KT-446-801

The Timing and Location of Major Ore Deposits
in an Evolving Orogen

Geological Society Special Publications
Society Book Editors
A. J. FLEET (CHIEF EDITOR)
P. DOYLE
F. J. GREGORY
J. S. GRIFFITHS
A. J. HARTLEY
R. E. HOLDSWORTH
A. C. MORTON
N. S. ROBINS
M. S. STOKER
J. P. TURNER

Special Publication reviewing procedures

The Society makes every effort to ensure that the scientific and production quality of its books matches that of its journals. Since 1997, all book proposals have been refereed by specialist reviewers as well as by the Society's Books Editorial Committee. If the referees identify weaknesses in the proposal, these must be addressed before the proposal is accepted.

Once the book is accepted, the Society has a team of Book Editors (listed above) who ensure that the volume editors follow strict guidelines on refereeing and quality control. We insist that individual papers can only be accepted after satisfactory review by two independent referees. The questions on the review forms are similar to those for *Journal of the Geological Society*. The referees' forms and comments must be available to the Society's Book Editors on request.

Although many of the books result from meetings, the editors are expected to commission papers that were not presented at the meeting to ensure that the book provides a balanced coverage of the subject. Being accepted for presentation at the meeting does not guarantee inclusion in the book.

Geological Society Special Publications are included in the ISI Index of Scientific Book Contents, but they do not have an impact factor, the latter being applicable only to journals.

More information about submitting a proposal and producing a Special Publication can be found on the Society's web site: www.geolsoc.org.uk.

It is recommended that reference to all or part of this book should be made in one of the following ways:

BLUNDELL, D. J., NEUBAUER, F. & VON QUADT, A. (eds) 2002. *The Timing and Location of Major Ore Deposits in an Evolving Orogen*. Geological Society, London, Special Publications, **204**.

BONI, M., MUCHEZ, P. & SCHNEIDER, J. 2002. Permo-Mesozoic multiple fluid flow and ore deposits in Sardinia: a comparison with Post-Variscan mineralization of Western Europe. *In*: BLUNDELL, D. J., NEUBAUER, F. & VON QUADT, A. (eds) 2002. *The Timing and Location of Major Ore Deposits in an Evolving Orogen*. Geological Society, London, Special Publications, **204**, 1–353.

GEOLOGICAL SOCIETY SPECIAL PUBLICATION NO. 204

The Timing and Location of Major Ore Deposits in an Evolving Orogen

EDITED BY

D. J. BLUNDELL
Royal Holloway, University of London, UK

F. NEUBAUER
University of Salzburg, Austria

and

A. VON QUADT
ETH-Z, Zurich, Switzerland

2002
Published by
The Geological Society
London

THE GEOLOGICAL SOCIETY

The Geological Society of London (GSL) was founded in 1807. It is the oldest national geological society in the world and the largest in Europe. It was incorporated under Royal Charter in 1825 and is Registered Charity 210161.

The Society is the UK national learned and professional society for geology with a worldwide Fellowship (FGS) of 9000. The Society has the power to confer Chartered status on suitably qualified Fellows, and about 2000 of the Fellowship carry the title (CGeol). Chartered Geologists may also obtain the equivalent European title, European Geologist (EurGeol). One fifth of the Society's fellowship resides outside the UK. To find out more about the Society, log on to www.geolsoc.org.uk.

The Geological Society Publishing House (Bath, UK) produces the Society's international journals and books, and acts as European distributor for selected publications of the American Association of Petroleum Geologists (AAPG), the American Geological Institute (AGI), the Indonesian Petroleum Association (IPA), the Geological Society of America (GSA), the Society for Sedimentary Geology (SEPM) and the Geologists' Association (GA). Joint marketing agreements ensure that GSL Fellows may purchase these societies' publications at a discount. The Society's online bookshop (accessible from www.geolsoc.org.uk) offers secure book purchasing with your credit or debit card.

To find out about joining the Society and benefiting from substantial discounts on publications of GSL and other societies worldwide, consult www.geolsoc.org.uk, or contact the Fellowship Department at: The Geological Society, Burlington House, Piccadilly, London W1J 0BG: Tel. +44 (0)20 7434 9944; Fax +44 (0)20 7439 8975; E-mail: enquiries@geolsoc.org.uk.

For information about the Society's meetings, consult *Events* on www.geolsoc.org.uk. To find out more about the Society's Corporate Affiliates Scheme, write to enquiries@geolsoc.org.uk.

Published by The Geological Society from:
The Geological Society Publishing House
Unit 7, Brassmill Enterprise Centre
Brassmill Lane
Bath BA1 3JN, UK
(*Orders*: Tel. +44 (0)1225 445046
        Fax +44 (0)1225 442836)
Online bookshop: http://bookshop.geolsoc.org.uk

The publishers make no representation, express or implied, with regard to the accuracy of the information contained in this book and cannot accept any legal responsibility for any errors or omissions that may be made.

© The Geological Society of London 2002. All rights reserved. No reproduction, copy or transmission of this publication may be made without written permission. No paragraph of this publication may be reproduced, copied or transmitted save with the provisions of the Copyright Licensing Agency, 90 Tottenham Court Road, London W1P 9HE. Users registered with the Copyright Clearance Center, 27 Congress Street, Salem, MA 01970, USA: the item-fee code for this publication is 0305-8719/02/$15.00.

British Library Cataloguing in Publication Data

A catalogue record for this book is available from the British Library.

ISBN 1-86239-122-X

Typeset by Keytec Typesetting Ltd., Bridport, Dorset, UK.

Printed by Antony Rowe, Chippenham, UK

Distributors

USA
AAPG Bookstore
PO Box 979
Tulsa
OK 74101-0979
USA
Orders: Tel. +1 918 584-2555
        Fax +1 918 560-2652
        E-mail bookstore@aapg.org

India
Affiliated East-West Press PVT Ltd
G-1/16 Ansari Road, Daryaganj,
New Delhi 110 002
Orders: Tel. +91 11 327-9113
        Fax +91 11 326-0538
        E-mail affiliat@nda.vsnl.net.in

Japan
Kanda Book Trading Co.
Cityhouse Tama 204
Tsurumaki 1-3-10
Tama-shi
Tokyo 206-0034
Japan
Orders: Tel. +81 (0)423 57-7650
        Fax +81 (0)423 57-7651

Contents

CONTENTS

Preface

When asked in 1996 what he thought was the major problem in geology still unresolved, Professor Rudolph Trümpy replied 'find the connection between mountain building processes and ore deposit formation'. His remark inspired the European Science Foundation to set up a five-year scientific programme in 1998 to investigate geodynamics and ore deposit evolution, GEODE, on a European scale. The GEODE programme was organised into five main projects, based on metallogenic provinces in Europe, supplemented by studies of metallogeny in the tectonically active regions of South America and the SW Pacific. A global comparison of major volcanic-hosted massive sulphide (VMS) deposits was also initiated. Metallogenesis has been related to geodynamic processes operating on a wide range of scales in space and time, involving large-scale processes conducive to the generation of magmatic and mineral fluids within the lithosphere–asthenosphere system, processes transporting and transforming those fluids through interactions with their surroundings in an ever changing thermal and pressure regime, and processes concentrating and depositing metals within the ore deposit regime progressively over time. To gain an understanding of these processes it is essential to examine them where they have been recently, or are currently, active and where information about the structure and characteristic properties of the lithosphere, and how they are evolving, can be gained from geophysical, geochemical, geochronological and other observations. As a consequence of previous European Science Foundation programmes, such as the European Geotraverse and EUROPROBE, the properties, structure and evolutionary history of the lithosphere of Europe is better known than almost anywhere else in the world. Metallogenic provinces in Europe range in time from the Archaean–Early Proterozoic to the Palaeozoic, to the Cenozoic and, in particular, in the active region of the Alpine–Balkan–Carpathian–Dinaride (ABCD) belt, to the Neogene. The latter, in comparison with the regions of South America and the SW Pacific, affords the opportunity to investigate metallogeny in a variety of modern tectonic contexts.

The contents of this book reflect these ideas. They are based on a symposium organised by GEODE and SGA (Society for Geology Applied to Mineral Deposits) held during the European Union of Geosciences Assembly in March 2001 entitled 'the timing and location of major ore deposits in an evolving orogen'. Although the majority of papers relate to Europe, their findings have a global significance for metallogenesis. Figure 1 provides a key to their locations. An introductory paper by **Blundell** to set the scene is followed by an account by **Allen & Weihed** of global comparisons of VMS districts from the first findings of a new international project. They conclude that the main VMS ore deposits take less than a few million years to form and generally occur near the top of a succession of felsic volcanic rocks within an extensional environment. A series of papers then examine various aspects of modern orogenic systems, starting with two (**Barley et al.** and **Macpherson & Hall**) on the SE Asia/SW Pacific region that demonstrate the speed of tectonic processes and the short duration of magmatic and mineralizing events. The latter are shown to relate to transient effects in a subduction complex, often through plate reorganization, rather than to steady state subduction. Changes in the balance between recycled fluxes of slab-derived fluid and sediment melt exert an important control on the chemical composition of arc lavas and, consequently, on their content of economically important metals. Arc-related magmatism in unusual tectonic settings produced the most abundant and largest deposits in SE Asia, the vast majority of which have formed since 5 Ma. It appears that whilst the timing of magmatic and metallogenic events can be explained in relation to tectonic changes, it is much more difficult to explain the tectonic controls on the locations of ore deposits. Papers by **Lips** and **Neubauer** identify the connections between collisional tectonics and ore deposit evolution in SE Europe and the ABCD belt, particularly relating to slab rollback in earlier stages and slab tear and detachment subsequently. **Lips** points out that roll-back of subducted lithosphere, restoration of orogenic wedge geometry and slab detachment are all scenarios that favour extension and the transfer of heat to relatively shallow lithospheric levels. Within the context of near-continuous subduction in the ABCD region over the past 100 Ma, **Neubauer** recognizes two periods of short-lived, late-stage collisional events that led to calc-alkaline magmatism and mineralization, in the Late Cretaceous and Oligocene–Neogene. In both periods, the type of mineralization and its timing changes progressively along the strike of the magmatic/metallogenic belt, which Neubauer links with the process of lithospheric slab tear observed independently in the

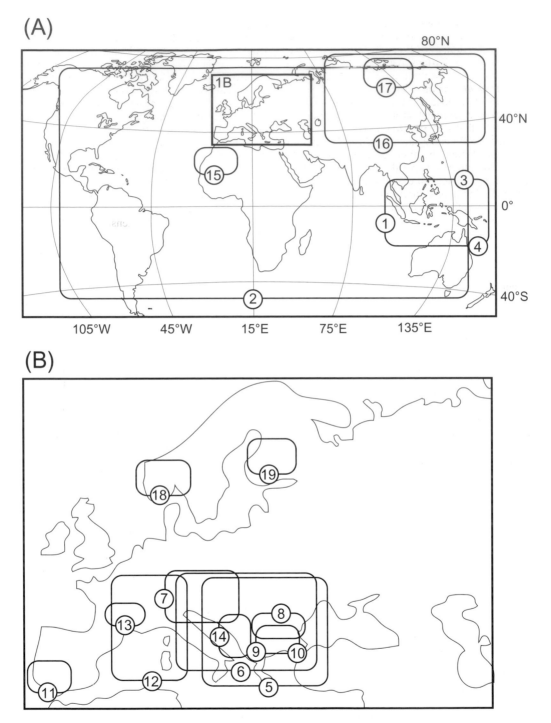

Fig. 1. Key maps showing general locations relating to the papers. (**a**) World map: 1, Blundell; 2, Allen & Weihed; 3, Barley *et al.;* 4, Macpherson & Hall; 15, Chauvet *et al.*; 16, Yakubchuk; 17, Fridovsky & Prokopiev. (**b**) Map of Europe: 5, Lips; 6, Neubauer; 7, Amann *et al.;* 8, von Quadt *et al.;* 9, Marchev & Singer; 10, Krohe & Mposkos; 11, Tornos *et al.;* 12, Boni *et al.* 13, Cuney *et al.;* 14, Jurković & Palinkaš; 18, Stein & Bingen; 19, Rajavuori & Kriegsman.

area from seismic tomography and other evidence. Papers by **Amann et al.** and **von Quadt et al.** explain the timing and genesis of specific ore deposits in the Eastern Alps and the Bulgarian Srednogorie zone of the Carpathians, respectively. **Amann et al.** examine late-tectonic gold mineralization of Oligocene–Miocene age related to a complex transtensional shear regime of conjugate strike-slip faults. **Von Quadt et al.** use high precision U–Pb dating of individual zircons from dykes bracketing the time of formation of the Elatsite porphyry Cu–Au deposit to demonstrate that the high-temperature ore forming process was constrained to a very short period within the Late Cretaceous collisional event, between 92.1 ± 0.3 and 91.84 ± 0.31 Ma ago. A paper by **Marchev & Singer** looking at the timing and nature of magmatism and hydrothermal activity in the eastern Rhodope region is complemented by an analysis of the structural evolution of the Rhodope mountains by **Krohe & Mposkos**. **Marchev & Singer** show how short the duration of volcanism and hydrothermal activity can be in a single ore district. They find that volcanism in the Madjarovo volcanic complex and ore district in Bulgaria began at 32.7 Ma and finished by 32.2 Ma, at which time the hydrothermal activity and fault-controlled base/precious metal mineralization occurred. This happened during a period, shown by **Krohe & Mposokos**, when the Rhodope region suffered pervasive deformation and granitoid intrusion, being uplifted, extended and exhumed through a series of inter-linked detachment faults.

Turning to older orogenic systems, one of the world's major metallogenic provinces is the Iberian Pyrite Belt of Southwest Iberia, set within the larger province of the western Variscides. **Tornos et al.** show how mineralization evolved as a consequence of oblique collisional plate tectonic processes. They find that most of the mineralization is related to localized extension in shear zones, pull-apart basins and escape structures within a regional transpressional regime in which deep, crustal-scale structures developed. Based on a detailed study of Sardinia, **Boni et al.** demonstrate the importance of large-scale, and persistent, fluid flow systems in the continental crust as a mechanism for ore deposit formation. They envisage a 'crustal-scale hydrothermal palaeofield' as applicable to much of the post-orogenic (Variscan) mineralization across western and southern Europe. **Cuney et al.** investigate the causes and effects of orogenic gold mineralization in the Massif Central, France. They distinguish two short-lived metallogenic episodes, related to emplacement of leucogranites. The first, c. 325 Ma, related to syn-collisional extension and the second, c. 310 Ma, related to general extension and rapid

exhumation of the Variscan belt, possibly due to lithosphere delamination. **Jurković & Palinkaš** tackle the task of discriminating between Variscide or Alpide (Triassic) origins of ore deposits in the Dinarides, in Palaeozoic and Permo–Triassic allochthonous units thrust over Mesozoic–Palaeogene rocks during Alpine deformation. With radiometric dating not possible, they successfully developed a package of discriminatory criteria, involving (^{34}S isotope values, $SrSO_4$ content and fluid inclusion properties. **Chauvet et al.** explain the structural controls on copper mineralization in the High Atlas of Morocco, a southerly extension of the Variscides to north Africa. They relate the mineralization to a ductile tectonic event associated with granite emplacement some 20 Ma earlier. To the east, and on a grand scale, the formation of very large ore deposits in relation to the evolution of orogenic collages across eastern Asia through the Mesozoic and Cenozoic is explained by **Yakubchuk**. He is able to demonstrate that the timing of mineralization is related to plate reorganization and oroclinal bending of magmatic arcs. Interestingly, the location of major gold deposits is correlated with marine shelf and platform sequences containing black shales. Complementing this work is a paper by **Fridovsky & Prokopiev** describing gold mineralization on the eastern margin of the north Asia craton, an area rich in black shales.

Stein & Bingen demonstrate the immense value of Re–Os dating techniques for obtaining direct information about the ages of mineral occurrences in southern Norway, which enable them to pinpoint the timing of deformation and metamorphic change associated with the mineral occurrences to a duration of just 30 Ma, between 1047 and 1017 Ma. The strength of their approach, as they say, is that 'small ore occurrences in molybdenum-endowed regions of the Earth's crust are capable of unleashing important age information that bears on the metamorphic history and tectonic assembly of major orogens'. Finally, **Rajavuori & Kriegsman** point out the role of fluorine as a pathfinder in metamorphic hydrothermal volcanogenic massive sulphides. Their study of F in Zn–Cu–Pb deposits in Finland and Australia examines the role of F in the fluid transport system of magmatic–hydrothermal ore formation.

The overall conclusion to be drawn from this collection of papers is that ore formation in magmatic–hydrothermal systems occurs over a short period of time, probably less than a million years, usually at or close to the end of a magmatic event. The timing relates to a transient effect of plate reorganization, which creates heat and a particular style of magmatism conducive to the generation of mineralizing fluids. Specific scenarios have been proposed. The location of ore deposits is difficult to pinpoint but is often controlled by localized extensional structures developed within a regional transpressional regime. But what controls the amount of mineralization produced and what determines where it is concentrated into very large ore deposits remain obscure.

All the papers in this book have benefited greatly from the critical assessments and helpful comments given by referees, who have given freely of their time and expertise. We would like to thank the referees, D. Alderton, M. Anderton, S. Cuthbert, T. Berza, A.-V. Bojar, H. de Boorder, K. Clark, R. Clayton, N. Cook, L. Diamond, M. Economou-Eliopoulos, M. Faure, H. Fritz, Ch. Gauert, J. Glodny, R. Hall, R. Handler, W. Halter, Y. Kostitsyn, D. Kozelj, W. Kurz, R. Large, J. Lexa, E. Marcoux, K. McClay, S. McCutcheon, A. Mogessier, R. Moritz, L. Nakov, W. Nokleberg, S. Petersen, T. Pettke, J. Raith, J. Relvas, J. Ridley, J. Schneider, S. Sengör, C. Spöttl, M. Tichimorova, F. Tornos, P. Tropper, N. White, E. Willingshofer and M. Wortel.

D. Blundell,
F. Neubauer
& A. von Quadt

The timing and location of major ore deposits in an evolving orogen: the geodynamic context

DEREK J. BLUNDELL

Department of Geology, Royal Holloway, University of London, Egham, Surrey TW20 0EX, UK (e-mail: d.blundell@gl.rhul.ac.uk)

Abstract: Although it is possible to identify the potential controls on mineralization, the problem remains to identify the critical factors. Very large mineral deposits are rare occurrences in the geological record and are likely to have resulted from the combination of an unusual set of circumstances. When attempting to understand the mineralization processes that occurred to form a major ore deposit in the geological past, especially the reasons why the deposit formed at a particular time and location within an evolving orogenic system, it is instructive to look at mineralization in modern, active subduction complexes. There it is possible to measure and quantify the rates at which both tectonic and mineralizing processes occur. In a complex subduction system, regions of extension develop. For example, subduction hinge retreat is a process that creates extension and generates heat from the upwelling of hot asthenosphere ahead of the retreating slab, producing partial melting, magmatism and associated mineralization. Seismic tomography not only images mantle as it is now, but subduction slab anomalies can be interpreted in terms of the past history of subduction. This can be used to test tectonic plate reconstructions. Tectonic and magmatic events occur rapidly and are of short duration so that many are ephemeral and will not be preserved. Furthermore, they can be diachronous as is the case with the lithospheric slab tear clockwise around the Carpathian Arc during the Neogene.

If the tectonic setting is paramount in determining the onset of the mineralization process and generation of mineralizing fluids, the fluid transport system that localizes the mineralization in space and time and concentrates the metal charge is the key to finding when and where the ore deposits occur. Fault and fracture networks in the crust provide various mechanisms for the localized expulsion of fluid in pulses of short duration. Excess surface water flow following large earthquakes in the Basin and Range region of USA offers a modern analogue to quantify fluid flow related to extensional faulting. Evidence from the Woodlark basin, east of Papua New Guinea, suggests that similar conditions pertain in the oceanic environment. Whilst there are limits to the use of regions of active tectonism as modern analogues to explain the mineralization of ancient orogenic systems, they do provide the best opportunity to understand the mechanisms of mineral processes and the controls on the location and timing of major ore deposits.

The formation of a large, world class ore deposit is a relatively rare event in geological history and requires the concurrence of a particular set of circumstances in space and time. These are likely to operate on a range of scales from the major tectonic context on a lithosphere scale, through the crustal-scale structural context down to the deposit and microscopic scales that determine the mode of mineralization. There are various approaches that can be made to determine the key factors conducive to the formation of large ore deposits. Ideally, a systems approach advocated by Ord *et al.* (1999) recognizes common crustal-scale fluid flow systems in ore deposit formation that are amenable to modelling, regardless of the diversity of tectonic settings. 'Soft' modelling of the common factors provides a set of input parameters for 'hard' modelling to quantify and predict the likely occurrences of ore deposits prescribed by a specific set of circumstances. However, in order to do this successfully, an essential pre-requisite is to understand the geodynamic processes involved so as to constrain the mechanisms incorporated into the modelling. Another approach has been used recently by Goldfarb *et al.* (2001) to produce a global synthesis of orogenic gold deposits. In this, they began by examining the characteristic features of the ore deposits and their tectonic settings, before making a detailed synthesis of their occurrence through geological time from the Archaean to the Present, within a succession of orogenic systems. This

From: BLUNDELL, D.J., NEUBAUER, F. & VON QUADT, A. (eds) 2002. *The Timing and Location of Major Ore Deposits in an Evolving Orogen*. Geological Society, London, Special Publications, **204**, 1–12.
0305-8719/02/$15.00 © The Geological Society of London 2002.

emphasises the long term variation through time of the amount of mineralization that has occurred. However, to examine the timing and location of major ore deposits within an evolving orogen, the best starting point is an appraisal of modern orogenic systems, in which the dynamically changing tectonic activity and lithospheric structure responsible for metallogeny can be related through the use of geophysical and geochemical observations. These systems can be used as modern analogues to substantiate those geodynamic processes and structures that, together, control the timing and location of large ore deposits within an evolving orogen.

Regardless of deposit type or form of mineralization, there are certain common factors in ore genesis (equivalent in hydrocarbon parlance to source, migration pathway and trap).

(1) *A source region within the Earth to provide the metal charge.* This can range from a region of partial melting within the upper mantle (or possibly deeper if generated by a mantle plume) to a region within the crust that is scavenged by hot brine within some form of hydrothermal system. The size of the source region is likely to be large in comparison with that of the ore deposit, so that a high degree of concentration of metals within the fluid between source and ore deposit is essential.

(2) *A fluid system that provides the mechanisms of transport and concentration between source region and ore deposit.* This is controlled by the evolving thermo-tectonic setting of the orogen and the rheological properties of its location within the lithosphere.

(3) *A localized structural/stratigraphic setting and chemical regime* that is conducive to the precipitation of a large quantity of metals in ore minerals within a deposit. Furthermore, conditions for the preservation and possible exhumation of the deposit are required subsequently for it to be at or near surface at the present day and thus exploitable as a resource.

Taking these factors into account, modern orogenic systems offer insights into processes that are ephemeral and allow rates and duration of process to be quantified. These can be used to interpret the key factors in past orogenies that have run to completion, where evidence of processes active during their evolution are no longer preserved.

A modern orogenic system: the SE Asia–SW Pacific region

The SE Asia–SW Pacific region forms an active subduction complex that provides excellent examples of most of the geodynamic processes in-

volved in an evolving orogen. Across the very large area depicted in Figure 1, three major plates are interacting; the dominantly continental Eurasian plate in the NW, the India–Australia plate in the south containing both old oceanic lithosphere and the Australian continent, and the oceanic Pacific plate in the east. In between are smaller plates containing continental fragments and young oceanic lithosphere that form a collage of rapidly changing and deforming tectonic units. Present day horizontal motions of structural elements have been quantified from analysis of GPS data to deduce relative velocity vectors. Rates of uplift or subsidence are measured by various means. Earthquake hypocentre and focal mechanism determinations define in three dimensions the configuration of subduction slabs in the upper mantle and quantify the state of stress. Their configurations are confirmed by seismic tomographic images of seismic velocity anomalies in the upper mantle, along with those of various other features (Bijwaard et al. 1998; Spakman & Bijwaard 1998). Gravity, seismic and heat flow measurements also provide information about lithospheric structure and physical properties at depth. All these provide a snapshot of this huge subduction complex at the present time.

In addition, palaeomagnetic data combined with high resolution dating provide evidence of plate movements in the past, both from ocean floor basalt anomalies and from rotations and relative translations of continental fragments. From these and other data, the rates of relative motions between the plates have been measured and their evolution over the past 50 Ma has been tracked in a succession of plate reconstructions at 1 Ma intervals (Hall 1996, 2002). This evolutionary model plus measurements of currently active processes provide the framework for attempting to understand the conditions required to generate large ore deposits. In this paper, just a few examples are presented to illustrate the value of this approach.

(1) Within a plate system of long-term convergence a large amount of extension has occurred, especially across the large region of Sundaland and the Banda Sea behind the Banda Arc. Many of the subduction zones defined by earthquake activity are steeply dipping and, together with seismic tomographic imagery, offer ample evidence of subduction hinge retreat, or 'rollback'. One effect of rollback is to introduce additional heat in front of the retreating slab, giving rise to partial melting in the mantle (see Macpherson & Hall 2002). Rollback also results in the development of back-arc and intra-arc extensional basins which, as Hutchinson (1973) first pointed out, are amongst the range of tectonic settings conducive

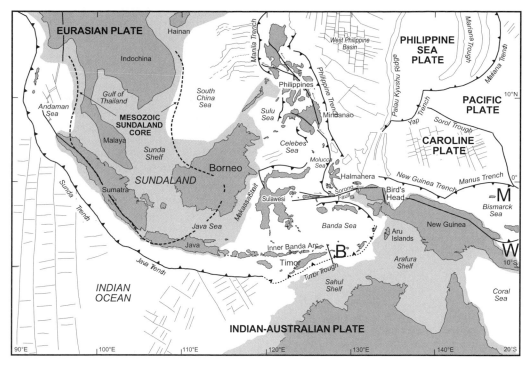

Fig. 1. Main plate tectonic features of the SE Asia–SW Pacific region, based on Hall (2002), reproduced with his permission, showing locations of Manus basin (M), Woodlark basin (W) and Banda Arc cross-section (B) of Figure 2.

to the formation of VHMS deposits. These are characterized by bimodal volcanics with rhyolitic and basaltic compositions.

In areas where the subduction hinge has advanced, volcanism often ceases as the mantle wedge is not continually being replenished by fresh mantle material and fewer conduits for fluid flow to the surface are available. Magmatism instead leads to the formation of intrusions within the crust. According to Sillitoe (1999), this is a more likely setting for porphyry copper deposits, characterized by cylindrical stocks related to intrusions, associated with andesite dominated magmatism where arcs are not in extension. Examining the tectonic settings of mineral deposits in Indonesia, Carlile & Mitchell (1994) noted that all known mineral deposits lie within magmatic arcs and formed during or shortly after the magmatic activity. But of 15 Cenozoic magmatic arcs recognized, only six were known to contain significant mineralization. The mineralization is dominated by porphyry Cu–Au and epithermal Au deposits, all founded on continental crust, the former occurring on both island arc and continental settings and the latter best developed in continental arcs. In contrast with the settings of the porphyry deposits, epithermal mineralization

is commonly linked to extensional movements on low angle detachment faults at depth. Within the same arc, epithermal deposits may be younger than the porphyry deposits and have formed through a different mechanism. Most of the major deposits are of Pliocene age, associated with mid-Miocene to late Pliocene magmatism. Carlile & Mitchell (1994) suggest that uplift and erosion have removed all the older mineral deposits but have been insufficient to reveal those of Quaternary age. Macpherson & Hall (2002) suggest instead that the timing of magmatism and style of mineralization relates to a change in plate dynamics from a period of hinge advance of the Banda arc between 25 and 15 Ma to one of hinge retreat since 10 Ma. Barley *et al.* (2002) relate these plate reorganizations to periods of gold mineralization. They note that both the largest and the largest number of ore deposits, associated with arc magmatism in unusual tectonic settings, formed in the past 5 million years during a major period of changing motions between the India–Australia and Pacific plates.

(2) The northward movement of the India–Australia plate at 60 mm a^{-1} relative to the Banda Sea plate has resulted in the collision of the NW Australian shelf with the Banda Arc in the region

of Timor. The collisional history appears to have begun 8 Ma ago when an outlier of thin continental crust reached the north-dipping subduction zone and transferred from the lower (India–Australia) plate to the upper (Banda Sea) plate when subduction jumped to the south of it. Oceanic lithosphere was subducted until 2.4 Ma ago when the NW Australian shelf arrived. A section of continental crust was subducted, to around 100 km depth, until relatively recently when subduction jumped to the north of the volcanic arc and changed polarity. Thus Timor and the adjacent section of the volcanic arc are currently moving north as part of the India–Australia plate (McCaffrey 1996). To the north, the Banda Sea is a back-arc basin currently being overthrust at the south-dipping Wetar Fault. The present cross-sectional configuration is sketched as a cartoon in Figure 2 (Richardson & Blundell 1996), showing the thickened continental mass below Timor, wedged between the Australian shelf continental mass to the south and the Banda Sea plate to the north, laterally shortened by over 50% and currently being uplifted at up to 2 mm a^{-1}.

The Banda Arc–Australia collision has been used by Brown & Spadea (1999) as a modern analogue for the evolution of the Uralides during the late Devonian, Figure 3, involving the collision of the Magnitogorsk Arc with the East European continental mass. It also offers the basis of an explanation for the location and timing of VHMS deposits in the southern Urals. However, the Uralide orogeny continued beyond arc–continent collision to complete continent–continent collision, granite magmatism and post-orogenic extensional collapse in the early Carboniferous. VHMS deposits in the southern Urals are found in well-preserved mounds (Zaykov 2000), whose modern counterparts are linked with black smokers, such as the 'black smoker' recovered from 2 km depth on the seafloor of the SE Manus basin. This is 2.7 m long, 80 cm in diameter at the base,

teeming with bacteria and archaea, and provides clear evidence of current mineralization in the area (Binns *et al.* 2002).

(3) A mantle plume has been recognized beneath the Manus basin in the Bismark Sea (location, Fig. 1) on the basis of the petrology, geochemistry and He isotope ratios of active volcanics (Macpherson *et al.* 1998). It is confirmed by the presence of a cylindrical-shaped seismic tomographic low velocity anomaly in the upper mantle (Spakman & Bijwaard 1998) with a diameter of 700 km. Investigating the widespread occurrence of Mg- and Si-rich boninite magmatism in the Izu–Bonin–Mariana forearc of the Philippine Sea plate during the middle Eocene, Macpherson & Hall (2001) have proposed that this is due to the presence of a mantle plume. They point out that the generation of boninite magma requires excess heat in order to melt residual peridotite within the mantle portion of an oceanic lithospheric plate overriding a subducted plate. Additionally, a hydrous fluid flux from the subducted lithosphere has to be incorporated into the melt. Using a plate reconstruction from 50 Ma (Hall 1996), they demonstrate that the boninite volcanics would have been located within a circle of 1,500 km diameter. Continuing with subsequent reconstructions they have been able to track the path of this plume to its present-day location in the Manus basin, where VHMS style mineralization has been found (Binns & Scott 1993). Thus the presence of a mantle plume beneath a subduction complex may be relatively long lived and can interact with a range of subduction processes as a result of the ever changing configurations of microplates within the complex. The additional thermal energy provided by the plume can thus promote substantial magmatic activity–the volume of boninite volcanics is comparable with other plume-related volcanism such as the Tertiary igneous province of NW Britain or the Columbia River basalts. The characteristic style of the middle Eocene boninite volcanism provides a Cenozoic analogue that can be recognized within the geological record, such as the oldest volcanic unit found in the Magnitogorsk forearc in the southern Urals (Brown & Spadea 1999) or the late Archaean Abitibi Belt in Canada (Wyman 1999).

(4) A key observation of the SE Asia–SW Pacific subduction complex is the speed at which the various tectonic processes take place and the short duration of tectonic and magmatic events. For this reason, the single most important requirement for a better understanding of mineralization is to have high precision, high accuracy dating of suites of ore minerals. Dynamic changes in plate motions, which are dependent upon time-variable parameters, are the events most likely to control

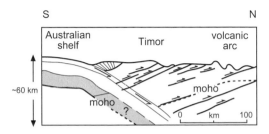

Fig. 2. Cartoon cross-section of the Banda Arc–NW Australian shelf collision zone interpreted from BIRPS deep seismic sections (Richardson & Blundell 1996); stippled area Australian lower crust.

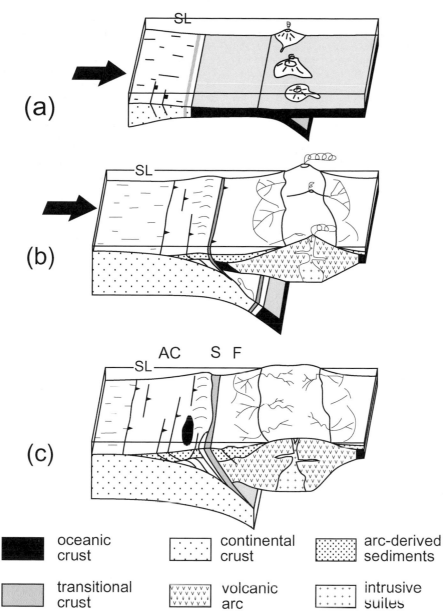

Fig. 3. Schematic model of arc–continent collision, redrawn from Brown & Spadea (1999), reproduced with permission from the Geological Society of America, based on modern analogues from the western Pacific, to explain the evolution of the southern Uralides: (**a**) Intra-oceanic subduction of young, hot oceanic lithosphere, with volcanic arcs of tholeiite and boninite composition during the early to mid-Devonian: SL is sea level. (**b**) As continental crust arrives at the subduction zone during the late Devonian, collision-related processes predominate, with calc-alkaline suites indicating a mature island arc setting. (**c**) In the Early Carboniferous, when collision is nearly complete, the accretionary complex (AC) and forearc (F) areas attain their final architecture; S is the suture.

the timing and location of mineralization. Sillitoe (1997) observed that various unusual arc settings can be conducive to mineralization, such as subduction polarity reversals, changes of plate boundaries from convergent to transform, or the cessation of subduction resulting from arc–continent collision. He found that some of the largest gold deposits in the circum-Pacific region occurred in short-lived, areally restricted events that resulted from partial melting of stalled subduction slabs conducive to the oxidation of mantle sulphides and release of gold. The changes in palaeogeography evident in Hall's plate reconstructions over the past 50 Ma underline the ephemeral nature of many of the subduction elements present today. For example, the Woodlark basin only began to develop as an oceanic ridge started to rift, driven by slab pull, just over 7 Ma ago. It is currently being subducted as fast as it is growing so that no trace of it will remain after another 2–3 Ma. A region of oceanic lithosphere flooring the Molucca Sea north of Halmahera was subducted between 11 Ma and 2.5 Ma, both to the east and to the west (Hall 1999), and is imaged by seismic tomography as an inverted V in cross-section (see Macpherson & Hall 2002, fig. 3). The forearc wedges of the upper plates to the east and west of it have now collided, with the western one overriding the other. Two arcs are now in contact with an accretionary complex sandwiched between. Meanwhile the inverted V of oceanic lithosphere is sinking through the upper mantle. Geologists in a few million years time will have great difficulty in distinguishing this arc-collisional history and in recognizing the sunken plate.

Most volcanic episodes are over within 3–5 Ma, though they may recur subsequently in the same location, but under different stress conditions. The rates of uplift of many of the islands, such as Timor, at up to $2 \, \mathrm{mm \, a}^{-1}$ ensure rapid erosion and exhumation. Thus the epithermal copper and gold deposits on East Mindanau associated with the Philippine Fault (probably formed at depths around 5 km less than 5 Ma ago) and the VHMS deposits on Wetar (currently at surface but formed in deep water 4.7 Ma ago) could be gone in another 2–3 Ma, possibly to reappear later in the sedimentary record as placer deposits. In New Guinea, gold deposits are associated with thrusts and lateral ramps (Pubellier & Ego 2002). Syn-tectonic porphyry Cu was injected between 7 and 2 Ma into thrust planes and then cut by later thrusts. The tectonic setting changed again after 2 Ma becoming wrenching at the front of the belt, the lateral ramps subsequently being re-activated as normal faults. Mineralization may come with one phase and be destroyed by the next one. These rapidly evolving systems show how essential it is to have accurate dating of events and underline the over-simplifications that we make when we try to understand ancient orogens.

Thus, although the SE Asia–SW Pacific subduction complex may provide examples of tectonic settings within an actively evolving orogen, much of what is observed at present is unlikely to be preserved after the orogeny has run its course through to continental collision and eventual collapse. Van Staal et al. (1998) illustrated this by modelling the SE Asia–SW Pacific complex forward into the future some 45 Ma, until the end of collision. Figure 4 replicates this model, which they used as an analogue for the Appalachian–Caledonian orogen, pointing out that it would be impossible to reconstruct the evolution of the orogeny because so much of the evidence had been lost. Thus there are bound to be severe limitations on interpretations of the tectonic settings of mineral deposits in ancient orogenic belts unless, like the Uralides, they are exceptionally well preserved.

A modern orogenic system: the Carpathian Arc

Forming part of the Alpine–Carpathian system, the Carpathian Arc has been evolving for the past 80–100 Ma, but is now in the late stage of continental collision when plate convergence has almost come to an end. This is one of the circumstances when slab detachment occurs, by lateral tear, particularly when continental lithosphere is incorporated into the subducting slab. Seismic tomographic images on radial cross-sections across the Carpathian Arc (Wortel & Spakman 2000) reveal the presence of a gap between the lithosphere and the down-going slab, apart from the Vrancea region in the SE where the slab is unbroken. The down-going slab turns to horizontal at depth, an indication of an earlier hinge retreat and rollback. The effect of a partly torn slab is to increase the weight of the slab where it is attached, resulting in subsidence at surface, and to unload the lithosphere where it has torn, resulting in surface uplift (Fig. 5). The history of slab tear clockwise around the Carpathian Arc since the Oligocene can be tracked by the migration of the null point between subsidence and uplift around the arc. Earthquake activity is strong in the Vrancea region where the slab is still attached. Working with numerical models, Wortel & Spakman (2000) have shown that detachment is dependent on the strength of the down-going slab: it can occur if the slab strength becomes less than the tensional stress from slab pull. The strength can be reduced, for example through a change from oceanic to continental lithosphere in the subducting slab, sufficient to initiate slab tear. Leech (2001) has pointed out that eclogitization is likely to be a major factor in weakening the lithosphere in these circumstances and that an influx of fluids into the subduction zone is

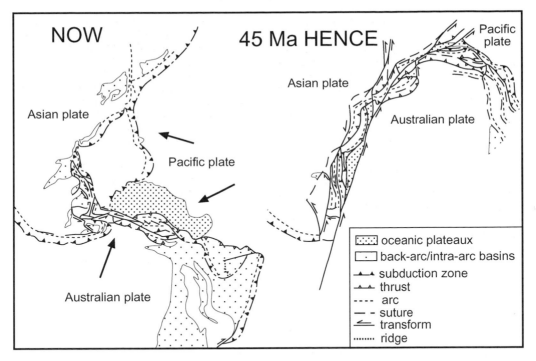

Fig. 4. Forward model of plate motions of the SE Asia–SW Pacific region 45 Ma into the future when continental collision occurs and relative plate motions cease (Van Staal *et al.* 1998).

required in the metamorphic processes that are involved. The actual depth of detachment is dependent upon the thermal structure and rheology of the subducting slab–the warmer and weaker it is, the shallower is the depth of detachment. Considering the thermal consequences, a shallow detachment can locally thin the lithosphere and bring hot asthenosphere closer to surface, with a consequent thermal pulse and the likelihood of partial melting. A slow convergence velocity results in a shallow detachment at a later time in the orogenic process, but a relatively small temperature jump. A minimum depth of detachment is at 35 km when very warm lithosphere is subducted, but the depth is greater for cooler lithosphere, around 70–80 km. Modelling indicates that partial melting is restricted to the plate contact region where the tear has formed. The overlying lithosphere, which is uplifting and extending laterally, is thus affected by a sudden flow of heat from the hot lithosphere. This can cause anatexis and granite magmatism, with related ore deposits. A further significance of this thermo-tectonic modelling is that it can be related to $P–T–t$ information from metamorphic minerals and the isotopic and geochemical signatures of the partial melts. Thus there is now a method of

analysis capable of predicting the tectonic settings of mineralization formed in the later stages of orogenesis. De Boorder *et al.* (1998) have shown that there is a close spatial link between Late Cenozoic deposits of Hg, Sb and Au, which have a mantle origin, and areas of lithospheric slab detachment beneath the Carpathian Arc picked out by seismic tomographic low velocity (i.e. hot) regions. They report a 2.2 Ma age difference between the volcanic host and the mineralization, which may indicate the time lag between slab detachment and mineralization. Neubauer (2002) finds that, within the long-term subduction complex of the Alpine–Balkan–Carpathian–Dinaride region that evolved over the past 120 Ma, two short-lived, late-stage collisional events occurred, one in the Late Cretaceous, the other of Oligocene–Neogene age. Plate reorganization led to magmatism and extensive mineralization in three distinct belts. Within these belts, the type of mineralization varies significantly along strike. In the Oligocene–Neogene Serbomacedonian–Rhodope belt, related to the Carpathian arc, the timing of magmatism and mineralization migrates along strike in keeping with the Carpathian arc, and so can be linked with the process of lithospheric slab tear.

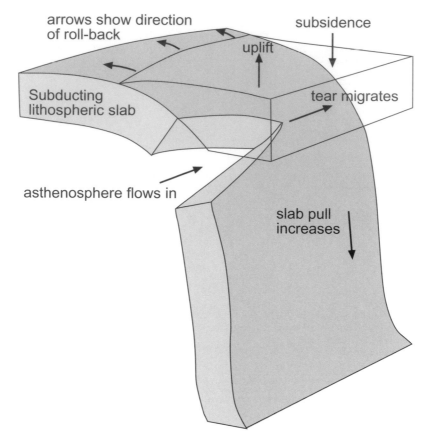

Fig. 5. Mechanism of slab tear proposed byWortel & Spakman (2000), reproduced with permission from *Science* **290**, 1910 (2000) fig.4. Copyright 2000 American Association for the Advancement of Science. The progress of the tear can be followed by the migration of the null point between subsidence and uplift around the arc. Slab tear enhances roll-back. Inflow of hot asthenoshere into tear gap creates specific conditions for short-lived magmatism and mineralization.

Modern fault systems and fluid flow

Whilst the tectonic settings just discussed determine the location of the source region and the timing of the generation of magmatic and/or hydrothermal mineralizing fluids, the transportation system to bring these fluids to the ore deposit location is of fundamental importance to ore genesis. The close association of ore deposits with faults and fracture systems implies that these are the prime components of the transport system, at least within the brittle upper crust. Evidence from seismic s-wave velocity anisotropy by Crampin (1994) has established the ubiquitous presence of vertical cracks in the crystalline crust. Deeper in the continental crust and upper mantle, anastomosing ductile shear zones hold the key in regions of extension, whilst crustal-scale shears, ramps and thrust anticlines dominate regions of compres-

sion. Figure 6 illustrates these differences, as imaged on deep seismic reflection profiles, which are a consequence of the differing rheological responses of rocks under tensile and compressional stress (Cloetingh & Banda 1992). In continental regions under extension, the seismogenic zone in which earthquakes occur is normally above a mid-crustal level of around 15–20 km but in regions of active thrusting during continental collision, such as the Himalayas, earthquakes occur at depths nearly down to the Moho.

A considerable body of research has established various possible mechanisms for fluid transport in the crust. Sibson (1996) showed that mesh structures involving faults interlinked with extensional vein-fracture systems can act as major conduits for large volume fluid flow. Fluid pressure provides the drive and the mechanism for opening elements in the mesh, allowing transient increase

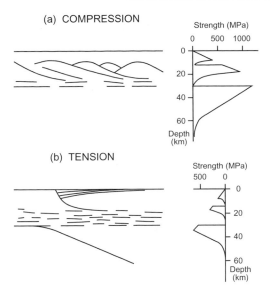

(a) COMPRESSION

Strength (MPa)

(b) TENSION

Strength (MPa)

Fig. 6. Cartoons based on deep seismic reflection profiles across continental lithosphere showing typical structural configurations in relation to strength profiles: (**a**) under compression where crustal-scale thrust ramps and anticlines dominate, (**b**) under tension where brittle conditions and faulting occur in the upper crust and ductile conditions with anastomosing shears pervade the lower crust. The Moho M can act as a detachment, below which, occasionally, dipping reflections are observed in the upper mantle.

in hydraulic conductivity, resulting in fluid flow and earthquake swarm activity. Sibson (1981, 1987) had earlier proposed an earthquake 'pumping' mechanism to explain the association of epithermal mineralization with extensional offsets between fault segments. Dilation in the relay zone between fault segments as the fault ruptures causes a reduction in fluid pressure so that fluid is sucked into the void space. This discharges subsequently as strain is reduced during the aftershock sequence. Sibson et al. (1988) proposed a similar mechanism to explain the association of mesothermal gold deposits with high-angle reverse faults at mid-crustal levels. In this case the brittle upper crust (see Fig. 6) acts as a seal to fluids trapped below the seismogenic zone until fluid pressure increases to a point when the fault fails and rupture occurs. The fault acts as a valve that opens from time to time, allowing pulses of high pressure fluid to be released upwards. However, this mechanism may have more to do with fault inversion events than crustal-scale thrusting. Sanderson & Zhang (1999) presented numerical models with coupled mechanical and hydraulic behaviour to demonstrate how fluid flow through a

network of interconnected cracks becomes highly organized and effective when fluid pressure reaches a critical point. Based on a random polygonal network of cracks, their model shows that hydraulic conductivity increases when rising fluid pressure reaches a critical value, less than but close to the lithostatic pressure. At this point, deformation of the network results in the formation of straighter, longer cracks, allowing slip, localized stress concentrations and localized opening of the more organized fractures. The sudden increase, locally, in hydraulic conductivity by several orders of magnitude allows the sudden upward expulsion of fluid from the system. Increasing fluid pressure can arise from loading due to tectonic stress or from heating–the latter a 'pressure cooker' effect. This mechanism is particularly applicable where flow localization is important, as in vein-hosted mineralization. Cox (1999) regards the crack networks as composed of through-going, or 'backbone' fractures linked to branching crack systems. Upstream, at depth, the branching elements act as tributaries that feed fluid into the system whereas downstream, at higher levels, the branching elements act as distributary structures and are the likely sites for mineral deposits. He emphasizes the transient nature of fluid flow, as well as its localization, as a result of the competition between processes creating crack growth and processes that close or seal the cracks. Continuing deformation regenerates crack growth to exceed crack closure and maintain a succession of fluid flow pulses.

Of particular significance are the findings of Muir-Wood & King (1993) from a study of the amounts of water expelled at surface due to earthquake-related fault movements. They showed that whilst normal faults in a continental setting expelled substantial quantities of water, strike-slip faults expelled less than a tenth of the water from normal faults and flow was variable, and reverse faults expelled virtually no water. They quantified their findings with detailed measurements of excess surface water flow from two earthquakes related to normal faults in the Basin and Range region of USA, at Hebgen Lake in 1959 and Borah Peak in 1983. In both cases, monitoring networks covering the complete catchment areas were operating at the time so that flow rates could be measured on a daily basis. Within a few days of each earthquake the flow rate peaked and an excess flow was recorded for nearly a year. In that time the total quantity of water expelled from a 20 km length of fault break due to a magnitude 7 earthquake was 0.5 km^3. To explain this flow, Muir-Wood & King (1993) proposed a model of coseismic strain illustrated in Figure 7. During the interseismic period, which would normally last

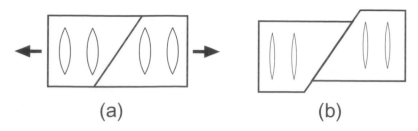

Fig. 7. Coseismic strain model redrawn from Muir-Wood & King (1993), reproduced with permission from the American Geophysical Union, of fluid flow related to earthquake-induced rupture of a normal fault. (**a**) Cracks open during the interseismic period under horizontal tensile stress. The cracks fill with fluid. (**b**) Cracks close at the time of an earthquake due to coseismic compressional rebound as the stress drops and strain is released. Fluid is expelled from the cracks and upwards via the fault.

100–150 years for earthquakes of this magnitude, horizontal tensile stress opens a network of vertical cracks, allowing water into the space. When the earthquake occurs and the fault breaks, the stress drops and the dilatational strain is released. The cracks close and the water is expelled. Since the fracture density is greatest close to the fault and the fault gouge provides a relatively high permeability pathway towards the surface, this is the prime route for the expulsion of water. Although the mechanism is similar to the 'seismic pumping' schemes of Sibson (1987) and Sibson et al. (1988), it links fault-related expulsion of fluids specifically with normal faulting and extensional tectonics, together with strike-slip fault systems where transtension occurs. It predicts the inverse effect for compressional faulting, in which cracks close during the interseismic periods, expelling water, but open when the earthquake causes fault rupture when water is drawn in. Of further significance for mineralization is the relatively long residence time of fluid in basement cracks with high aspect ratios under horizontal extension during interseismic periods. With a large surface area relative to volume, each crack is conducive to fluid–wall rock interaction and the possible scavenging of metals. The transport of mineralizing fluids from the crystalline basement to higher levels in the crust where they may concentrate in an ore body by this mechanism is by a large number of intermittent pulses of flow sustained over a significant period of time. For many large hydrothermal ore deposits, the total quantity of fluid required is between 10^3 and 10^4 km^3. At 0.5 km^3 per seismic event at 150 year intervals, a time span of between 200 000 and 2 million years is required.

A similar situation appears to hold in an oceanic setting. Floyd et al. (2001) report evidence of fault weakness and fluid flow from a low-angle normal fault in the Woodlark basin east of Papua New Guinea that is seismically active.

The fault is at the northern edge of the Moresby Seamount at the western tip of an active spreading ridge. The fault has been clearly imaged on a seismic reflection profile as a planar surface dipping north at 26°, visible to 9 km depth below the ocean floor. It developed 3.5 Ma ago and now has a displacement of 10 km. It is related to a number of magnitude 6 earthquakes with normal fault focal mechanisms. From a careful analysis of the reflection characteristics on the seismic section, Floyd et al. deduced that the fault gouge is 33 m in thickness with high (>30%) porosity, maintained by a high (near-lithostatic) fluid pressure. They suggest that high extensional stress and magmatic heat ahead of the ridge tip have created conditions for strain localization and hydrothermal fluid flow.

Conclusions

Currently active orogenic systems can provide modern analogues for mineralization in ancient orogens. Geodynamic processes can be observed and measured and their rates and duration quantified, both at the Earth's surface and throughout the lithosphere and upper mantle. In consequence, it is possible with modern systems to identify a variety of tectonic settings conducive to the initiation of mineralization at various stages within the evolution of an orogeny. Because geodynamic processes, such as subduction rollback and slab tear, can be modelled quantitatively, not only can the mechanism be understood but the consequential effects, such as localized heat production, partial melting, metamorphism and mineralization can be predicted. Furthermore, these models can be tested with observations on mineral deposits of Neogene age. Modern orogenic systems, such as the SE Asia–SW Pacific region, demonstrate the complexity, speed and short duration of many of the tectonic and magmatic processes involved and the ephemeral nature of many of their present features. Rapid uplift and erosion, as well as subduction can destroy many of the present features within a few million years. Much of the evidence of geodynamic processes taking place within an evolving orogen are not preserved by the time it has reached its end when relative plate movement and deformation have ceased. Thus, there are limitations to the use of modern systems to interpret what happened in ancient orogens that ran to a conclusion long ago.

As important to the formation of a large ore deposit as its tectonic setting is the mechanism for transporting mineralizing fluids to the site of ore deposition. The close association observed between the location of ore deposits and faults and fracture networks has led to the appreciation that these are the main elements of the transport system. Various mechanisms for generating upward fluid flow through networks of interconnected fracture systems and faults have been proposed, supported by numerical models, but observations of fluid flow generated by modern earthquake-related fault movements provide quantitative information. In particular, it has been observed that normal faults deliver much greater flows than strike-slip faults and that reverse faults produce practically no flow. This has clear implications for modelling mineralizing systems in the upper crust and provides quantitative information to constrain them.

References

BARLEY, M.E., RAK, P. & WYMAN, D. 2002. Tectonic controls on magmatic-hydrothermal gold mineralization in the magmatic arcs of SE Asia. *In:* BLUNDELL, D.J., NEUBAUER, F. & VON QUADT, A. (eds) *The Timing and Location of Major Ore Deposits in an Evolving Orogen.* Geological Society, London, Special Publications, **204**, 39–47.

BIJWAARD, H., SPAKMAN, W. & ENGDAHL, E.R. 1998. Closing the gap between regional and global travel time tomography. *Journal of Geophysical Research,* **103**, 30055–30078.

BINNS, R.A. & SCOTT, S.D. 1993. Actively forming polymetallic sulphide deposits associated with felsic volcanic rocks in the eastern Manus back-arc basin, Papua New Guinea. *Economic Geology,* **88**, 2226–2236.

BINNS, R.A., BARRIGA, F. & MILLER, J. 2002. Anatomy of an active felsic-hosted hydrothermal system, eastern Manus Basin, sites 1188-1191: introduction. ODP Initial Reports, 193, http://www-odp.tamu.edu/publications/193_IR/193ir.htm

BROWN, D. & SPADEA, P. 1999. Processes of forearc and accretionary complex formation during arc-continent collision in the southern Urals. *Geology,* **27**, 649–652.

CARLILE, J.C. & MITCHELL, A.H.G. 1994. Magmatic arcs and associated gold and copper mineralization in Indonesia. *Journal of Geochemical Exploration,* **50**, 91–142.

CLOETINGH, S. & BANDA, E. 1992. Europe's lithosphere – physical properties. *In:* BLUNDELL, D., FREEMAN, R. & MUELLER, S. (eds) *A continent revealed: the European Geotraverse.* Cambridge University Press, Cambridge, 80–91.

CRAMPIN, S. 1994. The fracture criticality of crustal rocks. *Geophysical Journal International,* **118**, 428–438.

COX, S.F. 1999. Deformational controls on the dynamics of fluid flow in mesothermal gold systems. *In:* MCCAFFREY, K.J.W., LONERGAN, L. & WILKINSON, J.J. (eds) *Fractures, Fluid Flow and Mineralization.* Geological Society, London, Special Publications, **155**, 123–140.

DE BOORDER, H., SPAKMAN, W., WHITE, S.H. & WORTEL, M.R. 1998. Late Cenozoic mineralization, orogenic collapse and slab detachment in the European Alpine Belt. *Earth and Planetary Science Letters,* **164**, 569–575.

FLOYD, J.S., MUTTER, J.C., GOODLIFFE, A.M. & TAYLOR, B. 2001. Evidence for fault weakness and fluid flow within an active low-angle normal fault. *Nature,* **411**, 779–783.

GOLDFARB, R.J., GROVES, D.I. & GARDOLL, S. 2001. Orogenic gold and geologic time: a global synthesis. *Ore Geology Reviews,* **18**, 1–75.

HALL, R. 1996. Reconstructing Cenozoic SE Asia. *In:* HALL, R. & BLUNDELL, D.J. (eds) *Tectonic Evolution of SE Asia.* Geological Society, London, Special Publications, **106**, 153–184.

HALL, R. 1999. Neogene history of collision in the Halmahera region, Indonesia. *Proceedings of the Indonesian Petroleum association,* **27**, 487–493.

HALL, R. 2002. Cenozoic geological and plate tectonic evolution of SE Asia and the SW Pacific: computer-based reconstructions, model and animations. *Journal of Asian Earth Sciences,* **20**, 353–434.

HUTCHINSON, R.W. 1973. Volcanogenic sulfide deposits and their metallogenic significance. *Economic Geology,* **68**, 1223–1246.

LEECH, M.L. 2001. Arrested orogenic development: eclogitization, delamination, and tectonic collapse. *Earth and Planetary Science Letters,* **185**, 149–159.

MACPHERSON, C.G. & HALL, R. 2001. Tectonic setting of Eocene boninite magmatism in the Izu-Bonin-Mariana forearc. *Earth and Planetary Science Letters,* **186**, 215–230.

MACPHERSON, C.G. & HALL, R. 2002. Timing and tectonic controls on magmatism and ore generation in an evolving orogen: evidence from SE Asia and the western Pacific. *In:* BLUNDELL, D.J., NEUBAUER, F. & VON QUADT, A. (eds) *The Timing and Location of Major Ore Deposits in an Evolving Orogen.* Geological Society, London, Special Publications, **204**, 49–67.

MACPHERSON, C.G., HILTON, D.R., SINTON, J.M., POREDA, R.J. & CRAIG, H. 1998. High ^{3}He/^{4}He ratios in the Manus Back-Arc Basin: implications for mantle mixing and the origin of plumes in the western Pacific Ocean. *Geology,* **26**, 1007–1010.

MCCAFFREY, R. 1996. Slip partitioning at convergent

plate boundaries of SE Asia. *In:* HALL, R. &
BLUNDELL, D.J. (eds) *Tectonic Evolution of SE
Asia.* Geological Society, London, Special Publica-
tions, **106**, 3–18.

MUIR-WOOD, R. & KING, G.C.P. 1993. Hydrological
signatures of earthquake strain. *Journal of Geophysical Research,* **98B**, 22035–22068.

NEUBAUER, F. 2002. Contrasting Late Cretaceous with
Neogene ore provinces in the Alpine-Balkan-Carpathian-Dinaride collision belt. *In:* BLUNDELL, D.J.,
NEUBAUER, F. & VON QUADT, A. (eds) *The Timing
and Location of Major Ore Deposits in an Evolving
Orogen.* Geological Society, London, Special Publications, **204**, 81–102.

ORD, A., WALSHE, J.L. & HOBBS, B.E. 1999. Geody-
namics and giant ore deposits. *In:* STANLEY, C.J. *ET
AL.* (eds) *Mineral deposits: processes to processing.*
A.A. Balkema, Rotterdam, 1341–1344.

PUBELLIER, M. & EGO, F. 2002. Anatomy of an escape
tectonic zone, Irian Jaya (Indonesia). *Tectonics,* **27**,
in press.

RICHARDSON, A.N. & BLUNDELL, D.J. 1996. Continental
collision in the Banda Arc. *In:* HALL, R. &
BLUNDELL, D.J. (eds) *Tectonic Evolution of SE
Asia.* Geological Society, London, Special Publications, **106**, 47–60.

SANDERSON, D.J. & ZHANG, X. 1999. Critical stress
localization of flow associated with deformation of
well-fractured rock masses, with implications for
mineral deposits. *In:* MCCAFFREY, K.J.W., LONE-
RGAN, L. & WILKINSON, J.J. (eds) *Fractures, Fluid
Flow and Mineralization.* Geological Society, London, Special Publications, **155**, 69–81.

SIBSON, R.H. 1981. Fluid flow accompanying faulting:
field evidence and models. *In:* SIMPSON, D.W. &
RICHARDS, P.G. (eds) *Earthquake Prediction.* AGU
Maurice Ewing Series, **4**, 593–603.

SIBSON, R.H. 1987. Earthquake rupturing as a mineralizing agent in hydrothermal systems. *Geology,* **15**,
701–704.

SIBSON, R.H. 1996. Structural permeability of fluid-
driven fault-fracture meshes. *Journal of Structural
Geology,* **18**, 1031–1042.

SIBSON, R.H., ROBERT, F. & PAULSEN, K.H. 1988. High
angle reverse faults, fluid pressure cycling and
mesothermal gold deposits. *Geology,* **16**, 551–555.

SILLITOE, R.H. 1997. Characteristics and controls of the
largest porphyry copper-gold and epithermal gold
deposits in the circum-Pacific region. *Australian
Journal of Earth Sciences,* **44**, 373–388.

SILLITOE, R.H. 1999. VMS and porphyry copper deposits: Products of discrete tectono-magmatic settings.
In: STANLEY, C.J. *ET AL.* (eds) *Mineral deposits:
processes to processing.* A.A. Balkema, Rotterdam,
7–10.

SPAKMAN, W. & BIJWAARD, H. 1998. Mantle structure
and large-scale dynamics of South-East Asia. *In:*
WILSON, P. & MICHEL, G.W. (eds) *The Geodynamics of S and SE Asia (GEODYSSEA) Project.*
GeoForschingsZentrum, Potsdam, 313–339.

VAN STAAL, C.R., DEWEY, J.F., MAC NIOCIALL, C. &
MCKERROW, W.S. 1998. The Cambrian–Silurian
tectonic evolution of the northern Appalachian and
British Caledonides: history of a complex, west and
soutwest Pacific-type segment of Iapetus. *In:* BLUN-
DELL, D.J. & SCOTT, A.C. (eds) *Lyell: the Past is
the Key to the Present.* Geological Society. London,
Special Publications, **143**, 199–242.

WORTEL, M.J.R. & SPAKMAN, W. 2000. Subduction and
slab detachment in the Mediterranean–Carpathian
region. *Science,* **290**, 1910–1917.

WYMAN, D.A. 1999. A 2.7 Ga depleted tholeiite suite:
evidence of plume-arc interaction in the Abitibi
Greenstone Belt, Canada. *Precambrian Research,*
97, 27–42.

ZAYKOV, V. 2000. Mineralisation in the Urals - VHMS
focus. *In:* HERRINGTON, R. (ed.) *Report of GEODE
workshop, London, April 2000.*

Global comparisons of volcanic-associated massive sulphide districts

RODNEY L. ALLEN[1,2], PÄR WEIHED[3] & THE GLOBAL VMS RESEARCH PROJECT TEAM[4]

[1]*Research Institute of Materials and Resources, Akita University, 1-1 Tegata-gakuen-cho, Akita 010-8502, Japan*

[2]*Present address: Volcanic Resources Limited, Guldgatan 11, 936 32 Boliden, Sweden (e-mail: rodallen@algonet.se)*

[3]*CTMG, Division of Applied Geology, Luleå University of Technology, 971 87 Luleå, Sweden (e-mail: par.weihed@sb.luth.se)*

[4]*Main contributors: Derek Blundell, Tony Crawford, Gary Davidson, Alan Galley, Harold Gibson, Mark Hannington, Richard Herrington, Peter Herzig, Ross Large, David Lentz, Valery Maslennikov, Steve McCutcheon, Jan Peter and Fernando Tornos.*

Abstract: Although volcanic-associated massive sulphide (VMS) deposits have been studied extensively, the geodynamic processes that control their genesis, location and timing remain poorly understood. Comparisons among major VMS districts, based on the same criteria, have been commenced in order to ascertain which are the key geological events that result in high-value deposits. The initial phase of this global project elicited information in a common format and brought together research teams to assess the critical factors and identify questions requiring further research. Some general conclusions have emerged.

(1) All major VMS districts relate to major crustal extension resulting in graben subsidence, local or widespread deep marine conditions, and injection of mantle-derived mafic magma into the crust, commonly near convergent plate margins in a general back-arc setting.

(2) Most of the world-class VMS districts have significant volumes of felsic volcanic rocks and are attributed to extension associated with evolved island arcs, island arcs with continental basement, continental margins, or thickened oceanic crust.

(3) They occur in a part of the extensional province where peak extension was dramatic but short-lived (failed rifts). In almost all VMS districts, the time span for development of the major ore deposits is less than a few million years, regardless of the time span of the enclosing volcanic succession.

(4) All of the major VMS districts show a coincidence of felsic and mafic volcanic rocks in the stratigraphic intervals that host the major ore deposits. However, it is not possible to generalize that specific magma compositions or affinities are preferentially related to major VMS deposits world-wide.

(5) The main VMS ores are concentrated near the top of the major syn-rift felsic volcanic unit. They are commonly followed by a significant change in the pattern, composition and intensity of volcanism and sedimentation.

(6) Most major VMS deposits are associated with proximal (near-vent) rhyolitic facies associations. In each district, deposits are often preferentially associated with a late stage in the evolution of a particular style of rhyolite volcano.

(7) The chemistry of the footwall rocks appears to be the biggest control on the mineralogy of the ore deposits, although there may be some contribution from magmatic fluids.

(8) Exhalites mark the ore horizon in some districts, but there is uncertainty about how to distinguish exhalites related to VMS from other exhalites and altered, bedded, fine grained tuffaceous rocks.

(9) Most VMS districts have suffered fold-thrust belt type deformation, because they formed in short-lived extensional basins near plate margins, which become inverted and deformed during inevitable basin closure.

(10) The specific timing and volcanic setting of many VMS deposits, suggest that either the felsic magmatic–hydrothermal cycle creates and focuses an important part of the ore solution, or that specific types of volcanism control when and where a metal-bearing geothermal solution can be focused and expelled to the sea floor, or both.

From: BLUNDELL, D.J., NEUBAUER, F. & VON QUADT, A. (eds) 2002. *The Timing and Location of Major Ore Deposits in an Evolving Orogen*. Geological Society, London, Special Publications, **204**, 13–37. 0305-8719/02/$15.00 © The Geological Society of London 2002.

This and other questions remain to be addressed in the next phase of the project. This will include in-depth accounts of VMS deposits and their regional setting and will focus on an integrated multi-disciplinary approach to determine how mineralisation, volcanic evolution and extensional tectonic evolution are interrelated in a number of world-class VMS districts.

Volcanic-associated massive sulphide (VMS) deposits are one of the world's major sources of zinc, copper, lead, silver and gold (Fig. 1). These deposits form an important part of the metal-mining industry in Australia, Canada and Europe (Sweden, Spain, Portugal and Russia). Although studied extensively at the deposit and district scale, many questions remain about the fundamental geodynamic processes that control the genesis, deposit characteristics and the timing and location of major VMS deposits. The fundamental question of what controls the distribution and timing of world-class VMS ore deposits and districts cannot be answered by considering just one deposit, or even one mining district, but requires global comparisons among districts to determine which key geological events were common to districts with high-value deposits and/or numerous economic deposits. These comparisons need to be carried out using similar criteria and expertise in each district. The project reported here seeks to carry

out these global comparisons. No such study has been attempted previously.

In most VMS districts there is one main stratigraphic position (or time line) on which the major VMS deposits are developed (Fig. 2). At the local scale, this time line commonly corresponds to a particular stage in the evolution of individual submarine volcanoes (e.g. Horikoshi 1969) and at the sub-regional scale it may represent a particular magmatic event in the evolution of the volcanic succession. At the mining district scale, this time line in turn represents a particular stage in the tectonic evolution of the whole region (e.g. Allen 1992). Defining the key magmatic/volcanic events and tectonic stage(s) that relate directly to major massive sulphide development, and exploring the connection between them from the local to regional scales, is critically important from both an ore genesis and mineral exploration perspective. By studying this problem in a series of global volcanic belts that host major VMS deposits, it should

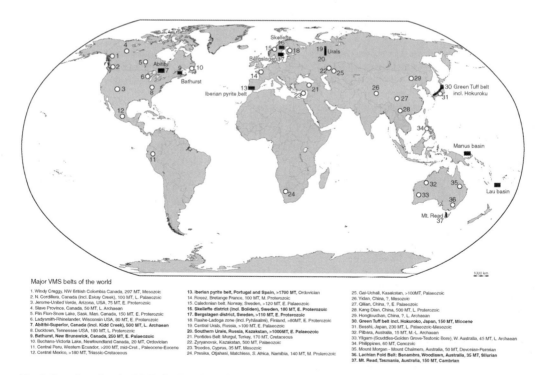

Major VMS belts of the world

1. Windy Craggy, NW British Colombia Canada, 297 MT, Mesozoic
2. N. Cordillera, Canada (incl. Eskay Creek), 100 MT, L. Palaeozoic
3. Jerome-United Verde, Arizona, USA, 75 MT, E. Proterozoic
4. Slave Province, Canada, 50 MT, L. Archaean
5. Flin Flon-Snow Lake, Sask. Man. Canada, 150 MT, E. Proterozoic
6. Ladysmith-Rhinelander, Wisconsin USA, 80 MT, E. Proterozoic
7. **Abitibi-Superior, Canada (incl. Kidd Creek), 500 MT, L. Archaean**
8. Ducktown, Tennessee USA, 180 MT, L. Proterozoic
9. **Bathurst, New Brunswick, Canada, 250 MT, E. Palaeozoic**
10. Buchans-Victoria Lake, Newfoundland Canada, 20 MT, Ordovician
11. Central Peru, Western Ecuador, >200 MT, mid-Cret., Paleocene-Eocene
12. Central Mexico, >180 MT, Triassic-Cretaceous

13. **Iberian pyrite belt, Portugal and Spain, >1700 MT, Ordovician**
14. Rouez, Bretange France, 100 MT, M. Proterozoic
15. Caledonian belt, Norway, Sweden, >120 MT, E. Palaeozoic
16. **Skellefte district (incl. Boliden), Sweden, 180 MT, E. Proterozoic**
17. **Bergslagen district, Sweden, >110 MT, E. Proterozoic**
18. Raahe-Ladoga zone (incl. Pyhäsalmi), Finland, >80MT, E. Proterozoic
19. Central Urals, Russia, >100 MT, E. Palaeozoic
20. **Southern Urals, Russia, Kazakstan, >1000MT, E. Palaeozoic**
21. Pontides Belt: Murgul, Turkey, 170 MT, Cretaceous
22. Zyryanovsk, Kazakstan, 500 MT, Palaeozoic
23. Troodos, Cyprus, 35 MT, Mesozoic
24. Presika, Otjahasi, Matchless, S. Africa, Namibia, 140 MT, M. Proterozoic

25. Gai-Uchali, Kasakstan, >100MT, Palaeozoic
26. Yidan, China, ?, Mesozoic
27. Qilian, China, ?, E Palaeozoic
28. Kang Dian, China, 500 MT, L. Proterozoic
29. Hongtouchan, China, ?, L. Archaean
30. **Green Tuff belt incl. Hokuroko, Japan, 150 MT, Miocene**
31. Besshi, Japan, 230 MT, L. Palaeozoic-Mesozoic
32. Pilbara, Australia, 15 MT, M.-L. Archaean
33. Yilgarn (Scuddles-Golden Grove-Teutonic Bore), W. Australia, 45 MT, L. Archaean
34. Philippines, 60 MT, Cenozoic
35. Mount Morgan - Mount Chalmers, Australia, 50 MT, Devonian-Permian
36. **Lachlan Fold Belt: Benambra, Woodlawn, Australia, 35 MT, Silurian**
37. Mt. Read, Tasmania, Australia, 150 MT, Cambrian

Fig. 1. Location of major VMS districts, including (in black) districts involved in the Global VMS research project.

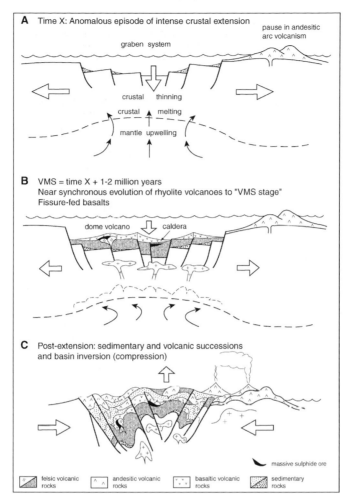

Fig. 2. Schematic tectonic-volcanic model encompassing the possible relationships between crustal-scale tectonism, volcanism and formation of VMS deposits. The model provides a theme that links the component studies of the Global VMS research project.

be possible to identify and compare these key events, and thus significantly improve our understanding of massive sulphide formation, and mineral exploration targeting. A further reason for choosing this approach is that the mining industry now operates with a global perspective and mining companies are interested in global scientific appraisals of important ore deposits. Consequently an aim of this project is to determine if it is possible to produce better genetic models for VMS deposits that can be applied globally.

This paper describes the start of the 'Global VMS research project' and provides preliminary results presented at the 'Volcanic Environments and Massive Sulphide Deposits' conference and associated workshop, Hobart, Tasmania in November 2000. In this contribution, emphasis is placed

on comparing the tectonic and volcanic setting of VMS districts and deposits. Figure 2 provides the general context for the settings and evolution of many VMS districts, and serves as the theme that links our VMS studies.

A database of major VMS districts

For the reasons described above, the European Science Foundation GEODE (Geodynamics and Ore Deposit Evolution) programme and CODES (Centre for Ore Deposit Studies, University of Tasmania) set up a project in 1999, with a global perspective, to compare and contrast major VMS ore deposits and their settings. This brought together research teams conversant with the VMS provinces of Abitibi and the Bathurst Camp in

Canada, the Mount Read Volcanics in Tasmania, the Iberian Pyrite Belt (IPB) of Portugal and Spain, the Bergslagen and Skellefte districts in Sweden, and the southern Urals in Russia. Active plate tectonics, hydrothermal processes and mineralization in the Manus and Lau basins in the SW Pacific were included to provide modern analogues.

A database was developed by means of a questionnaire, drawn up by Professor Large (CODES). The questionnaire comprised a set of questions concerned with the tectonic and structural setting, timing and location of mineralization, volcanic architecture, styles of ore deposits, ore deposit characteristics, favourable horizons, exhalites, alteration facies, and mechanisms of ore genesis involving the metal and fluid sources, plumbing and fluid circulation, fluid–rock interactions and the thermo-mechanical driving forces. Each team responded with the relevant information, based on the current state of knowledge, to create a database with which to compare features in a consistent fashion. The database is accompanied by a bibliography of the key publications relating to each of the VMS districts. The data have been compiled on CD-ROM and are freely available from CODES. They provide systematic information on what is known about the VMS deposits and also indicate where there are gaps in knowledge. The teams met at a workshop held in Hobart, Tasmania in November 2000, to appraise the data. A summary of the results of the workshop is provided in this paper.

Brief descriptions of some major VMS districts

Eleven regions with major VMS deposits are briefly described below. They are arranged according to their stratigraphy and interpreted tectonic setting, from island arc terranes and rifted continental margin arcs (Green Tuff Belt, Skellefte, Urals), to rifted continental margins without thick andesitic arc successions (Lachlan Fold Belt, Bathurst, Mount Read Volcanics), to continental margin regions with anomalously intense felsic volcanism (Bergslagen) and relatively minor volcanism (Iberian Pyrite Belt), to an Archaean province (Abitibi) and two young back-arc basins that contain VMS mineralization (Lau and Manus Basins). The young age (Miocene) and extensive research of the Green Tuff belt have enabled the construction of a more detailed and reliable record of this belt's evolution than most, if not all, other VMS districts. Consequently, the Green Tuff belt is described in more detail than the other regions.

Green Tuff Belt, Japan

The Japan volcanic arc is built on 30 km thick, Palaeozoic–Oligocene continental crust and was part of a continental margin arc along the eastern margin of the Eurasian continent from at least the Cretaceous to the Early Miocene (about 70–24 Ma) (Taira *et al.* 1989). Japan separated from the Eurasian continent during back-arc extension in the Late Oligocene–Middle Miocene (28–14 Ma) (Otofuji *et al.* 1985; Tamaki *et al.* 1992). The back-arc basin is mainly floored by extended continental crust. New oceanic crust formed by spreading is known in only a part of the basin. The Green Tuff belt (Fig. 3) represents the marine volcanic and sedimentary succession that formed along the eastern, arc-side of the back-arc basin in response to rifting (Ohguchi *et al.* 1989). The belt is 1500 km long and contains a complex, 1–5 km thick, Lower Miocene to Lower Pliocene (24–4 Ma) volcanic stratigraphy. The belt contains many massive sulphide (kuroko) districts. However, the most famous and important is the Hokuroku district in northern Honshu (Fig. 3), which contains eight main clusters of massive sulphide lenses and several other scattered smaller deposits (Ishihara *et al.* 1974; Ohmoto & Skinner 1983).

Along the far western side of Japan, including Oga Peninsula (Fig. 3), the type area of the Green Tuff, alkaline to calc-alkaline andesite and lesser basalt and rhyolite were erupted in a subaerial to littoral environment during the Late Oligocene to Early Miocene (35–24 Ma) (Ohguchi 1983; Ohguchi *et al.* 1989, 1995). This volcanism migrated eastwards towards the trench from 30–20 Ma (Ohguchi *et al.* 1989) and appears to have been a precursor to major back-arc rifting, which commenced at 28–21 Ma (Otofuji *et al.* 1985; Tamaki *et al.* 1992). Rifting peaked at 16–15 Ma, when there was a strong rotation of the Japanese arc (Otofuji *et al.* 1985), a peak in the marine transgression, and rapid fault-controlled subsidence of up to 3 km in 1 million years, which formed deep marine grabens along the western side of Japan (within the Green Tuff belt) (Yamaji & Sato 1989; Sato & Amano 1991). Intense basalt-dominant bimodal (basalt–rhyolite) submarine volcanism closely followed this rapid subsidence along the present Japan Sea Coast of northern Honshu, and intense rhyolite-dominant bimodal submarine volcanism and the kuroko mineralization followed the subsidence in the present Backbone Ranges belt to the east. The basalts are tholeiitic to transitional in composition, whereas the rhyolites are calc-alkaline (Konda 1974; Dudás *et al.* 1983). The crustal stress regime at this time, determined from the orienta-

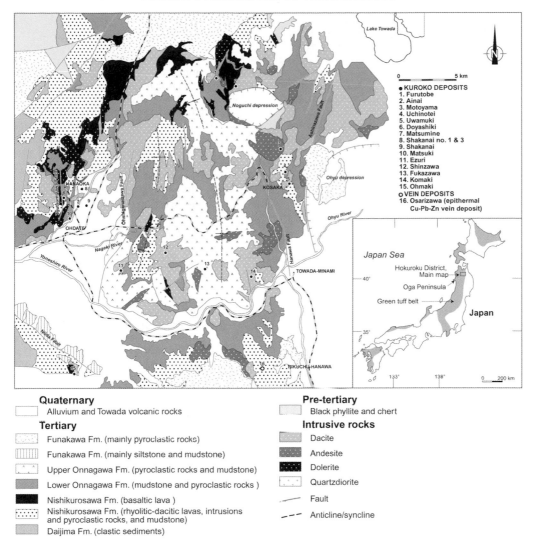

Fig. 3. Summary geological map of the Hokuroku VMS district. Modified from Tanimura *et al.* (1983).

tion of dykes, faults and veins, comprised east–west to SE–NW extension (Yamagishi & Watanabe 1986; Sato 1994).

The intense early Mid Miocene (16–15 Ma) subsidence was followed (Mid–Late Miocene, 14–10 Ma) by a transitional stress field and slow (thermal) subsidence (Sato & Amano 1991; Sato 1994). Bimodal volcanism continued from about 16–12 Ma, then changed to more typical arc-composition calc-alkaline andesite–dacite–rhyolite volcanism (Konda 1974; Okamura 1987). In the Japan Sea coast area, the slow subsidence continued into the Early Pliocene (14–3 Ma) and a thick, deep marine mudstone–turbidite succession with subordinate volcanic rocks accumulated. However to the east, the present Backbone

Ranges, including the major Kuroko districts, began to uplift during the Late Miocene–Early Pliocene (10–3 Ma) and numerous felsic caldera volcanoes formed in terrestrial to shallow marine environments. This uplift and volcanism may have been caused by igneous underplating (Sato 1994). A compressional stress regime and reactivation of the Miocene normal faults as thrusts commenced in the Late Pliocene, and resulted in widespread terrestrial conditions and the onset of the present arc regime (Yamagishi & Watanabe 1986; Sato 1994).

The kuroko deposits occur in a narrow time-stratigraphic interval (Mid Miocene 15–12 Ma) within the deep marine, bimodal (rhyolite–basalt) volcanic successions. The Miocene strata are mainly gently dipping, but complex volcanic facies architecture and abundant normal faults make it difficult to correlate stratigraphy from one submarine volcano or fault block to the next. Where the volcanic rocks are interbedded with mudstones, fossil control (mainly foraminifera, nannofossils) enables resolution of 2–3 Ma time-stratigraphic intervals. In the Hokuroku district, the volcanic succession comprises a lower, 300–1000 m thick, massive rhyolite–basalt complex and an overlying 300–1000 m thick, more strati-fied succession dominated by alternating tuffaceous mudstone and thick subaqueous mass flow units of felsic pyroclastic debris. The lower volcanic complex comprises several overlapping volcanic centres dominated by felsic lava domes, intrusive domes and related autoclastic and pyroclastic rocks (Horikoshi 1969; Ishihara *et al.* 1974). The kuroko deposits are Zn–Pb–Cu–Ag–Au type VMS deposits that formed on and below the seafloor, mainly in the upper, proximal (near vent) part of these volcanoes (Horikoshi 1969).

The name 'Green Tuff' is derived from the various shades of green of the volcanics due to marine diagenetic and hydrothermal alteration, and the tuffaceous (pyroclastic) origin ascribed to many of the rocks. The regional diagenetic altera-tion is characterized by a vertical zonation of clays (especially montmorillonite and saponite) and zeolites (Iijima 1974), formed under high geother-mal gradients of over $100\,°C\,km^{-1}$ (Utada 1991). The kuroko deposits are enclosed by local hydrothermal alteration that overprints and inter-fingers with the diagenetic alteration, and is zoned from the ore deposit to the margin as follows: quartz–K-feldspar, kaolinite, sericite–chlorite, montmorillonite–illite, montmorillonite and ana-lcite–calcite (Iijima 1974; Utada 1991).

Skellefte district, northern Sweden

The Skellefte district (Fig. 4) is a 120×30 km Early Proterozoic (1.90–1.88 Ga) magmatic re-gion that contains over 80 massive sulphide deposits (Rickard 1986; Weihed *et al.* 1992). The district lies between a region of continental, mainly felsic, volcanic rocks of similar to slightly younger age to the north, and a large region of deep marine sedimentary and subordinate mafic volcanic rocks, intruded by numerous granitoids, to the south and east. The marine sedimentary succession south of the Skellefte district appears to span an age range from older than to younger than the Skellefte district volcanic rocks (Lund-qvist *et al.* 1998).

The Skellefte district contains a >7 km thick stratigraphy of calc-alkaline basalt–andesite–da-cite–rhyolite, tholeiitic basalt–andesite–dacite, high Mg (komatiitic) basalt and subordinate sedi-mentary rocks, and is intruded by syn- and post-volcanic granitoids (Fig. 4; Vivallo & Claesson 1987; Allen *et al.* 1996*b*). The rocks are generally strongly deformed, steeply dipping and are meta-morphosed from greenschist to amphibolite facies. Primitive isotopic signatures suggest that magmas were mainly mantle-derived (Billström & Weihed 1996). The stratigraphy is very complex, laterally variable, diachronous, and marker horizons are rare. The only consistent regional stratigraphic pattern is a first order cycle comprising a lower >3 km thick marine volcanic complex (*c.* 1882–1890 Ma), overlain diachronously by a >4 km thick, mixed sedimentary and volcanic sequence. The lower volcanic complex consists of interfin-gering and overlapping rhyolite, dacite–andesite and basalt–andesite–dacite volcanoes (Allen *et al.* 1996*b*). However, about 50% of the volcanic rocks are rhyolites. Subaqueous lava, intrusion, autoclas-tic and pyroclastic facies are all common. Deep subaqueous (below wave base) depositional envir-onments are dominant throughout the lower volca-nic complex, which indicates that strong extension and subsidence preceded and/or accompanied the volcanism (Allen *et al.* 1996*b*). The overlying mixed sedimentary and volcanic sequence records uplift, erosion and renewed rifting, and includes medial–distal facies of the voluminous continental felsic magmatism (1877 Ma) that occurred directly north of the Skellefte district. The stratigraphic architecture, range of volcanic compositions and abundance of rhyolites suggest that the Skellefte district is a remnant of a strongly extensional intra-arc region that developed on continental or mature arc crust.

Most VMS deposits in the Skellefte district occur in near-vent facies associations at the top of local volcanic cycles (volcanoes), especially rhyo-litic dome-tuff cone volcanoes (Allen *et al.* 1996*b*). Regionally, these VMS deposits occur on at least two stratigraphic levels, but the highest concentration occurs near the upper contact of the main volcanic complex. The VMS deposits span a wide range in composition, geometry and altera-tion patterns: the main compositional types are Au–As–Cu–Zn, pyritic Zn–Pb–Cu–Au–Ag and pyritic Zn–Cu–Au–Ag deposits. Several deposits are Au-rich. The deposits are associated with strong quartz–sericite, chlorite (or phlogopite–cordierite), andalusite–muscovite and carbonate alteration.

Urals

The Urals orogenic belt extends for 2500 km and is up to 200 km wide. The main VMS deposits are confined to the southern half of the belt (Fig. 5).

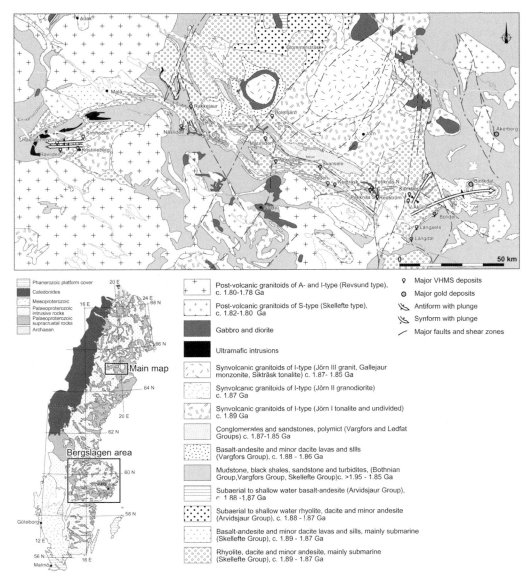

Fig. 4. Summary geological map of the Skellefte VMS district. Modified from Allen *et al.* (1996b).

They fall into two main age groups, Ordovician–Silurian and Devonian. The former group occurs west of the Main Urals Fault, which marks the suture representing the closure of the Palaeozoic Urals palaeo-ocean, and the latter group occurs to the east of the suture (Koroteev *et al.* 1997; Puchkov 1997). Altogether the various types of VMS deposits have a pre-mining tonnage of over 1000 Mt (Prokin & Buslaev 1999; Herrington 1999; Herrington *et al.* 2000).

The southern Urals can be further divided from west to east into the Sakmara Zone (SZ), the Main Uralian Fault Zone (MUFZ) and the Mednogorsk

Island Arc System (MIAS), all sub-parallel to the Main Uralian Fault (Fig. 5; Puchkov 1997; Koroteev *et al.* 1997; Herrington 1999; Herrington *et al.* 2000). The Urals palaeo-ocean was initiated at *c.* 460 Ma. The first known island arc terrane comprises Silurian volcanics in the SZ and these rocks host the oldest VMS deposits. The initiation of the Magnitogorsk island arc, located east of the MUFZ, started around 390 Ma with Caledonian collision between Baltica and Siberia. The arc sequence collided with the East European Continent diachronously, at around 360 Ma in the southern Urals, and in the Carboniferous in the

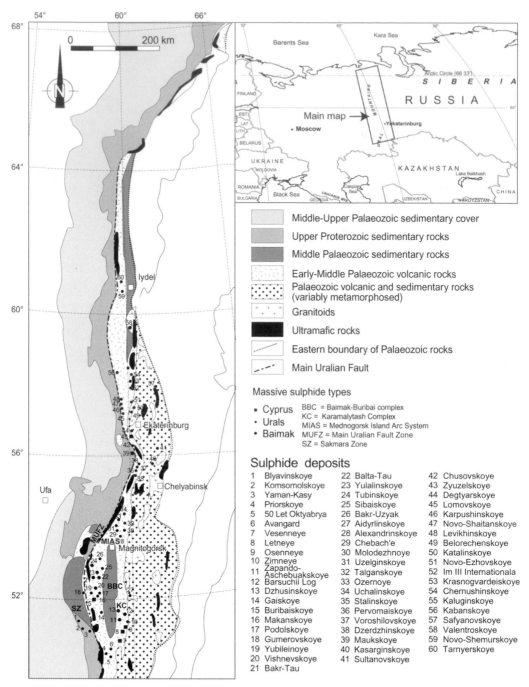

Fig. 5. Summary geological map of the southern Urals VMS district. Modified from Prokin *et al.* (1998).

central Urals. Subsequently, there was an easterly shift and an apparent switch in polarity of subduction prior to final continent–continent collision. At the western margin of the MUFZ, 400 Ma old boninites record the onset of eastward intraoceanic subduction, followed by an eastward temporal and spatial progression of several arc-volcanic complexes. Approach of the East European continental margin to the subduction zone, resulted in arc–continent collision (Puchkov 1997; Brown & Spadea 1999).

Arc-related VMS deposits are confined to the Baimak–Buribai Complex (BBC) and the Karamalytash Complex (KC) (Koroteev, *et al.* 1997; Prokin & Buslaev 1999; Herrington 1999; Herrington *et al.* 2000). The VMS deposits span a range in types and compositions but are dominated by Cu–Zn, Cu and Pyrite types. The allochthonous oceanic fragments of the Sakmara Zone contain the Cu–Zn VMS deposits of the Mednogorsk district, associated with areas of andesite-felsic rocks. The MUFZ contains several small uneconomic serpentinite- and basalt-hosted massive sulphides of debatable origins. The VMS deposits of the MIAS are associated with intermediate-felsic successions in the western MIAS and with felsic–mafic successions in the eastern MIAS.

Coherent lava flows are interpreted to be much more abundant than volcaniclastic rocks in the vicinity of the VMS deposits and many deposits are associated with rhyolitic–dacitic domes. The roofs of plagiogranite intrusions have been drilled below some of the VMS deposits. The deposits have extensive footwall alteration and most show clear evidence of having formed on the seafloor. The largest deposits are thought to be concentrated on two or three stratigraphic levels within each terrane (Herrington 1999; Herrington *et al.* 2000).

Lachlan Fold Belt, SE Australia

The Lachlan Fold Belt contains a >10 km thick Cambrian to Carboniferous stratigraphy that records a complex history of basin development, waves of diachronous deformation, magmatism, accretion and continental growth at the eastern margin of Gondwana (Cas 1983; Gray *et al.* 1997). Although there is no preserved distinct andesitic volcanic arc, the pattern of sedimentation, magmatism and deformation is attributed to plate convergence and subduction.

During the mid-late Silurian, a series of graben basins and widespread shallow and deep marine environments developed over an area of at least 600 by 400 km in response to transtension along the continental margin (Powell 1983). An enormous volume of S- and I-type granitoid magma was also intruded to high crustal levels in this area from Late Silurian–Early Devonian. The basins and granitoids are interpreted to have formed in a segment of continental crust between converging oceanic crust to the east and a marginal basin to the west (Gray *et al.* 1997). The mid-late Silurian basins thus have an essentially ensialic intra-arc

rift setting (Cas & Jones 1979). The basins have broadly similar stratigraphies, suggesting a common evolutionary cycle (Allen 1992): (1) crustal-derived felsic magmatism formed volcanic piles up to 3 km thick in terrestrial to shallow marine environments with fringing carbonate reefs and platforms, then (2) graben subsidence to deep marine conditions during the late stage of felsic volcanism, (3) accumulation of deep marine sediments and local eruption of rhyolite and mantle-derived basalt ± andesite–dacite within the grabens and finally (4) closure and uplift (structural inversion) of the basins between the end of the Silurian and Mid-Devonian. The Silurian basin successions were strongly deformed and metamorphosed to greenschist facies at this time.

VMS deposits formed in the grabens directly after initial deep subsidence. They formed in the vent areas of rhyolite dome volcanoes near the top of the felsic successions, at the time and place where mantle derived basalt ± andesite–dacite first erupted (Allen 1992; Stolz *et al.* 1997). The deep marine volcanic rocks are mainly lavas and their autoclastic facies, shallow sills, and mass flow units of pyroclastic debris that were shed into the basins from the basin margins. The VMS deposits are associated with strong quartz–sericite and chlorite alteration.

Bathurst

The Bathurst mining camp in eastern Canada comprises a 100 × 75 km area of complexly deformed Ordovician (480–457 Ma) volcanic and sedimentary units intruded by syn-volcanic plutons (Fig. 6). The region is interpreted as an ensialic back-arc basin that was strongly deformed and metamorphosed to upper greenschist facies during closure of the basin in the Late Ordovician to Late Silurian (van Staal 1987).

The district hosts about 35 VMS deposits of Pb–Zn–Cu–Ag type with a total tonnage of over 250 Mt (McCutcheon 1992). The deposits are associated with a bimodal, rhyolite–rhyodacite dominated, marine volcano-sedimentary sequence. The felsic rocks are attributed to partial melting of the continental basement, whereas the mafic rocks are tholeiitic to alkaline basalts (Lentz 1999). In the Brunswick belt, the most productive part of the district, the VMS deposits formed directly after a major episode of felsic pyroclastic volcanism (Nepisiguit Falls Formation of the Tetagouche Group) and before the deposition of overlying rhyolitic lavas, hyaloclastites, tuffs and tuffaceous sedimentary rocks. The large volume of juvenile pyroclastic rocks in the footwall to the ore deposits suggests that the ores may have formed in calderas following climactic eruption and cal-

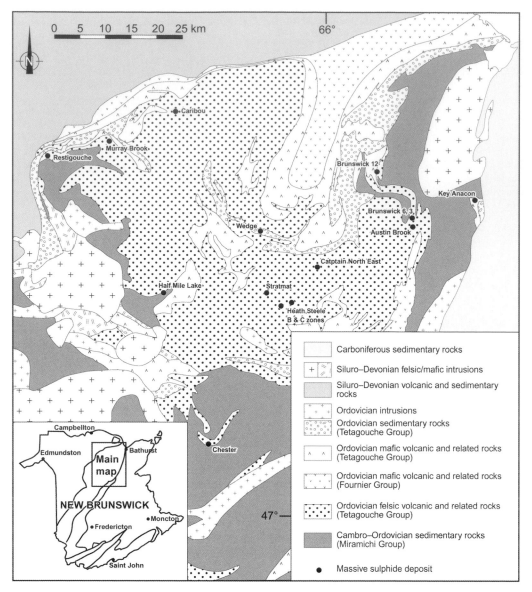

Fig. 6. Summary geological map of the Bathurst VMS district. Modified from Lentz (1999).

dera subsidence. However, details of the volcanic architecture are not known.

The VMS deposits are mainly of stratiform-type, are hosted by fine-grained volcaniclastic and sedimentary rocks, and are in part associated with iron formation exhalites that generally extend <1 km along strike, but locally up to 5 km (Saif 1983; Peter & Goodfellow 1996). The deposits are interpreted to have formed in deep to very deep water. Some other deposits are stratabound and occur in first or second-cycle felsic fragmental rocks.

Mount Read Volcanics, Tasmania

This region comprises a 200×20 km area of Cambrian, moderately to strongly deformed, mainly felsic volcanic rocks and volcano-sedimentary successions, lying on and between a series of Precambrian basement blocks near a margin of the Gondwanan continent (Fig. 7; Corbett 1992). The district contains a series of rich VMS deposits that span a wide range in deposit styles (Green *et al.* 1981; Large *et al.* 1988; Gemmell & Large 1992; Large 1992; Halley & Roberts 1997; Solomon &

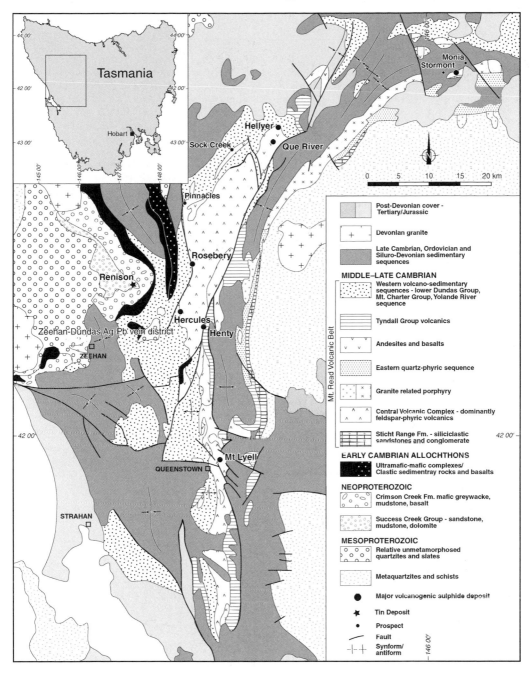

Fig. 7. Summary geological map of the Mount Read Volcanic VMS district. Modified from Corbett (1992).

Khin Zaw 1997; Corbett 2001). Metamorphic grade is mainly lower greenschist facies.

In the central, best-known part of the district, the lowest exposed stratigraphic unit is a >1 km thick, marine tholeiitic basalt–andesite volcanic complex. This is overlain by the 3 km thick Central Volcanic Complex (CVC), which is a mass of interfingering calc-alkaline dacite–rhyolite lavas, pyroclastic facies, shallow intrusions and local medium- to high-K calc-alkaline andesite, all emplaced in a shallow to deep marine environment (Corbett 1992). The CVC is on-lapped to the west and north by a >3 km thick, deep marine, mixed volcanic–sedimentary succession (Mt Charter Group) with calc-alkaline andesites, dacites and rhyolites, and medium-K to shoshonitic basalt–andesite. The CVC is overlain to the east by a 1 km thick, shallow marine to subaerial, rhyolitic succession with lesser basalt (Tyndall Group). Cambrian granites, contemporaneous with the volcanic rocks, occur in a belt along the eastern side of the Mount Read Volcanics (Large *et al.* 1996). Regional stratigraphic relationships are complex, laterally variable, and correlations are difficult to demonstrate (McPhie & Allen 1992; Corbett 1992). The Mount Read Volcanics are attributed to a period of extension on the Gondwanan continental margin, but the details are still debated. Crawford *et al.* (1992) argued that extension was related to crustal collapse after an intra-oceanic arc and fore-arc complex collided with, and was overthrust onto, the Gondwanan passive margin.

VMS deposits are interpreted to lie in two main stratigraphic settings: (1) at the top of the CVC in vent areas of major rhyolite–dacite volcanoes (Rosebery, Hercules, Mt Lyell deposits) and (2) in proximal (near-vent) facies associations at the top of andesite–dacite volcanoes in the mixed volcanic–sedimentary succession (Hellyer, Que River deposits). Strong quartz–sericite, chlorite, silicification and carbonate alteration are the main alteration types associated with the VMS deposits.

Bergslagen, central Sweden

Bergslagen (Fig. 4) is the intensely mineralized part of a 280 × 300 km, Early Proterozoic (1.90–1.87 Ga) felsic magmatic region of mainly medium to high metamorphic grade. The volcanic succession is 1.5 km thick and overlies turbiditic metasedimentary rocks in the east, and is over 7 km thick with no exposed base in the west (Lundström 1987; Allen *et al.* 1996a). Basement is interpreted to be Precambrian continental crust.

The volcanic succession is overwhelmingly (90%) calc-alkaline rhyolite with minor calc-alkaline dacite and andesite, and chemically unrelated, probably tholeiitic basalts. Strong K-feldspar and albite alteration occur on a regional scale, whereas Mg-alteration (talc, chlorite, phlogopite, cordierite, skarn) is more local and spatially closely associated with mineralization (Lagerblad & Gorbatschev 1985). The stratigraphy commonly follows the pattern. (1) Lower 1–5 km thick, poorly stratified felsic complex, dominated by the proximal–medial facies of interfingering and overlapping large caldera volcanoes, and minor interbedded limestone. (2) Middle 0.5–2.5 km thick, well stratified interval dominated by medial–distal juvenile volcaniclastic facies and limestone sheets. (3) Upper >3 km thick post-volcanic argillite–turbidite sequence (Baker *et al.* 1988; Allen *et al.* 1996a). Depositional environments fluctuated mainly between shallow and moderately deep subaqueous throughout accumulation of the lower and middle stratigraphic intervals, then became consistently deep subaqueous in the upper interval.

The supracrustal succession has been intruded by an enormous volume of syn- and post-volcanic granitoids and has been strongly deformed, such that it now occurs as scattered, tightly folded outliers, enveloped by granitoids. The stratigraphy reflects an evolution from intense magmatism, thermal doming and crustal extension, followed by waning extension, waning volcanism and thermal subsidence, then reversal from extension to compressional deformation and metamorphism. The region is interpreted as an intra-continental, or continental margin back-arc, extensional region (Baker *et al.* 1988; Allen *et al.* 1996a).

Bergslagen has a diverse range of ore deposits, including banded iron formation, magnetite–skarn, manganiferous skarn- and limestone-hosted iron ore, apatite–magnetite iron ore, stratiform and stratabound Zn–Pb–Ag– (Cu–Au) sulphide ores, and W skarn (Hedström *et al.*1989; Sundblad 1994; Allen *et al.* 1996a). The massive and semi-massive base-metal sulphide ores occur in interbedded limestone and volcaniclastic rocks, especially near the top of the volcanic succession. These ores occur close to proximal felsic volcanic facies and anomalous concentrations of basalt intrusions and lavas.

Iberian Pyrite Belt

The Iberian Pyrite Belt (IPB) is a 250 × 60 km belt of Upper Devonian–Lower Carboniferous sedimentary and volcanic rocks (Fig. 8), which hosts about 90 massive sulphide deposits, including eight extremely large deposits with more than 100 Mt of massive sulphide (Schermerhorn 1975; Carvalho *et al.* 1997; Leistel *et al.* 1998). It is part of the South Portuguese Zone (SPZ), the southernmost fold and thrust terrane of the Variscan orogen in Europe (Silva *et al.* 1990; Quesada 1998).

Most authors attribute the IPB to crustal extension and related magmatism, that were triggered by oblique collision of the continental SPZ with the active margin of the Iberian block to the north (Quesada 1991, 1998; Leistel *et al.* 1998; Tornos *et al.* 2002). Pull-apart graben basins with bimodal volcanic successions and VMS deposits formed in

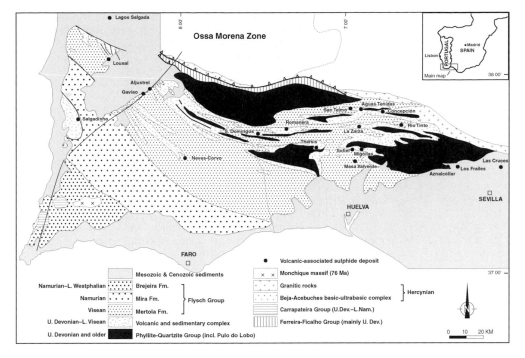

Fig. 8. Summary geological map of the Iberian Pyrite Belt VMS. Modified from Carvalho *et al.* (1997).

the IPB during this oblique collision. Continued collision ultimately resulted in obduction of the Iberian margin onto the SPZ, inversion of the extensional basins, and southward propagation of the SPZ fold and thrust belt.

The IPB has a relatively simple regional stratigraphy that can be divided into three distinct intervals. From base to top these are: (1) >1000 m of Upper Devonian terrigenous siliciclastic rocks and minor limestone, deposited on a shallow marine continental platform. (2) A 20–1000 m thick, mixed volcanic–sedimentary interval, comprising grey-black deep-water mudstones with intercalated rhyolitic–dacitic, basaltic, and lesser andesitic rocks. Basalts are tholeiitic whereas the other volcanic rocks are mainly calc-alkaline and interpreted to be partial melts of continental crust (Thieblemont *et al.* 1998). Shallow sills and associated autoclastic facies appear to be the most abundant volcanic facies, followed in abundance by lava domes/lobes and associated autoclastic facies and pyroclastic facies (Boulter 1993; Soriano & Martí 1999; Tornos *et al.* 2002). The facies pattern indicates a regionally extensive sill–lava lobe complex without distinct, large constructional volcanoes. VMS deposits and Jasper and Mn–Fe formations occur in the upper part of the volcano-sedimentary interval. In the southern part

of the IPB, a prominent 5–40 m layer of red, oxidized, shallow-water volcaniclastic sediments occurs near the top of the volcano-sedimentary interval, and grades up into (3) >3 km of first purple, then grey and black shales and turbidites (Schermerhorn 1975; Allen 2000).

Plutonic rocks of similar composition to the volcanic rocks are found in the northern part of the IPB, in a zone that probably represents the roots of the thrust complex.

The IPB has regional-scale, weak to moderate hydrothermal alteration (attributed to seafloor metamorphism) and local strong quartz–sericite and chlorite alteration zones related to VMS mineralization. The VMS deposits typically occur in grey-black mudstones, above, or intercalated with the felsic volcanic rocks. In the southern part of the belt, the VMS deposits are seafloor-type deposits and fossil control indicates that they formed at the same time (Upper Strunian) over an extensive area. In the northern part of the belt, the deposits are interpreted to have formed by replacement of mainly felsic volcanic rocks.

Abitibi

The Archaean Abitibi greenstone belt (Fig. 9) is approximately 500 × 200 km in size and origin-

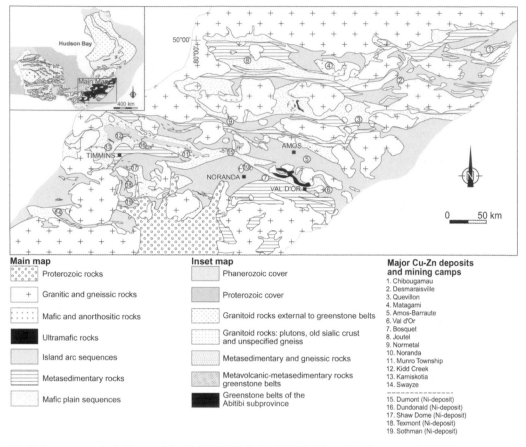

Main map

| | | **Inset map** | | **Major Cu-Zn deposits and mining camps** |

○○○ Proterozoic rocks — Phanerozoic cover

\+ Granitic and gneissic rocks — Proterozoic cover

⋯ Mafic and anorthositic rocks — Granitoid rocks external to greenstone belts

■ Ultramafic rocks — Granitoid rocks: plutons, old sialic crust and unspecified gneiss

Island arc sequences — Metasedimentary and gneissic rocks

Metasedimentary rocks — Metavolcanic-metasedimentary rocks greenstone belts

Mafic plain sequences — Greenstone belts of the Abitibi subprovince

1. Chibougamau
2. Desmaraisville
3. Quevillon
4. Matagami
5. Amos-Barraute
6. Val d'Or
7. Bosquet
8. Joutel
9. Normetal
10. Noranda
11. Munro Township
12. Kidd Creek
13. Kamiskotia
14. Swayze
- - - - - - - - - - - - - - - - -
15. Dumont (Ni-deposit)
16. Dundonald (Ni-deposit)
17. Shaw Dome (Ni-deposit)
18. Texmont (Ni-deposit)
19. Sothman (Ni-deposit)

Fig. 9. Summary geological map of the Abitibi VMS district. Modified from Hannington *et al.* (1999).

ally contained over 675 Mt of volcanic-associated massive sulphide. Most VMS deposits are associated with bimodal volcanic successions dominated by tholeiitic and komatiitic basalts, and tholeiitic–calc alkaline, high silica rhyolites (Barrie *et al.* 1993; Prior *et al.* 1999*a*; Barrett & MacLean 1999). The VMS deposits mainly occur in successions that contain over 150 m of felsic volcanic rocks.

The mineralized successions can be divided into three types: (1) bimodal, tholeiitic basalt–andesite and high silica rhyolite successions that are host to more than 50% of the VMS deposits by tonnage, but comprise only 10% of the areal distribution of volcanic rocks, (2) bimodal, transitional tholeiitic to calc-alkaline andesite and rhyolite successions that are host to about 30% of the VMS deposits by tonnage but again comprise only 10% of the areal distribution of volcanic rocks and (3) a minor calc-alkaline andesite–rhyolite assemblage that is host to only one deposit (Selbaie). Barren volcanic assemblages include

calc-alkaline basaltic andesite to rhyodacite and mafic to felsic alkalic volcanic rocks.

The different assemblages are interpreted to have formed in different tectonic settings. The first group is interpreted to have formed in thickened oceanic rift suites, similar to the Galapagos spreading centre or the Iceland rift zones. The rhyolites are thought to have formed by partial melting of mafic crust, reflecting high heat flow or mantle plumes (Barrie *et al.* 1993; Prior *et al.* 1999*b*). The second group is similar to rifted island arcs (e.g. Hokuroku) and the third group may be comparable to continental arc suites.

The mafic volcanic rocks are mainly lava flows and the felsic rocks include subaqueous mass flow units of pyroclastic debris, and lava flows and domes, including their autoclastic facies (de Rosen-Spence *et al.* 1980; Dimroth 1982). The VMS deposits are interpreted to have formed in proximal volcanic settings in mainly deep water environments. They are commonly associated with rhyolite domes, even though domes form a minor part of the successions. The deposits include both seafloor and sub-seafloor types (Kerr & Gibson 1993; Galley *et al.* 1996) and generally have strong chlorite and sericite footwall alteration zones (Riverin & Hodgson 1980; Barrett *et al.* 1991). Many of the deposits have at least a spatial association with mafic and/or felsic subvolcanic intrusions. The Rouyn–Noranda VMS camp, one of the best studied areas, is interpreted as a large shield volcano with a large central cauldron (Gibson & Watkinson 1990). Most VMS deposits occur above syn-volcanic faults within the cauldron.

Lau and Manus Basins

The Lau and Manus Basins are modern intra-oceanic back-arc basins in which sea floor sulphide deposits have recently been discovered (Hawkins 1995; Binns & Scott 1993; Fouquet *et al.* 1993). These basins are included in the Global VMS research project to facilitate comparisons between ancient and modern VMS mineralization. The Lau and Manus Basins are especially relevant to understanding the settings of major ancient VMS districts because these basins contain complex bimodal volcanic areas dominated by andesite, dacite and rhyolite (Vallier *et al.* 1991; Binns *et al.* 1996).

The VMS deposits are polymetallic mounds and chimney complexes that occur in proximal volcanic positions at the axis and flanks of spreading

ridges (Binns & Scott 1993; Herzig *et al.* 1993; Gemmell *et al.* 1999). Several of the deposits are associated with andesitic and dacitic volcanic centres at these spreading ridges. Water depths at the sites of the VMS deposits range from 1650–2500 m, and active research is in progress to sample the volcanic host rocks, the sulphide bodies and the hydrothermal fluids that are forming them.

Tectonic setting

From these summary descriptions of VMS districts and the consensus view at the Hobart workshop, we conclude that all major VMS districts relate to major crustal extension resulting in graben subsidence, marine transgression, development of local or widespread deep marine environments, and injection of mantle-derived mafic magma into the crust (Fig. 2 and Table 1). The tectonic setting is largely confined to extensional basins near convergent plate margins, commonly in a general back-arc setting. This contrasts with many of the modern systems studied, which occur at mid-ocean spreading ridges. Back-arc to intra-arc rifts and pull-apart basin settings such as the Miocene Japan arc-back arc system and the modern Okinawa trough and Lau and Manus Basins probably provide the nearest young analogues of the settings of major Proterozoic and Phanerozoic world-class VMS districts. However, several VMS districts (e.g. Bergslagen, IPB,

Table 1. *Comparison of tectonic setting and stratigraphy of six Proterozoic and Palaeozoic VMS Belts (modified after Allen 2000)*

	GTB	Skell	MRV	LFB	Berg	IPB
Deep graben subsidence → marine incursion	✔	✔	✔	✔	✔	✔
Thick, complex volcanic stratigraphy	✔	✔	✔	✔	✔	X
'Arc-like' volcanic compositions and architecture	✔X	✔X	X	X	X	X
Continental basement	✔	?	✔	✔	?✔	✔
Mature arc basement	✔X	?✔	X	X	X	X
Volcanic arc nearby (<200 km)	✔	✔	X	X	X	X
Plate margin nearby (<500 km)	✔	✔	✔	✔	?	✔
Regional stratigraphic marker horizons	X	X	X	X	✔X	✔
Extensive platform/shelf facies	X	X	X	✔	✔	✔
Coincidence of mantle-derived mafic and crustal-derived felsic magmas	✔	?✔	✔	✔	✔	✔
Bimodal volcanism regionally dominant	X	X	✔X	✔X	✔	✔
Ores mostly related to bimodal interval	✔	✔X	X	✔	✔	X
Small (S) or large volume (L) felsic volcanic centres	S	S	L	L	L	?S
VMS mainly associated with proximal felsic volcanic facies associations	✔	✔	✔	✔	?✔	?X
Most VMS related to specific volcano-type	✔	✔	X	✔	?✔	X
Timing late in a magmatic-hydrothermal cycle	✔	✔	✔	✔	✔	?✔
Pyroclastic felsic volcanism dominant	X	X	X	X	✔	X
Lava and/or shallow intrusions dominant	✔	✔	✔	✔	X	✔

GTB, Green Tuff Belt; Skell, Skellefte district; MRV, Mt Read Volcanics; LFB, Lachlan Fold Belt; Berg, Bergslagen; IPB, Iberian Pyrite Belt.

Mount Read Volcanics) do not seem to have close young analogues in terms of both tectonic setting and regional stratigraphy. The Archaean districts have similarities with some young rifted oceanic and arc terrains, but again it has not been possible to find specific very close young analogues.

In detail it seems that a variety of settings exist from extensional regions at continental margins to those associated with oceanic arcs. However, most of the world-class VMS districts have significant volumes of felsic volcanic rocks and are attributed to extensional regions associated with moderately evolved island arcs (Urals?), island arcs with continental basement, especially those that formed by the rifting of continental margins and development of limited marginal basins (Japan, Silurian stage of Lachlan Fold Belt, Bathurst?, Skellefte?) or continental margin arcs (Bergslagen?), rather than extensional regions in primitive oceanic crust. The Iberian Pyrite Belt and the Mount Read Volcanics may be examples of more complex continental margin extensional settings, in which extension and VMS mineralization occurred on a continental passive margin during, or following, collision with an arc terrane. This scenario may also be relevant to parts of the southern Urals where the giant VMS deposits occur in the latest volcanic phase, shortly after the East European continent had collided with the Urals volcanic arcs.

One key feature is that in almost all VMS districts, the actual time span for the development of all the major deposits is only a few million years or less, regardless of the time span of the enclosing volcanic succession. This short time interval of major VMS formation may result from the deposits being related to specific episodes of anomalous extension, subsidence and magmatism (Fig. 2). The overwhelming impression obtained from this appraisal is that the precise setting of the deposit (e.g. mid-ocean ridge, island-arc, continental-margin etc.) is not critical but the concurrence of a set of specific tectonic and magmatic processes related to crustal extension near a plate margin may hold the key (cf. Lentz 1998).

A criticism of recent work in ancient VMS terranes is that the determination of tectonic setting has relied too much on geochemical tectonic discrimination diagrams and too little on careful, detailed documentation of the regional stratigraphy, facies architecture, igneous petrology (including geochemistry) and structural evolution.

Critical areas that need addressing to improve understanding of the tectonic setting of VMS deposits include:

1 high resolution dating of the key tectonic, magmatic and metamorphic events;

2 detailed documentation of the stratigraphic–volcanic evolution of the host settings, especially periods of extension;

3 palinspastic reconstruction of the facies architecture of ancient VMS belts;

4 integrated, detailed studies of stratigraphy, igneous petrology and igneous geochemistry so that the evolution of magmatic suites, changes from one magmatic suite to another, and changes in geochemical affinity can be tied to stratigraphy and tectonic evolution.

Volcanic facies

All of the major VMS districts show a coincidence of felsic and mafic volcanic rocks in the stratigraphic intervals that host the major ore deposits, regardless of whether the whole province is dominated by mafic, intermediate or felsic rocks. Furthermore, the VMS districts are characterized by an anomalous combination of abundant felsic rocks and widespread relatively deep marine environments compared with broadly similar tectonic settings that do not host a major VMS district (e.g. compared with most mature oceanic volcanic arcs and continental margin arcs). In this respect the major VMS districts are not typical volcanic arcs, but reflect an important dramatic extensional tectonic–magmatic event superimposed on an arc or other type of plate margin. Examples at two extremes of the spectrum of settings are Bergslagen, which is a major thick rhyolitic magmatic province that anomalously maintained subaqueous depositional environments throughout most of its development, and the Archaean Abitibi belt, which is a dominantly oceanic terrane, but the deposits themselves are associated with local thick rhyolite sequences.

The mafic rocks most closely associated with the VMS deposits are mantle-derived basalts and basaltic andesites. They are generally tholeiitic or transitional in affinity, although in the Silurian Cowombat rift of southeastern Australia they are part of a mantle-derived calc-alkaline basalt–andesite–dacite suite (Stolz *et al.* 1997). The felsic volcanic rocks associated with the major VMS deposits are calc-alkaline or tholeiitic dacite–rhyolite and are mainly interpreted to have been derived by crustal melting (Fig. 2). Intermediate andesite–dacite suites are abundant in some of the VMS regions, but they are not preferentially associated with the major VMS deposits. Consequently, it is not possible to generalize that specific magma compositions or affinities are preferentially related to major VMS deposits world-wide, although this may be the case in individual VMS districts. However, the coincidence in time and space of felsic volcanism and mantle-derived mafic magma (in an extensional marine basin) appears critical.

Understanding the facies architecture of the host rocks is a key to understanding the setting in which the VMS deposits formed. In most districts (with the possible exception of parts of the IPB), the major VMS deposits occur within or adjacent to proximal (near-vent) rhyolitic facies associations. Submarine rhyolite dome complexes with associated pyroclastic facies are probably the most common host volcano type (Hokuroku, Skellefte, Abitibi, Silurian successions in the Lachlan Fold Belt). However, calderas dominated by rhyolitic pyroclastic facies or lava flows are also an important host (Rosebery–Hercules in the Mount Read Volcanics, Bathurst?, Noranda). Individual dome volcanoes that host VMS deposits may generally be components of larger multi-vent volcanoes or volcanic systems. However, these larger volcanic systems are at present generally poorly defined and understood. An exception is the Rouyn–Noranda VMS camp where mapping shows that rhyolite dome complexes are components of a large shield volcano and cauldron (Gibson & Watkinson 1990).

Volcaniclastic rocks are abundant in the host sequences to VMS deposits and are generally diagenetically and hydrothermally altered, and in the case of ancient terranes, also deformed and metamorphosed. Consequently, primary rock textures are partly obscured and it is no simple matter to map and interpret the origin of the rocks. Until recently, most clastic felsic volcanic rocks were assumed to be pyroclastic rocks. However, recent studies show that in many VMS districts, the abundance of pyroclastic rocks has been greatly overestimated and that many of the apparent clastic rocks are hyaloclastites, autobreccias, debris flow deposits, and even altered coherent lavas, all generated by non-explosive fragmentation mechanisms (de Rosen-Spence et al. 1980; Yamagishi 1987; Allen 1988). Furthermore, there is no guarantee that thick beds of pyroclastic debris in marine basins were erupted locally within the deep water part of the basin. In many cases these beds are subaqueous mass flow deposits of pyroclastic debris transported into deep water environments from subaerial and shallow-water vents (e.g. significant parts of the Mount Read Volcanics and Green Tuff successions). However, locally erupted felsic submarine pyroclastic rocks are still regarded as abundant (Bathurst) or locally abundant (Mount Read Volcanics, Green Tuff) in some VMS districts. The details of these submarine pyroclastic eruptions, such as the location of vents, water depths of the vent areas, eruption styles and mechanisms, and relationship to evolu-

tion of the magmatic and hydrothermal system, have rarely been carefully documented. Large pyroclastic caldera volcanoes were proposed in the 1970–80's as the host to many VMS districts (e.g. Ohmoto 1978). However, several of those interpretations were partly based on the superficial circular shape of some mining districts (Bathurst, Hokuroku), now known to be an artefact of younger deformation and the position of cover sequences, and partly to the erroneous interpretation that most felsic volcaniclastic rocks are pyroclastic deposits.

Volcaniclastic rocks are also significant in that a simple plot of large tonnage ore bodies against volumes of volcaniclastic rocks shows a correlation, which suggests a causal relationship (H. Gibson, 2000 Hobart workshop presentation). Volcaniclastic sequences appear to promote formation of massive sulphides by infiltration and replacement below the sea floor; a situation that may trap more of the total metal budget. In volcanic sequences dominated by lava flows and intrusions, most VMS ores form at the sea floor and more of the metal budget might potentially become dispersed from hydrothermal vents into the basin waters.

In addition to being located in proximal volcanic facies, most major VMS deposits show evidence of having formed at a particular stage in the evolution of the enclosing volcanic succession. In many VMS districts the main VMS deposits are concentrated at the top, or in the upper part of the major syn-rift felsic volcanic unit (IPB, Green Tuff belt, Skellefte, Mount Read Volcanics, Silurian SE Australia, Bergslagen, Bathurst). Horikoshi (1969) also showed that at the more local scale, each of the Kuroko deposits of the Kosaka area in the Hokuroku district, formed at the end of a specific rhyolite volcano cycle, comprising (1) extrusion of large domes and flows, accompanied by pyroclastic eruptions of gas-rich magma, (2) continued extrusion and intrusion of degassed magma to form smaller domes and cryptodomes, (3) small phreatic ('steam') eruptions from the lava domes, followed by (4) effusion of ore solutions to the sea floor to form massive sulphides. This volcanic cycle is analogous to the classic subaerial dome eruption cycle (see Cole 1970). Other VMS deposits have also been shown to be preferentially associated with the late stage of evolution of specific rhyolite volcanoes (e.g. Allen et al. 1996b). Furthermore, the VMS deposits and their host rhyolite volcanoes are commonly followed by a significant change in the pattern, composition and intensity of volcanism and sedimentation, especially a pause in felsic volcanism, deposition of mudstone, and fissure eruption of mantle-derived basalts (Mount Read Volcanics,

Silurian SE Australia, Skellefte, Green Tuff belt, Abitibi, Bathurst, Bergslagen). These basalts commonly form the mafic component of the coincident felsic and mafic volcanism related to VMS deposits.

The formation of many major VMS deposits at specific stratigraphic positions, and at specific stages in the evolution of individual rhyolite volcanoes indicate that either the felsic magmatic–hydrothermal cycle (and it's interplay with mafic magmatism) creates and focuses an important part of the ore solution, or that specific types of volcanism focus when and where a metal-bearing geothermal solution can be concentrated and expelled to the sea floor, or both. A major challenge now is to determine which of these possibilities is correct and why, and to determine in what way these links are recorded in the regional tectonic–volcanic evolution of the basin, and evolution of the hydrothermal system. These are major aims of the Global VMS research project.

Critical areas that need addressing include:

1 the need for more facies mapping in VMS districts to build up models of volcanic facies architecture, volcanic evolution and palaeoenvironment;

2 what is the connection between the position of VMS deposits in the evolutionary cycle of their host volcanoes and the position of the deposits in the regional tectonic and basin evolution and what are the common features from district to district;

3 can the stratigraphic position (ore horizon) of VMS deposits be recognized in the regional stratigraphy away from known ore deposits, and if so, how;

4 the role of basalts in the evolution of the felsic volcanoes that host VMS deposits and in the formation of the VMS deposits themselves;

Structure

The VMS belts and ore deposits under consideration span an enormous range in style and degree of deformation and metamorphism. Most of the successions, except the modern Pacific basins, the young Green Tuff belt and parts of the Archaean Abitibi, are strongly deformed. Fold-thrust belt type deformation patterns are particularly common. A likely explanation for this deformation is that VMS districts are mainly related to extensional basins near plate margins, and these basin successions generally become inverted and strongly deformed during the inevitable closure of

the basins. Basin closure probably in most cases results from arc–arc, arc–continent, or continent–continent collision at the plate margin.

Despite the extent of deformation in most VMS districts and the important role of structural studies in unravelling the stratigraphic successions, the amount of detailed, high-quality structural work that has been carried out varies greatly. Problems include the difficulty in carrying out structural analysis in complex volcanic successions without prominent marker units and the paucity of structural geologists with experience of volcanic stratigraphy. However, some districts have been the focus of excellent structural studies that have demonstrated their use in understanding the geological evolution of the district and in ore discovery (Bathurst, Mount Read Volcanics). In the late 1980s and early 1990s the Mount Read Volcanics were the focus of a government regional mapping campaign, and this work has greatly increased the knowledge of the district and laid the foundation for subsequent research studies and mineral exploration. In contrast, the intensely deformed Bergslagen district is just now receiving its first modern regional structural interpretation. In the strongly deformed and glacial sediment covered Skellefte district, electrical geophysical exploration methods have until recently had enormous success in locating ores. This success has contributed to the generally poor knowledge of regional stratigraphy and structure. The recent success of gravity surveys in locating large ore bodies in the IPB has to some extent also retarded studies of stratigraphy and structure in the IPB, despite the fact that the IPB has enormous potential for the discovery of blind ore bodies in fold–thrust structures.

It can be concluded that good quality regional mapping of stratigraphy and structure is required in order to make meaningful interpretations of regional basin evolution and the setting of VMS ores, and for a foundation for local detailed research projects.

Critical areas that need addressing include:

1 integrated mapping and research studies of volcanic stratigraphy and structure at the regional and ore deposit scale.

Ore deposit characteristics

Despite the criticism that in-depth ore deposit studies are just collecting data, it is acknowledged that very few modern detailed accounts of VMS deposits and districts have been published. Furthermore, such studies provide the integrated data sets on which realistic detailed interpretation can be based. Recent studies have been made on the Noranda, Kidd Creek and Bathurst districts. The Urals has a great deal of primary data in Russian that is in need of compilation. Regional facies mapping has commenced in the Skellefte and Bergslagen districts, but comprehensive modern descriptions of the ore deposits are lacking. Detailed ore deposit descriptions and facies mapping are advanced in parts of the Mount Read Volcanics, but are not yet sufficient to make correlations between ore deposit characteristics, volcanic setting, depositional environment and basin evolution. Recent work in several VMS districts suggests that there are relationships between the facies architecture of the host volcanic succession (lava pile, lithic volcaniclastic rocks, pumiceous volcaniclastic rocks) and the style (sea floor exhalative, subsea-floor replacement), shape (mound, sheet, multiple lenses or tongues) and size of VMS deposits. However, these relationships have not been explored in detail.

Many VMS districts have numerous small–medium sized deposits and just one or two giants (e.g. Abitibi–85 deposits around 1 to 2 million tonnes, Kidd Creek and Horne around 100 Mt; Skellefte–over 80 deposits, the 52 largest have a median size of 1 million tonnes, and only Kristineberg and Rakkejaur are over 20 Mt). However, the largest deposits are often not the richest (e.g. Skellefte, Mount Read Volcanics). Studies should focus on large, well-exposed deposits and the smaller rich deposits so that the characteristics of the most economically viable deposits are fully documented. Compilations of the settings and characteristics of small deposits and low value deposits are a second priority but are also important for the purposes of comparison.

Critical areas that need addressing include:

1 more high quality, integrated, multi-disciplinary studies of major VMS deposits;
2 a better descriptive classification of the styles (types) of massive sulphide deposits;
3 improved criteria for the genetic classification of deposit types–current nomenclature is too rigid and is based on limited data;
4 investigation of the relationships between host stratigraphy (facies architecture) and the style (geometry, seafloor versus subsea-floor position, size, metal zonation) of VMS deposits;
5 more information on the mineralogy and ore mineral textures of deposits (e.g. a database on mineral textures) complemented by modern chemical and isotopic studies;
6 fluid-inclusion data and other information on the salinity of ore fluids, fluid compositions,

fluid sources, pathways and effects such as alteration and ore mineral deposition;
7 modelling of fluid-rock reactions in VMS hydrothermal systems.

Exhalites

Some VMS districts have prominent sulphide or oxide facies exhalites that mark the ore-horizon away from the deposit (e.g. Bathurst, Noranda), whilst others do not (Mount Read Volcanics, Skellefte). There remains uncertainty about which are true hydrothermal exhalites, which may relate to higher than average heat-flow, which are reworked clastic weathering products and which are regional weathering phenomena of volcanic rocks. Some VMS deposits, such as sub-sea floor replacement deposits, may possibly have no sea-floor expression. However, this seems difficult to envisage, as the hydrothermal fluid has to escape from the mineralizing system somewhere.

An excellent database has been constructed for the Bathurst exhalites and documents extensive iron-formations of both volcanic-association (Algoma-type) and possibly starved sediment sources (Superior-type). However, some other districts that have prominent chemical sediment units and exhalites, such as Bergslagen, have no modern, systematic, comprehensive data on these rocks.

Critical areas that need addressing include:

1 how to recognize exhalites in the field and are exhalites, or the chemical and mineralogical changes resulting from their interaction with the sea floor substrate, common in VMS districts but largely unrecognized;
2 a compilation of exhalite mineralogy and geochemistry in several VMS districts;
3 how to distinguish exhalites related to VMS mineralizing systems from other exhalites (e.g. iron formations) not related to VMS systems;
4 the possibility that some iron formations in VMS camps may not be exhalites–e.g. Bergslagen;
5 the reasons why some exhalites extend beyond the sulphide deposit whereas others do not.

Mineralogy, metal zonation and alteration

The chemistry of the footwall volcanic rocks appears to be the biggest control on the mineralogy of the ore deposits. This may simply reflect the mineralogy of the volcanic rocks that are leached during hydrothermal alteration, e.g. break-

Table 2. *Relationships between the composition of massive sulphide mineralization, host rock type and tectonic setting, suggested from exploration of the modern sea floor*

	Pb, As, Sb, Au			
	Very high	High	Low	Absent
Setting	Continental back arc	Oceanic back arc	Sedimented mid-ocean ridge	Sediment free mid-ocean ridge
Host rocks	Rhyolite/dacite	Dacite/andesite	Turbidite sequences, MORB	MORB
Examples	Okinawa Trough	Lau and Manus basins	Mid-ocean ridges	Mid-ocean ridges

down of mafic minerals–copper and zinc, breakdown of felsic minerals–lead, barium etc. However, some workers attribute part of the metal composition of the ore deposits to primary magmatic contributions to the hydrothermal solution (Urabe 1987).). Exploration of the modern sea floor suggests the relationships shown in Table 2 between the composition of VMS ores, host rock type and tectonic setting:

Exploration of the modern sea floor suggests the relationships shown in Table 2 between the composition of VMS ores, host rock type and tectonic setting.

Metal zonation appears to be largely related to the temperature of the hydrothermal system within the actively forming VMS deposit, and the pattern of temperature decrease between the high temperature core of the deposit and the low temperature margins (e.g. high temperature–copper, low temperature–zinc, lead, barium). Metal zonation may be complicated by changes in the hydrothermal fluid flow pattern and temperature gradients with time.

Several factors (e.g. rock composition, temperature, fluid composition, salinity) affect hydrothermal alteration mineralogy; but fluid pH appears to have the largest effect. At very low pH (below 3.5), kaolinite and pyrophyllite become the stable phases instead of muscovite. Footwall rocks provide the buffer to fluid chemistry. Low pH fluids might indicate a closer association with a magmatic source, i.e. fluids more out of equilibrium with wall-rocks. The mineralogy of ores, gangue minerals and footwall alteration could be used better to help interpret the evolution of VMS deposits. Lithogeochemistry can be used to map out chemical zonations and mass changes in the host succession caused by hydrothermal alteration, and to locate the main hydrothermal conduits (upflow zones) and recharge zones (Barrett *et al.* 1991; Barrett & MacLean 1999). New ideas of the complex electro-chemical interaction between oxidized seawater, precipitated sulphide assemblages and the volcano-sedimentary host rocks could help

to explain some of the variations in deposits seen in ancient systems.

Critical areas that need addressing include:

1 systematic data on the mineral chemistry of ore and alteration minerals at a selection of different VMS deposit types;

2 correlate chemistry of alteration minerals with whole rock geochemistry and volcanicsedimentary facies;

3 synthesis of the geometry, mineralogical zonation and chemical zonation of hydrothermal alteration (footwall and hanging-wall) at a range of different types of VMS deposits.

The way forward

The discussion above provides a summary and some highlights of the GEODE VMS database and the Hobart workshop. This also chronicles the background and birth of a global VMS research project. The ultimate scope and success of the project depends on the enthusiasm and imagination of the research teams and the level of funding that can be achieved. We are now organizing the main research stage of the project to answer some of the important scientific questions raised above, and to address the following general objectives.

(1) Create an international network of experienced researchers with a range of different skills and develop a collaborative international research programme in which research teams work on common goals and with significant exchange of researchers, skills, ideas and data (Fig. 10). The project plan includes teams with scientists from seven European nations (Sweden, Germany, Spain, Portugal, United Kingdom, France, Switzerland) and a Russian, Canadian, Australian, Japanese and possibly an Andean-Swiss team.

(2) Using an integrated multi-disciplinary approach, determine how mineralization, volcanic evolution, and extensional tectonic evolution are interrelated in a number of world-class VMS

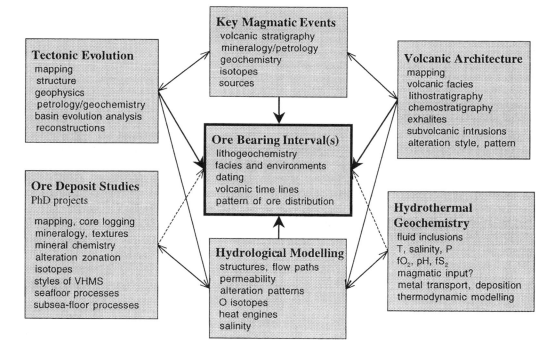

Fig. 10. Framework for the integration of component studies of the Global VMS research project.

districts. Field-oriented research (stratigraphy, volcanology, structural geology) will be integrated with laboratory studies (petrology, geochemistry, isotope geochemistry, analysis of geophysical data) as illustrated in Figure 10. Mining districts under consideration for inclusion in the project include Skellefte and Bergslagen districts, Sweden, Southern Urals, Russia, Iberian Pyrite Belt, Spain and Portugal, Mount Read Volcanics, Australia, Bathurst, Abitibi and Flin Flon districts, Canada, the Andean Belt, South America, Green Tuff belt, Japan and Manus and Lau Basins, Pacific Ocean (Fig. 1).

(3) Develop criteria to recognize these relationships in the field so that it becomes possible to identify the volcanic intervals that are likely to host major ore deposits in any given district.

(4) Organize and guide the studies in the various mining districts and bring them together in a global synthesis. Specifically, compare and contrast the results from each district and thereby distinguish those features that are essential to formation of world-class ore deposits (i.e. occur in all districts), from those that are not essential but are locally important (i.e. occur only in specific districts).

In this paper we have synthesized data, ideas and suggestions from many colleagues, especially scientists in the Global VMS research teams. We wish to thank all

these colleagues for the great spirit of collaboration that forms the core of the project. Rod Allen and Pär Weihed also thank the Research Institute of Materials and Resources (RIMR) at Akita University, Japan, the Centre for Applied Ore Geology Studies (CTMG) at Luleå University of Technology, Sweden, the European Science Foundation programme for Geodynamics and Ore Deposit Evolution (GEODE) and the GEORANGE research programme, Sweden for supporting this work.

References

ALLEN, R.L. 1988. False pyroclastic textures in altered silicic lavas, with implications for volcanic-associated mineralization. *Economic Geology*, **83**, 1424–1446.

ALLEN, R.L. 1992. Reconstruction of the tectonic, volcanic and sedimentary setting of strongly deformed Cu–Zn massive sulfide deposits at Benambra, Victoria. *Economic Geology*, **87**, 825–854.

ALLEN, R.L. 2000. Comparison of tectonic and stratigraphic setting of ores in six VMS Belts: Green Tuff Belt, Skellefte district, Bergslagen, Mt Read Volcanics, Iberian Pyrite Belt and Lachlan Fold Belt. *In:* GAMMEL, B.B. & PONGRATZ, J. (eds) *Volcanic environments and massive sulfide deposits. Program with abstracts.* Codes Special Publication, **3**, 3–5.

ALLEN, R.L., LUNDSTRÖM, I., RIPA, M., SIMEONOV, A. & CHRISTOFFERSON, H. 1996a. Facies Analysis of a 1.9 Ga, Continental Margin, Back-Arc, Felsic Caldera Province with Diverse Zn-Pb-Ag-(Cu-Au) Sulfide and Fe Oxide Deposits, Bergslagen Region, Sweden. *Economic Geology*, **91**, 979–1008.

ALLEN, R.L., WEIHED, P. & SVENSON, S-Å. 1996b. Setting of Zn-Cu-Au-Ag massive sulfide deposits in the evolution and facies architecture of a 1.9 Ga marine volcanic arc, Skellefte district, Sweden. *Economic Geology*, **91**, 1022–1053.

BAKER, J.H., HELLINGWERF, R.H. & OEN, I.S. 1988. Structure, stratigraphy and ore-forming processes in Bergslagen: implications for the development of the Svecofennian of the Baltic Shield. *Geologie en Mijnbouw*, **67**, 121–138.

BARRETT, T.J. & MACLEAN, W.H. 1999. Volcanic sequences, lithogeochemistry, and hydrothermal alteration in some bimodal volcanic-associated massive sulfide systems. *In:* BARRIE, C.T. & HANNINGTON, M.D. (eds) *Volcanic-Associated Volcanic Massive Sulfide Systems: Processes and Examples in Modern and Ancient Settings.* Reviews in Economic Geology, **8**, 101–131.

BARRETT, T.J., CATTALANI, S. & MACLEAN, W.H. 1991. Massive sulfide deposits of the Noranda area, Quebec; I., The Horne Mine. *Canadian Journal of Earth Sciences*, **28**, 465–488.

BARRIE, C.T., LUDDEN, J.N. & GREEN, T.H. 1993. Geochemistry of volcanic rocks associated with Cu-Zn and Ni-Cu deposits in the Abitibi subprovince. *Economic Geology*, **88**, 1341–1358.

BILLSTRÖM, K. & WEIHED, P. 1996. Age and provenance of host rocks and ores of the Paleoproterozoic Skellefte District, northern Sweden. *Economic Geology*, **91**, 1054–1072.

BINNS, R.A. & SCOTT, S.D. 1993. Actively forming polymetallic sulfide deposits associated with felsic volcanic rocks in the eastern Manus back-arc basin, Papua New Guinea. *Economic Geology*, **88**, 2226–2236.

BINNS, R.A., WATERS, J.C., CARR, G.R. & WHITFORD, D.J. 1996. A submarine andesite-dacite lineage of arc affinity, Pual Ridge, eastern Manus Basin, Papua New Guinea. *EOS Transactions, American Geophysical Union*, **77**, 119–120.

BOULTER, C.A. 1993. High level peperitic sills at Rio Tinto, Spain: implications for stratigraphy and mineralization. *Transactions of the Institute of Mining and Metallurgy*, **102b**, 30–38.

BROWN, D. & SPADEA, P. 1999. Processes of forearc and accretionary complex formation during arc-continent collision in the southern Ural Mountains. *Geology*, **27**, 649–652.

CARVALHO, D., BARRIGA, F. & MUNHÁ, J. 1997. The Iberian Pyrite Belt of Portugal and Spain: Examples of bimodal–siliciclastic systems. *In:* BARRIE, C.T. & HARRINGTON, M.D. (eds) *Volcanic-Associated Massive Sulfide Deposits: Processes and Examples in Modern and Ancient Settings.* Reviews in Economic Geology, **8**, 375–402.

CAS, R.A.F. 1983. A review of the paleogeographic and tectonic development of the Palaeozoic Lachlan Fold Belt of southeastern Australia. *Geological Society of Australia, Special Publications*, **10**.

CAS, R.A.F. & JONES, J.G. 1979. Palaeozoic interarc basin in eastern Australia and a modern New

Zealand analogue. *New Zealand Journal of Geology and Geophysics*, **22**, 71–81.

COLE, J.W. 1970. Structure and eruptive history of the Tarawera Volcanic Complex. *New Zealand Journal of Geology and Geophysics*, **13**, 879–902.

CORBETT, K.D. 1992. Stratigraphic-volcanic setting of massive sulphide deposits in the Cambrian Mt Read Volcanics, Tasmania. *Economic Geology*, **87**, 564–586.

CORBETT, K.D. 2001. New mapping and interpretations of the Mount Lyell mining district, Tasmania: A large hybrid Cu-Au system with an exhalative top. *Economic Geology*, **96**, 1089–1122.

CRAWFORD, A.J., CORBETT, K.D. & EVERARD, J.L. 1992. Geochemistry of the Cambrian volcanic-hosted massive sulfide-rich Mount Read Volcanics. *Economic Geology*, **87**, 597–619.

DE ROSEN-SPENCE, A.F., PROVOST, G., DIMROTH, E., GOCHNAVER, K. & OWEN, V. 1980. Archean subaqueous felsic flows, Rouyn-Noranda, Quebec, Canada, and their Quaternary equivalents. *Precambrian Research*, **12**, 43–77.

DIMROTH, E., IMREII, L., ROCHELEAU, M. & GOULET, N. 1982. Evolution of the south-central part of the Archean Abitibi belt, Quebec. Part I: Stratigraphy and paleogeographic model. *Canadian Journal of Earth Sciences*, **19**, 1729–1758.

DUDÁS, F.Ö., CAMPBELL, I.H. & GORTON, M.P. 1983. Geochemistry of igneous rocks in the Hokuroku district, Northern Japan. *In:* OHMOTO, H. & SKINNER, B.J. (eds) *The Kuroko and related volcanogenic massive sulphide deposits.* Economic Geology Monographs, **5**, 115–133.

FOUQUET, Y., VON STACKELBERG, U., CHARLOU, J.L., ERZINGER, J., HERZIG, P.M., MÜHE, R. & WIEDICKE, M. 1993. Metallogenesis in back-arc environments: the Lau Basin example. *Economic Geology*, **88**, 2154–2181.

GALLEY, A.G., WATKINSON, D.H., JONASSON, I.R. & RIVERIN, G. 1996. The subsea-floor formation of volcanic-hosted massive sulfide: Evidence from the Ansil deposit, Rouyn-Noranda, Canada. *Economic Geology*, **90**, 2006–2017.

GEMMELL, J.B. & LARGE, R.R. 1992. Stringer system and alteration zones underlying the Hellyer volcanogenic massive sulphide deposit, Tasmania, Australia. *Economic Geology*, **87**, 620–649.

GEMMELL, J.B., BINNS, R.A. & PARR, J.M. 1999. Submarine, high sulfidation alteration within DESMOS caldera, Manus Basin, PNG. *In:* STANLEY, C.J. (ed.) *Mineral Deposits: Processes to Processing.* A.A.Balkeme, Rotterdam, 503–506.

GIBSON, H.L. & WATKINSON, D.H. 1990. Volcanogenic massive sulphide deposits of the Noranda cauldron and shield volcano, Quebec. *In:* RIVE, M., VERPAELST, P., GAGNON, Y., LULIN, J.-M., RIVERIN, G. & SIMBRA, A. (eds) *The Northwestern polymetallic belt; a summary of 60 years of mining exploration; Proceedings of the Rouyn–Noranda 1990 Symposium.* Canadian Institute of Mining and Metallurgy Special Volumes, **43**, 119–132.

GRAY, D.R., FOSTER, D.A. & BUCHER, M. 1997. Recognition and definition of orogenic events in the Lachlan Fold Belt. *Australian Journal of Earth*

Science, **44**, 489–501.

GREEN, G.R., SOLOMON, M. & WALSHE, J.L. 1981. The formation of the volcanic-hosted massive sulphide deposit at Rosebery, Tasmania. *Economic Geology*, **76**, 304–338.

HALLEY, S.W. & ROBERTS, R.H. 1997. Henty: a shallow-water gold-rich volcanogenic massive sulfide deposit in western Tasmania. *Economic Geology*, **92**, 438–447.

HANNINGTON, M.D., BARRIE, C.T. & BLEEKER, W. 1999. Preface and Introduction. *In:* HANNINGTON, M.D., BARRIE, C.T. & BLEEKER, W. (eds) *The Giant Kidd Creek Volcanogenic Massive Sulfide Depsit, Western Abitibi Subprovince, Canada.* Economic Geology Monographs, **10**, 1–30.

HAWKINS, J.W. 1995. The geology of the Lau Basin. *In:* TAYLOR, B. (eds) *Back-Arc Basins: Tectonics and Magmatism.* Plenum Publishing, New York, 63–138.

HEDSTRÖM, P., SIMEONOV, A. & MALMSTRÖM, L. 1989. The Zinkgruvan ore deposit, south-central Sweden: A Proterozoic, proximal Zn-Pb-Ag deposit in distal volcanic facies. *Economic Geology*, **84**, 1235–1261.

HERRINGTON, R.J. 1999. *Volcanic hosted massive sulphide deposits of the southern Urals.* Unpublished guidebook for the 1999 SGA-IAGOD Fieldtrip, August 1999.

HERRINGTON, R., ARMSTRONG, R., ZAYKOV, V. & MASLENNIKOV, V. 2000. Volcanic-hosted massive sulfide deposits of the southern Urals, their ore facies and geological settings. *In:* GEMMELL, J.B. & PONGRATZ, J. (eds) *Program with abstract: Volcanic environments and massive sulfide deposits.* CODES Special Publications, **4**, 81–82.

HERZIG, P.M., HANNINGTON, M.D., FOUQUET, Y., VON STACKELBERG, U. & PETERSEN, S. 1993. Gold-rich polymetallic sulfides from the Lau back-arc and implications for the geochemistry of gold in seafloor hydrothermal systems in the Southwest Pacific. *Economic Geology*, **88**, 2182–2209.

HORIKOSHI, E. 1969. Volcanic activity related to the formation of the Kuroko-type deposits in the Kosaka District, Japan. *Mineralium Deposita*, **4**, 321–345.

IIJIMA, A. 1974. Clay and zeolitic alteration zones surrounding Kuroko deposits in the Hokuroku District, northern Akita, as submarine hydrothermal-diagenetic alteration products. *In:* ISHIHARA, S., KANEHIRA, K., SASAKI, A., SATO, T. & SHIMAZAKI, Y. (eds) *Geology of the Kuroko deposits.* Mining Geology, Special Issue, **6**, 267–289.

ISHIHARA, S., KANEHIRA, K., SASAKI, A., SATO, T. & SHIMAZAKI, Y. (EDS) 1974. *Geology of the Kuroko deposits.* Mining Geology, Special Issue, **6**.

KERR, D.J. & GIBSON, H.L. 1993. A comparison between the Horne volcanogenic massive sulfide deposit and intracauldron deposits of the mine sequence, Noranda, Quebec. *Economic Geology*, **88**, 1419–1442.

KONDA, T. 1974. [Bimodal volcanism in the Northeast Japan arc]. *Journal of the Geological Society of Japan*, **80**, 81–89. [in Japanese].

KOROTEEV, V.A., DE BOORDER, H., NETCHEUKIN, V.M. & SAZONOV, V.N. 1997. Geodynamic setting of the mineral deposits of the Urals. *Tectonophysics*, **276**, 291–300.

LAGERBLAD, B. & GORBATSCHEV, R. 1985. Hydrothermal alteration as a control of regional geochemistry and ore formation in the central Baltic Shield. *Geologische Rundschau*, **74**, 33–49.

LARGE, R.R. 1992. Australian volcanic-hosted massive sulfide deposits: features, styles, and genetic models. *Economic Geology*, **87**, 471–510.

LARGE, R.R., MCGOLDRICK, P.J., BERRY, R.F. & YOUNG, C.H. 1988. A tightly folded, gold-rich, massive sulfide deposit: Que River mine, Tasmania. *Economic Geology*, **83**, 681–693.

LARGE, R., DOYLE, M., RAYMOND, O., COOKE, D., JONES, A. & HEASMAN, L. 1996. Evaluation of the role of granites in the genesis of world class VMS deposits in Tasmania. *Ore Geology Reviews*, **10**, 215–230.

LEISTEL, J.M., MARCOUX, E., THIEBLEMONT, D., QUESADA, C., SÁNCHEZ, A., ALMODOVAR, G.R., PASCUAL, E. & SÁEZ, R. 1998. The volcanic-hosted massive sulphide deposits of the Iberian Pyrite Belt. Review and preface to the special issue. *Mineralium Deposita*, **33**, 2–30.

LENTZ, D.R. 1998. Petrogenetic evolution of felsic volcanic sequences associated with Phanerozoic volcanic-hosted massive sulphide systems: the role of extensional geodynamics. *Ore Geology Reviews*, **12**, 289–327.

LENTZ, D.R. 1999. Petrology, geochemistry, and oxygen isotope interpretative of felsic volcanic and related rocks hosting the Brunswick 6 and 12 massive sulfide deposits (Brunswick Belt), Bathurst Mining Camp, New Brunswick, Canada. *Economic Geology*, **94**, 57–86.

LUNDQVIST, T., VAASJOKI, M. & PERSSON, P.-O. 1998. U-Pb ages of plutonic and volcanic rocks in the Svecofennian Bothnian Basin, central Sweden, and their implications for the Palaeoproterozoic Evolution of the Basin. *GFF*, **120**, 357–363.

LUNDSTRÖM, I. 1987. Lateral Variations in Supracrustal Geology within the Swedish part of the Southern Svecokarelian Volcanic Belt. *Precambrian Research*, **35**, 353–365.

MCCUTCHEON, S.R. 1992. Base-metal deposits of the Bathurst-Newcastle district: characteristics and depositional models. *Exploration and Mining Geology*, **1**, 105–119.

MCPHIE, J. & ALLEN, R.L. 1992. Facies architecture of mineralised submarine sequences: Cambrian Mount Read Volcanics, Western Tasmania. *Economic Geology*, **87**, 587–596.

OHGUCHI, T. 1983. Stratigraphical and petrographical study of the late Cretaceous to Early Miocene volcanic rocks in northeast inner Japan. *Journal of the Mining College Akita University, Series A*, **6**, 189–258.

OHGUCHI, T., YOSHIDA, T. & OKAMI, K. 1989. Historical change of the Neogene and Quaternary volcanic field in the Northeast Honshu Arc, Japan. *In:* KITAMURA, N., OTSUKI, K. & OHGUCHI, T. (eds) *Cenozoic geotectonics of northeast Houston Arc.* Geological Society of Japan, Memoirs, **32**, 431–455. [in Japanese with English Abstract].

OHGUCHI, T., HAYASHI, S., KOBAYASHI, N., ITAYA, T. &

YOSHIDA, T. 1995. K-Ar dating unravels the volcanic history of the Kuguriiwa Lava and Kamo Lava Members, Oligocene Monzen Formation, Oga Peninsula, NE Japan. Geological Society of Japan, Memoirs, **44**, 39–45. [in Japanese with English abstract].

OHMOTO, H. 1978. Submarine calderas: A key to the formation of volcanogenic massive sulphide deposits? *Mining Geology*, **28**, 219–231.

OHMOTO, H. & SKINNER, B.J. (EDS) 1983. *The Kuroko and related volcanogenic massive sulfide deposits.* Economic Geology Monographs, **5**.

OKAMURA, S. 1987. Geochemical variation with time in the Cenozoic volcanic rocks of southwest Hokkaido. Japan. *Journal of Volcanology and Geothermal Research*, **32**, 161–176.

OTOFUJI, Y., MATSUDA, T. & NOHDA, T. 1985. Paleomagnetic evidence for the Miocene counter-clockwise rotation of northeast Japan. *Earth and Planetary Science Letters*, **75**, 265–278.

PETER, J.M. & GOODFELLOW, W.D. 1996. Mineralogy, bulk and rare earth element geochemistry of massive sulphide-associated hydrothermal sediments of the Brunswick horizon, Bathurst Mining Camp, New Brunswick. *Canadian Journal of Earth Sciences*, **33**, 252–283.

POWELL, C.M.A. 1983. Tectonic relationship between the Late Ordovician and Late Silurian palaeogeographies of southeastern Australia. *Journal of the Geological Society of Australia*, **30**, 353–373.

PRIOR, G.J., GIBSON, H.L., WATKINSON, D.H. & COOK, R.E. 1999a. Anatomy, lithogeochemistry and emplacement mechanisms for the QP rhyolite, Kidd Creek Mine, Timmins, Ontario. *In:* HANNINGTON, M.D., BARRIE, C.T. & BLEEKER, W. (eds) *The Giant Kidd Creek Volcanogenic Massive Sulfide Deposit, Western Abitibi Subprovince, Canada.* Economic Geology Monographs, **10**, 123–142.

PRIOR, G.J., GIBSON, H.L., WATKINSON, D.H., COOK, R.E. & HANNINGTON, M.D. 1999b. Rare earth and high field strength element geochemistry of the Kidd Creek rhyolites, Abitibi greenstone belt, Canada: Evidence for Archaean felsic volcanism and volcanogenic massive sulphide formation in an Iceland-style rift environment. *In:* HANNINGTON, M.D., BARRIE, C.T. & BLEEKER, W. (eds) *The Giant Kidd Creek Volcanogenic Massive Sulfide Deposit, Western Abitibi Subprovince, Canada.* Economic Geology Monographs, **10**, 471–498.

PROKIN, V.A. & BUSLAEV, F.P. 1999. Massive copper-zinc sulphide deposits in the Urals. *Ore Geology Reviews*, **14**, 1–69.

PROKIN, V.A., BUSLAEV, F.P. & NASEDKIN, A.P. 1998. Types of massive sulphide deposits in the Urals. *Mineralium Deposita*, **34**, 121–126.

PUCHKOV, V.N. 1997. Sructure and geodynamics of the Uralian Orogen. *In:* BURG, J.-P. & FORD, M. (eds) *Orogeny through Time.* Geological Society, London, Special Publications, **121**, 201–236.

QUESADA, C. 1991. Geological constraints on the Paleozoic tectonic evolution of tectonostratigraphic terranes in the Iberian Massif. *Tectonophysics*, **185**, 225–245.

QUESADA, C. 1998. A reappraisal of the structure of the Spanish segment of the Iberian Pyrite Belt. *Mineralium Deposita*, **33**, 31–44.

RICKARD, D. (ED.) 1986. *The Skellefte Field.* Sveriges Geologiska Undersökning Ca, **62**.

RIVERIN, G. & HODGSON, C.J. 1980. Wall-rock alteration at the Millenbach Cu-Zn mine, Noranda, Quebec. *Economic Geology*, **75**, 424–444.

SAIF, S-I. 1983. Petrographic and geochemical characteristics of iron-rich rocks and their significance in exploration for massive sulfide deposits, Bathurst, New Brunswick, Canada. *Journal of Geochemical Exploration*, **19**, 705–721.

SATO, H. 1994. The relationship between late Cenozoic tectonic events and stress field and basin development in northeast Japan. *Journal of Geophysical Research*, **99B**, 22261–22274.

SATO, H. & AMANO, K. 1991. Relationship between tectonics, volcanism and basin development, late Cenozoic, central part of northern Honshu, Japan. *Sedimentary Geology*, **74**, 323–343.

SCHERMERHORN, L.J.G. 1975. Spilites, regional metamorphism and subduction in the Iberian Pyrite Belt: some comments. *Geologie en Mijnbouw*, **54**, 23–35.

SILVA, J.B., OLIVEIRA, J.T. & RIBEIRO, A. 1990. Structural outline of the South Portuguese Zone. *In:* DALLMEYER, R.D. & MARTINEZ GARCÍA, E. (eds) *Pre-Mesozoic Geology of Iberia.* Springer Verlang, Berlin, 348–362.

SOLOMON, M. & KHIN ZAW, 1997. Formation on the seafloor of the Hellyer volcanogenic massive sulfide deposit. *Economic Geology*, **92**, 686–695.

SORIANO, C. & MARTÍ, J. 1999. Facies analysis of volcano-sedimentary successions hosting massive sulfide deposits in the Iberian Pyrite Belt, Spain. *Economic Geology*, **94**, 867–882.

STOLZ, A.J., DAVIES, G.R. & ALLEN, R.L. 1997. The importance of different types of magmatism in VMS mineralization: evidence from the geochemistry of the host volcanic rocks to the Benambra massive sulphide deposits, Victoria, Australia. *Mineralogy and Petrology*, **59**, 251–286.

SUNDBLAD, K. 1994. A genetic reinterpretation of the Falun and Åmmeberg ore types, Bergslagen, Sweden. *Mineralium Deposita*, **29**, 170–179.

TAIRA, A., TOKUYAMA, H. & SOH, W. 1989. Accretion tectonics and evolution of Japan. *In:* BEN-AVRAHAM, Z. (ed.) *The evolution of the Pacific Ocean margins.* Oxford University Press, Oxford, 100–123.

TAMAKI, K., SUYEHIRO, K., ALLAN, J., INGLE, C. & PISCIOTTO, K.A. 1992. Tectonic synthesis and implications of Japan Sea ODP drilling. *Proceedings of the Ocean Drilling Program, Scientific results*, **127/128**(2), 1333–1348.

TANIMURA, S., DATE, J., TAKAHASHI, T. & OHMOTO, H. 1983. Geological setting of the Kuroko deposits, Japan. Part II. Stratigraphy and structure of the Hokuroko Distric. *In:* OHMOTO, H. & SKINNER, B.J. (eds) *The Kuroko and related volcanogenic massive sulfide deposits.* Economic Geology Monographs, **5**, 24–38.

THIEBLEMONT, D., PASCUAL, E. & STEIN, G. 1998. Magmatism in the Iberian Pyrite Belt: petrological constraints on a metallogenic model. *Mineralium Deposita*, **33**, 98–110.

TORNOS, F., CÉSAR CASQUET, C., JORGE, M.R.S., RELVAS, J.M.R.S., BARRIGA, F.J.A.S. & REINALDO SÁEZ, R. 2002. The relationship between ore deposits and oblique tectonics: the southwestern Iberian Variscan Belt. *In:* BLUNDELL, D.J., NEUBAUER, F. & VON QUADT, A. (eds) *The Timing and Location of Major Ore Deposits in and Evolving Orogen.* Geological Society, London, Special Publications, **204**, 179–195.

URABE, T. 1987. Kuroko deposit modelling based on magmatic hydrothermal theory. *Mining Geology*, **37**, 159–176.

UTADA, M. 1991. Zeolitization in the Neogene formations of Japan. *Episodes*, **14**, 242–245.

VALLIER, T.L., JENNER, G.A., FREY, F.A., GILL, J.B., DAVIS, A.S., VOLPE, A.M., HAWKINS, J.W., MORRIS, J.D., CAWOOD, P.A., MORTON, J.L., SCHOLL, D.W., RAUTENSCHLEIN, M., WHITE, W.M., WILLIAMS, R.W., STEVENSON, A.J. & WHITE, L.D. 1991. Subalkaline andesite from Valu Fa Ridge, a back-arc spreading center in southern Lau Basin: petrogenesis, comparative chemistry, and tectonic implications. *Chemical Geology*, **91**, 227–256.

VAN STAAL, C.R. 1987. Tectonic setting of the Tetagouche Group in northern New Brunswick: implications for plate tectonic models in the northern Appalachians. *Canadian Journal of Earth Sciences,*

24, 1329–1351.

VIVALLO, W. & CLAESSON, L-Å. 1987. Intra-arc rifting and massive sulphide mineralization in an early Proterozoic volcanic arc, Skellefte district, northern Sweden. *In:* PHARAOH, T.C., BECKINSALE, R.D. & RICKARD, D. (eds) *Geochemistry and Mineralization of Proterozoic Volcanic Suites.* Geological Society, London, Special Publications, **33**, 69–79.

WEIHED, P., BERGMAN, J. & BERGSTRÖM, U. 1992. Metallogeny and tectonic evolution of the early Proterozoic Skellefte District, northern Sweden. *Precambrian Research*, **58**, 143–167.

YAMAGISHI, H. 1987. Studies on the Neogene subaqueous lavas and hyaloclastites in southwest Hokkaido. *Report of the Geological Survey, Hokkaido*, **59**, 55–117.

YAMAGISHI, H. & WATANABE, Y. 1986. Change of stress field of late Cenozoic southwest Hokkaido, Japan. *In:* Association for the Geological Collaboration in Japan, Monographs, **31**, 321–331. [in Japanese with English abstract].

YAMAJI, A. & SATO, H. 1989. Miocene subsidence of the Northeast Honshu Arc and its mechanism, Japan. *In:* KITAMURA, N., OTSUKI, K. & OHGUCHI, T. (eds) *Cenozoic geotectonics of northeast Houston Arc.* Geological Society of Japan, Memoirs, **32**, 339–349. [in Japanese with English abstract].

Tectonic controls on magmatic–hydrothermal gold mineralization in the magmatic arcs of SE Asia

M. E. BARLEY[1], P. RAK[1] & D. WYMAN[2]

[1]Centre for Global Metallogeny, Department of Geology and Geophysics, The University of Western Australia, 35 Stirling Highway, Crawley, WA, 6009, Australia
(e-mail: mbarley@geol.uwa.edu.au)

[2]School of Geosciences, Edgeworth David Building FO5, University of Sydney, Sydney, NSW 2006, Australia

Abstract: The magmatic arcs of SE Asia contain some of the world's major gold deposits. These are mainly related to magmatic–hydrothermal activity and include epithermal, porphyry Cu–Au and skarn deposits. Most gold deposits in SE Asian arcs formed during three intervals of tectonic reorganization rather than during periods of normal or steady-state subduction. These plate reorganizations and periods of gold mineralization were caused initially by the collision of the Australian craton with the Philippine Sea plate arc at 25 Ma. A second Mid-Miocene period of mineralization accompanied plate reorganization following maximum rotation or extrusion of Indochina and the cessation of spreading of the South China Sea at 17 Ma. However, the vast majority and largest deposits formed since 5 Ma in a broad belt from Taiwan to the Solomon Islands during an important period of tectonic reorganization. This tectonic reorganization accompanied a postulated change in the relative motion between the Indian–Australian and Pacific plates between 5 and 3.5 Ma following collision of the Philippine arc and the Eurasian plate in Taiwan. Arc-related magmatism in unusual tectonic settings produced the most abundant and largest deposits with many deposits associated with relatively rare high-K calc-alkaline, shoshonite, adakite and alkaline magmatism. In particular peak mineralization appears related to melting of mantle that had been previously modified by subduction. During large-scale plate reorganization this can occur at the end of a period of normal subduction, following the cessation of subduction, following arc collision or accompanying subduction reversal at approximately the same time in different parts of a complex system of magmatic arcs such as those in SE Asia. Such tectonic controls may be impossible to recognize in older orogenic belts where recognition of a major change in tectonic style and unusual magma types may be the best guides to mineralization.

The Cenozoic magmatic arcs that surround the Pacific Ocean are richly endowed with magmatic–hydrothermal gold deposits such as high- and low-sulphidation epithermal deposits, porphyry Cu–Au deposits and skarn Cu–Au deposits. Although a spatial and temporal link between these types of metal deposit and subduction-related magmatism has been recognized for some time (Mitchell & Garson 1972, 1976; Sillitoe 1972, 1989), the deposits are most abundant within specific arc sectors and during specific periods (e.g. Sillitoe 1989, 1997). This strongly suggests that tectonic factors other than normal or steady state subduction of oceanic lithosphere are important for the formation and localization of magmatic–hydrothermal gold deposits in magmatic arcs.

The magmatic arcs of SE Asia formed during the Cenozoic as a result of the convergence of the Indian–Australian, Philippine–Pacific and Eurasian plates (Fig. 1). This area contains some of the world's largest and richest magmatic–hydrothermal gold deposits and has experienced relatively rapid changes in plate configuration and rates of tectonic processes during the Cenozoic. It is thus a key area for assessing the effects of regional tectonics on the spatial and temporal distribution of magmatic–hydrothermal gold deposits. In this study a database of gold deposit ages and styles was superimposed on an animated tectonic reconstruction of SE Asia, presented by Hall (1996, 1998) and available from http://www.gl.rhul.ac.uk/seasia/welcome.html/. A version of this reconstruction, which includes the locations of gold deposits is available from *http://www.cgm.uwa.edu.au/*.

Examination of the spatial and temporal distribution of gold deposits relative to the regional

From: BLUNDELL, D.J., NEUBAUER, F. & VON QUADT, A. (eds) 2002. *The Timing and Location of Major Ore Deposits in an Evolving Orogen*. Geological Society, London, Special Publications, **204**, 39–47. 0305-8719/02/$15.00 © The Geological Society of London 2002.

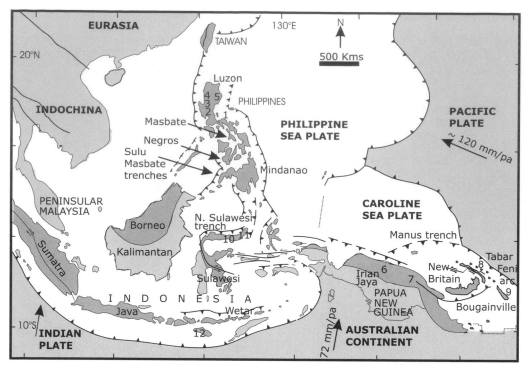

Fig. 1. Map of SE Asia, showing tectonic plates (Eurasian, Indian–Australian and Pacific) in pale grey, and Cenozoic magmatic arcs in dark grey. Major gold districts and deposits are numbered: 1, Camarines Norte; 2, Batangas; 3, Baguio (including Santo Tomas II); 4, Mankayan (including Lepanto, Far SE); 5, Isabella-Dipidio; 6, Ertsberg–Grasberg; 7, Ok Tedi–Porgera; 8, Ladolam; 9, Panguna; 10, Tombulilato–Gorontalo;11, Kotamobagu–Ratatotok;12, Batu Hijau.

tectonic model shows that most gold deposits formed during major tectonic reorganizations (Fig. 2). The first of these followed the collision of the Australian continent with the Philippine Sea plate arc around 25 Ma. This was followed by a Mid-Miocene period of mineralization that accompanied and followed the maximum rotation or extrusion of Indochina and cessation of spreading in the South China Sea at around 17 Ma (Sibuet *et al.* 2002). The majority of (and largest) deposits formed in a broad belt from Taiwan to the Solomon Islands during tectonic reorganization that followed the collision of the Philippine arc with the Eurasian plate at 5 Ma (Hall 1996). This period of plate reorganization and mineralization accompanies a postulated change in the relative motions of the Indian–Australian and Pacific plates (Cox & Engebretson 1985; Harbert & Cox 1989).

Gold in island arcs

Major SE Asian, magmatic–hydrothermal gold deposits (containing more than 10 tonnes of gold)

are dominantly high- or low-sulphidation epithermal deposits or porphyry Cu–Au deposits. Skarn, sediment-hosted, and gold-bearing carbonate-hosted base metal deposits are less abundant.

Epithermal Au deposits are mainly veins and disseminated deposits that possess textural and mineralogical characteristics of shallow emplacement (< 2 km) with many showing clear evidence of having formed in fossil hot spring and maar–diatreme systems in active volcanic environments. Low-sulphidation (adularia–sericite) epithermal Au deposits are characterized by alteration assemblages of quartz–adularia–sericite–carbonate formed from low temperature (150 to 300 °C) near-neutral hydrothermal fluids (Bonham 1986; White & Hedenquist 1995). These deposits are typically volcanic-hosted veins and breccias with native Au (± gold tellurides), silver sulphides and sulphosalts, high Ag/Au ratios and low base metal contents. They are the most numerous Au deposit type in the SE Asian arcs and generally contain less than 100 tonnes Au, although Ladolam (Tabar–Feni-arc) and Baguio (Philippines) (Fig. 1) are huge deposits of this type (Sillitoe 1997). In contrast, high-sulphidation epithermal Au deposits are characterized by intense argillic and advanced argillic alteration, especially quartz–alunite assemblages and ore minerals such as enargite characterized by their high sulphidation state (Bonham 1986; White & Hedenquist 1995). In several deposits of this type such as Lepanto (Mankayan district Philippines) (Fig. 1) and Wafi (Papua New Guinea) there is a clear relationship between high-sulphidation epithermal Au mineralization and intrusion-related porphyry-style mineralization with a substantial overlap in deposit characteristics.

Porphyry Cu–Au deposits are low-grade, large (>100 million tonnes) ore deposits associated with high-level porphyritic calk-alkaline to alkaline intrusions. Ore is characterized by veinlet or disseminated iron–copper sulphides distributed through a large volume of hydrothermally altered rock by high volume hydrothermal systems driven by heat from the igneous intrusion. Many of the porphyry Cu deposits of SE Asia are gold-rich (>0.1 ppm Au) by world standards with deposits such as Grasberg (Irian Jaya) and Panguna (Bougainville) containing more than 700 tonnes of gold, and Santo Tomas II (Baguio district Philippines), Far SE (Mankayan district Philippines), Ok Tedi (Papua New Guinea) and Batu Hijau (Indonesia) (Fig. 1) containing more than 200 tonnes of gold (Sillitoe 1997). Gold-rich porphyry Cu deposits are typically distinguished from other porphyry deposits by an abundance of hydrothermal magnetite in zones of Au- and Cu-bearing K-silicate alteration (Sillitoe 1979, 1997).

Copper and gold-rich skarn deposits are contact metasomatic deposits that form as a result of recrystallization and metasomatism of calcareous sedimentary rocks and causative intrusions. Important examples include the Ertsberg, Ertsberg East and Wabu skarns associated with the Grasberg porphyry deposit in Irian Jaya (Mertig et al. 1994; Meinert et al. 1997) (Fig. 1). Sediment-hosted Au, and carbonate-hosted base metal gold deposits are less common variants of the porphyry–skarn–epithermal suite of magmatic–hydrothermal gold deposits (Sillitoe & Bonham 1990; Leach & Corbett 1994).

The magmatic rocks associated with magmatic–hydrothermal gold mineralization in SE Asian arcs include island-arc tholeiite, calc-alkaline, high-K calc-alkaline (including shoshonite), adakite and alkaline suites. These magmas have a complex petrogenesis (Pearce & Peate 1995), initiated by dehydration of subducted oceanic crust and sediment at depth which releases an oxidized $H_2O–CO_2–Cl$ fluid capable of removing incompatible elements from the subducted slab and mobilizing metals both from the slab and in

the mantle. In the simplest case (many island arc tholeiite and calc-alkaline suites) these fluids initiate melting of the overlying mantle and are incorporated into the resulting basalt melt that then rises to the base of the crust. In the crust, a combination of fractional crystallization, assimilation and mixing with crustal melts results in the wide range of magma types found in magmatic arcs. During their emplacement in the upper crust and eruption, these relatively H_2O-rich magmas set up the hydrothermal systems that form epithermal, porphyry and skarn deposits. Because these magmas are associated with Cu–Au mineralization in both oceanic and continental arcs it is generally thought that the metals are derived from the mantle source of the magmas, rather than a local crustal source.

A large number of the gold deposits are associated with the high-K calc-alkaline, shoshonite, adakite and alkaline suites (Muller & Groves 1993; Richards & Kerrich 1993; Sillitoe 1997; Sajona & Maury 1998). Melting of sub-arc mantle that has been extensively metasomatized by fluids derived from ongoing, or earlier, episodes of subduction is thought to be important in the genesis of these suites (McInnes & Cameron 1994; Sajona et al. 1994; McInnes et al. 1999).

Gold deposits and the evolution of SE Asia

All principal deposits in SE Asian magmatic arcs formed after 25 Ma with the majority dated between 6 and 1 Ma (Late Miocene to Pleistocene) (Fig. 2). It has been suggested that the preponderance of young deposits reflects increased likelihood of erosion with increasing age (Sillitoe 1989). Though erosion will certainly remove older near-surface metal deposits, pre-Pliocene volcanic and high-level intrusive rocks are common in SE Asian arcs with little evidence that they were richly mineralized. Hence it is likely that the dominance of Pliocene gold deposits relative to other periods is not caused solely by the removal by erosion of older deposits, but may be related to the tectonic evolution of the arcs.

At around 25 Ma, the arrival of the Australian craton at the Philipine Sea plate subduction zone (Fig. 3a) caused major tectonic reorganization in the SE Asian region (Hall 1996). The direct effects were a change in the tectonic regime in the Papua New Guinea region from near orthogonal subduction to sinistral strike-slip and the commencement of clockwise rotation of the Philippine Sea plate. As the Philippine Sea plate is central to the region, this change in motion affected neighbouring terranes. Other effects may have included a reversal of subduction beneath the Philippines from eastward subduction on the proto-Manila

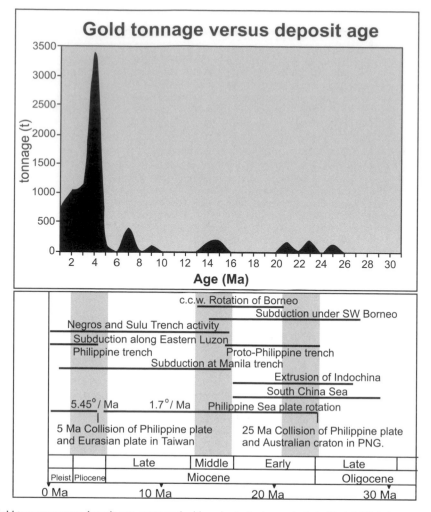

Fig. 2. Gold tonnage versus deposit age, compared with major tectonic events that affected SE Asian magmatic arcs. Grey shading on lower diagram indicates periods of plate reorganization.

trench to westward subduction at the proto-Philippine trench (Hayes & Lewis 1984). This period coincided with gold deposits in the Isabella–Didipio district (Luzon) and in New Britain (Mitchell & Leach 1991; Lindley 1998) (Figs 1 and 3a).

During the Mid-Miocene, the island arcs experienced more rapid convergence and plate rotation following the maximum extrusion, or rotation, of Indochina (e.g.Tapponnier *et al.* 1982; England & Molnar 1990) and the opening of the South China Sea (Fig. 3b). A tectonic response to spreading in the South China Sea was southwards directed subduction under Borneo. Cessation of subduction between 18 and 15 Ma and associated crustal thickening may be

related to the formation of Central Kalimantan gold deposits such as Kelian with >200 tonnes of contained gold, between 24 and 18 Ma (Simmons & Browne 1992; van Leeuwen *et al.* 1990) (Figs 1 and 3a). As maximum spreading of the South China Sea was reached at around 17 Ma, a subduction reversal occurred in the Philippines, from the proto-Philippine trench in the east to the modern Manila and Sulu–Masbate trenches in the west (Hayes & Lewis 1984) (Fig. 1). This subduction reversal coincided with gold deposits in the Camarines Norte district, eastern Philippines, and the Negros, Masbate, Batangas and Baguio districts along the western side of the archipelago (Mitchell & Leach 1991) (Figs 1 and 3b).

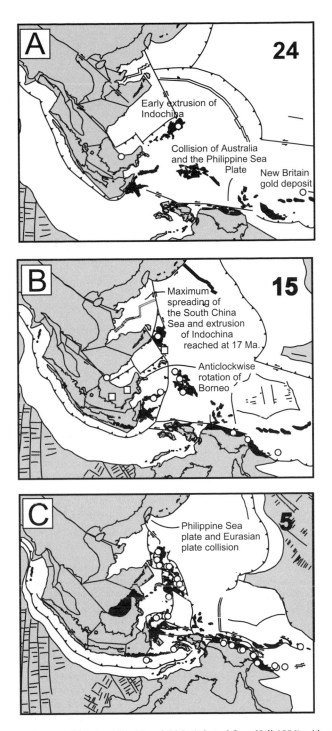

Fig. 3. Tectonic reconstructions of SE Asia at 24, 15 and 5 Ma (adapted from Hall 1996) with superimposed gold deposits and districts. (**a**) All gold deposits formed between 24 and 25 Ma. (**b**) Circles: all gold deposits formed between 12 and 15 Ma. Squares: those formed between 16 and 24 Ma. (**c**) All gold deposits formed between 0 and 6 Ma

Mid-Miocene gold deposits (15 to 12 Ma) also formed in the Maramuni arc of northern New Guinea (Russell 1990; Tau-Loi & Andrews 1998). This period of mineralization followed collision of the Philippine–Caroline arc with the extended margin of New Guinea (Abbott 1995; Crowhurst *et al.* 1996) and the initial collision of the Ontong–Java Plateau with the Solomons arc (Hill 2000).

A major tectonic reorganization has occurred in the SE Asian region since about 5 Ma, approximately the postulated time of change in the direction of relative movement between the Indian–Australian and Pacific plates (Cox & Engebretson 1985). This also coincided with collision of the Philippine arc with the Eurasian plate in Taiwan (Hall 1996). At this time the rate of rotation of the Philippine Sea plate increased from 1.7° to 5.4° per Ma and its pole of rotation changed from 15°N 160°E to 48°N 157°E (Hall 1996). The consequent effects on neighbouring terranes were widespread. In the Philippines, subduction appears to be transferring from the Manila trench, where subduction has slowed since 5 Ma, to the modern Philippine trench, where subduction commenced at around 5 Ma (Fig. 1). Significant gold deposits formed in the eastern Philippines, and particularly the Baguio and Mankayan districts in Luzon as well as in Mindanao (Mitchell & Leach 1991) (Figs 1 and 3c).

Further afield, the changes in plate movement affected Sulawesi, Irian Jaya and Papua New Guinea. Subduction was initiated in the Pliocene at the north Sulawesi trench, also associated with subduction reversal. Associated gold deposits include those in the Tombulilato/Gorontalo and Ratatotok/Kotamobagu districts (Figs 1 and 3c). In Irian Jaya and Papua New Guinea, subduction was initiated at the New Guinea trench and Pliocene magmatism was associated with significant gold deposits, including Ertsberg–Grasberg (MacDonald & Arnold 1994; Mertig *et al.* 1994), Ok Tedi (Rush & Seegers 1990) and Porgera (Richards & Kerrich 1993) in the New Guinea Orogen. Subduction was also initiated on the Eastern New Britain–San Cristobal–Vanuatu trenches following slowing down and cessation of subduction on the Kilinailau trench. The Panguna deposit on Bougainville, and the Ladolam deposit on Lihir Island in the Tabar–Feni arc, formed at this time (Figs 1 and 3c).

Discussion

There is a clear relationship between the age of gold deposits in SE Asian magmatic arcs and the tectonic reorganizations caused by changes in the regional tectonic regime (Figs 2 and 3). The most important of these started at about 5 Ma. This corresponded with a relative change in direction of movement of the Australian and Pacific plates between *c.* 5 and 3.5 Ma (Cox & Engebretson 1985; Harbert & Cox 1989) and followed the collision of the Philippine arc with Eurasia (Hall 1996). Further tectonic reorganization which started at about this time included the collision of eastern and western Mindanao and reversal of subduction in the Philipines and Sulawesi (Hall 1996). In Irian Jaya and Papua New Guinea rapid convergence between the Indian–Australian and Philippine Sea–Caroline plates resulted in southwards propagation of the New Guinea Orogen (Crowhurst *et al.* 1996). Cessation of subduction on the Kilinailau trench and obduction of the southern margin of the Ontong Java Plateau (Yan & Kroenke 1993) were accompanied by initiation of subduction on the New Britain–San Cristobal–Vanuatu trenches and the opening of the Woodlark Basin. Further afield, clockwise rotation of the Vanuatu arc (Yan & Kroenke 1993) and the change from transtension to transpression along the Alpine Fault in New Zealand (Norris *et al.* 1990) also occurred at about this time.

The resulting intensely mineralized belt extends from Taiwan through the Philippines, Sulawesi, eastern Indonesia, Papua New Guinea to the Solomon Islands and Fiji and contains many of the world's largest magmatic–hydrothermal Au deposits. Deposits with over 200 tonnes of contained gold include Santo Tomas II, Far Southeast (Sillitoe 1997) and Baguio (Cooke *et al.* 1996) in the Philippines, Batu Hijau (Irianto & Clark 1995) and Ertsberg–Grasberg (MacDonald & Arnold 1994) in Indonesia, and Ok Tedi (Rush & Seegers 1990), Porgera (Richards & Kerrich 1993), Ladolam (Moyle *et al.* 1990) and Panguna (Clark 1990) in Papua New Guniea.

How tectonic reorganizations control the location of mineralized districts is less certain. Sillitoe (1989, 1997) observed that all mineralized districts occurred in arcs with evidence for active extension or uplift at the time of mineralization, with the largest deposits associated with unusual rather than normal arc settings and magma types. Furthermore, many mineralized districts with high gold content (e.g. Baguio/Mankayan Luzon, east Mindanao, north Sulawesi, Panguna Bougainville) occur in volcanic arcs following a reversal in subduction polarity (Solomon 1990). In such a setting, local extension may result from slowing subduction (and possibly rollback of the slab) on one side of the volcanic arc and incipient subduction on the other. Magmatism may result from either dehydration or melting of the slab, or inflow of hot mantle as subduction slows, or during slab rollback. Solomon (1990) also suggested that this would induce melting of sub-arc mantle that had been both metasomatized and previously melted by earlier episodes of subduction and that such magmas may be intrinsically gold rich.

Melting of metasomatized (subduction modified) mantle may generate fluid-rich highly oxidized magmas as well as destabilizing mantle sulphides to release Cu and Au (McInnes & Cameron 1994). Important new evidence from veined peridotite xenoliths sampling the mantle beneath the Tabar–Feni arc, which hosts the Ladolam deposit, shows that they are strongly enriched in Cu, Au, Pt and Pd relative to surrounding depleted arc mantle (McInnes *et al.* 1999). These peridotites also have similar Os isotopic compositions to the Ladolam gold ores, indicating the primary source of the metals was the subduction-modified mantle (McInnes *et al.* 1999).

Gold deposits that follow subduction reversal occur in Late Oligocene, Mid-Miocene and Pliocene rocks of Luzon and Pliocene rocks of Mindanao, north Sulawesi and Bougainville. Further possible districts with this setting are Romang–Wetar islands in Indonesia, New Britain, New Ireland and the Solomon Islands. Subduction cessation and reversal appear to be a common regime for formation magmatic–hydrothermal gold deposits. However, in Mindanao, it is unlikely that the Celebes Sea slab had penetrated to sufficient depth to provide a source for shoshonitic and adakitic magmas associated with Pliocene mineralization (Sajona *et al.* 1994; Sajona & Maury 1998; Macpherson & Hall 1999). Also the Pliocene magmas associated with gold deposits of the New Guinea Orogen including the prodigious Ertsberg–Grasberg district (the highest gold tonnage in the region), Ok Tedi and Porgera follow arc accretion and cannot be easily linked to either coeval subduction, or subduction reversal. Thermal or tectonic reactivation of subduction-modified mantle beneath a thickened arc or orogen, with melting in the mantle and lower crust, seems a more likely cause for magmatism in both cases.

In all cases magmas associated with magmatic–hydrothermal gold mineralization in SE Asian magmatic arcs have geochemical features indicating their origins are linked to subduction processes. However, in several important provinces, including many of those with the largest gold deposits, magmatism is not directly linked to an actively subducting slab. Examination of the spatial and temporal distribution of gold deposits relative to the regional tectonic model for SE Asia shows that most gold deposits did not form during periods of normal or steady state subduction. Rather periods of plate reorganization characterized by magmatism in unusual arc-related settings

produced the most abundant and largest deposits. In particular, peak mineralization appears related to melting of mantle that has been previously modified by subduction. During large-scale plate reorganization this can occur at the end of a period of normal subduction, following the cessation of subduction, following arc collision, or accompanying subduction reversal at approximately the same time in different parts of a complex system of magmatic arcs such as those in SE Asia. Such tectonic controls may be impossible to recognize in older orogenic belts where recognition of a major change in tectonic style and unusual magma types may be the best guides to mineralization.

We thank R. Hall for permission to use his tectonic model. P. Rak compiled the gold deposit data-base and tectonic animations as part of a B.Sc. (Honours) thesis at the University of Western Australia.

References

ABBOTT, L.D. 1995. Neogene tectonic reconstruction of the Adelbert-Finistrerre-New Britain collision, northern Papua New Guinea. *Journal of Southeast Asian Earth Sciences*, **11**, 33–51.

BONHAM, H.F. 1986. Models for volcanic-hosted epithermal precious metal deposits: a review. *Proceedings of International Volcanological Congress Symposium*, **5**, 13–17.

CLARK, G.H. 1990. Panguna copper-gold deposit. *In:* HUGHES, F.E. (eds) *Geology of the Mineral Deposits of Australia and Papua New Guinea.* The Australasian Institute of Mining and Metallurgy Monographs, **14**, 1807–1816.

COOKE, D.R., MCPHAIL, D.C. & BLOOM, M.S. 1996. Epithermal gold mineralization, Acapan, Baguio district, Philippines: geology, mineralization, alteration, and the thermochemical environment of ore deposition. *Economic Geology*, **91**, 243–272.

COX, A. & ENGEBRETSON, D. 1985. Change in motion of the Pacific plate at 5 Myr BP. *Nature*, **313**, 472–474.

CROWHURST, P.V., HILL, K.C., FOSTER, D.A. & BENNETT, A.P. 1996. Thermochronological and geochemical constraints on the tectonic evolution of northern Papua New Guinea. *In:* HALL, R. & BLUNDELL, D.J. (eds) *Tectonic Evolution of Southeast Asia.* Geological Society, London, Special Publications, **106**, 525–537.

ENGLAND, P.C. & MOLNAR, P. 1990. Right-lateral shear and rotation and rotation as the explanation for strike-slip faulting in eastern Tibet. *Nature*, **344**, 140–142.

HALL, R. 1996. Reconstructing Cenozoic SE Asia. *In:* HALL, R. & BLUNDELL, D.J. (eds) *Tectonic Evolution of Southeast Asia.* Geological Society, London, Special Publications, **106**, 153–184.

HALL, R. 1998. The plate tectonics of Cenozoic SE Asia and the distribution of land and sea. *In:* HALL, R. & HOLLOWAY, J.D. (eds) *Biogeography and Geologi-*

cal Evolution of SE Asia. Backhuys Publishers, Leiden, 99–131.

HARBERT, W. & COX, A. 1989. Late Neogene motion of the Pacific Plate. *Journal of Geophysical Research, 94B,* 3052–3064.

HAYES, D.E. & LEWIS, S.D. 1984. A geophysical study of the Manila Trench, Luzon, Philippines: 1, Crustal structure, gravity, and regional tectonic evolution. *Journal of Geophysical Research,* **89B,** 9171–9195.

HILL, K.C. 2000. Mesozoic-Tertiary evolution of the northern Australian margin. *Geological Society of Australia Abstracts,* **59,** 223.

IRIANTO, B. & CLARK, G.H. 1995. The Batu Hijau porphyry copper-gold deposit, Sumbawa Island, Indonesia. *Proceedings PACRIM,* **99,** 299–304.

LEACH, T.M. & CORBETT, G.J. 1994. Porphyry-related carbonate-base-metal gold systems in the southwest Pacific: characteristics. *In:* ROGERSON, R. (ed.) *Proceedings of the PNG Geology, Exploration and Mining Conference, 1994.* Australasian Institute of Mining and Metallurgy, 84–91.

LINDLEY, I.D. 1998. Mount Sinivit gold deposits. *In:* BERKMAN, D.A. & MACKENZIE, D.H. (eds) *Geology of Australian and Papua New Guinean Mineral Deposits.* The Australasian Institute of Mining and Metallurgy, 821–826.

MACPHERSON, C.G. & HALL, R. 1999. Tectonic controls of geochemical evolution in arc magmatism of SE Asia. *Proceedings PACRIM,* **99,** 359–367.

MACDONALD, G.D. & ARNOLD, L.C. 1994. Geological and geochemical zoning of the Grasberg igneous complex, Irian Jaya, Indonesia. *Journal of Geochemical Exploration,* **50,** 143–178.

MCINNES, B.I.A. & CAMERON, E.M. 1994. Carbonated, alkaline hybridizing melts from a sub-arc environment: mantle wedge samples from the Tabar-Lihir-Tanga-Feni arc Papua New Guinea. *Earth and Planetary Science Letters,* **122,** 125–141.

MCINNES, B.I.A., MCBRIDE, J.S., EVANS, N.J., LAMBERT, D.D. & ANDREW, A.S. 1999. Osmium isotope constraints on ore metal recycling in subduction zones. *Science,* **286,** 512–516.

MEINERT, L.D., HEFTON, K.K., MAYES, D. & TASIRAN, I. 1997. Geology, zonation and fluid evolution of the Big Gossan Cu-Au skarn deposit, Ertsberg District, Irian Jaya. *Economic Geology,* **92,** 509–534.

MERTIG, H.J., RUBIN, N.J. & KYLE, J.R. 1994. Skarn Cu-Au orebodies of the Gunung Bijih (Ertsberg) district Irian Jaya, Indonesia. *Journal of Geochemical Exploration,* **50,** 172–202.

MITCHELL, A.H.G. & GARSON, M.S. 1972. Relationship of porphyry copper and circum-Pacific tin deposits to palaeo-Benioff zones. *Transactions of the Institute of Mining and Metallurgy,* **81,** B10–B25.

MITCHELL, A.H.G. & GARSON, M.S. 1976. Mineralization at Plate boundaries. *Minerals Science Engineering,* **8,** 129–169.

MITCHELL, A.H.G. & LEACH, T.M. 1991. *Epithermal Gold in the Philippine.* Academic Press, London.

MOYLE, E.H., DOYLE, B.J., HOOGVLIET, H. & WARE, A.R. 1990. Ladolam gold deposit, Lihir Island. *In:* HUGHES, F.E. (ed.) *Geology of the Mineral Deposits of Australia and Papua New Guinea.* The Australa-

sian Institute of Mining and Metallurgy Monographs, **14,** 1793–1805.

MULLER, D. & GROVES, D.I. 1993. Direct and indirect associations between potasic igneous rocks, shoshonites and copper-gold mineralization at a convergent plate margin. *Ore Geology Reviews,* **8,** 383–406.

NORRIS, R.J., KOONS, P.O. & COOPER, A.F. 1990. The obliquely-convergent plate boundary in the South Island of New Zealand: implications for ancient collision zones. *Journal of Structural Geology,* **12,** 715–725.

PEARCE, J.A. & PEATE, D.W. 1995. Tectonic implications of the composition of island arc magmas. *Annual Reviews of Earth and Planetary Sciences,* **23,** 251–285.

RICHARDS, J.P. & KERRICH, R. 1993. The Porgera gold mine, Papua New Guinea: magmatic hydrothermal to epithermal evolution of an alkalic-type precious metal deposit. *Economic Geology,* **88,** 1017–1052.

RUSH, P.M. & SEEGERS, H.J. 1990. Ok Tedi copper gold deposits of Papua New Guinea. *In:* HUGHES, F.E. (eds) *Geology of the Mineral Deposits of Australia and Papua New Guinea.* The Australasian Institute of Mining and Metallurgy Monographs, **14,** 1747–1754.

RUSSELL, P. 1990. Woodlark Island gold deposits. *In:* HUGHES, F.E. (eds) *Geology of the Mineral Deposits of Australia and Papua New Guinea.* The Australasian Institute of Mining and Metallurgy Monographs, **14,** 1735–1739.

SAJONA, F.G., BELLON, H., MAURY, R.C., PUBELLIER, M., COTTEN, J. & RANGIN, C. 1994. Magmatic response to abrupt changes in geodynamic settings: Pliocene-Quaternary calc-alkaline and Nb-enriched lavas from Mindanao (Philippines). *Tectonophysics,* **237,** 47–72.

SAJONA, F.G. & MAURY, R.C. 1998. Association of adakites with gold and copper mineralization in the Philippines. *Comptes Rendus de l'Academie des Sciences, Paris,* **326,** 27–34.

SIBUET, J-C., HSU, S-K., LE PICHON, X., LE FORMAL, J-P., REED, D., MOORE, G. & LIU, C-S. 2002. East Asia plate tectonics since 15 Ma: constraints from the Taiwan region. *Tectonophysics,* **134,** 103–134.

SILLITOE, R.H. 1972. Relation of metal provinces in western America to subduction of oceanic lithosphere. *Bulletin of the Geological Society of America,* **83,** 813–818.

SILLITOE, R.H. 1979. Some thoughts on gold-rich porphyry copper deposits. *Mineralium Deposita,* **14,** 161–174.

SILLITOE, R.H. 1989. Gold deposits in western Pacific island arcs: the magmatic connection. *In:* KEAYS, R., RAMSAY, R. & GROVES, D. (eds) *The Geology of Gold Deposits: The Perspective in 1988.* Economic Geology Monographs, **6,** 274–291.

SILLITOE, R.H. 1997. Characteristics and controls of the largest porphyry copper-gold and epithermal gold deposits in the circum Pacific region. *Australian Journal of Earth Sciences,* **44,** 373–388.

SILLITOE, R.H. & BONHAM, H.F. 1990. Sediment-hosted gold deposits: distal products of magmatic hydrothermal systems. *Geology,* **18,** 157–161.

SIMMONS, S.F. & BROWNE, P.R.L. 1992. Mineralogic,

alteration and fluid inclusion studies of epithermal gold veins at the Mount Muro Prospect, central Kalimantan (Borneo), Indonesia. *Journal of Geochemical Exploration*, **35**, 63–103.

SOLOMON, M. 1990. Subduction, arc reversal, and the origin of porphyry copper-gold deposits in island arcs. *Geology*, **18**, 630–633.

TAPPONNIER, P., PELTZER, G., LE DAIN, A.Y., ARMIJO, R. & COBBOLD, P. 1982. Propogating extrusion tectonics in Asia: new insights from simple experiments with plasticine. *Geology*, **7**, 171–174.

TAU-LOI, D. & ANDREWS, R.L. 1998. Wafi copper-gold deposit. *In:* BERKMAN, D.A. & MACKENZIE, D.H. (eds) *Geology of Australian and Papua New Gui-nean Mineral Deposits*. The Australasian Institute of Mining and Metallurgy, 827–831.

YAN, C.Y. & KROENKE, L.W. 1993. A plate reconstruction of the southwest Pacific, 0-100 Ma. *In:* BERGER, W.H., KROENKE, L.W. & MAYER, L.A. (eds) *Proceedings of the Ocean Drilling Program Scientific Results*. **130**, 697–709.

VAN LEEUWEN, T., LEACH, T., HAWKE, A.A. & HAWKE, M. 1990. The Kelian disseminated gold deposit, East Kalimantan, Indonesia. *Journal of Geochemical Exploration*, **35**, 1–61.

WHITE, N.C. & HEDENQUIST, J.W. 1995. Epithermal gold deposits: styles characteristics and exploration. *Society of Economic Geologists Newsletter*, **23**, 9–13.

Timing and tectonic controls in the evolving orogen of SE Asia and the western Pacific and some implications for ore generation

COLIN G. MACPHERSON[1] & ROBERT HALL[2]

[1]*Department of Geological Sciences, University of Durham, South Road, Durham DH1 3LE,UK (e-mail: colin.macpherson@durham.ac.uk)*

[2]*SE Asia Research Group, Department of Geology, Royal Holloway University of London, Egham, Surrey TW20 0EX, UK*

Abstract: SE Asia lies at the convergence of the Eurasian, Pacific and Australian plates. The region is made up of many active arcs, extensional basins, and the remnants of similar tectonic environments developed throughout the Cenozoic. There are many important hydrothermal mineral deposits and prospects in SE Asia but their formation is often poorly understood due to the complicated tectonic history of this region and the knowledge of relationships between mineralization and tectonics. Plate reconstruction offers a framework to integrate geological and geochemical data that can be used to unravel the large-scale tectonic processes that affected mineralized provinces. We present examples of the information that can be derived from this approach and discuss the implications for understanding the origin of some hydrothermal mineral deposits in SE Asia.

Formation of an economic hydrothermal ore deposit is the culmination of many different processes, some of which are intrinsically linked to the tectonic setting of the deposit. The recurrence of particular styles of mineralization in certain tectonic settings indicates a strong relationship between geodynamics and mineralization and has led to the development of models in which ore generation is linked to plate tectonic processes (e.g. Solomon 1990; Sillitoe 1999). Some models aim to establish links between magmas and mineralization, implying that mineralization style may depend either on the mechanism and source of melting (e.g. Sajona & Maury 1998) or the ability of magmas to transport metals (Feiss 1978; Sillitoe 1997). Verification of associations between mineralization and either tectonic setting or magmatic characteristics would provide a powerful exploration tool in the search for new mineral deposits.

SE Asia and the west Pacific host several of the world's most important ore deposits, which include a range of mineralization styles and metallogenic provinces. Many of these deposits are relatively young and therefore their current tectonic environment provides a good approximation of that in which they formed. Such relationships provide a basis for interpreting other ore deposits and assessing the likelihood that economic reserves may be identified in particular locations or may have formed at certain times (Solomon 1990;

Sillitoe 1997). This contribution examines the importance of some tectonic processes that have been proposed as possible controls on ore formation for the SE Asian and SW Pacific regions.

Tectonic setting of SE Asia and western Pacific

SE Asia and the SW Pacific are dominated by convergence between the Eurasian, the Indian–Australian and the Pacific plates (Fig. 1). In addition, the Philippine Sea plate, which lies between the Pacific and Eurasian plates, has played a crucial role in development of the region. There are at present a large number of other small plates and plate fragments, and there was probably a similar complexity throughout the Cenozoic (Hall 1996, 2002). The magmatic provinces of SE Asia and the western Pacific can be attributed to four major zones of convergence (Fig. 1).

The first lies between the Eurasian and Indian–Australian plates and extends from Myanmar in the west to the Banda arc in the east and results from subduction of Indian Ocean lithosphere beneath Sundaland. Within this zone the Sunda arc comprises the islands of the Andamans, Sumatra, Java and Nusa Tenggara. Mesozoic magmatism is known in Sumatra, Java (Katili 1975; Rock *et al.* 1982; McCourt *et al.* 1996) and Sumba (e.g. Rutherford *et al.* 2001) although in

From: BLUNDELL, D.J., NEUBAUER, F. & VON QUADT, A. (eds) 2002. *The Timing and Location of Major Ore Deposits in an Evolving Orogen*. Geological Society, London, Special Publications, **204**, 49–67. 0305-8719/02/$15.00 © The Geological Society of London 2002.

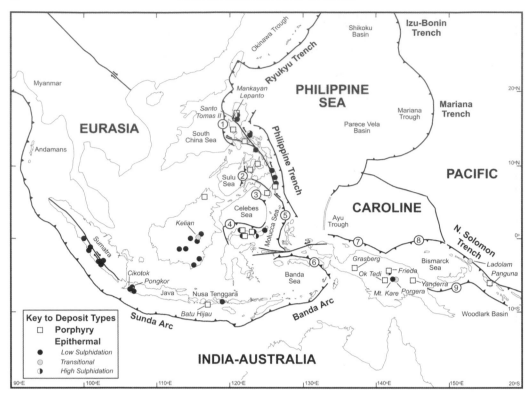

Fig. 1. Map of SE Asia highlighting major tectonic plates and plate boundaries and the location of important hydrothermal ore deposits. Plate names are shown in boldest font and subduction zones are indicated by lines with triangles on the overriding plate. Key to subduction zones not noted in figure: 1, Manila Trench; 2, Negros Trench; 3, Cotobato Trench; 4, North Sulawesi Trench; 5, Halmahera Trough (also frontal thrust of Sangihe arc); 6, Seram Trough; 7, New Guinea Trench; 8, Manus Trench; 9, New Britain Trench. Key to hydrothermal mineral deposit types is shown and major deposits are labelled in italics.

most islands of Nusa Tenggara the oldest known magmatic rocks are Miocene (e.g. Abbott & Chamalaun 1978; Barberi *et al.* 1987). The eastern part of this convergent zone is the Banda arc where the Australian continental margin has been subducted towards the north (Fig. 2: Carter *et al.* 1976; Vroon *et al.* 1993). In the west, mineralization in Sumatra and west Java is mainly made up of low-sulphidation epithermal gold deposits (Basuki *et al.* 1994; Marcoux & Milési 1994; Jobson *et al.* 1994;Carlile & Mitchell 1994). Well-developed mineralization is largely unrecognized in central and eastern Java but Nusa Tenggara hosts one major porphyry copper deposit at Batu Hijau in Sumbawa (Meldrum *et al.* 1994) and several epithermal prospects (Carlile & Mitchell 1994). The easternmost mineralization in this zone is comprised mainly of submarine exhalative Ag–Au–barite Volcanic Massive Sulphide deposits (VMS) such as those on Wetar Island (Sewell & Wheatley 1994) with or without base metals

(Carlile & Mitchell 1994). For most deposits in the Sunda–Banda arc, a Late Miocene to Pliocene mineralization age has been determined or is inferred.

A second zone of convergence occurs where lithosphere of the Pacific Ocean is subducted towards the west. Subduction extends from Japan, through the Izu–Bonin and Mariana trenches to the northern end of the Ayu Trough. Magmatism has occurred in this arc since the Eocene (Cosca *et al.* 1998) and involved the opening of several marginal basins within the Philippine Sea plate: the Parece Vela Basin, thc Shikoku Basin and the Mariana Trough (Fig. 1). Actively forming VMS deposits have recently been identified within the Izu–Bonin arc (Iizasa *et al.* 1999).

The third zone of convergence runs south from Taiwan, through the Philippine Islands to the Molucca Sea region in the south. This is a complex region with subduction zones dipping both east and west beneath the Philippine Islands

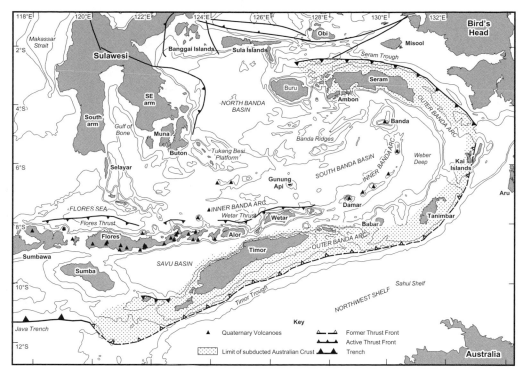

Fig. 2. Geographical features of the Banda Sea and surrounding regions. Small black filled triangles are volcanoes from the Smithsonian database (http://www.nmnh.si.edu/gvp/), and bathymetry is from the GEBCO digital atlas (IOC, IHO, BODC 1997). Bathymetric contours at 1000 m, 2000 m and 4000 m.

linked by strike-slip faults (Mitchell *et al.* 1986). The Philippine Sea plate is subducted towards the west at the Philippine Trench. Eastward subduction at the Manila, Negros and Cotobato trenches consumes oceanic lithosphere of the South China, the Sulu and Celebes Seas respectively, which formed in Cenozoic extensional and back-arc basins at the eastern margin of Sundaland. The Celebes Sea is also being subducted towards the south beneath the northern arm of Sulawesi.

The Philippine Islands host a huge number of ore bodies and prospects (Sillitoe & Gappe 1984; Mitchell & Leach 1991). Porphyry-style and epithermal mineralization occur in close proximity to one another in several parts of the archipelago. The Central Cordillera and Camarines Norte District of Luzon and eastern Mindanao host many porphyry-style and epithermal mineral deposits that are rich in gold. In each of these cases the deposits lie close to the present trace of the Philippine Fault although the connection between faulting and mineralization remains unclear. Reported radiometric ages, obtained by whole-rock K–Ar dating, have a bimodal distribution with many porphyry-related intrusions of Early to Mid-Miocene age (20–12 Ma) whereas epithermal rocks and some porphyry-related intrusions are Plio-Pleistocene. Mitchell & Leach (1991) used this evidence to question the link between porphyry and epithermal mineralization. However, the Far South East porphyry and Lepanto epithermal deposits in the Central Cordillera of Luzon appear to have formed over a short interval of <0.5 Ma (Arribas *et al.* 1995). This suggests that the bimodal distribution of mineralization ages adjacent to the Philippine Fault could reflect either two distinct mineralization events or problems of K–Ar dating. Many rocks are susceptible to alteration and weathering, particularly in tropical latitudes, which could reset the K–Ar clock leading to a spuriously young age (Dickin 1995), or could add potassium, and therefore also ^{40}Ar, to the rock resulting in an overestimation of the age. Unlike some dating techniques which employ materials that are more resistant to weathering (e.g. mineral separates for Rb–Sr) or may identify K and Ar mobility (e.g. $^{40}Ar/^{39}Ar$), the effects of alteration on the K–Ar isotope systematics cannot be quantified. Reliable dating of mineral deposits remains one of the main barriers to improving our

C. G. MACPHERSON & R. HALL

understanding of the geodynamic setting of ore generation.

At the southern end of the Philippines the Molucca Sea plate is being subducted to the east and west beneath Halmahera and Sangihe arcs, respectively (Fig. 3: Cardwell *et al.* 1980). The greatest ages yet obtained for volcanism associated with subduction of the Molucca Sea are Mid-Miocene (Baker & Malaihollo 1996; Elburg & Foden 1998), although Late Cretaceous–Eocene and Oligocene phases of subduction magmatism are recorded in the Halmahera region (Hakim & Hall 1991; Hall *et al.* 1995). In the case of Halmahera, the Miocene to Recent igneous activity appears to have gone through periods of greater and lesser intensity (Baker & Malaihollo 1996). We also include the north arm of Sulawesi in this zone, where the Celebes Sea is being subducted towards the south (Silver *et al.* 1983; Walpersdorf *et al.* 1998). Hydrothermal mineralization is relatively uncommon in the Sangihe Islands and Halmahera (e.g. Oldberg *et al.* 1999), but many porphyry and, primarily low sulphidation, epithermal deposits and prospects are known from northern Sulawesi (Carlile & Mitchell 1994). Much of

this mineralization is Mid-Miocene or younger, thus predating the initiation of the current southward subduction of the Celebes Sea Plate (Silver *et al.* 1983; Surmont *et al.* 1994).

A final convergence zone extends east to the young subduction systems and oceanic basins that result from complex convergence and strike-slip faulting linking the New Guinea margin to the Melanesian arcs at the boundary of the Australian and Pacific plates. Among the many hydrothermal occurrences in this region are Grasberg, Ok Tedi and Porgera on New Guinea. The Ladolam deposit on Lihir and the Panguna deposit on Bougainville lie in segments of the Vitiaz–West Melanesian arc that collided with the Ontong Java Plateau during the Mid-Miocene (Kroenke 1984; Phinney *et al.* 1999) leading to tectonic quiescence during the Mid-Miocene (Hilyard & Rogerson 1989; Stewart & Sandy 1988). Subsequent tectonic activity on these islands, including mineralization, was associated with north and westward subduction of the Solomon Sea. Like the Philippines and Molucca Sea, there is a mixture of both porphyry and epithermal styles of mineralization through this zone and most of these have ages that are younger

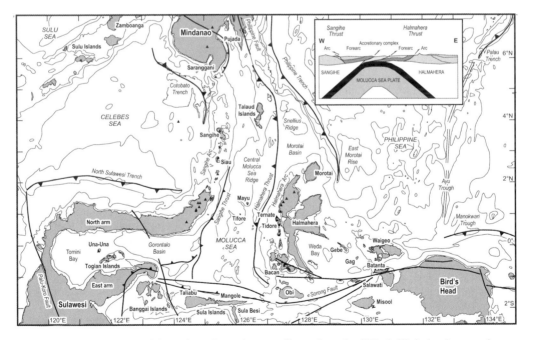

Fig. 3. Geographical features of the Molucca Sea and surrounding regions. Small black filled triangles are volcanoes from the Smithsonian database. Bathymetric contours at 200 m, 2000 m, 4000 m, and 5000 m from the GEBCO digital atlas (IOC, IHO, BODC 1997). Large barbed lines are subduction zones and small barbed lines are thrusts. Note that the direction of surface thrusting in the Molucca Sea is correctly shown. Thrusts on each side of the Molucca Sea are directed outwards towards the adjacent arcs although the subducting Molucca Sea plate dips east beneath Halmahera and west below the Sangihe arc. The Halmahera arc is being overridden by the Sangihe arc as shown in the inset figure.

than Mid-Miocene (Sillitoe, 1997). Ladolam is distinctive in that it is a telescoped deposit in which the sector collapse of a volcano resulted in the overprinting of a porphyry body by epithermal mineralization (Sillitoe 1994). A further notable feature of this region is the active formation of VMS deposits in the eastern Bismarck Sea, which are associated with extensional magmatism in the Manus back-arc basin. This basin is related to northward subduction of the Solomon Sea (Binns & Scott 1993; Scott & Binns 1995).

Tectonic controls on mineralization

A hydrothermal mineral deposit will form when fluids circulating through the Earth's crust concentrate an element, or elements, in a particular location. Sufficient concentration of the ore to create an economic body relies on an appropriate balance between several variables. The lifetime of the system, the chemical environment provided by the country rocks, and the porosity and permeability of the country rocks probably play a significant role in determining how elements will be distributed within a hydrothermal system. While these factors will influence the size and/or grade of hydrothermal deposits, to a greater or lesser degree, they are largely independent of tectonic processes.

In contrast, the crustal heat flow, the fluid pathways permitted by the crustal structure hosting the system, the chemistry of the hydrothermal fluids (Hedenquist & Lowenstern 1994; Sillitoe 1997) and the composition of the magmatism associated with a deposit can be directly related to the tectonic forces exerted on the crust hosting a hydrothermal system. Heat flow and crustal structure in convergent margins will depend on the lithospheric thickness and the stress regime imposed by plate interactions. Similarly, the chemical composition of magmas (and the fluids they exsolve) depend on melting conditions. These, in turn, result from chemical and physical factors, such as composition, pressure, temperature and volatile content of the mantle or crustal source (Pearce & Peate 1995). Therefore, the processes that generate magmatism and affect the lithospheric structure at convergent margins can influence the timing and location of mineralization in several ways.

The association between subduction and magmatism is clear. The recognition that volcanism in island arcs is related to subduction was quickly incorporated into tectonic models once plate tectonic theory was accepted at the end of the 1960s but since then research has led to revision of many ideas about the causes of melting, and the relative importance of tectonic processes. Early models (e.g. Oxburgh & Turcotte 1970) emphasized magmatism due to frictional heating on the Benioff zone, melting of the subducting slab and compressional shortening of the overriding plate in the arc–trench gap. In contrast, melts are now seen as originating primarily in the mantle wedge above the subduction zone and are thought to result from the input of volatiles that lower the mantle solidus (Fig. 4a). Furthermore, extension of the overriding plate is now recognized as a common feature in arc settings (Hamilton 1995). Tectonics may influence magmatism at regional and local scales, for example, by inducing melting during extension, by providing pathways for magmas, or by varying the rate of supply of volatile components. The character of magmatism is of particular interest in tectonic interpretation if the geochemistry of igneous rocks can be used reliably to infer the tectonic setting (e.g. Pearce & Peate 1995).

The location of subduction zone magmatism is related to the mechanisms that cause subduction. Hamilton (1988, 1995) emphasizes the necessity of viewing subduction in a dynamic way. Subduction zones have often been portrayed as 'static' systems with the subducting slab rolling over a stationary hinge and sliding down a slot fixed in the mantle (Fig. 4b). Viewed dynamically, Hamilton (1995) argues that subduction is most commonly driven by gravity acting on the subducting slab (Fig. 4c) and, therefore, most subduction zones are characterized by retreat of the hinge with time, or slab rollback (Fig. 4d). As a result the mantle wedge between the two plates can be replenished by an inflow of hot, undepleted mantle (Andrews & Sleep 1974; Furukawa 1993). This flow may also lead to ablation of the base of the overriding plate (Furukawa 1993; Iwamori 1997; Rowland & Davies 1999) and the mantle wedge that upwells into the ablated volume will be susceptible to melting by decompression. In this way steady-state arc magmatism at a relatively constant distance from the trench can be maintained over prolonged periods. Subduction zones may also be characterized by fixed hinges or by hinge advance. A fixed hinge may result when the system is locked or when the velocity of the whole system is exactly opposite to the rate of hinge retreat. Similarly, hinge advance may in part result if the two plates are both moving in the same direction as the slab is being subducted.

Geochemical investigation of tectonic processes

Many mineral deposits in SE Asia can be linked to magmatism at convergent margins. Subduction zone magmatism frequently displays geochemical

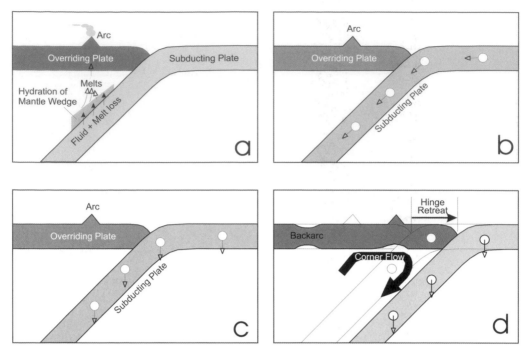

Fig. 4. Illustration of the different concepts of static and dynamic subduction. (**a**) Schematic cross section showing relationship between subducted and overriding plates and the mantle wedge between them. The distinctive geochemistry of subduction zone magmatism results from dehydration and melting of the subducted plate. (**b**) 'Conveyor belt' subduction inferred in many textbook models of subduction. (**c**) Schematic representation of gravitational forces acting on a subducted slab at any point in time as discussed by Hamilton (1995). (**d**) Result of subduction driven by gravity from condition shown in (c) with initial condition shown by faint lines. As the subducted plate moves down in the mantle, mantle circulates through the mantle wedge. The resulting migration of the trench at the Earth's surface is known as hinge retreat.

characteristics that are distinct from melts gener-ated in other tectonic environments, particularly mid-ocean ridge basalts (MORB). Compared to MORB, the relative and absolute concentrations of trace elements that are not mobile in fluids, such as the high field strength elements (HFSEs), are often depleted in lavas from island arcs (Ewart & Hawkesworth 1987; McCulloch & Gamble 1991). Such depletion suggests that the mantle wedge has been stripped of a basaltic component to varying degrees prior to the present phase of magmatism (Woodhead *et al.* 1993) and demonstrates that the wedge exerts important controls on general geo-chemical characteristics of arc lavas (e.g. Woodhead 1989). HFSE depletion is particularly noticeable in arcs paired with back-arc spreading centres, suggesting that the peridotite which melts to yield arc lavas may have been processed during its passage through the source region of back-arc magmatism (Woodhead *et al.* 1993). This is consistent with the concept of corner flow in the mantle wedge (Andrews & Sleep 1974; Furukawa 1993).

As a consequence of the cooling effect of the subducting slab and the refractory nature of peridotite in some mantle wedges, slab-derived fluid or volatile-rich melt is probably necessary to induce melting beneath many arc fronts (Morris *et al.* 1990; Ryan & Langmuir 1993). Any portion of the sedimentary cover that is subducted with the lithosphere may also be prone to melting at, or close to, the depths at which the mantle wedge melts. Therefore, the recycled component may have two different provenances within individual arcs (Elliott, *et al.* 1997; Turner *et al.* 1996). Subducted crustal material within the source re-gion of arc magmas provides a reservoir not present during MORB genesis. The nature of this material and the mechanisms of mass transport between slab and wedge can be resolved by examining the variable enrichment of trace ele-ments compared to MORB and characteristic departures of isotope ratios from upper mantle values (Tatsumi & Eggins 1995; Turner *et al.* 1996; Hawkesworth *et al.* 1997). For example, fluid transport from slab to wedge can induce chemical fractionation between elements such as La and Nb that otherwise behave similarly during magmatic processes, but have significantly different mobility in the presence of hydrous fluids or brines (Tatsumi *et al.* 1986; Thirlwall *et al.* 1994; Brennan *et al.* 1995; Keppler 1996; You *et al.* 1996). Taken together, observations of this sort have led to continual refinements of models that describe the nature of sources and the mechanisms of melting at subduction zones (e.g. Woodhead 1989; Hawkesworth *et al.* 1991; Elliott *et al.* 1997).

Following generation, melts must traverse the plate that overrides the subduction zone prior to eruption. This transport provides an opportunity for chemical differentiation, either through melt–solid–vapour partitioning, or through open system processes, such as magma-mixing, crustal contamination or assimilation with fractional crystallization (AFC). The contaminants involved in these processes can resemble the materials being subducted and care must be taken to fully account for the influence of the overriding plate if recycled fluxes are to be accurately constrained (Davidson & Harmon 1989; Macpherson *et al.* 1998).

Studies leading to new theories of melt generation in subduction zones have largely focused on active, or recently active, arcs. Much progress has been made by investigating spatial variations within individual arcs that allow evaluation of changes in the balance of different processes. Major factors influencing petrogenesis, such as fluid and melt fluxes from the subducted plate and the temperature, geometry and fertility of the mantle wedge, also have the potential to vary in time (e.g. Plank & Langmuir 1988; McCulloch & Gamble 1991; Turner & Hawkesworth 1997; Elburg & Foden 1998). Therefore, geochemical studies have potential to yield information on the evolution of tectonic events at convergent margins. If dynamic changes in subduction zones are a major factor in the generation of large mineralized systems (Sillitoe 1997), then it is important to understand and quantify the role of these events so that other factors influencing ore genesis at particular sites can be understood in their regional context. This is particularly relevant to SE Asia where there have been relatively rapid changes in plate configuration and rates of tectonic processes during the Cenozoic.

There are several properties that can vary to produce changes in the character of magmatism at subduction zones. Changes in the rate of subduction could affect both the rate at which the mantle wedge is replenished and the fluxes of material (fluids and melts) derived from the down-going slab. The nature of subducted material may influence magmatism and may, in some circumstances,

provide a chemical tracer of tectonic history. If magmas interact with the crust, the character of crust in the overriding plate will be important. For example, in east Indonesia it is possible to identify changes in the character of crust from the geochemistry of the Neogene volcanic rocks (see below). In the case of oblique subduction, there may be changes in the character of crust where there are trench-parallel strike-slip faults as in Sumatra and the Philippines. The following examples show how geochemical investigations of arc lavas have been employed to identify changes of this type in a single subduction system, the Molucca Sea Collision Zone.

Evolution of recycled components

Both the nature of recycled materials and recycling processes can evolve during the lifetime of a subduction zone. For example, Elburg & Foden (1998) have suggested that the recycled component in southern Sangihe arc lavas was derived mainly from altered oceanic crust during the Mid-Miocene, but is presently dominated by a sediment-derived component. This evolution can be compared with that of Neogene (Forde 1997) and Quaternary (Morris *et al.* 1983) lavas in the Halmahera arc. An orthogonal collision between the Sangihe and Halmahera arcs is consuming the Molucca Sea plate (Fig. 3). Therefore, the crust and sediments being subducted beneath both arcs may be similar.

Figure 5 shows the changes in Zr/Nb in the Halmahera lavas during differentiation of suites of lavas from Halmahera (Fig. 3). There is consider-

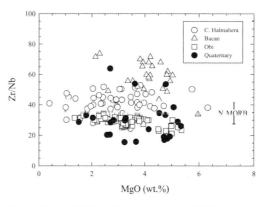

Fig. 5. Plot of MgO (wt%) content versus Zr/Nb for Neogene and Quaternary to Recent volcanics from the Halmahera Arc. Zr/Nb ratios higher than the value for MORB infer residual mantle wedge while low values can reflect incorporation of subducted sediment. Halmahera data from Forde (1997) and Morris *et al.* (1983) and N-MORB from Sun & McDonough (1989).

able variation in Zr/Nb between the different Neogene centres investigated by Forde (1997), but the range is restricted within each centre. Magmas from each location evolved from basic magmas (with MgO > 6 wt%) with Zr/Nb ratios that were similar to, or higher than, N-MORB (Fig. 5). None of the phenocryst phases present in Halmahera lavas are able to produce significant changes in the Zr/Nb ratio during fractional crystallization. Lavas from Obi have relatively constant Zr/Nb close to the MORB range whereas ratios increase successively through Central Halmahera and Bacan. High values for Zr/Nb ratios occur in melts derived from mantle that has experienced previous melt extraction (Woodhead *et al.* 1993) and probably require a relatively high fluid flux to induce melting. Thus, the Neogene data for Halmahera are consistent with a mantle wedge that was variably depleted, with respect to the MORB source, being fluxed by a slab-derived fluid.

The majority of Quaternary lavas from Halmahera also have Zr/Nb ratios that are similar to or higher than N-MORB (Fig. 5). This suggests that, like Neogene magmas, the majority of these lavas were derived from mantle that was similar to or more depleted than the MORB source. However, several Quaternary lavas display Zr/Nb lower than any observed from the Neogene arc. Normal mantle processes, such as previous melt extraction or fractional melting, are unable to produce melts with low Zr/Nb ratios suggesting a significant input from a reservoir characterized by low Zr/Nb. Sediments provide such a reservoir, and partial melting of sediment, as opposed to bulk additions to the mantle, may further lower the ratio (Elliott *et al.* 1997). The Zr/Nb ratios are consistent with Quaternary arc lavas containing a more significant contribution from subducted sediment than lavas erupted during the Neogene (Elburg & Foden 1998). In this respect, the Neogene to present day evolution of the Halmahera and Sangihe arcs has been similar. However, unlike the Elburg & Foden model for the Sangihe arc, Quaternary lavas that record increased sediment recycling in the Halmahera arc are contemporaneous with lavas that record recycling of fluid-dominated components derived from altered oceanic crust (Fig. 5).

Figure 6 compares Pb and Nd isotope data for the arcs colliding in the Molucca Sea region. This figure also illustrates the less extreme time dependence of mass transfer processes of the Halmahera arc compared to Sangihe. The younger (Pliocene to Quaternary) Sangihe volcanics are successively displaced closer to the field of sediments than their Neogene counterparts (Elburg & Foden 1998). In the Halmahera arc there is considerable overlap between lavas of different ages, although the Quaternary rocks are again more concentrated

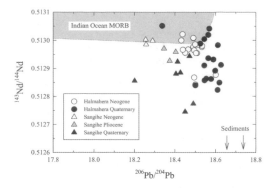

Fig. 6. $^{206}Pb/^{204}Pb$ versus $^{143}Nd/^{144}Nd$ in lavas from the Molucca Sea collision zone. Lavas from the Sangihe arc show a clear temporal progression towards lower $^{143}Nd/^{144}Nd$. While there is more overlap between Neogene and Quaternary arc lavas in the Halmahera data a similar overall progression exists. Halmahera data from Forde (1997), Sangihe data from Elburg & Foden (1998) and Indian Ocean MORB from Michard *et al.* (1986), Ito *et al.* (1987) and references therein.

towards the field of sediments. The most conspicuous feature of this plot is the distinct offset between the fields for the two arcs; the Halmahera lavas possess consistently higher $^{206}Pb/^{204}Pb$. This may be interpreted in several ways. It may be incorrect to assume that similar sediment is subducted under each arc and that sediment has higher $^{206}Pb/^{204}Pb$ in the east. Alternatively, the same sediment may enter both subduction zones but mixes with two isotopically distinct mantle wedges. The isotopic characteristics of both arcs, throughout their Neogene to Recent history, favour derivation from mantle similar to the source of Indian Ocean (I-)MORB. I-MORB has a sufficiently large isotopic range to accommodate the range of sources required to account for the differences between Halmahera and Sangihe lavas (Fig. 6). A final possibility is that the same sediment is being subducted beneath both arcs where it is contaminating mantle wedges that have similar isotopic characteristics but different Pb/Nb ratios, resulting in greater sensitivity to Pb contamination in the low Pb/Nd wedge (Halmahera).

Evolution of the mantle wedge

Progressive melting of the mantle wedge might be expected to result in lavas from increasingly depleted sources being erupted in the later stages of an arc's activity. However, lavas from the Quaternary Halmahera arc show no significant evidence for greater depletion of the mantle wedge when compared to the Neogene arc (Fig. 5). This suggests that fresh material has been available to maintain the fertility of the zone of melting beneath the Halmahera arc since the Neogene.

Upwelling of asthenosphere from the back-arc region is a significant control on the depletion of the mantle wedge. Mantle can be 'processed' through the back-arc region before being advected into the mantle wedge by corner flow (Fig. 4d; McCulloch & Gamble 1991). Where melting is minimal, or absent, in the back-arc region this will allow the mantle wedge to be replenished with fertile peridotite. Therefore, arcs that are tectonically susceptible to strong corner flow may be less likely to show a progressive depletion in wedge fertility. Upwelling will be an inevitable tectonic response in subduction zones when hinge retreat occurs (Hamilton 1995). Figure 7a shows that the boundary between the Molucca Sea plate and the Philippine Sea plate, to the west of Halmahera, migrated relatively rapidly westwards through the Neogene resulting in substantial hinge retreat west of the Halmahera arc. Peridotite from beneath the Philippine Sea plate or deeper in the asthenosphere must have been drawn into the mantle wedge during this process. Recharging of the mantle wedge by asthenospheric peridotite could reduce, or even suppress, the overall wedge depletion throughout the lifetime of an arc.

This process may also play a factor in the relative volumes of arcs. Volumetrically, the Halmahera arc is considerably larger than the Sangihe arc on the opposing side of the Molucca Sea collision zone. During the Neogene there was relatively little movement of the subduction hinge at the Sangihe arc, separating the Molucca Sea plate and the Eurasian plate (Fig. 7a). This situation would not favour upwelling of fresh fertile peridotite beneath the Sangihe arc, thus limiting the volume of magmatism. Replenishment of the Halmahera mantle wedge would provide a greater reservoir of fertile mantle from which to construct a larger arc.

Evolution of crustal interaction

The composition of basement upon which many island arcs are constructed is relatively poorly known. To a large extent, this is due to the poor exposure of basement rocks that have often been blanketed by the products of the arc. Geophysical techniques can be used to indirectly examine the nature of arc basements, but interactions between the crust and passing melts can produce distinctive geochemical signatures in the melts. Where local basement rocks have been geochemically characterized, such signatures can be exploited to understand the degree and processes of interaction (e.g. Macpherson et al. 1998). Conversely, where the basement is not exposed, suites of lavas

displaying geochemical trends that can be attributed to crustal interactions may reveal something of the nature of the arc basement. Where the age of magmatism is known this may also provide tectonic constraints on the development of crustal domains (Vroon et al. 1996).

The island of Bacan, in the Halmahera arc, provides an example. Morris et al. (1983) noted that dacitic lavas in southern Bacan, erupted near areas of exposed metamorphic basement, possess exceptionally radiogenic strontium and lead isotopic ratios ($^{87}Sr/^{86}Sr = 0.7198-0.7240$, $^{206}Pb/^{204}Pb \approx 40.2$). In conjunction with elevated alkali contents in the lavas, Morris et al. (1983) used these isotopic ratios to suggest that the dacites were contaminated by crust of continental origin. Neogene dacites displaying elevated isotopic ratios indicate that this continental material was present in the arc crust by the Late Miocene (Vroon et al. 1996; Forde 1997). However, the locally exposed basement (Sibela Metamorphic Complex) has lead and strontium isotopic ratios that are too low to produce the values found in the dacites requiring older basement material to be available in the crust (Forde 1997). In the western part of Bacan isotopic ratios of Late Miocene basaltic andesites show $^{87}Sr/^{86}Sr$ and $^{143}Nd/^{144}Nd$ ratios typical of oceanic arcs, whereas Pliocene lavas in the same part of the island have higher $^{87}Sr/^{86}Sr$ and lower $^{143}Nd/^{144}Nd$ that are consistent with contamination by the rocks of the Sibela Complex. This implies either (a) a change in magma dynamics leading to interaction between melts and parts of the crust that were previously unaffected by magmatism or (b) tectonic movement–possibly associated with the Sorong Fault Zone (Hall et al. 1991)–introducing previously unavailable material. The latter scenario suggests active strike-slip faulting in the region during the late Neogene as also implied by palaeomagnetic studies (Ali & Hall 1995).

Slab melting

The role of slab melting, as opposed to melting of peridotite in the mantle wedge, has received an increasing amount of attention during the last decade (Defant & Kepezhinskas 2001). Slab melting was initially proposed as a mechanism for generating the majority of andesites at convergent margins (Green & Ringwood 1968), but was soon deemed to be negligible in most arcs due to the basaltic parents calculated for the majority of arc lava suites (Gill 1981). However, Defant & Drummond (1990) published a compilation of data for lavas that appeared to possess the characteristics of slab melts that were erupted at margins where young oceanic crust was being subducted. Under certain circumstances, numerical models show it

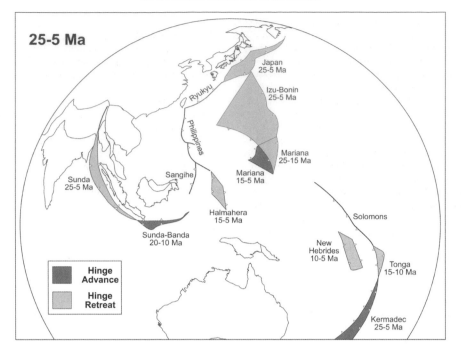

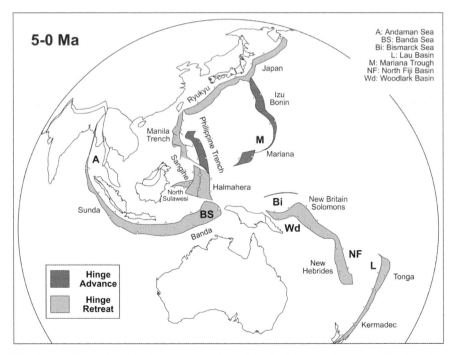

Fig. 7. (**a**) Regions and intervals of significant movement of subduction hinges in the Neogene based on the reconstructions of Hall (1998). The map represents the interval between the regional plate reorganizations of 25 Ma and 5 Ma. Major continental outlines are shown at 25 Ma for reference. For subduction zones shown without shading there was no significant movement of hinge. (**b**) Regions of significant movement of subduction hinges since 5 Ma based on the reconstructions of Hall (1998). Bold letters indicate areas of young marginal basin formation. Major continental outlines are shown at 5 Ma for reference.

is possible to produce limited melting of the mafic parts of subducted oceanic crust that is younger than 25 Ma at the time of subduction, since younger crust will still retain some heat from accretion and may be hot enough to cross the solidus during subduction (Peacock *et al.* 1994). Melting would occur during the amphibolite to eclogite transition and yield an evolved melt for which Defant & Drummond (1990) coined a geochemical name: adakite. High Sr concentrations in adakites are attributed to plagioclase breakdown, while Y retention in residual garnet and hornblende leads to low concentrations (and consequently high La/Y and Sr/Y). Defant & Drummond (1990) suggested the association of young subducting crust and the distinctive volcanic rocks provided evidence for slab melting.

Adakitic lavas have been identified in various parts of SE Asia, most notably the Philippines, and attributed to slab melting (Sajona *et al.* 1993, 1994, 1997). However, many of the Philippines adakites are unlikely to have been generated in this way as in most cases the age of the subducted slab is much greater than permitted by slab melting models. For example, adakitic lavas in Mindanao have been interpreted to be the result of subduction at the Cotobato trench (Sajona *et al.* 1994, 1997). However, the age of the subducted Celebes Sea crust is Eocene. These adakites are older than 1 Ma but the subducted plate currently only penetrates to depths of about 80 km. At the time of magmatism the slab would have been even shallower. Thus, the link between adakites and the subducted Celebes Sea lithosphere is questionable. Like southern Mindanao, adakites in eastern Mindanao are associated with the earliest stages of subduction. This suggests that rather than melting the subducted slab, adakites may be produced by melting of the mantle wedge and/or lower crust due to an influx of hot mantle above the foundered crust on the west side of the Philippines (Fig. 7b). In both these parts of Mindanao, adakites are associated with strike-slip faults which may have contributed additional heat by ductile shearing in the lower crust or upper mantle as in other major strike-slip zones (e.g. Thatcher & England 1998) or may have acted simply as conduits for melts.

The origin of adakitic magmas has been given added economic significance since a number of authors have claimed that adakites may be closely related to porphyry copper mineralization and may, therefore, provide a useful exploration tool (Thiéblemont *et al.* 1997; Sajona & Maury 1998; Defant & Kepezhinskas 2001). Sillitoe (1997) suggested that the mantle may become oxidized by the passage of slab melts, which would break down sulphides and make gold and copper available for transport into the overriding plate. However, mantle wedge peridotite is normally considered to be more oxidized than the average upper mantle (Brandon & Draper 1996 and references therein; Parkinson & Pearce 1998) so that slab melts need not be required to decrease sulphide stability.

The association between adakitic magmatism and hydrothermal mineralization requires further exploration with a focus on the analysis of relevant specimens from mineralized areas. Since the term was coined in 1990, adakites have been found in several locations where the subducting crust is too old to melt under the conditions permitted by thermal modelling. While particular tectonic factors have been identified in each of these cases that *may* permit older slab crust to melt (Defant & Drummond 1990; Defant *et al.* 1992; Sajona *et al.* 1993; Yogodzinski *et al.* 1995; Gutscher *et al.* 2000), an alternative explanation is that adakitic magma can be generated from mafic reservoirs in the overriding plate, as described above for the Philippines. This is particularly important where previous phases of subduction zone magmatism have modified the overriding plate. Development of plumbing systems to facilitate transport of arc magmas through arc crust occurs with elevated heat flow in the overriding plate (Furukawa 1993) and hydrous arc magmas will pond at a range of levels within the crust and mantle beneath an island arc. As any single phase of arc magmatism ceases such material will become frozen into the plate until the geothermal gradient is raised sufficiently to cause remelting. Hydrous arc magmas frozen into the lithosphere at $\geq$ 10kbar will either crystallize directly or re-equilibrate isobarically to garnet-bearing mafic assemblages (De Bari & Coleman 1989). During subsequent heating these assemblages would partially melt to yield a magma similar to those generated by melting of hydrated basalt in subducted oceanic crust.

The mechanism proposed above bears some similarities to Atherton & Petford's (1993) model in which hot basalt recently accreted to the base of arc crust remelts to yield tonalitic magma. The difference is that mafic reservoirs need not be related to the active phase of magmatism but remnants of previous arcs could reside in the lower crust or lithospheric mantle for many millions of years. Arc tholeiites, produced by hydrous melting of peridotite in the mantle wedge and frozen into the upper mantle or lower crust in this way, would originally have been emplaced beneath the arc front. Later perturbation of the geotherm could result in partial melting of these frozen mafic segregations, which may originally have formed individual conduits and veins or were akin to the MASH zones of Hildreth & Moorbath (1988), and would produce intermediate magmas with adakitic characteristics reflecting the stability of garnet in the residue. Therefore, the distribution and geochemistry of adakitic lavas need not vary systematically with respect to a currently subducting slab since the location of the mafic source within the mantle wedge or lower crust would result from the geometry of the previous phase of arc magmatism. Cenozoic adakitic magmatism commonly occurred during the first magmatic event following either a quiescence in subduction zone magmatic activity (Defant et al. 1992) or a change in subduction zone architecture (Sajona et al. 1993). The model outlined above is entirely consistent with each of these tectonic scenarios. In the earliest stages of subduction, melting of hydrous crustal or mantle domains would be one of the first processes to occur as the geothermal gradient increased in the overriding plate.

Relevance of subduction evolution to mineralization

Magmatism provides heat to generate hydrothermal systems and also releases fluids that redistribute metals within the crust. Furthermore, stress patterns around intrusions may induce fractures that act as fluid pathways, which can then become sites for ore deposition. Finally, a proportion of the metals in ore bodies may ultimately be derived from the same magma bodies that provide the heat and fluid to drive the system. Therefore, both the sources that contribute to subduction zone magmatism and the processes that cause these to melt are important factors in the origin of hydrothermal mineral deposits.

Sources of ore metals

One of the most important reasons for exploring the link between magmatism and mineralization is to determine the source of ore metals. Fluid inclusion and stable isotope data indicate that hydrothermal systems incorporate fluids derived from magmatic and non-magmatic sources (Hedenquist & Lowenstern 1994) but there has been less certainty about the relative contributions of metals from mantle-derived magmas compared to metals stripped from the arc crust. Early attempts to resolve the provenance of metals focused on elemental and isotopic ratios that are commonly used to investigate the origin and evolution of magmatic rocks, such as isotope ratios of the lithophile elements Sr, Nd and Pb. For, example, initial $^{87}Sr/^{86}Sr$ ratios of porphyry-related magmatic rocks in the Philippines are uniformly low, suggesting a source of the magmas dominated by mantle contributions (Divis 1980). However, magmatism in western New Guinea, as at Grasberg and Ertsberg, is characterized by Sr, Nd and Pb isotope ratios suggesting the presence of enriched mantle and crustal contaminants in addition to a depleted mantle component (Housh & McMahon 2000). Furthermore, the relevance of such data to ore deposits is unclear since lithophile elements may experience processes that ore metals escape, while the mobilization of ore metals may not be recorded by lithophile tracers.

Osmium is now recognized as a potentially powerful tracer of metallogenesis because, as a siderophile element, its geochemical behaviour is similar to that of the ore metals in porphyry deposits (Freydier et al. 1997; McInnes et al. 1999a). Recent application of the osmium isotope system to ore bodies and contemporaneous rocks has indicated that the mantle wedge may provide the major fraction of ore metal in some porphyry deposits. McInnes et al. (1999a) found that mantle nodules associated with the Ladolam deposit on Lihir Island are highly enriched in ore metals and interpreted their unradiogenic nature as evidence that this enrichment occurred within the mantle, from where the magmas associated with the ore body inherited their inventory of osmium and other metals. This conclusion is compatible with results of Os isotope studies in South America (Freydier et al. 1997), and Mathur et al. (2000a) suggest that in South America higher copper contents occur in porphyry deposits that record Os isotope ratios closest to the value for the upper mantle. Such findings emphasize the need to understand the geodynamic setting of modern ore deposits to fully assess the mineral potential of ancient and other contemporary margins. However, like the isotope ratios of lithophile elements, osmium isotope data for the Grasberg deposit suggest that remobilization of metals from crustal sources may be an important process in some terranes (Mathur et al. 2000b). Such processes will probably be more important in collisional settings, like western New Guinea, or on continental margins where hydrothermal systems may develop in the vicinity of suitable crustal contaminants, and where crustal thickening may encourage crustal melting.

Tectonic controls on magmatism

The generation of many mineral deposits in SE Asia, particularly the largest bodies, appears to be associated with changes in the geometry of the subduction zone, such as stalling (and subsequent melting) of subducted lithosphere, slab steepening and subduction reversal (Solomon 1990; Sillitoe

1997). The spatial and temporal distribution of igneous rocks provides a means to understand the changes in dynamics associated with such events (see above). To investigate the relationship between magmatism and tectonics further we compiled a database of the location and timing of magmatism in SE Asia based on an extensive literature review. The database contains information on radiometric age determinations of rock samples, the locations from which these samples were collected and the methods employed to measure the ages. Information can be accessed using Microsoft Excel and Filemaker Pro software, but MapInfo Professional is the preferred medium as this also allows the data to be viewed graphically and integrated with other GIS data. The database is available through the internet at http://www.gl.rhul.ac.UK/seasia. In general the measurements included in the database were made using isotopic techniques although a number of fission track ages are also included. Public domain data were compiled from the scientific literature and all other available sources. Exhaustive literature searches were conducted for Indonesia, the Philippines and Papua New Guinea to ensure maximum coverage. The geographically located data have been integrated with plate tectonic reconstructions (Hall 1996, 2002). Below, we employ these reconstructions to discuss the relationships between tectonic processes, magmatism and mineralization.

In SE Asia most new Cenozoic subduction systems appear to have been initiated close to the boundary between thick and thin crust after collision events or plate reorganizations. For the period since 25 Ma, most subduction zones in SE Asia and the western Pacific experienced significant arc volcanism during hinge retreat (Fig. 7), which in many cases was accompanied by marginal basin formation. In contrast, reduction or cessation of volcanic activity occurred during hinge advance. The eastern Sunda and Banda arcs provide excellent examples of these contrasting situations. In the eastern Sunda arc in Nusa Tenggara there was hinge advance from about 25 Ma to 15 Ma and there is little evidence of magmatic activity (Fig. 8a). From about 10 Ma there was significant hinge retreat accompanied by an increase in arc magmatism and back-arc spreading. Increased magmatic activity during periods of hinge retreat can be explained by the mantle flow that is induced to conserve mass beneath the arc as the slab descends (Andrews & Sleep 1974). This corner flow introduces fertile mantle into the mantle wedge and may also possess a component of flow towards the surface (Andrews & Sleep 1974; Tokosöz & Hsui 1978; Furukawa 1993; Elkins Tanton et al. 2001) resulting in a component of decompression

melting. Both the supply of fertile peridotite and upwelling will be subdued if the hinge does not retreat and thus the volume of magma produced will diminish. Based on current tectonic models it appears that hinge movement is a first order factor in (the control of) volcanism and ore formation. We expect that further improvements in tectonic models will allow examination of the tectonic history of different parts of the region in greater detail.

Evidence for links between major regional tectonic events and mineralization also comes from the particular composition of lavas associated with mineralization (see Barley et al. 2002). For example, Housh & McMahon (2000) identified rocks with lamproitic affinities throughout the mineralized region of western New Guinea. While an association between lamproitic magmatism and gold mineralization has previously been postulated for Papua New Guinea (e.g. Rock & Finlayson 1990), Richards et al. (1991) point out the extremely limited volume of lamproite at Porgera in Papua New Guinea. Instead, Richards et al. (1991) suggest that the Porgera deposit formed in association with alkali-rich basaltic magmatism. A similar association is favoured by Müller & Groves (2000) and may be reliant on the volatile characteristics of such lavas. We are currently compiling a database of chemical analyses of igneous rocks from throughout SE Asia that will be coupled with radiometric age data and plate reconstructions to investigate such associations further. A particular target is identifying the role of mantle that has previously been modified by subduction and which subsequently melts through other mechanisms, such as decompression or heating (Macpherson & Hall 1999). This is especially relevant to western New Guinea where plate reconstruction and tomography provide no evidence for subduction having occurred during the inferred period of mineralization.

Evolution of lithosphere thickness

Exposure of several very young deposits in SE Asia has been used as evidence for a direct link between rapid uplift and mineralization. During rapid uplift of the crust, induced by changes in slab dip, slab break-off or collision between the arc and continental crust, depressurization of magma could facilitate fluid release and encourage formation of hydrothermal systems (Sillitoe 1997). Rapid uplift may also explain the relative paucity of economic hydrothermal deposits in the oldest and youngest arcs of the region. Carlile & Mitchell (1994), noting the restriction of most major Indonesian ore bodies to the Neogene, advocated a favourable balance between uplift and

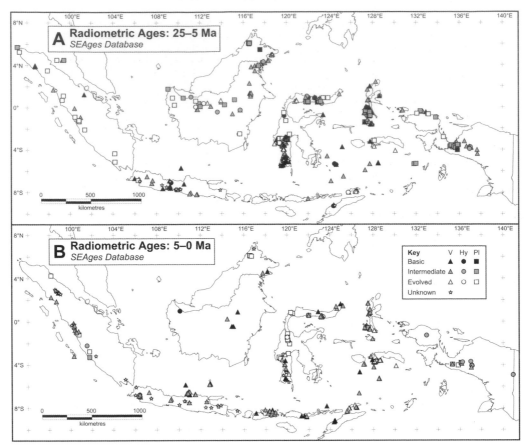

Fig. 8. Maps of Indonesia showing the location of radiometric age determinations for *in situ* igneous rocks. Symbol shades designate the relative silica content of the rocks: black, basic; grey, intermediate; white, evolved; star; unknown. Symbol shapes represent the level of cooling: triangles, volcanic rocks; circles, hypabyssal intrusive rocks; squares, plutonic intrusive rocks. (**a**) Period from 25 to 5 Ma. (**b**) Period from 5 Ma until Recent.

denudation for the exposure of the most economic levels of the arc crust. This mechanism may also explain the difference between the ages of major phases of mineralization in arcs at substantially different latitudes (Sillitoe 1997). Barley *et al.* (2002), however, question the role of uplift as a primary control on the distribution of economic deposits and relate peak mineralization to remelting of subduction-modified mantle (Macpherson & Hall 1999). Recent advances in thermochronometric dating of topography development (e.g. McInnes *et al.*, 1999*b*) provide a tool that could be integrated with plate reconstructions to assess the relative importance of uplift in different arc systems throughout SE Asia.

Although there are young mountain ranges constructed at the convergent margins of New Guinea, the Philippines and east Indonesia not all hydrothermal deposits in the region formed in contractional systems. Figure 7 shows that during the last 5 Ma much of the Sunda–Banda arc system, host to several porphyry and epithermal deposits, has experienced hinge retreat, a condition likely to minimize contraction or even lead to extension (Hamilton 1995). In New Guinea and the Philippines there are important strike-slip faults which, depending on their geometry, may be releasing or restraining. Therefore, the existence of a contractional regime should not, in itself, be taken as a condition for generation of a hydrothermal ore deposit. This, however, does not preclude uplift of arc crust that is under extension. Crust thinned during extension will experience a component of isostatic uplift. Furthermore, extension of the overriding plate is most likely to occur in association with hinge retreat, which will contribute to uplift in two ways. First, corner flow resulting from hinge retreat will cause hot, buoyant mantle to be drawn into the mantle wedge and will elevate the overriding crust (Gvirtzman & Nur, 1999). Second, circulation in the mantle wedge will ablate the lower side of the overriding plate thus enhancing the isostatic uplift of the arc crust. Similar phenomena may also be induced in areas of post-collisional extension, such as those identified in Indonesia by Charlton (1991).

Summary

The timing and location of hydrothermal mineralization in SE Asia are often related to major evolutionary events at plate boundaries, for example in New Guinea, northern Luzon, eastern Mindanao and the Sunda–Banda arc, and are ultimately driven by the same processes that cause mantle melting. The spatial and temporal distribution of magmatism, plate reconstruction and the geochemistry of magmatic rocks can be used to investigate processes occurring in these settings. Hinge migration is important in controlling the volume of melt produced at an active margin and the dynamic nature of subduction zones requires that this process encourages mantle circulation through the mantle wedge. Changes in the balance between recycled fluxes of slab-derived fluid and sediment melt exert an important control on the chemical composition of arc lavas and, therefore, on their inherent content of, and ability to transport, economically important metals. Deformation of the overriding plate leads to changes in the crustal rocks available to interact with melts during transport to the site of eruption. Plate reconstruction models provide a means to relate the location and age of mineral deposits to data derived from geological and geochemical investigation of lavas.

Financial support has been provided by NERC, the Royal Society, the London University Central Research Fund, and the Royal Holloway SE Asia Research Group. We thank J.A.F. Malaihollo for valuable assistance in compiling the radiometric age and mineralization databases. H. De Boorder, E. Marcoux and D.J. Blundell provided valuable comments on an earlier draft.

References

ABBOTT, M.J. & CHAMALAUN, F.H. 1978. *New K/Ar age data for Banda arc volcanics*. Institute for Australian Geodynamics Publications, **78/5**.

ALI, J.R. & HALL, R. 1995. Evolution of the plate boundary between the Philippine Sea Plate and Australia: palaeomagnetic evidence from eastern Indonesia. *Tectonophysics*, **251**, 251–275.

ANDREWS, D.J. & SLEEP, N.H. 1974. Numerical modelling of tectonic flow behind island arcs. *Geophysical Journal of the Royal Astronomical Society*, **38**, 237–251.

ARRIBAS, A., HEDENQUIST, J.W., ITAYA, T., OKADA, T., CONCEPCION, R.A. & GARCIAL, J.S. 1995. Contemporaneous formation of adjacent porphyry and epithermal Cu-Au deposits over 300ka in northern Luzo Philippines. *Geology*, **23**, 337–340.

ATHERTON, M.P. & PETFORD, N. 1993. Generation of sodium-rich magmas from newly underplated basaltic crust. *Nature*, **362**, 144–146.

BARBERI, F., BIGIOGGERO, B., BORIANI, A., CATTANEO, M., CAVALLIN, A., CIONI, R., EVA, C., GELMINI, R., GIORGETTI, F., IACCARINO, S., INNOCENTI, F., MARINELLI, G., SLEIJKO, D. & SUDRADJAT, A. 1987. The island of Sumbawa: a major structural discontinuity in the Indonesian arc. *Bulletin of the Geological Society of Italy*, **106**, 547–620.

BAKER, S. & MALAIHOLLO, J. 1996. Dating of Neogene igneous rocks in the Halmahera region: arc initiation and development. *In:* HALL, R. & BLUNDELL, D.J. (eds) *Tectonic evolution of Southeast Asia*. Geological Society, London, Special Publications, **106**, 499–509.

BARLEY, M.E., RAK, P. & WYMAN, D. 2002. Tectonic controls on magmatic-hydrothermal gold mineralization in the magmatic arcs of SE Asia. *In:* BLUNDELL, D.J., NEUBAUER, F. & VON QUADT, A. (eds) *The Timing and Location of Major Ores Deposits within an Evolving Orogen*. Geological Society, London, Special Publications, **204**, 39–49.

BASUKI, A., SUMANAGARA, D.A. & SINAMBELA, D. 1994. The Gunung Pongkor gold-silver deposit, west Java, Indonesia. *Journal of Geochemical Exploration*, **50**, 371–391.

BINNS, R.A. & SCOTT, S.D. 1993. Actively forming polymetallic sulfide deposits associated with felsic volcanic rocks in the eastern Manus back-arc basin, Papua New Guinea. *Economic Geology*, **88**, 2226–2236.

BRANDON, A.D. & DRAPER, D.S. 1996. Constraints on the origin of the oxidation state of mantle overlying subduction zones: An example from Simcoe, Washington, USA. *Geochimica et Cosmochimica Acta*, **60**, 1739–1749.

BRENNAN, J., SHAW, H.F., PHINNEY, D.L. & RYERSON, J.F. 1995. Mineral-aqueous fluid partitioning of trace elements at 900°C and 2.0Gpa: Constraints on the trace element chemistry of mantle and deep crustal fluids. *Geochimica et Cosmochimica Acta*, **59**, 333–3350.

CARDWELL, R.K., ISACKS, B.L. & KARIG, D.E. 1980. The spatial distribution of earthquakes, focal mechanism solutions and subducted lithosphere in the Philippine and northeastern Indonesian islands. *In:* HAYES, D.E. (ed.) *The Tectonic and Geologic Evolution of Southeast Asian Seas and Islands*. American Geophysical Union Geophysical Monograph Series, **23**, 1–36.

CARLILE, J. & MITCHELL, A.H.G. 1994. Magmatic arcs and associated gold and copper mineralization in Indonesia. *Journal of Geochemical Exploration*, **50**, 91–142.

CARTER, D.J., AUDLEY-CHARLES, M.G. & BARBER, A.J. 1976. Stratigraphical analysis of island arc-continental margin collision in eastern Indonesia. *Journal of the Geological Society, London*, **132**, 179–189.

CHARLTON, T.R. 1991. Postcollisional extension in arc-continental collision zones, eastern Indonesia. *Geology*, **19**, 28–31.

COSCA, M.A., ARCULUS, R.J., PEARCE, J.A. & MITCHELL, J.G. 1998. $^{40}Ar/^{39}Ar$ and K-Ar geochronological age constraints for the inception and early evolution of the Izu-Bonin-Mariana arc system. *The Island Arc*, **7**, 579–595.

DAVIDSON, J.P. & HARMON, R.S. 1989. Oxygen isotope constraints on the petrogenesis of volcanic arc magmas from Martinique, Lesser Antilles. *Earth and Planetary Science Letters*, **95**, 255–270.

DE BARI, S.M. & COLEMAN, R.G. 1989. Examination of deep levels of an island arc: evidence from the Tonsia ultramafic assemblage, Tonsia, Alaska. *Journal of Geophysical Research*, **94**, 4373–4391.

DEFANT, M.J. & DRUMMOND, M.S. 1990. Derivation of some modern arc magmas by melting of young subducted lithosphere. *Nature*, **347**, 662–665.

DEFANT, M.J. & KEPEZHINSKAS, P. 2001. Evidence suggests slab melting in arc magmas. *EOS*, **82**, 65–69.

DEFANT, M.J., JACKSON, T.E., DRUMMOND, M.S., DE-BOER, J.Z., BELLON, H., FEIGENSON, M.D., MAURY, R.C. & STEWART, R.H. 1992. The geochemistry of young volcanism throughout western Panama and southeastern Costa Rica: an overview. *Journal of the Geological Society, London*, **149**, 569–579.

DICKIN, A.P. 1995. *Radiogenic Isotope Geology*. Cambridge University Press, Cambridge, Cambridge University Press, Cambridge.

DIVIS, A.F. 1980. The petrology and tectonics of recent volcanism in the central Philippine Islands. *In:* HAYES, D.E. (ed.) *The Tectonic and Geologic Evolution of Southeast Asian Seas and Islands*. American Geophysical Union Geophysical Monograph Series, **23**, 127–144.

ELBURG, M.A. & FODEN, J.D. 1998. Temporal changes in arc magma geochemistry, north Sulawesi, Indonesia. *Earth and Planetary Science Letters*, **163**, 381–398.

ELKINS TANTON, L.T., GROVE, T.L. & DONNELLY-NOLAN, J. 2001. Hot, shallow melting under the Cascades volcanic arc. *Geology*, **29**, 631–634.

ELLIOT, T., PLANK, T., ZINDLER, A., WHITE, W. & BOURDON, B. 1997. Element transport from slab to volcanic front at the Mariana arc. *Journal of Geophysical Research*, **102**, 14991–15019.

EWART, A. & HAWKESWORTH, C.J. 1987. The Pleistocene-Recent Tonga-Kermadec arc lavas: interpretation of new isotopic and rare earth data in terms of a depleted mantle source model. *Journal of Petrology*, **28**, 495–530.

FEISS, P.G. 1978. Magmatic sources of copper in porphyry copper deposits. *Economic Geology*, **73**, 397–404.

FORDE, E. 1997. *The geochemistry of the Neogene Halmahera Arc, eastern Indonesia*. PhD Thesis, University of London.

FREYDIER, C., RUIZ, J., CHESLEY, J., McCANDLESS, T. & MUNIZAGA, F. 1997. Re-Os isotope systematics of sulphides from felsic igneous rocks: Application to base metal porphyry mineralization in Chile. *Geology*, **25**, 775–778.

FURUKAWA, Y. 1993. Depth of the decoupling plate interface and thermal structure under arcs. *Journal of Geophysical Research*, **98**, 20005–20013.

GILL, J.B. 1981. *Orogenic Andesites and Plate Tectonics*. Berlin, Springer-Verlag, Berlin.

GREEN, T.H. & RINGWOOD, A.E. 1968. Genesis of calc-alkaline igneous rock suite. *Contributions to Mineralogy and Petrology*, **18**, 105–162.

GUTSCHER, M.-A., MAURY, R., EISSEN, J.-P. & BOURDON, E. 2000. Can slab melting be caused by flat subduction? *Geology*, **28**, 535–538.

GVIRTZMAN, Z. & NUR, A. 1999. Plate detachment, asthenosphere upwelling and topography across subduction zones. *Geology*, **27**, 563–566.

HAKIM, A.S. & HALL, R. 1991. Tertiary volcanic rocks from the Halmahera arc, eastern Indonesia. *Journal of Southeast Asian Earth Sciences*, **6**, 271–287.

HALL, R. 1996. Reconstructing Cenozoic SE Asia. *In:* HALL, R. & BLUNDELL, D.J. (eds) *Tectonic evolution of Southeast Asia*. Geological Society, London, Special Publications, **106**, 153–184.

HALL, R. 1998. The plate tectonics of Cenozoic SE Asia and the distribution of land and sea. *In:* HALL, R. & HOLLOWAY, J.D. (eds) *Biogeography and Geological Evolution of SE Asia*. Backhuys Publishers, Leiden, The Netherlands, 99–131.

HALL, R. 2002. Cenozoic geological and plate tectonic evolution of SE Asia and the SW Pacific: computer-based reconstructions, model and animations. *Journal of Asian Earth Sciences*, **20**, 353–434.

HALL, R., ALI, J.R., ANDERSON, C.D. & BAKER, S.J. 1995. Origin and motion history of the Philippine Sea Plate. *Tectonophysics*, **251**, 229–250.

HALL, R., NICHOLS, G., BALLANTYNE, P., CHARLTON, T. & ALI, J. 1991. The character and significance of basement rocks of the southern Molucca Sea region. *Journal of Southeast Asian Earth Sciences*, **6**, 249–258.

HAMILTON, W.B. 1988. Plate tectonics and island arcs. *Geological Society of America Bulletin*, **100**, 1503–1527.

HAMILTON, W.B. 1995. Subduction systems and magmatism. *In:* SMELLIE, J.L. (ed.) *Volcanism associated with Extension at consuming plate margins*. Geological Society, London, Special Publications, **81**, 3–28.

HAWKESWORTH, C.J., HERGT, J.M., ELLAM, R.M. & McDERMOTT, F. 1991. Element fluxes associated with subduction related magmatism. *Philosophical Transactions of the Royal Society of London*, **A335**, 393–405.

HAWKESWORTH, C.J., TURNER, S.P., McDERMOTT, F., PEATE, D.W. & van CALSTEREN, P. 1997. U-Th isotopes in arc magmas: implications for element transfer from the subducted crust. *Science*, **276**, 551–555.

HEDENQUIST, J.W. & LOWENSTERN, J.B. 1994. The role of magmas in the formation of hydrothermal ore deposits. *Nature*, **370**, 519–527.

HILDRETH, W. & MOORBATH, S. 1988. Crustal contributions to arc magmatism in the Andes of Central Chile. *Contributions to Mineralogy and Petrology*, **98**, 455–489.

HILYARD, D. & ROGERSON, R. 1989. Revised stratigraphy of Bougainville and Buka Islands, Papua New Guinea. *In:* Marlow, M.S., Dadisman, S.V. & Exon, N.R. (eds) *Geology and offshore resource of Pacific island arcs - New Ireland and Manus region, Papua New Guinea.* Circum-Pacific Council for Energy and Mineral Resources Earth Science Series, **9**, 87–92.

Housh, T. & McMahon, T.P. 2000. Ancient isotopic characteristics of Neogene potassic magmatism in western New Guinea (Irian Jaya, Indonesia). *Lithos,* **50**, 217–239.

Iizasa, K., Fiske, R.S., Ishizuka, O., Yuasa, M., Hashimoto, J., Ishibashi, J., Naka, J., Horii, Y., Fujiwara, Y., Imai, A. & Koyama, S. 1999. A Kuroko-type polymetallic sulphide deposit in a submarine silicic caldera. *Science,* **283**, 975–977.

IOC, IHO, BODC, GEBCO-97 1997. *The 1997 Edition of the GEBCO Digital Atlas,* British Oceanographic Data Centre.

Ito, E., White, W.M. & Göpel, C. 1987. The O, Sr, Nd and Pb isotope geochemistry of MORB. *Chemical Geology,* **62**, 157–176.

Iwamori, H. 1997. Heat sources and melting in subduction zones. *Journal of Geophysical Research,* **102**, 14803–14820.

Jobson, D.H., Boulter, C.A. & Foster, R.P. 1994. Structural controls and genesis of epithermal gold-bearing breccias at the Lebong Tandai mine, western Sumatra, Indonesia. *Journal of Geochemical Exploration,* **50**, 409–428.

Katili, J.A. 1975. Volcanism and plate tectonics in the Indonesian island arcs. *Tectonophysics,* **26**, 165–188.

Keppler, H. 1996. Constraints from partitioning experiments on the composition of subduction-zone fluids. *Nature,* **380**, 237–240.

Kroenke, L.W. 1984. *Cenozoic tectonic development of the Southwest Pacific.* CCOP/SOPAC Technical Bulletin, **6**.

Macpherson, C.G. & Hall, R. 1999. Tectonic controls of geochemical evolution in arc magmatism of SE Asia. *Proceedings PACRIM 99 Congress, Australian Institute of Mining and Metallurgy Publication Series,* **4/99**, 359–368.

Macpherson, C.G., Gamble, J.A. & Mattey, D.P. 1998. Oxygen isotope geochemistry of lavas from an oceanic to continental arc transition, Kermadec–Hikurangi margin, S.W. Pacific. *Earth and Planetary Science Letters,* **160**, 609–621.

Marcoux, E. & Milési, J.-P. 1994. Epithermal gold deposits in west Java, Indonesia: geology, age and crustal source. *Journal of Geochemical Exploration,* **50**, 393–408.

Mathur, R., Ruiz, J. & Munizaga, F. 2000. Relationship between copper tonnage of Chilean base-metal porphyry deposits and Os isotope ratios. *Geology,* **28**, 555–558.

Mathur, R., Ruiz, J., Titley, S., Gibbins, D. & Margotomo, W. 2000. Different crustal sources for Au-rich and Au-poor ores of the Grasberg Cu-Au porphyry deposit. *Earth and Planetary Science Letters,* **183**, 7–14.

McCourt, W.J., Crow, M.J., Cobbing, E.J. & Amin, T.C. 1996. Mesozoic and Cenozoic plutonic evolu-

tion of SE Asia: evidence from Sumatra, Indonesia. *In:* Hall, R. & Blundell, D.J. (eds) *Tectonic Evolution of SE Asia.* Geological Society, London, Special Publications, **106**, 321–335.

McCulloch, M.T. & Gamble, J.A. 1991. Geochemical and geodynamical constraints on subduction zone magmatism. *Earth and Planetary Science Letters,* **102**, 358–374.

McInnes, B.I.A., McBride, J.S., Evans, N.J., Lambert, D.D. & Andrew, A.S. 1999a. Osmium isotope constraints on ore metal recycling at subduction zones. *Science,* **286**, 512–516.

McInnes, B.I.A., Farley, K.A., Sillitoe, R.H. & Kohn, B.P. 1999b. Application of apatite (U-Th)/He thermochronometry to the determination of the sense and amount of vertical fault displacement at the Chuquicamata porphyry copper deposit Chile. *Economic Geology,* **94**, 937–947.

Meldrum, S.J., Aquino, R.S., Gonzales, R.I., Burke, R.J., Suyadi, A., Irianto, B. & Clarke, D.S. 1994. The Batu Hijau porphyry copper-gold deposit, Sumbawa Island, Indonesia. *Journal of Geochemical Exploration,* **50**, 203–220.

Michard, A., Montigny, R. & Schlich, R. 1986. Geochemistry of mantle beneath the Rodriguez Triple Junction and the Southeast Indian Ridge. *Earth and Planetary Science Letters,* **78**, 104–114.

Mitchard, A.H.G., Hernandez, F. & dela Cruz, A.P. 1986. Cenozoic evolution of the Philippine archipelago. *Journal of Southeast Asian Earth Science,* **1**, 3–22.

Mitchell, A.H.G. & Leach, T.M. 1991. *Epithermal Gold in the Philippines· Island arc Metallogenesis, Geothermal Systems and Geology.* Geothermal Systems and Geology. Academic Press, Academic Press, London.

Morris, J.D., Jezek, P.A., Hart, S.R. & Gill, J.B. 1983. The Halmahera Arc, Molucca Sea Collision Zone, Indonesia: A geochemical Survey. *In:* Hayes, D.E. (ed.) *The Tectonic and Geologie Evolution & Southeast Asian Seas and Islands, Part 2.* American Geophysical Union Monographs, **27**, 373–387.

Morris, J.D., Leeman, W.P. & Tera, F. 1990. The subducted component in island arc lavas: constraints from Be isotopic and B-Be systematics. *Nature,* **344**, 31–36.

Müller, D. & Groves, D.I. 2000. *Potassic rocks and associated gold-copper mineralisation, 3rd edition.* Lecture Notes in Earth Sciences, Springer, Heidelberg, **56**.

Oldberg, D.J., Rayner, J., Langmead, R.P. & Coote, J.A.R. 1999. Geology of the Gosowng epithermal gold deposit, Halmahera, Indonesia. *In: Proceedings of the PACRIM 99 Congress.* Australian Institute of Mining and Metallurgy Publication Series, **4/99**, 359–368.

Oxburgh, E.R. & Turcotte, D.L. 1970. Thermal structure of island arcs. *Geological Society of America Bulletin,* **81**, 1665–1688.

Parkinson, I.J. & Pearce, J.A. 1998. Peridotites from the Izu-Bonin-Mariana forearc (ODP Leg 125): Evidence for mantle melting and melt-mantle interaction in a supra-subduction zone setting. *Journal of Petrology,* **39**, 1577–1618.

PEACOCK, S.M., RUSHMER, T. & THOMPSON, A.B. 1994.
   Partial melting of subducting oceanic crust. *Earth
   and Planetary Science Letters*, **121**, 227–244.
PEARCE, J.A. & PEATE, D.W. 1995. Tectonic implica-
   tions of the composition of volcanic arc magmas.
   Annual Reviews of Earth and Planetary Sciences,
   23, 251–285.
PHINNEY, E.J., MANN, P., COFFIN, M.F. & SHIPLEY,
   T.H. 1999. Sequence stratigraphy, structure and
   tectonic history of the southwest Ontong Java
   Plateau adjacent to the North Solomon Trench and
   Solomon Island arc. *Journal of Geophysical Re-
   search*, **104**, 20449–20466.
PLANK, T. & LANGMUIR, C.H. 1988. An evaluation of
   the global variations in the major element chemistry
   of arc basalts. *Earth and Planetary Science Letters*,
   90, 349–370.
RICHARDS, J.P., CHAPPELL, B.W., McCULLOCH, M.T. &
   McDOUGALL, I. 1991. The Porgera gold deposit,
   Papua New Guinea I: Association with alkalic
   magmatism in a continent-island arc collision zone.
   In: LADEIRA, E.A. (ed.) *Brazil Gold '91: the
   economics, geology, geochemistry and genesis of
   gold deposits*. Balkema, Rotterdam, 307–312.
ROCK, N.M.S. & FINLAYSON, E.J. 1990. Petrological
   affinities of intrusive rocks associated with the giant
   mesothermal gold deposit at Porgera, Papua New
   Guinea. *Journal of Southeast Asian Earth Science*,
   4, 247–257.
ROCK, N.M.S., SYAH, H.H., DAVIS, A.E., HUTCHISON,
   D., STYLES, M.T. & LENA, R. 1982. Permian to
   Recent volcanism in northern Sumatra, Indonesia: a
   preliminary study of its distribution, chemistry and
   peculiarities. *Bulletin Volcanologique*, **45**, 127–152.
ROWLAND, A. & DAVIES, J.H. 1999. Buoyancy rather
   than rheology controls the thickness of the over-
   riding mechanical lithosphere at subduction zones.
   Geophysical Research Letters, **26**, 3037–3040.
RUTHERFORD, E., BURKE, K. & LYTWYN, J. 2001.
   Tectonic history of Sumba Island, Indonesia, since
   the Late Cretaceous and its rapid escape into the
   forearc in the Miocene. *Journal of Asian Earth
   Sciences*, **19**, 453–479.
RYAN, J.G. & LANGMUIR, C.H. 1993. The systematics of
   boron abundances in young volcanic rocks. *Geochi-
   mica et Cosmochimica Acta*, **57**, 1489–1498.
SAJONA, F.G. & MAURY, R.C. 1998. Association of
   adakites with gold and copper mineralisation in the
   Philippines. *Comptes Rendus de l'Academie des
   Science Series II*, **326**, 27–34.
SAJONA, F.G., MAURY, R.C., BELLON, H., COTTON, J.,
   DEFANT, M.J. & PUBELLIER, M. 1993. Initiation of
   subduction and the generation of slab melts in
   western and eastern Mindanao, Philippines. *Geol-
   ogy*, **21**, 1007–1010.
SAJONA, F.G., BELLON, H., MAURY, R.C., PUBELLIER,
   M., COTTEN, J. & RANGIN, C. 1994. Magmatic
   response to abrupt changes in geodynamic settings;
   Pliocene Quaternary calc alkaline and Nb enriched
   lavas from Mindanao (Philippines). *Tectonophysics*,
   237, 47–72.
SAJONA, F.G., BELLON, H., MAURY, R.C., PUBELLIER,
   M., QUEBRAL, R.D., COTTEN, J., BAYON, F.E.,
   PAGADO, E. & PAMATIAN, P. 1997. Tertiary and

Quaternary magmatism in Mindanao and Leyte
   (Philippines): Geochronology, geochemistry and
   tectonic setting. *Journal of Asian Earth Sciences*,
   15, 121–154.
SCOTT, S.D. & BINNS, R.A. 1995. Hydrothermal processes and contrasting styles of mineralization in
   the western Woodlark and eastern Manus basins of
   the western Pacific. *In:* PARSON, L.M., WALKER,
   C.L. & DIXON, D.R. (eds) *Hydrothermal Vents and
   Processes*. Geological Society, London, Special
   Publications, **87**, 191–205.
SEWELL, D.M. & WHEATLEY, C.J.V. 1994. The Lerokis
   and Kali Kuning submarine exhalative gold-silver-
   barite deposits, Wetar Island, Maluku, Indonesia.
   Journal of Geochemical Exploration, **50**, 351–370.
SILLITOE, R.H. 1994. Erosion and collapse of volcanoes:
   Causes of telescoping in intrusion-centred ore
   deposits. *Geology*, **22**, 945–948.
SILLITOE, R.H. 1997. Characteristics and controls of the
   largest porphyry copper-gold and epithermal gold
   deposits in the circum-Pacific region. *Australian
   Journal of Earth Sciences*, **44**, 373–388.
SILLITOE, R.H. 1999. VMS and porphyry copper deposits:
   Products of discrete tectono-magmatic settings. *In:*
   STANLEY, C.J. *ET AL.* (eds) *Mineral Deposits: Pro-
   cesses to Processing*. Balkema, Rotterdam, 7–10.
SILLITOE, R.H. & GAPPE, I.M. 1984. *Philippine por-
   phyry copper deposits: geological setting and
   characteristics*. Committee for Co-ordination of
   Joint Prospecting for Mineral Resources in Asian
   Offshore Areas, Technical Publications, **14**.
SILVER, E.A., McCAFFREY, R. & SMITH, R.B. 1983.
   Collision, rotation and the initiation of subduction
   in the evolution of Sulawesi, Indonesia. *Journal of
   Geophysical Research*, **88**, 9407–9418.
SOLOMON, M. 1990. Subduction, arc reversal and the
   origin of porphyry copper-gold deposits in island
   arcs. *Geology*, **18**, 630–633.
STEWART, W.D. & SANDY, M.J. 1988. Geology of New
   Ireland and Djaul Islands, Northeastern Papua New
   Guinea. *In:* MARLOW, M.S., DADISMAN, S.V. &
   EXON, N.F. (eds) *Geology and offshore resource of
   Pacific island arcs - New Ireland and Manus region,
   Papua New Guinea*. Circum-Pacific Council for
   Energy and Mineral Resources Earth Science Ser-
   ies, **9**, 13–30.
SUN, S.S. & McDONOUGH, W.F. 1989. Chemical and
   isotopic systematics of oceanic basalts: implications
   for mantle composition and processes. *In:* SAUN-
   DERS, A.D. & NORRY, M.J. (eds) *Magmatism in the
   ocean basins*. Geological Society, London, Special
   Publications, **42**, 313–345.
SURMONT, J., LAJ, C., KISSAL, C., RANGIN, C., BELLON,
   H. & PRIADI, B. 1994. New paleomagnetic con-
   straints on the Cenozoic tectonic evolution of the
   North Arm of Sulawesi, Indonesia. *Earth and
   Planetary Science Letters*, **121**, 629–638.
TATSUMI, Y., HAMILTON, D.L. & NESBITT, R.W. 1986.
   Chemical characterisation of fluid phase released
   from a subducted lithosphere and origin of arc
   magmas: evidence from high-pressure experiments
   and natural rocks. *Journal of Volcanology and
   Geothermal Research*, **29**, 293–309.
TATSUMI, Y. & EGGINS, S. 1995. *Subduction Zone*

Magmatism. Blackwell Science, Oxford.

THATCHER, W. & ENGLAND, P. 1998. Ductile shear zones beneath strike-slip faults: implications for the thermomechanics of the San Andreas fault zone. *Journal of Geophysical Research*, **103**, 891–905.

THIÉBLEMONT, D., STEIN, G. & LESCUYER, J.-L. 1997. Epithermal and porphyry deposits: the adakite connection. *Comptes Rendus de l'Academie des Science Series II*, **325**, 103–109.

THIRLWALL, M.F., SMITH, T.E., GRAHAM, A.M., THEODOROU, N., HOLLINGS, P., DAVIDSON, J.P. & ARCULUS, R.J. 1994. High field strength element anomalies in arc lavas: Source or process? *Journal of Petrology*, **35**, 819–838.

TOKOSÖZ, M.N. & HSUI, A.T. 1978. Numerical studies of back-arc convection and the formation of marginal basins. *Tectonophysics*, **50**, 177–196.

TURNER, S. & HAWKESWORTH, C.J. 1997. Constraints on flux rates and mantle dynamics beneath island arcs from Tonga-Kermadec lava geochemistry. *Nature*, **389**, 568–573.

TURNER, S., HAWKESWORTH, C., VAN CALSTEREN, P., HEATH, E., MACDONALD, R. & BLACK, S. 1996. U-series isotopes and destructive plate margin magma genesis in the Lesser Antilles. *Earth and Planetary Science Letters*, **142**, 191–207.

VROON, P.Z., VAN BERGEN, M.J., WHITE, W.M. & VAREKAMP, J.C. 1993. Sr-Nd-Pb isotope systematics of the Banda Arc, Indonesia: Combined subduction and assimilation of continental material. *Journal of Geophysical Research*, **98**, 22349–22366.

VROON, P.Z., VAN BERGEN, M.J. & FORDE, E.J. 1996. Pb and Nd isotope constraints on the provenance of tectonically dispersed continental fragments in east Indonesia. *In:* HALL, R. & BLUNDELL, D.J. (eds) *Tectonic evolution of Southeast Asia*. Geological Society, London, Special Publications, **106**, 445–453.

WALPERSDORF, A., RANGIN, C. & VIGNY, C. 1998. GPS compared to long-term geologic motion of the north arm of Sulawesi. *Earth and Planetary Science Letters*, **159**, 47–55.

WOODHEAD, J.D. 1989. Geochemistry of the Mariana arc (western Pacific): source composition and processes. *Chemical Geology*, **76**, 1–24.

WOODHEAD, J.D., EGGINS, S. & GAMBLE, J.A. 1993. High field strength and transition element systematics in island arc and back-arc basin basalt: evidence for multi-phase melt extraction and a depleted mantle wedge. *Earth and Planetary Science Letters*, **114**, 491–504.

YOGODZINSKI, G.M., KAY, R.W., VOLYNETS, O.N., KOSLOSKOV, A.V. & KAY, S.M. 1995. Magnesian andesite in the western Aleutian Komandorsky region: implications for slab melting and processes in the mantle wedge. *Geological Society of America Bulletin*, **107**, 505–519.

YOU, C.-F., CASTILLO, P.R., GIESKES, J.M., CHAN, L.H. & SPIVAK, A.J. 1996. Trace element behaviour in hydrothermal experiments: implications for fluid processes at shallow depths in subduction zones. *Earth and Planetary Science Letters*, **140**, 41–52.

Correlating magmatic–hydrothermal ore deposit formation over time with geodynamic processes in SE Europe

ANDOR L. W. LIPS

[1]*Vening Meinesz Research School of Geodynamics, Utrecht University, the Netherlands*
[2]*Present address: BRGM, Mineral Resources Division, REM/VADO, 3 av C. Guillemin, BP 6009, 45060 Orléans cedex 2, France (e-mail: a.lips@brgm.fr)*

Abstract: Numerous magmatic–hydrothermal metal deposits in SE Europe occur in relatively narrow belts of magmatic rocks that have been classically related to subduction. In contrast to the hundred million years of convergence and consumption of lithosphere at the subduction zones in the region, the belts have been produced in discrete time intervals. It has also become recognized that the dominant phases of mineralization in the belts were accompanied by extension. Research focusing on the evolution of the lithosphere at convergent margins identifies secondary thermal and mechanical processes that accompany the primary process of subduction. Such secondary processes have been identified as the possible trigger of the magmatism, regional crustal extension, and enhanced flow of heat and circulation of fluids. The lithosphere dynamics may thus have played a vital role in the formation and localization of the mineralization. Roll-back of the subducted lithosphere, or restoration of the orogenic wedge geometry by changes in internal friction, set by variations in rates of the Africa–Eurasia convergence; and gravitational collapse (e.g. involving slab detachment or root delamination), are the scenarios that favour extension and the transfer of heat to relatively shallow lithospheric levels. Analysis of the temporal and spatial constraints of these geodynamic processes, together with a refining of the timing of the main phases of mineralization is the first step to discriminate between the geodynamic causes, and to determine their effects in relatively short-lived regional igneous and hydrothermal activities and the formation of related mineralization.

The interrelationship between plate tectonics and mineralization was first recognized about 30 years ago, with the appreciation of a strong correlation between plate margins and ore deposit distribution (Pereira & Dixon 1971; Mitchell & Garson 1972; Sillitoe 1972*a*, *b*). The initial correlations were based on the circum-Pacific region but were extended to the Alpine–Himalayan chain by Dixon & Pereira (1974), who showed that the location of a variety of mineral deposits in specific tectonic settings could be related to the development of the Tethyan region. Jankovic (1977) further extended this relationship, emphasizing the genetic link between copper mineralization and a dominantly calc-alkaline volcanic-intrusive belt with volcano-sedimentary intervals.

Mineralization at convergent margins has been seen as part of a continuous process that covers the entire period of convergence. It had become traditional to interpret the convergent margin mineralization in terms of the dip of the subduction zone and the build-up of the orogenic wedge during convergence, which upon reaching maturity then underwent melting and consequently released the fluids assisting in the melting and mineralizing processes (Sillitoe 1991). Whilst such a model broadly explains mineralization in terms of the convergence process, it does not explain why the ore deposits preferentially occur in areas of the convergent margin that are undergoing transtensional or extensional tectonics at the time of mineralization (as originally raised by Izawa & Urashima 1988). More recently, Mitchell (1992, 1996), Mitchell & Carlile (1994) and Jankovic (1997) emphasized the link between the position of the volcanic-intrusive belt and the distribution of dominantly porphyry and epithermal deposits in the Carpathian–Balkan system (Fig. 1). They related the generation of volcanic-intrusive belts at the European convergent margin to the subduction of oceanic lithosphere and associated extension in the related back-arc.

This paper examines the formation of magmatic–hydrothermal deposits, in contrast to the orogenic, or mesothermal, gold deposits (which, for example, have been documented in the nearby Alps by Pettke *et al.* 2000). These two classes of deposits vary significantly in terms of their depth of formation, their fluid source, and their tectonic position with respect to the convergent margin.

From: BLUNDELL, D.J., NEUBAUER, F. & VON QUADT, A. (eds) 2002. *The Timing and Location of Major Ore Deposits in an Evolving Orogen*. Geological Society, London, Special Publications, **204**, 69–79. 0305-8719/02/$15.00 © The Geological Society of London 2002.

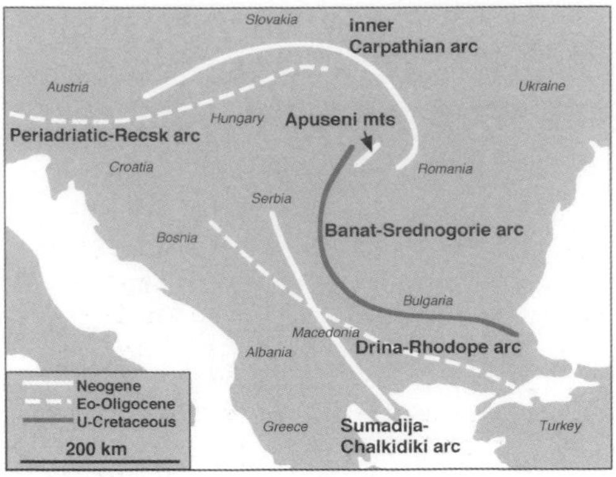

Fig. 1. Post-mid-Cretaceous mineralized volcanic-intrusive belts in the Carpathian–Balkan–Aegean region as presented by Mitchell (1996; after Jankovic 1977).

Magmatic–hydrothermal deposits are commonly associated with a back-arc location, while orogenic gold deposits form at the position of the growing fore-arc (Groves *et al.* 1998). A synthesis of orogenic gold occurrences in the circum-Pacific by Goldfarb *et al.* (2001) summarizes lithosphere-scale scenarios that have resulted in the formation of this class of deposits. Most scenarios involve regional thermal anomalies and extension (Goldfarb *et al.* 2001, fig. 10b–f) in contrast to the traditional view of plate subduction, plate thickening, and formation of a magmatic arc and associated mesothermal deposits (Goldfarb *et al.* 2001, fig. 10a). Despite the marked differences between the two classes of deposits, similar scenarios may have influenced the development of the back-arc, and thus also the formation of the magmatic–hydrothermal deposits.

Together with the presence of regional thermal anomalies and extension during phases of mineralization, the timing of the mineralization has become better constrained, indicating that the mineralizing phases are restricted to narrow time intervals in the convergence process. The refined compilation by Mitchell (1996, following Jankovic 1977, fig. 1) emphasizes that the individual mineralized belts in the Carpathian–Balkan region may have formed within less than 10 million years and indicates the formation of the mineralized belts in SE Europe at latest Cretaceous, Eo-Oligocene, and Neogene times, respectively. It is clear that formation of an entire magmatic belt in a short period of time is in contrast with the continuous consumption of subducted lithosphere and the continuously evolving active margin at the southern edge of Europe associated with over 100 million years of convergence between Africa and Eurasia.

Although academic research and industrially related metal exploration have added substantial data about mineralization in SE Europe in the past decades, two fundamental questions have remained. They form the foundation of the current research, which investigates the potential links between plate-scale geodynamics and mineralization (and forms the backbone of GEODE research).

(1) Why do we observe discrete igneous activity with associated mineralizing events along hundreds of kilometres along the strike of the Tethyan belt, over the past 100 million years?

(2) Why is the mineralization located at a particular place, in a particular geometry (at a particular time)?

In the context of potential plate-scale controls on the mineralization in the Carpathian–Balkan sector of the Alpine–Himalayan chain this paper evaluates relevant geodynamic mechanisms in relationship to the spatial and temporal distributions of the mineralized belts of SE Europe. This paper emphasizes the limits in the timing and duration of various geodynamic mechanisms and highlights their potential contributions (thermally and mechanically) to magmatism, fluid infiltration and extension in the crust. In conclusion, the paper will propose the most likely geodynamic mechanisms that have operated during the formation of the main regional phases of magmatic–hydrothermal mineralization in SE Europe.

Timing: quality, relevance, significance

To examine further the regional crustal features and the geodynamic mechanisms involved, the respective mineralization needs to be placed within the context of the regional tectonic history. A vital factor in this projection is an accurate age determination of the mineralizing event and a temporally well-resolved tectonic history. In principle, the sensitivity of modern age analyses should help to pinpoint any remarkable happening in the evolution of an orogen within a resolution of a few million years. In practice the timing of a combined physical and chemical anomaly like an ore deposit is difficult to establish. To fully appreciate the age information produced it is important to be familiar with the limits of the dating techniques to reduce the potential obscuring effect of misidentified absolute ages. A critical look at dating information shows that, as well as the analytical challenge, three key questions have to be resolved: what is the timing of a specific mineralization, what is the duration of a mineralizing event, and is it possible to identify different stages of mineralization in the same district. The integration of several observations is essential to understand the geological significance of the age information. Ideally this includes temporal information of the mineralization itself (timing of mineralization, timing of host rock deposition, duration of the mineralizing event, duration of associated volcanism, duration of associated hydrothermal activity, spatial variations of temporal information), but also involves information associated with the regional tectonics (e.g. timing of deformation in surrounding basement rocks, timing of basement denudation, timing of basin formation and sedimentation, timing of fault development). To apply the best possible age dataset it is important to have reliable relative age relationships (e.g. to establish overprinting mineralizing events, to discriminate between mineralization which is syn-genetic or epigenetic to the host rock), to appreciate existing age information and carry out strategic absolute age dating, to examine the analytical quality, and to check the geological relevance, before the regional significance can be established (N.B. it is the integration of the above that is currently in progress in the ABCD-GEODE-related Geographical Information System by BRGM http://giseurope.brgm.fr).

The timing of the mineralization in SE Europe has often been derived by interpolation of absolute or stratigraphic ages of host rocks. The most frequently applied age dating techniques do not produce a direct absolute age for the mineralization itself. With the advent of Re–Os dating applied to molybdenite (e.g. Stein et al. 1998) or Rb–Sr dating on sphalerite (Nakai et al. 1990, Pettke & Diamond 1996), the age information obtained directly for an individual mineral deposit has been greatly improved. Rb–Sr and ^{40}Ar/^{39}Ar isotopic dating techniques consequently help to pinpoint the timing and/or duration of the hydrothermal activity (e.g. Groff et al. 1997; Clark et al. 1998; Walshaw & Menuge 1998), along with a further refinement of host rock age information (also established by U–Pb dating). The geological relevance of the absolute ages produced is constrained by the effect of post ore chemical alteration and/or reheating on recrystallization and element diffusion characteristics. Considering this, existing datasets available from K–Ar dating are still applicable for the investigation of the volcanic host-rock deposition, but should be treated with caution when the data are derived from chemically altered samples. The closure temperature ranges of K–Ar, ^{40}Ar/^{39}Ar and Rb–Sr dating allow the investigation of the regional thermal conditions of the crust to project the mineralization and/or host rock in its respective regional crustal environment (e.g. Lips 1998; Lips et al. 1998). Re–Os and U–Pb dating are characterized by high closure temperatures and allow the pre-

servation of geologically relevant ages during subsequent thermal disturbances.

Timing of the mineralization in SE Europe

Examining the mineralized belts in SE Europe at Latest Cretaceous, Eo-Oligocene, and Neogene times (Fig. 1) in the light of recently published absolute ages (Fig. 2) shows that they are not equally well resolved. The 12–0.2 Ma Inner Carpathian Arc is best constrained by extensive K–Ar dating programmes (e.g. Pécskay et al. 1995a, b), with mineralization dated around 12–8 Ma (e.g. Lang et al. 1994; Kraus et al. 1999). K–Ar dating in the Apuseni mountains has indicated the timing of volcanic activity at 15–7 Ma (Pécskay et al. 1995a; Rosu et al. 1997, 2001), whilst the timing of mineralization is poorly constrained by limited 10–11 Ma ^{40}Ar/^{39}Ar laserprobe data on the Valea Morii porphyry alteration (Lips, unpublished). The 25–20 Ma Sumadija–Chalkidiki belt is only constrained in considerable detail at the Kassandra mining area in Greece (Gilg & Frei 1994). The 36–27 Ma Periadriatic–Recsk belt has been confined to a limited extent (e.g.von Blankenburg & Davies 1995) and younger volcanic rocks have been reported along strike (e.g. Pamic & Pécskay 1996). The 40–30 Ma Drina–Rhodope Belt is fairly well constrained in its Bulgarian part (e.g. Harkovska et al. 1989; Pécskay et al. 1992), and benefits from recent dating campaigns (e.g. Marchev & Singer 1999; Marchev et al. 2000). Its extension into Macedonia and Serbia has virtually

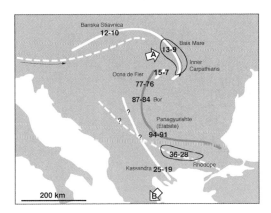

Fig. 2. Recently published absolute age datasets related to ore districts projected on the compilation by Mitchell (1996). Regional dating studies in the Inner Carpathian and Rhodopian domains are indicated by area outlines. White arrows indicate (outward) migration directions of the overriding plate of the subduction systems (A, Carpathian; B, Aegean). Black arrows along the Periadriatic and Inner Carpathian arcs indicate proposed propagation directions of detaching slabs.

no absolute age information. The Banat–Srednog-orie arc has been ascribed to the Upper Cretaceous based on limited absolute age information (e.g. see references in *Berza et al.* 1998). Recently obtained age information suggests an heterogeneous set of ages along the arc, as indicated by 76–77 Ma U–Pb and Re–Os ages obtained from SW Romania (Ciobanu *et al.* 2000), by 84–87 Ma $^{40}Ar/^{39}Ar$ ages (Lips *et al.* unpublished) and similar to slightly younger K–Ar ages (Banjesevic *et al.* 2001) from Serbia, and by 91–94 Ma U–Pb ages from western Bulgaria (von Quadt *et al.* 2001). The above information shows the necessity of an extensive compilation of all the absolute age information produced, along with continued strategic dating campaigns.

In summary (Fig. 2), the main phases of magmatism and associated mineralization are temporally best constrained at 95–75 Ma (Banat–Srednogorie), 40–30 Ma (Rhodope), 25–20 Ma (Chalkidiki), and 12–8 Ma (Inner Carpathians). In addition to the periods during which the main phases of mineralization occurred, numerous studies have reported shifts in age of emplacement and chemical characteristics of the magmatism or have related the role of extension to the onset of magmatic activity in the region (e.g. Jones *et al.* 1992; Seyitoglu & Scott 1992; Pirajno 1995; Karamata *et al.* 1997; Berza *et al.* 1998; Rosu *et al.* 2001).

Geodynamic settings

With the periods of mineralization established in the previous section, it is now appropriate to consider the tectono-magmatic evolution of the respective regions during the relevant time periods, in order to determine the geodynamic mechanisms that might have operated. Originally the mechanisms have been put forward as lithosphere-scale causes of extension in the crust and/or the magmatism. In principle, they may have had a control on the timing of the mineralization. This approach aims to identify the geodynamic mechanisms involved in the development of the Carpathian–Balkan region in the past hundred million years, as follows.

Subduction

Classically, the primary process of subduction has been taken to cause the back-arc magmatism and associated mineralization. Examination of the existing dataset on the African–Eurasian convergence (Müller & Roest 1992, fig. 3) shows that for about the past 100 million years convergence between the African and Eurasian plates was active at variable rates, but was slow compared with e.g. Pacific convergence rates. The present Aegean subducting slab appears to be the most western part of the Neo-Tethys subduction (e.g. Wortel & Spakman 2000), as seismic tomography studies have illustrated the long history of northward subduction of lithosphere at the southern margin of Europe (Spakman *et al.* 1988). The Carpathian subduction system appears to be of a similar age (e.g.Wortel & Spakman 2000).

Initiation of back-arc volcanism may have followed the onset of Africa–Eurasia with a time lag of several million years. Based on normal convergence at a constant rate of 20 km Ma^{-1} and a subduction angle of 45°, there is a time lag of more than 7 million years after subduction was initiated before the subducting slab reaches a depth of *c.* 100 km and enters the magma generation window (after Mason *et al.* 1998). This time lag would be greater when subduction is slower, or is oblique to the subduction zone and/or when the angle of subduction is smaller. Calculation shows that although northward subduction commenced around 100 Ma, the associated back-arc volcanism may have initiated since 90 Ma.

Roll-back

A review of the Tertiary subduction history in the Mediterranean and Carpathian region (Wortel & Spakman 2000) concludes that roll-back of subducted lithosphere, set by the Africa–Eurasian convergence, is a primary geodynamic mechanism which has shaped the present-day configuration. It operated from *c.* 30 Ma to the present in the Carpathian and Aegean arcs (e.g. see references in Wortel & Spakman 2000). In detail the Africa–Eurasia convergence rates (Fig. 3) show an alternation of phases of extremely slow convergence (i.e. 6–8 km Ma^{-1}) and relatively faster convergence (i.e. 15–20 km Ma^{-1}) in variable directions. The two phases of extremely slow convergence are identified for the periods around 20–0 Ma, and 75–55 Ma. During the past 30 million years, oceanic lithosphere in the subducting slab has been old (and cold) enough to promote roll-back of the slab (Fig. 4; Royden 1993) and, as a consequence, accentuate the curvature of the arc (Wortel & Spakman 2000). A similar process affecting the region in the 75–55 Ma period is still questionable as the subducted oceanic lithosphere was younger, and thus less dense, at that time and there is no direct evidence from seismic tomography. When roll-back occurs it will enhance extension in the back-arc of the overriding plate and allow upwelling of asthenosphere to shallow mantle levels (becoming a potential source of heat driving melting in the lithosphere),

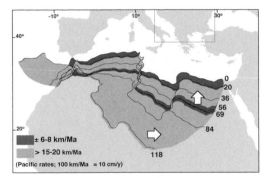

Fig. 3. Summary of overall convergence rates of African and Eurasian plates projected on the present-day outline of Europe and Africa (data extrapolated from Müller & Roest 1992), indicating varying convergence directions and two episodes of slow convergence (c. 75–55 Ma and post-20 Ma), alternating with episodes of slightly faster convergence (pre c. 75 Ma and c. 55–20 Ma). Square in upper part of figure indicates the region considered in this paper.

at which it may induce asthenosphere-dominated magmatism.

Critical wedges

Another process related to plate convergence is caused by associated changes in internal friction of the active margin wedge (based on critical wedge theories of Davis et al. 1983 and Dahlen 1990). Drastic changes in the convergence rates, and thus in internal friction of the wedge, can affect the potential energy distribution in the orogenic wedge at the margin of the overriding

plate. Faster convergence, causing an increase in friction, leads to a narrower and thicker orogenic wedge. Similarly, when the internal friction in an orogenic wedge is lowered, or lost, the wedge will force the redistribution of material to restore its wider and flatter critical wedge geometry. Such a process has been identified for the 15–5 Ma period in the East Carpathians, where rapid morphological and tectonic changes in the orogenic wedge may be directly related to roll-back (Sanders 1998; Sanders et al. 1999). A similar process, but preceding the roll-back, has been interpreted for the Aegean arc (Lips 1998) with the transformation from a high friction wedge to a low friction wedge geometry at c. 45 Ma. The additional interaction with the roll-back has caused an accelerated back-arc extension and allowed collapse of the Rhodope region in the internal parts of the orogen (Figs 5 and 6; Lips 1998; Lips et al. 2000). It is emphasized that the above scenario is seen as a crustal process, rather than a process operating at the scale of the whole lithosphere, when evaporite horizons are involved as the decollement-forming lithology. Deep-seated low-friction lithologies (e.g. at the base of the crust) may introduce the whole lithosphere to the restoration process. Its potential role in the formation of mineralization is expected in the combined extension and fluid infiltration in the rear of the wedge, generated at elevated thermal gradients.

Collapse and slab detachment

The terminal collapse of an orogen is classically interpreted to occur following the removal of the thickened lithospheric root of the orogen (Platt &

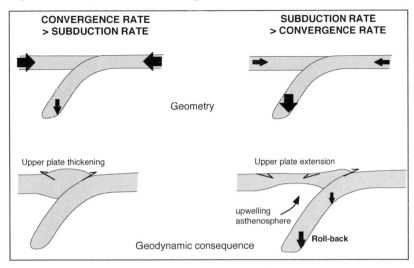

Fig. 4. The concept of roll-back (Royden 1993).

(instantaneous) evaporite decollement formation

overthrusting of wedge underlain by evaporite decolleme

outward flow of thrust wedge

extensional doming of internal basement

tilting of internal evaporite decollement surfaces
(inducing gravitational sliding)

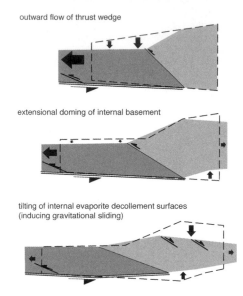

Fig. 5. Schematic representation of the development of a wedge after initiation of a decollement fault in evaporites (double lines in basal fault zone), based on the theoretical critical taper geometries in a non-cohesive wedge as proposed by Davis *et al.* (1983) and Dahlen (1990). Top figure; restoration of the wedge geometry after instantaneous evaporite decollement formation and transformation of a high friction decollement (narrow and high wedge; dashed outline) to a state when the basal decollement experiences no friction (wide and flat wedge). (based on Mohr diagrams of the basal state of effective stress). Lower three figures represent a schematized plausible scenario for the evolution of the Aegean orogenic wedge; restoration of the wedge geometry after accretion of material overlying an evaporite decollement, resulting in internal doming of overthrust material by outward escape of the wedge, and continuing doming and associated tilting of the decollement horizon, which will initiate further gravitational sliding along the decollement.

England 1993). One mechanism for doing this is the process of slab detachment (Wortel & Spakman 1992, 2000). This final state in the subduction history is achieved when tearing of the subducted slab is initiated. This most probably occurs at the point when continental lithosphere in the upper plate reaches the trench and starts to get subducted, following behind denser, stronger oceanic lithosphere or, alternatively, when subduction becomes stagnant (e.g. von Blanckenburg & Davies 1995; Wong a Ton & Wortel 1997; van de Zedde & Wortel 2001). Lateral migration of a tear in a subducted slab along the strike of the subduction zone can enhance slab roll-back and arc migration where the slab is not yet detached, initiate orogenic collapse, and change the signatures of arc volcanism. In the region it has been interpreted to operate along the Periadriatic Lineament from *c.* 45–40 Ma (von Blanckenburg & Davies 1995) and in the Carpathians from 15 Ma onwards (Mason *et al.* 1998), whilst in the Hellenic arc it may be very young (Pliocene) or absent (Wortel & Spakman 2000 and references therein). Slab detachment has been proposed as a possible mechanism involved in the formation of the Neogene mineralized belt (Fig. 7; de Boorder *et al.* 1998). It is very well constrained by independent studies in the inner Carpathian arc (e.g. Mason *et al.* 1998; Nemcok *et al.* 1998), but is not constrained for the Sumadija–Chalkidiki belt. Berza *et al.* (1998) inferred slab detachment to have occurred during the formation of the Banat–Srednogorie arc, although the regional operation of this mechanism in Late Cretaceous times has not been substantiated. A phase of lithospheric delamination is currently considered as likely to have controlled the formation of the Eo-Oligocene mineralized Rhodope belt (Marchev pers. comm.).

Roll-back and detachment: (transient) sources of heat and extension

The above mechanisms have different thermal and mechanical effects, in space and time, on the crust. These differences allow a first order discrimination between the regional scale sources of heat and extension. In both the cases of roll-back and slab detachment, cold mantle lithosphere is replaced by hot mantle asthenosphere and the thermo-mechanical effect on the crust may appear more or less similar. The two processes may differ considerably in their temporal and spatial characteristics. In the case of slab roll-back, the asthenosphere associated source of heat may cover a broad zone parallel to the subduction zone, theoretically migrating over time towards the subduction trench (e.g. Sillitoe 1991) with the induced volcanism following a similar pattern (e.g. Tatsumi *et al.* 1989). Related to the laterally migrating tear in a subducted slab, other heterogeneities in the

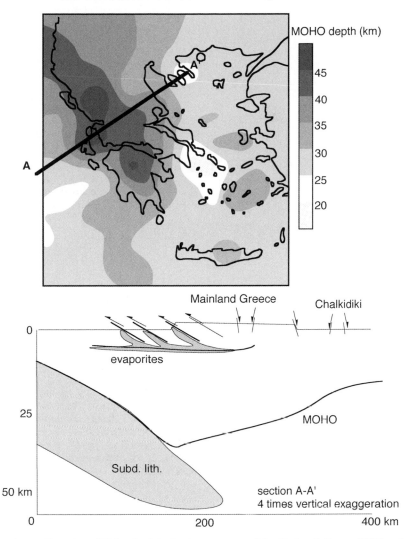

Fig. 6. Present-day configuration of Moho depth in the Aegean system (after Tsokas & Hansen 1997) and cross-section across the Greek mainland, indicating elevated Moho levels in the Aegean back-arc system, initiated by lithospheric delamination or wedge restoration and subsequent roll-back.

subducted lithosphere may become the locus of a rupture generating space for upwelling asthenosphere. The orientation of these ruptures or 'windows' with respect to the subduction direction and their opening/widening characteristics will be reflected in temporal and spatial variations while convergence continued (e.g. Thorkelson & Taylor 1989). Such 'windows' have been recognized in recent studies and have been temporally and genetically associated with back-arc extension, mafic, alkalic, volcanism (Hole *et al.* 1991), migration of igneous activities (Terakado & Nohda 1993), and with different classes of mineraliza-tion (Haeussler *et al.* 1995; Goldfarb *et al.* 1998; de Boorder *et al.* 1999; Goldfarb *et al.* 2001).

The lateral detachment of subducted lithosphere as proposed for the Balkan–Carpathian region is likely to have generated a point source of heat propagating with the direction of tearing parallel to the subduction zone (e.g.von Blanckenburg & Davies 1995;Wortel & Spakman 2000). The best recognized example in the region, in the Inner Carpathians, shows the generation of volcanism in the overriding plate with increasing mantle affinities, propagating over 150 km in 10 Ma and affecting individual locations over 2–4 Ma

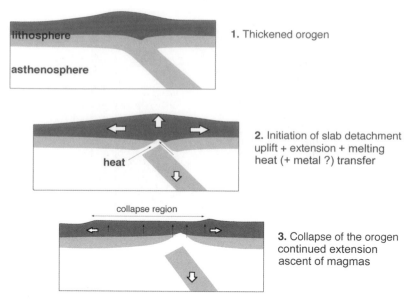

Fig. 7. The hypothetical role of orogenic collapse by slab detachment in the formation of ore deposits (de Boorder *et al.* 1998).

(Mason *et al.* 1998). The thermal implication of shallow slab detachment has been further investigated by van de Zedde & Wortel (2001) and shows that a transient thermal anomaly is created with a temperature increase in the overlying material up to 500 °C, as the rising asthenosphere fills the gap created by the sinking slab. The resulting pulse of heat allows for partial melting of the asthenosphere and the remaining metasomatized lithosphere for several million years.

Conclusions

In summary, the following geodynamic scenarios are proposed for the respective mineralized belts.

The Banat–Srednogorie belt appears to have been generated by subduction related magmatism (as traditionally proposed) that followed the onset of Africa–Eurasia convergence since *c.* 110 Ma. The relatively low convergence rates caused a time-lag of 10–20 million years before the melting process was initiated. Variations in convergence directions with respect to the geometry of the active margin may have caused the temporal variation in the magmatism (and mineralization), resulting in the occurrence of 95–90 Ma, 85–90 Ma and 75–80 Ma systems in the west Bulgarian (Panagyurishte), east Serbian (Bor) and SW Romanian (Ocna de Fier) parts of the belt, respectively. It is unlikely that the system was affected by detachment of the subducted slab at that time (e.g. Berza *et al.* 1998), as the relatively low density of the young slab proposed, and its

limited length, would have offered unfavourable conditions.

The Rhodope mineralized province was most probably controlled by gravitational collapse produced by lithospheric delamination (Marchev pers comm), or by the reconfiguration of the accretionary wedge after the loss of internal friction in its frontal parts (Lips 1998). The spatial and temporal characteristics of both scenarios suggest the introduction of a regional heat source over an area with an unknown geometry. Based on the above scenarios the identification of the Drina–Rhodope arc is premature and a regional extension of the Rhodope province into former Yugoslavia and into Turkey is still unfounded and requires further attention.

The Adriatic–Recsk belt appears to have been controlled by slab detachment following Alpine subduction as proposed by von Blanckenburg & Davies (1995). Although the propagation direction of the heat source from west to east is plausible, the exact timing and propagation rate are not resolved by the geochronological dataset. Further constraints on the timing and localization of individual mineral occurrences requires an improved temporal resolution of the geodynamic process.

The Chalkidiki region suffered from progressive collapse of the Aegean orogenic wedge (Lips 1998), assisted by roll-back of the subducted lithosphere. This initiated around 25–20 Ma and produced extension, magmatism, and mineralization in the rear of the orogenic wedge. Further extension towards the north(west) and (south)east may not concern a single belt but a wide region, trending roughly parallel to the regional outcrop patterns, as the Miocene–Recent collapse has affected the whole region extensively.

Detachment of the subducted slab is the most plausible scenario for magma generation in the Inner Carpathians and the propagation of the magma source from NW to SE from 12 Ma to 0.2 Ma. In the Inner Carpathians it is best constrained by independent studies and techniques. The timing of the mineralization is however still remarkable and its generation should remain under further consideration as the main ore districts (e.g. Banska Stiavnica and Baia Mare) were produced in the same time interval around 12–10 Ma. Also little mineralization appears to be associated with the 10–0 Ma history of the detachment process.

The evolution of the Apuseni Mountains may be partially considered together with the Inner Carpathians as it may well represent a fragment of the proto-Carpathians that has become dismembered by detachment and enhanced roll-back of the southeastern portion of the Carpathian subducted slab. The dominant phase of magmatism and associated mineralization may be similar to the 13–9 Ma mineralized belt of the Inner Carpathians. Supported by the palaeomagnetic and geochemical results of the Apuseni (Rosu et al. 2001) and of the Inner Carpathians (Mason et al. 1998) it is speculated that a synchronous formation of the mineralization across the Carpathians is controlled by the timing of a regional fragmentation of the crust. The fragmentation may have occurred when the overriding lithosphere responded to a combined ENE-directed roll-back and a gradual SE-directed detachment of the subducted lithosphere. The process caused the development of the present geometry of the Carpathians, thinning of the lithosphere, the dismembering of existing crustal fragments, and magma ascent along the dominant crustal fractures. A careful restoration of the crust and mantle lithosphere around the Carpathians at 13–9 Ma is required to further examine this speculation.

The investigations reported here were supported by the Vening Meinesz Research School of Geodynamics (VMSG). Discussion with Paul Mason, Hugo de Boorder, Rinus Wortel, Stan White and Derek Blundell, and reviews by Wolfgang Frisch and Kurt Stüwe on a previous version have been appreciated. Critical comments from reviews by Thomas Pettke and Robert Handler have helped to improve the manuscript.

References

BANJESEVIC, M., COCIC, S. & RADOVIC, M. 2001. Petrology and K/Ar ages of volcanic rocks for widen Bor zone as the part of the Timok magmatic complex (east Serbia). In: Abstracts of ABCD-GEODE workshop, Vata Bai Romania, Romanian Journal of Mineral Deposits, **79**, 39–40.

BERZA, T., CONSTANTINESCU, E. & VLAD, S.-N. 1998. Upper Cretaceous magmatic series and associated mineralzation in the Carpathian-Balkan Orogen. Resource Geology, **48**, 291–306.

CIOBANU, C.L., STEIN, H. & COOK, N.J. 2000. A Re-Os age for molybdenite, Dognecea skarn deposit, Banat, Romania. Abstracts of ABCD-GEODE workshop, Borovets Bulgaria, 19.

CLARK, A.H., ARCHIBALD, D.A., LEE, A.W., FARRAR, E. & HODGSON, C.J. 1998. Laser probe ^{40}Ar/^{39}Ar ages of early- and late-stage alteration assemblages, Rosario porphyry copper-molybdenum deposit, Collahuasi district, I Region, Chile. Economic Geology, **93**, 326–337.

DAHLEN, F.A. 1990. Critical taper models of fold-and-thrust belts and accretionary wedges. Annual Review Earth and Planetary Science, **18**, 55–99.

DAVIS, D., SUPPE, J. & DAHLEN, F.A. 1983. Mechanics of fold-and-thrust belts and accretionary wedges. Journal of Geophysical Research, **88**, 1153–1172.

DE BOORDER, H., SPAKMAN, W., WHITE, S.H. & WORTEL, M.J.R. 1998. Late Cenozoic mineralisation, orogenic collapse and slab detachment in the European Alpine Belt. Earth and Planetary Science Letters, **164**, 569–575.

DE BOORDER, H., SPAKMAN, W., WHITE, S.H., LIPS, A.L.W., WORTEL, M.J.R. & KRONEMEIJER, W.M. 1999. Late orogenic mineralization: tomography and geodynamic setting. In: STANLEY, C.J. ET AL. (eds) Mineral Deposits: Processes to Processing. Balkema, Rotterdam, 1305–1306.

DIXON, C.J. & PEREIRA, J. 1974. Plate tectonics and mineralization in the Tethyan region. Mineralium Deposita, **9**, 185–198.

GILG, H.A. & FREI, R. 1994. Chronology of magmatism and mineralization in the Kassandra mining area, Greece: the potentials and limitations of dating hydrothermal illites. Geochimica et Cosmochimica Acta, **58**, 2107–2122.

GOLDFARB, R.J., PHILLIPS, G.N. & NOKLEBERG, W.J. 1998. Tectonic setting of synorogenic gold deposits of the Pacific Rim. Ore Geology Reviews, **13**, 185–218.

GOLDFARB, R.J., GROVES, D.I. & GARDOLL, S. 2001. Orogenic gold and geological time: a global synthesis. Ore Geology Reviews, **18**, 1–75.

GROFF, J.A., HEITZLER, M.T., MCINTOSH, W.C. & NORMAN, D.I. 1997. 40Ar/39Ar dating and mineral paragenesis for Carlin-type gold deposits along the Getchell trend, Nevada: Evidence for Cretaceous and Tertiary gold mineralization. Economic Geology, **92**, 601–622.

GROVES, D.I., GOLDFARB, R.J., GEBRE-MARIAM, M., HAGEMANN, S.G. & ROBERT, F. 1998. Orogenic gold deposits: a proposed classification in the context of their crustal distribution and relationship to other gold deposit types. Ore Geology Reviews, **13**, 7–27.

HAEUSSLER, P.J., BRADLEY, D., GOLDFARB, R.J., SNEE, L. & TAYLOR, C. 1995. Link between ridge subduction and gold mineralization in southern Alaska. *Geology*, **23**, 995–998.

HARKOVSKA, A., YANEV, Y. & MARCHEV, P. 1989. General features of the Paleogene orogenic magmatism in Bulgaria. *Geologica Balcanica*, **19**, 37–72.

HOLE, M.J., ROGERS, G., SAUNDERS, A.D. & STOREY, M. 1991. Relation between alkalic volcanism and slab-window formation. *Geology*, **19**, 657–660.

IZAWA, E. & URASHIMA, Y. 1988. Quaternary gold mineralization and its geological environment in Kyushu, Japan. *In: Bicentennial gold, Geological Society Australia Abstracts*, **22**, 177–182.

JANKOVIC, S. 1977. The copper deposits and geotectonic setting of the Tethyan Eurasian metallogenic belt. *Mineralium Deposita*, **12**, 37–47.

JANKOVIC, S. 1997. The Carpatho-Balkanides and adjacent area: a sector of the Tethyan Eurasian metalogenic belt. *Mineralium Deposita*, **32**, 426–433.

JONES, C.E., TARNEY, J., BAKER, J.H. & GEROUKI, F. 1992. Tertiary granitoids of Rhodope, northern Greece: magmatism related to extensional collapse of the Hellenic Orogen? *Tectonophysics*, **210**, 295–314.

KARAMATA, S., KNEZEVIC, V., PÉCKSAY, Z. & DJORDJEVIC, M. 1997. Magmatism and metallogeny of the Ridanj-Krepoljin belt (eastern Serbia) and their correlation with northern and eastern analogues. *Mineralium Deposita*, **32**, 452–458.

KRAUS, I., CHERNYSHEV, I.V., SUCHA, V., KOVALENKER, V.A., LEBEDEV, V.A. & SAMAJOVA, E. 1999. Use of illite for K/Ar dating of hydrothermal precious and base metal mineralization in central Slovak Neogene volcanic rocks. *Geologica Carpathica*, **50**, 353–364.

LANG, B., EDELSTEIN, O., STEINITZ, G., KOVACS, M. & HALGA, S. 1994. Ar-Ar dating of adularia - A tool in understanding relations between volcanism and mineralization: Baia Mare area (Gutii Mountains) Northwestern Romania. *Economic Geology*, **89**, 174–180.

LIPS, A.L.W. 1998. Temporal constraints on the kinematics of the destabilization of an orogen; syn- to post-orogenic extensional collapse of the Northern Aegean region. *Geologica Ultraiectina*, **166**, 1–224.

LIPS, A.L.W., WHITE, S.H. & WIJBRANS, J.R. 1998. 40Ar/39Ar laserprobe direct dating of discrete deformational events, a continuous record of Early Alpine tectonics in the Pelagonian Zone, NW Aegean area, Greece. *Tectonophysics*, **298**, 133–153.

LIPS, A.L.W., WHITE, S.H. & WIJBRANS, J.R. 2000. Middle-Late Alpine thermotectonic evolution of the southern Rhodope Massif, Greece. *Geodynamica Acta*, **13**, 281–292.

MARCHEV, P. & SINGER, B. 1999. Timing of magmatism, alteration-mineralization, and caldera evolution in the Spahievo ore field, Bulgaria, from laser-fusion 40Ar/39Ar dating. *In:* STANLEY, C.J. *ET AL* (eds) *Mineral Deposits: Processes to Processing*. Balkema, Rotterdam, 1271–1274.

MARCHEV, P., RAICHEVA, R., SINGER, B., DOWNES, H., AMOV, B. & MORITZ, R. 2000. Isotopic evidence for the origin of Palaeogene magmatism and epithermal ore deposits of the Rhodope Massif. *In:*

Abstracts of ABCD-GEODE workshop, Borovets Bulgaria, 47.

MASON, P.R.D., SEGHEDI, I., SZÁKACS, A. & DOWNES, H. 1998. Magmatic constraints on geodynamic models of subduction in the East Carpathians, Romania. *Tectonophysics*, **297**, 157–176.

MITCHELL, A.H.G. 1992. Andesitic arcs, epithermal gold and porphyry-type mineralization in the western Pacific and eastern Europe. *Institution of Mining and Metallurgy Transactions*, **B101**, 125–138.

MITCHELL, A.H.G. 1996. Distribution and genesis of some epizonal Zn-Pb and Au provinces in the Carpathian-Balkan region. *Institution of Mining and Metallurgy Transactions*, **B105**, 127–138.

MITCHELL, A.H.G. & CARLILE, J.C. 1994. Mineralization, antiforms and crustal extension in andesitic arcs. *Geological Magazine*, **131**, 231–242.

MITCHELL, A.H.G. & GARSON, M.S. 1972. Relationship of porphyry copper and circum-Pacific tin deposits to palaeo-Benioff zones. *Institution of Mining and Metallurgy Transactions*, **B81**, 10–25.

MÜLLER, R.D. & ROEST, W.K. 1992. Fracture zones from combined Geosat and Seasat data. *Journal of Geophysical Research*, **97**, 3337–3350.

NAKAI, S., HALLIDAY, A.N., KESLER, S.E. & JONES, H.D. 1990. Rb-Sr dating of sphalerites from Tenessee and the genesis of Mississippi Valley type ore deposits. *Nature*, **346**, 354–357.

NEMCOK, M., POSPISIL, L., LEXA, J. & DONELICK, R.A. 1998. Tertiary subduction and slab break-off model of the Carpathian-Pannonian region. *Tectonophysics*, **295**, 307–340.

PAMIC, J. & PÉCSKAY, Z. 1996. Geological and K-Ar ages of Tertiary volcanic formations from the southern part of the Pannonian Basin in Croatia - based on surface and subsurface data. *Nafta*, **47**, 195–202.

PÉCSKAY, Z., BALOGH, K. & HARKOVSKA, A. 1992. K-Ar dating of the Perelik volcanic massif (central Rhodopes, Bulgaria) . *Acta Geologica Hungarica*, **34**, 101–110.

PÉCSKAY, Z., LEXA, J., SZAKÁCS, A., BALOGH, K., SEGHEDI, I., KONECNY, V., KOVACS, M., MARTON, E., KALICIAK, M., SZEKY-FUX, V., POKA, T., GYARMATI, P., EDELSTEIN, O., ROSU, E. & ZEC, B. 1995a. Space and time distribution of Neogene-Quaternary volcanism in the Carpatho-Pannonian region. *Acta Vulcanologica*, **7**, 15–28.

PÉCSKAY, Z., EDELSTEIN, O., SEGHEDI, I., SZAKÁCS, A., KOVACS, M., CRIHAN, M. & BERNAD, A. 1995b. Recent K-Ar datings of Neogene-Quaternary calc-alkaline volcanic rocks in Romania. *Acta Vulcanologica*, **7**, 53–61.

PEREIRA, J. & DIXON, C.J. 1971. Mineralization and plate tectonics. *Mineralium Deposita*, **6**, 404–405.

PETTKE, T. & DIAMOND, L.W. 1996. Rb-Sr dating of sphalerite based on fluid inclusion-host mineral isochrons: a clarification why it works. *Economic Geology*, **91**, 951–956.

PETTKE, T., DIAMOND, L.W. & KRAMERS, J.D. 2000. Mesothermal gold lodes in the north-western Alps: a review of genetic constraints from radiogenic isotopes. *European Journal of Mineralogy*, **12**, 213–230.

PIRAJNO, F. 1995. Volcanic-hosted epithermal systems in

northwest Turkey. *South African Journal of Geology*, **98**, 13–24.

PLATT, J.P. & ENGLAND, P.C. 1993. Convective removal of lithosphere beneath mountain belts: thermal and mechanical consequences. *American Journal of Science*, **293**, 307–336.

ROSU, E., PÉCSKAY, Z., STEFAN, A., POPESCU, G., PANAIOTU, C. & PANAIOTU, C.E. 1997. The evolution of the Neogene volcanism in the Apuseni Mountains (Rumania): constraints from new K-Ar data. *Geologica Carpathica*, **48**, 353–359.

ROSU, E., SZAKACS, A., DOWNES, H., SEGHEDI, I., PÉCSKAY, Z. & PANAIOTU, C. 2001. The origin of Neogene calc-alkaline and alkaline magmas in the Apuseni Mountains, Romania: the adakite connection. *Romanian Journal of Mineral Deposits*, **79**, 3–23.

ROYDEN, L.H. 1993. Evolution of retreating subduction boundaries formed during continental collision. *Tectonics*, **12**, 629–638.

SANDERS, C.A.E. 1998. *Tectonics and erosion; Competitive forces in a compressive orogen - A fission track study of the Romanian Carpathians*. PhD thesis, Vrije Universiteit Amsterdam.

SANDERS, C.A.E., ANDRIESSEN, P.A.M. & CLOETINGH, S.A.P.L. 1999. Life cycle of the East Carpathian orogen: Erosion history of a doubly vergent critical wedge assessed by fission track thermochronology. *Journal of Geophysical Research*, **104**, 29095–29112.

SEYITOGLU, G. & SCOTT, B.C. 1992. Late Cenozoic volcanic evolution of the northeastern Aegean region. *Journal of Volcanological and Geothermal Research*, **54**, 157–176.

SILLITOE, R.H. 1972a. Formation of certain massive sulphide deposits at sites of sea floor spreading. *Institution of Mining and Metallurgy Transactions*, **B81**, 141–148.

SILLITOE, R.H. 1972b. A plate tectonic model for the origin of porphyry copper deposits. *Economic Geology*, **67**, 184–197.

SILLITOE, R.H. 1991. Gold metallogeny of Chile - an introduction. *Economic Geology*, **86**, 1187–1205.

SPAKMAN, W., WORTEL, M.J.R. & VLAAR, N. 1988. The Hellenic subduction zone: a tomographic image and its geodynamic implications. *Geophysical Research Letters*, **15**, 60–63.

STEIN, H.J., MORGAN, J.W., MARKEY, R.J. & HANNAH, J.L. 1998. An introduction to Re-Os - what's in it for the mineral industry? *SEG Newsletter*, **32**, 7–15.

TATSUMI, Y., OTOFUJI, Y., MATSUDA, T. & NOHDA, S. 1989. Opening of the Sea of Japan back-arc basin by asthenospheric injection. *Tectonophysics*, **166**, 317–329.

TERAKADO, Y. & NOHDA, S. 1993. Rb-Sr dating of acidic rocks from the middle part of the Inner Zone of southwst Japan: tectonic implications for the migration of the Cretaceous to Paleogene igneous activity. *Chemical Geology*, **109**, 69–87.

THORKELSON, D.J. & TAYLOR, R.P. 1989. Cordilleran slab windows. *Geology*, **17**, 833–836.

TSOKAS, G.N. & HANSEN, R.O. 1997. Study of the crustal thickness and the subducting lithosphere in Greece from gravity data. *Journal of Geophysical Research*, **102**, 20585–20597.

VON BLANCKENBURG, F. & DAVIES, J.H. 1995. Slab break-off: A model for syncollisional magmatism and tectonics in the Alps. *Tectonics*, **14**, 120–131.

VAN DE ZEDDE, D.M.A. & WORTEL, M.J.R. 2001. Shallow slab detachment as a transient source of heat at midlithospheric depths. *Tectonics*, **20**, 868–882.

VON QUADT, A., PEYCHEVA, I. & KAMENOV, B. 2001. The Elatsite porphyry copper metal deposit of the Panagyurishte corridor, Srednogorie zone, Bulgaria: U-Pb zircon and isotopic investigations for timing and ore genesis. *Journal of Conference Abstracts*, **6**, 000–001.

WALSHAW, R.D. & MENUGE, J.F. 1998. Dating crustal fluid flow by the Rb-Sr isotopic analysis of sphalerite: a review. *In:* PARNELL, J. (ed.) *Dating and duration of fluid flow and fluid-rock interaction*. Geological Society, London, Special Publications, **144**, 137–143.

WONG A TON, S.Y.M. & WORTEL, M.J.R. 1997. Slab detachment in continental collision zones; an analysis of controlling parameters. *Geophysical Research Letters*, **24**, 2095–2098.

WORTEL, M.J.R. & SPAKMAN, W. 1992. Structure and dynamics of subducted lithosphere in the Mediterranean region. *Proceedings of the Koninklijke Nederlandse Akademie voor Wetenschappen*, **95**, 325–347.

WORTEL, M.J.R. & SPAKMAN, W. 2000. Subduction and slab detachment in the Mediterranean–Carpathian region. *Science*, **290**, 1910–1917.

Contrasting Late Cretaceous with Neogene ore provinces in the Alpine–Balkan–Carpathian–Dinaride collision belt

FRANZ NEUBAUER

Institute of Geology and Palaeontology, University of Salzburg, Hellbrunner Str. 34, A-5020 Salzburg, Austria (e-mail:franz.neubauer@sbg.ac.at)

Abstract: Internal sectors of the Alpine–Balkan–Carpathian–Dinaride (ABCD) orogen comprise fundamentally different ore deposits along strike in three temporally and spatially distinct belts. These were formed by several short-lived, late-stage collisional processes (including slab break-off) during the Late Cretaceous and Oligocene to Neogene times. Reconstruction of Late Cretaceous (*c.* 92–65 Ma) collisional structures, magmatic features and mineralization reveals contrasting variations along strike, including the following.

(1a) The Late Cretaceous 'banatite' magmatic belt, which extends from the Apuseni mountains to the Balkans, associated mainly with porphyry Cu–Au, massive sulphide and Fe–Cu skarn mineral deposits. In respect to their country rocks and geodynamic setting, the magmatism is interpreted to represent either post-collisional or Andean-type calc-alkaline due to continuous subduction or break-off of the subducted lithosphere.

(1b) The Alpine–West Carpathian sector, which is characterized by strong Late Cretaceous metamorphic/deformational overprint, lack of magmatism and both syn- and late-orogenic formation of metasomatic and metamorphogenic talc, magnesite, siderite and vein- and shear zone-type Cu and As–Au due to the exhumation of metamorphic core complexes.

(2a) The Oligocene–Miocene Serbomacedonian–Rhodope metallogenic zone extends across several structural units from the Bosnian Dinarides to the Rhodopes and to Thrace. It includes both a belt with volcanic-hosted and vein-type Pb–Zn deposits and a belt of porphyry Cu–Au–Mo and epithermal Au mineralization, which is more common in the south. Both belts appear to relate to microcontinent collision and associated subsequent magmatism, again possibly due to slab break-off.

(2b) Different types of mineralization were also formed along the internal Inner Carpathian and Alpine sectors during Late Oligocene to Miocene collision. In the Alps, mineralization formed due to eastward extrusion of fault-bounded blocks into the Carpathian arc. Associated mineral deposits are always related to exhumation of metamorphic core complexes and include: sub-vertical mesothermal Au–quartz veins and replacement As–Ag–Cu ore bodies within the metamorphic core complex, fault-bounded mineralization (Pb–W–Au) along low-angle ductile normal faults along the upper margins of the metamorphic core complex, mineralized (Sb–Au) strike-slip faults and sub-vertical Au–Ag–Sb-bearing tension veins.

(2c) In contrast, nearly all Miocene ore deposits within the Carpathians are related to volcanic activity contemporaneous with the invasion of fault-bounded blocks into the Carpathian arc. These have been related to slab break-off and cessation of subduction. Mineral deposits include structurally controlled Au–Sb–Cu–Pb–Zn ore bodies within shallow volcanic edifices, with a preference for steep tension veins parallel to the motion direction of laterally escaping crustal blocks.

Many different types of mineralization are found in orogenic belts of various ages, including hydrothermal vein and porphyry deposits (e.g. Lehmann *et al.* 2000; Blundell 2002). Major questions concerning the formation of late stage orogenic hydrothermal and porphyry Cu ore deposits in orogenic belts include the origin of metals and fluids, the nature of the heat source driving hydrothermal processes, the composition of magmatic melts and the general geodynamic context within which ore formation occurs (e.g.

Hedenquist & Lowestern 1994; Heinrich *et al.* 1999; Ulrich *et al.* 1999; Lang & Baker 2001). Although the formation of hydrothermal and orogenic ore deposits includes many processes, only a few models exist which link these to specific large-scale geodynamic processes such as slab break-off (e.g. de Boorder *et al.* 1998) or to a change of the subduction angle (Oyarzun *et al.* 2001). For example, many hydrothermal lodes and porphyry Cu deposits are related to subduction of oceanic crust or slab windows within subducting

From: BLUNDELL, D.J., NEUBAUER, F. & VON QUADT, A. (eds) 2002. *The Timing and Location of Major Ore Deposits in an Evolving Orogen*. Geological Society, London, Special Publications, **204**, 81–102.
0305-8719/02/$15.00 © The Geological Society of London 2002.

slabs, as in the western USA (e.g. Haeussler *et al.*
1995) and the Andes (Dill 1998). Recent observations show that many orogenic ore deposits
formed rather by punctuated events and not by
continuous processes. The generation of distinct,
adakitic melts clearly plays a role in the formation
of giant deposits in a subduction zone environment (Oyarzun *et al.* 2001).

Several models have been proposed which
relate ore deposits to collisional and post-collisional processes within convergent continental
plates (e.g. Spencer & Welty 1986; Beaudoin *et
al.* 1991). Mineralization within collisional orogens involves a number of different processes.
These include mainly (1) Andean-type mineralization due to subduction of oceanic lithosphere
predating collision, (2) mineralization associated
with collisional granites and (3) late-stage orogenic mineralization resulting from hydrothermal
activity associated with orogen-scale faulting due
to adjustment of the upper, brittle crust. This is
often associated with calc-alkaline and subsequent
alkaline magmatism. The latter process is possibly
triggered by slab break-off (de Boorder *et al.*
1998). The Alpine–Balkan–Carpathian–Dinaride
(ABCD) orogen is a particularly interesting young
belt as late-stage orogenic mineral deposits (e.g.
Evans 1975; Petrascheck 1986) differ in style
along strike. A deep-seated heat source, visible in
mantle tomography, has been postulated recently
as the cause of the magmatism and associated
mineralization (de Boorder *et al.* 1998). Data from
the Alpine–Balkan–Carpathian–Dinaride orogen
may help to constrain models for the late-stage
orogenic mineralization.

The ABCD orogen has a geological history
characterized by complex interactions between
various microplates squeezed in between the African and Eurasian continents, a situation which is
basically similar to the present-day interactions
between Australia and SE Asia in the region of the
Indonesian island arc system (Milsom 2001;
Blundell 2002). For the ABCD orogen, only a few
attempts have been made so far to relate large-scale
tectonic processes with mineralization. These include, for example, the work of Petrascheck
(1986), Mitchell (1996), Serafimovski *et al.*
(1995), de Boorder *et al.* (1998), Vlad & Borcos
(1998).

Late-stage orogenic ore provinces in the ABCD belt

The ABCD belt is a complex, arcuate, double-
vergent orogen (Fig. 1) that formed during two
independent stages of continent-continent collision
during Mid–Late Cretaceous and Late Eocene–
Palaeogene. The present-day structure resulted

from the final collision of the stable European–
Moesian platform with the Adriatic hinterland,
deforming a number of continental microplates in
between. The Cretaceous orogen was heavily
deformed during Tertiary microplate movements,
as these moved towards the north against the
stable European continent, due to southward subduction of the Penninic ocean and its expansion
into the future Carpathian realm (e.g. Neugebauer
et al. 2001 and references therein). Mineralization
occurred during two distinct major orogenic stages
between the Late Cretaceous and the Neogene,
which therefore suggests a punctuated development of mineralization rather than a continuous
process during orogeny. The ABCD orogenic belt
is still one of the most active mining and exploration areas of Europe. Principal deposits are compiled in Table 1 and shown in Figure 1. Details of
economic importance can be found in Heinrich &
Neubauer (2002).

Mineral deposits are contained within the following three belts, which may, from the metallogenic point of view, be subdivided into further
zones. For the sake of simplicity, this is not done
here, but explained in the following. The principal
emphasis is on contemporaneous mineralization
along strike of distinct zones of the ABCD
orogen.

(1) The Late Cretaceous metallogenic banatite
belt is formed of volcanic and plutonic suites,
which include major porphyry Cu and massive
sulphide deposits, replacement ores and hydrothermal veins. Late Cretaceous magmatic rocks comprise high-level plutonic suites, collectively
referred to as banatites, and some subordinate
volcanic sequences (von Cotta 1864). The magmatic rocks and mineral deposits extend in a
continuous belt from Burgas (close to the Black
Sea) to the Apuseni Mountains. These mineral
deposits contrast with contemporaneous metamorphogenic siderite/magnesite/talc and hydrothermal
Cu vein deposits exposed in the Eastern Alps and
Western Carpathians, forming a spatially distinct
metallogenic sub-zone in the western lateral extension of the banatite belt.

(2) The Oligocene Serbomacedonian–Rhodope
metallogenic belt is mostly related to calc-alkaline
magmatism and includes numerous Pb–Zn,
mostly volcanic hosted, vein and replacement
deposits extending from Bosnia through Serbia
and Macedonia to southern Bulgaria and Greece.
A second type of mineralization is represented by
a number of porphyry Cu–Au deposits that are
more common in northern Greece and Macedonia.

(3) The Inner Carpathian–Alpine belt includes
Oligocene to Neogene volcanism with numerous
epithermal Pb–Zn–Ag deposits in the Inner Car-

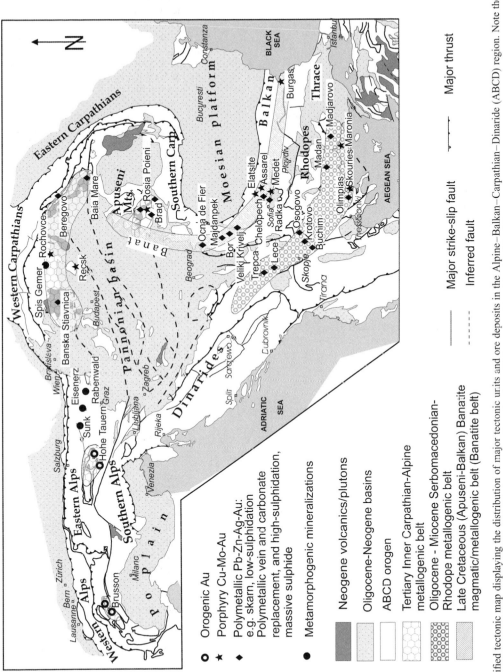

Fig. 1. Simplified tectonic map displaying the distribution of major tectonic units and ore deposits in the Alpine–Balkan–Carpathian–Dinaride (ABCD) region. Note the distribution of three Late Cretaceous to Miocene belts of orogenic mineralization which are considered in detail in this paper.

Table 1. *Major Cretaceous and Tertiary mineral resources of the ABCD belt—most of the deposits are in operation*

Deposit	Country	Metals/ mineral	Type	Age	Reference
Late Cretaceous magmatic–metallogenetic belt					
Burgas district	Bulgaria	Pb–Zn	Hydrothermal veins	Late Cretaceous–Paleocene ?	Berza *et al.* (1998)
Chelopech	Bulgaria	Cu–Au	Massive sulphide	Late Cretaceous	Strashimirov & Popov (2000)
Elatsite	Bulgaria	Cu–Au	Porphyry Cu	Late Cretaceous	Strashimirov & Popov (2000)
Radka	Bulgaria	Cu–Au	Massive sulphide	Late Cretaceous	Kouzmanov (2002)
Bor	Serbia	Cu–Au	Massive sulphide, porphyry Cu	Late Cretaceous	Karamata *et al.* (1997)
Majdanpek	Serbia	Cu–Au	Massive sulphide, porphyry Cu	Late Cretaceous	Karamata *et al.* (1997)
Veliki Krivelj	Serbia	Cu	Porphyry Cu	Late Cretaceous	Ciobanu *et al.* (2002)
Ocna de Fier/Dognecea	Romania	Fe–Cu–Pb–Zn	Skarn	Late Cretaceous	Nicolescu *et al.* (1999)
Rochovce	Slovakia	Mo–W	Porphyry Mo–W	Late Cretaceous	Grecula *et al.* (1995)
Spis–Gemer	Slovakia	Siderite	Metamorphogenic	Late Cretaceous	Grecula *et al.* (1995)
Jelsava, Dubrava	Slovakia	Magnesite	Metamorphogenic	Late Cretaceous	Grecula *et al.* (1995)
Gemerska Poloma	Slovakia	Talc	Metamorphogenic	Late Cretaceous	Grecula *et al.* (1995)
Eisenerz	Austria	Siderite	Metamorphogenic	Late Cretaceous	Weber *et al.* (1997)
Sunk/ Trieben	Austria	Magnesite	Metamorphogenic	Late Cretaceous	Weber *et al.* (1997)
Rabenwald	Austria	Talc	Metamorphogenic	Late Cretaceous	Weber *et al.* (1997)
Oligocene–Miocene Serbomacedonian–Rhodope metallogenetic belt					
Trepca	Serbia/Kosovo	Pb–Zn–Ag(–Au)	Breccia pipe, skarn, stratiform manto	Miocene	Serafimovski (2000)
Buchim	Macedonia	Cu–Au	Porphyry Cu	Oligocene	Serafimovski (2000)
Skouries	Greece	Cu–Au–Ag	Porphyry Cu	Oligocene–Miocene	Serafimovski (2000)
Maronia	Greece	Cu–Au–Mo	Porphyry Cu	Oligocene(?)	Melfòs *et al.* (2002)
Madjarovo	Bulgaria	Au–Ag–Pb–Zn	Hydrothermal veins	Oligocene	Marchev *et al.* (2002)
Madan	Bulgaria	Pb	Hydrothermal veins	Oligocene	Dimov *et al.* (2000)
Osogovo–Blagodat	Macedonia	Pb–Zn–Ag	Volcanic hosted	Oligocene	Naftali (2002)
Kratovo–Zletovo	Macedonia	Pb–Zn–Ag	Volcanic hosted	Oligocene	Serafimovski (2000)
Olimpias	Greece	Pb–Zn–Au–Ag	Breccia and volcanic hosted along regional normal fault	Oligocene	Kroll *et al.* (2002)
Kapaonik/Trepca	Serbia	Pb–Zn–Ag	Volcanic hosted	Oligocene	Naftali (2002)
Oligocene–Neogene Innercarpathian–Alpine extrusion–related deposits					
Rosia Montana	Romania	Au	Porphyry Cu	Late Miocene	Alderton & Fallick (2000)
Brad–Sacarimb	Romania	Au–Cu–Te	Porphyry Cu and veins	Late Miocene	Alderton & Fallick (2000)
Baia Mare	Romania	Pb–Zn–Au	Hydrothermal veins	Late Miocene	Lexa (1999)
Beregovo	Ukraina	Au–Ag (–Sb)	Epithermal stockwork and veins, volcanic hosted	Mid-Miocene	Lexa (1999)
Recsk–Lahoca	Hungary	Cu, Au	Porphyry Cu, high-sulphidation epithermal Au	Late Eocene	Gatter *et al.* (1999)
Banska Stiavnica	Slovakia	Pb–Zn–Ag	Volcanic hosted, epithermal stockwork and veins, porphyry Cu	Mid-Miocene	Lexa (1999); Lexa *et al.* (1999)
Hohe Tauern	Austria	Au(–Pb–Zn)	Hydrothermal veins	Miocene	Weber (1997)
Brusson	Italy	Au	Mesothermal hydrothermal veins	Oligocene	Pettke *et al.* (2000)

pathians and important Cu–Au–Te deposits in the Apuseni Mountains. These contrast with metallogenically distinct, mesothermal Au mineralization within and on top of metamorphic domes of the Western and Eastern Alps. From this, a subordinate zone may be distinguished extending from the Ivrea area to Recsk, associated with Late Eocene to Oligocene Periadriatic magmatic rocks, mainly intrusives. From these, only Recsk represents a major porphyry Cu mineralization, all other intrusives are remarkably poor in mineralization.

Cretaceous tectonic restoration

Before discussing the Cretaceous structures, a restoration of Tertiary structures is necessary in order to define the Cretaceous palaeogeographic relationships of various tectonic units. The details of Cretaceous and Tertiary deformation are still not fully understood but sufficient data exist to allow a reasonable reconstruction. Various reconstructions have been discussed by, for example, Burchfiel (1980), Csontos et al. (1992), Dercourt et al. (1993), Csontos (1995), Robertson & Karamata (1994), Channell & Kozur (1997), Stampfli & Mosar (1999), Neugebauer et al. (2001) and Willingshofer (2000). None of them appears to have solved all the problems, so that many open questions remain.

Figure 2 shows a reconstruction of the Cretaceous configuration that is mainly based on palaeomagnetic data from Late Cretaceous units as collected by Willingshofer (2000) and Neugebauer et al. (2001). These show that the Late Cretaceous units can be divided into (1) the ALCAPA (Alpine–Carpathian–Pannonian) block comprising the Austroalpine units in the Eastern Alps and Inner Western Carpathians, (2) the Tisia block extending from Zagreb in Croatia to the Apuseni Mountains, (3) the Dacia block which includes the Eastern und Southern Carpathians and the Balkan, (4) the Rhodope block and (5) the South Alpine/ Dinaric block (see Figs 1 and 2). None of these

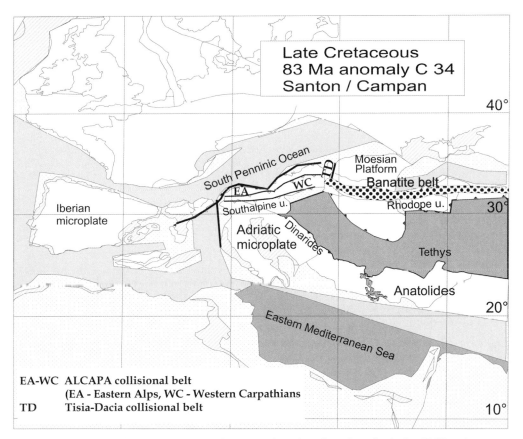

Fig. 2. Reconstruction of the Late Cretaceous palaeogeography and geodynamic setting in the ABCD region (modified after Neugebauer et al. 2001).

behaved as a rigid block during Tertiary times, but they were strongly deformed, mainly around their margins. Additionally, they were also partially rotated during their invasion into the future Alpine–Carpathian realm (for details, see below). The ALCAPA block displays a 50–90° counter-clockwise rotation during the Late Palaeogene–Early Miocene (e.g. Márton 1997 and references therein), whilst the Tisia block records a 50–90° clockwise rotation, mostly during the mid-Miocene (Panaiotu pers. comm.). A simple solution suggests that the ALCAPA and Tisia blocks invaded the Carpathian arc during the Tertiary, the Tisia block pushing at its front the western sectors of the Dacia block. The Moesian platform represents an indenter that is interpreted to have moved westwards during the Late Cretaceous and Palaeogene due to the opening of the West Black Sea oceanic basin. The essential result of this restoration is that the ALCAPA, Tisia and Dacia blocks together formed an east–west-trending, straight orogen during the critical time at c. 80 Ma when most of the so-called banatite magmatism occurred (Figs 1 and 2; see below). This view is also supported by palaeomagnetic data from banatites which call for post-Cretaceous bending and orocline formation of the banatite belt during its invasion into the Carpathian arc (Patrascu et al. 1992). In some interpretations (e.g. Neugebauer et al. 2001; Willingshofer 2000; Schmid et al.1998), the Balkan sector of the Dacia block was connected with the Moesian platform during the Late Cretaceous, and the Balkan block with the Rhodope block. Therefore, it is reasonable to assume that the Cretaceous orogen has separated a northern, South Penninic oceanic tract from the still open Tethys ocean in the south. Furthermore, the reconstruction by Neugebauer et al. (2001; Figure 2) shows an open oceanic tract between the Balkan zone and the Moesian platform that is not well constrained by data. The South Alpine–Dinaric belt was connected with the southern ALCAPA block when an open ocean was closed in a scissor-like manner due to convergence of the Dinarides towards the ALCAPA Tisia–Dacia–Rhodope continent.

In summary, reconstructions indicate open oceanic tracts, both to the north and south of the Late Cretaceous orogenic belt. This belt was attached to the Moesian platform in the east, and to the Adriatic microplate in the west. This leaves open the question as to which geodynamic process occurred within this belt during the Late Cretaceous, e.g. continuous subduction or collision along segments that are attached to continental blocks (Moesia–Europe) in the east, and the Adriatic block in the west. Orogenic polarity of the closure and nappe stacking was, respectively, to the north

and NW, rotating units back to their present-day positions (e.g. Ratschbacher et al. 1989, 1993; Schmid et al. 1998).

Overview of Cretaceous structural evolution

I now briefly discuss the principal Late Jurassic to Cretaceous orogenic structures that are common features for the whole Cretaceous ABCD belt. As has now been well established, between pre-Late Cretaceous collisional structures, mainly ductile thrusts, and Late Cretaceous collapse structures, sedimentary basins can be distinguished. These structures and basins are associated with banatite magmatism in southeastern sectors, together with ductile low-angle normal and high-angle strike-slip faults that relate to the basin formation.

Late Jurassic to early Late Cretaceous compressional structures

The Austroalpine units of the Eastern Alps and Western Carpathians represent a continental, basement–cover nappe pile which received its essential internal nappe structure during Early and early Late Cretaceous orogenic events (e.g. Ratschbacher 1986; Dallmeyer et al. 1996; Plasienka 1997; Neubauer et al. 2000). Collision between Austroalpine units in a footwall position and emplacement of overlying Meliata–Juvavic–Silice units in a hanging-wall position occurred during the early Late Cretaceous (e.g. Dallmeyer et al. 1996; Plasienka 1997; Willingshofer et al. 1999b). Nappe stacking was directed towards the NW and west, and probably migrated from hanging wall to footwall (e.g. Ratschbacher 1986; Neubauer et al. 2000). Upper Austroalpine units formed klippen with very low- to low-grade metamorphic overprint during Cretaceous continent–continent collision. These juxtaposed Middle Austroalpine units of the central sectors of the Eastern Alps with a metamorphic overprint that increases from greenschist facies conditions in the north to eclogite facies metamorphism in the south (e.g. Frey et al. 1999).

At approximately the same time within the Cretaceous, similar ductile nappe structures were formed in the Apuseni Mountains, Southern and Eastern Carpathians. In the Southern Carpathians, the Severin oceanic domain was closed during the early Late Cretaceous (Sandulescu 1984; Zacher & Lupu 1999). In the Apuseni Mountains, a ductile nappe stack formed at c. 120–110 Ma, and the Mures ophiolite was subsequently emplaced onto the underlying basement–cover nappes during the Late Cretaceous (Dallmeyer *et al.* 1996). Each of these present-day isolated mountain groups represents basement–cover nappes that formed during ductile thrusting at very low-grade to low-grade metamorphic conditions. The Southern Carpathians include, from footwall to hanging wall, the Danubian basement–cover nappe stack, the Jurassic Severin ophiolite and the Getic–Supragetic nappe complexes. This suggests the presence of an oceanic tract between the Moesian platform and the Dacian units. The whole nappe stack formed at *c.* 120–80 Ma, whilst post-orogenic collapse started slightly prior to *c.* 80 Ma (Bojar *et al.* 1998; Dallmeyer *et al.* 1996; Willingshofer *et al.* 2001).

The Bulgarian Balkan region is linked by the Serbian Timok zone with the extension of the Danubian nappe complex of the Southern Carpathians (Fig. 1). In Bulgaria, the Zone with specific Late Cretaceous sedimentary/volcanogenic basins is called the Srednogorie zone (Aiello *et al.* 1977; Bocaletti *et al.* 1974, 1978). The Srednogorie zone extends to the Black Sea and there it is also superposed onto the southerly adjacent Strandja zone (with mainly Late Jurassic tectonism; o). The basement of the Srednogrie zone experienced weak early Late Cretaceous deformation at *c.* 102–100 Ma (e.g. Velichkova *et al.* 2001), which is post-dated by volcano-sedimentary basins (see below).

Late Cretaceous, post-orogenic collapse structures

Late Cretaceous collapse basins sealed the ophiolite sutures (e.g. Meliata suture) and previously formed basement–cover nappe structures all over the area (e.g. Neubauer *et al.* 1995; Willingshofer *et al.* 1999*a*; Fig. 1). 'Gosau' refers to a locality in the Eastern Alps where this sort of basin was first described. Gosau-type basins can be traced from the Eastern Alps to the Srednogorie zone and represent a prominent feature of the region. The formation of the Gosau basins in the Eastern Alps was associated with sinistral wrenching along roughly ENE–WNW-trending faults, normal faulting at shallow crustal levels and exhumation of eclogite-bearing crust within Austroalpine units (Neubauer *et al.* 1995, 2000; Froitzheim *et al.* 1997; Willingshofer *et al.* 1999*a, b*). This led to the juxtaposition of eclogite–amphibolite-facies metamorphic rocks of the lower tectonic units with very low-grade to low-grade metamorphic rocks of the upper tectonic units along ductile low-angle normal faults in the Eastern Alps, Western Carpathians, Apuseni Mountains (Willingshofer *et al.* 1999*a*) and the Balkan–Rhodope

Mountains (Burg *et al.* 1990, 1993). Because of the calc-alkaline magmatism, the Srednogorie zone has been interpreted as an intra-arc basin (e.g. Aiello *et al.* 1977) or as a post-collisional rift-type basin (e.g. Popov 1987).

The Bulgarian Balkan region is an extension of the Danubian nappe complex in Southern Carpathians (Popov 1987), via the Timok zone (Serbia). However, it shows weak early Late Cretaceous deformation that is post-dated by volcano-sedimentary basins occurring within the Srednogorie Zone (Popov & Popov 2000). The Maritsa shear zone, which also in part includes sheared Late Cretaceous granites, separates the Srednogorie zone from the southerly adjacent Rhodope massif. In the Rhodope massif, the uppermost unit is a Cretaceous metamorphic unit that formed within Cretaceous amphibolite facies metamorphic conditions (e.g. Ricou *et al.* 1998). The Srednogorie Zone shows subsidence and formation of local sedimentary basins, volcanism and shallow granitoid intrusions. Furthermore, ductile structures suggest that the Maritsa shear zone and splays to the north (e.g. Iskar–Yavonitsa fault zone, Kamenitsa–Rakovitsa, etc.) can be regarded as a Late Cretaceous dextral wrench corridor that was active mainly under greenschist facies metamorphic conditions. Since igneous rocks (e.g. the Vitosha granite) intruded into the Maritsa shear zone were dated at *c.* 80 Ma (von Quadt & Peytcheva pers. comm.), initial exhumation and cooling of the uppermost tectonic units exposed within the Rhodope Mountains should also have an age of *c.* 80 Ma, contemporaneous with the main subsidence in the adjacent Srednogorie Basin.

Overview of Late Cretaceous magmatism and mineralization

Many different terms, such as the Apuseni–Srednogrie magmatic–metallogenic belt, have been proposed for this belt but, for simplicity, I here use 'banatite belt'. The banatites are associated with various types of mineralization including porphyry copper, massive sulphide and replacement ores (Popov 1987, 1996; Berza *et al.* 1998). Banatite magmatic rocks are exposed in an arcuate belt from the Apuseni Mountains to the Black Sea (Figs 1 and 2). In general, calc-alkaline suites largely predominate, although minor alkaline rocks are reported from the Bulgarian Srednogorie Zone (e.g. Bocaletti *et al.* 1974, 1978; Berza *et al.* 1998; von Quadt *et al.* 2001, 2002). In the Apuseni Mts, the western South Carpathians and the Banat (Romania/Serbia), banatites occur as intrusive rocks that are mainly high-level plutons of predominantly granodioritic composition. In Serbia and Bulgaria, andesites and dacites dominate, with subordinate intrusive rocks. Within the Panagyurishte region (Elatsite–Medet in Fig. 1) is a roughly north–south-trending zone with predominantly shallow intrusive bodies across the Srednogorie zone, which are associated with porphyry-Cu and massive sulphide mineralization (e.g. Strashimirov & Popov 2000).

In some areas, volcanic products are intercalated between fossil-bearing Late Cretaceous rocks, as in the Hateg basin of the Southern Carpathians (Willingshofer et al. 2001 and references therein) and in the Srednogorie Zone (Aiello et al. 1977; Popov & Popov 2000). Biostratigraphic data indicate a Cenomanian to Maastrichtian age (Aiello et al. 1977; Popov & Popov 2000; Willingshofer et al. 2001). Correlation of volcanics with nearby plutonic suites has not yet been carried out. It has also to be noted that beyond the present-day banatite belt, Late Cretaceous to Early Palaeogene (sub-)volcanic rocks and granites were recently reported from the Dinarides in Croatia (part of the Tisia block; Pamic et al. 2000) and Western Carpathians (Poller et al. 2001).

Recent geochronological data from banatitic volcanic and plutonic rocks are compiled in Table 2. Additional K–Ar ages of amphibole and biotite presented in some reports (Bleahu et al. 1984; Kräutner et al. 1986; Cioflica et al. 1992, 1994; Downes et al. 1995) generally lack descriptions of rocks and fabrics. These incomplete data make the assessment of their value for estimating the duration and timing of banatite magmatism difficult. In general, the Romanian and Serbian banatites appear to be younger than those in Bulgaria. In Romania, they range between c. 80 and 62 Ma, except for one locality with K–Ar ages of 89–83 Ma (Kräutner et al. 1986; Cioflica et al. 1992, 1994; Downes et al. 1995). The predominant K–Ar ages of dacite and andesite in the Serbian Timok zone are 74–70 Ma (Karamata et al. 1997) but ages of c. 60 Ma are also reported. In Bulgaria, geochronological ages range from c. 92 to 80 Ma (Table 2 with references). Three U–Pb zircon ages from Elatsite range from 92 to 91 Ma (von Quadt et al. 2002; Peycheva et al. 2001). Preliminary $^{40}Ar/^{39}Ar$ amphibole and biotite ages from sub-volcanic rocks of the Panagyurishte region are between c. 92 Ma (Elatsite) and c. 80 Ma (Velichkova et al. 2001; own unpubl. results).

Recently, Poller et al. (2001) reported a late Cretaceous U–Pb zircon age (75.6 ± 1.1 Ma) from the Rochovce granite of the Inner Western Carpathians which may thus represent the northernmost magmatic rock of the banatite suite. If true, these data could be taken together as evidence for an along-strike shift of magmatism from older magmatic activity in the SE to younger magmatism in the present-day north. Exceptions to this trend are a few K–Ar ages from Late Cretaceous granites of the Strandja zone in Turkey which are as young as 83 ± 3 and 78 ± 2 Ma (Okay et al. 2001 and references therein).

Banatite magmatism and mineralization took place contemporaneously with the formation of the Gosau-type collapse basins, and may be interpreted to represent either a product of continuous northward subduction or post-collisional I-type magmatism due to break-off of the subducted lithosphere. Except for the unique Rochovce granite, Cretaceous magmatism has not affected the Alpine to Inner Western Carpathian sectors of the ABCD belt. The lack of magmatism in the Alpine sectors has been an enigma until now. Possible explanations for this fact are: (1) the presence of only a small, subducting ocean (e.g. Evans 1975, but note the presence of an ocean at least 500 km wide: Wortmann et al. 2001) and (2) subduction of mainly water-poor carbonates which are not appropriate to trigger melting in the overlying lithospheric mantle wedge.

Late Cretaceous banatite magmatism is associated with rich and widespread Cu–Au, Fe–Cu, Pb–Zn mineralization, from the Apuseni Mountains in Romania to the Srednogorie Zone in Bulgaria (e.g. Sillitoe 1980; Karamata et al. 1997; Vlad 1997; Berza et al. 1998). These mineral deposits are numerous and include some of the largest known in Europe. The mineralization is highly diverse and includes well known skarn deposits in the Apuseni Mountains and Southern Carpathians (e.g. Baita Bihor, Ocna de Fier), mainly porphyry Cu(–Au) and andesite-hosted massive sulphide deposits in Serbia (Bor, Majdanpek; Kozelj & Jelenkovic 2001) and western Bulgaria (Panagyurishte corridor with Elatsite, Chelopech, Assarel and Medet) and Pb–Zn vein deposits (Burgas district) near the Black Sea (Figs 1 and 3a; Table 1). As a whole, the belt exposes deeper levels (e.g. skarn at the contact with intrusive rocks) in the north and shallower levels in the central southern and eastern sectors. An exception is the extensive subsurface porphyry-type Mo–W mineralization associated with the Rochovce granite of Inner Western Carpathians (Grecula et al. 1995). The age of mineralization postdates magmatism but, except for Elatsite, is generally not well dated in the case of plutonic host rocks. von Quadt et al. (2002) report an age of c. 91 Ma for Cu mineralization from Elatsite. Another exception is the recently reported Re–Os molybdenite age of 76.6 ± 0.3 Ma of Ocna de Fier in the Romanian Southern Carpathians (Ciobanu et al. 2002).

Table 2. *Geochronological data from banatite magmatic rocks*

Pluton	Rock	Method	Age ± error (Ma	Interpretation	Reference
Elatsite (B)	Porphyry dyke	U–Pb zircon	92.1 ± 0.3	Age of intrusion	von Quadt et al. (2002)
Elatsite (B)	'Late' dyke	U–Pb zircon	91.42 ± 0.15	Age of intrusion	von Quadt et al. (2002)
Plutons and sills,	Porphyry dyke, andesite	Ar–Ar	91.72 ± 0.70 to 80.21 ± 0.45	Age of cooling	Velichkova et al. (2001);
Panagyuriste region					Handler et al. (2002)
Vocin (C)	Basalt	K–Ar	72.8 ± 2.1	Age of extrusion	Pamic et al. (2000)
Vocin (C)	Basalt	K–Ar	62.1 ± 1.8	Age of extrusion	Pamic & Pécskay (1994)
Cadavica (C)	Basalt	K–Ar	71.6 ± 3.0, 69.4 ± 2.5	Age of extrusion	Pamic et al. (2000)
Mt Pozeska Gora (C)	Granite	Rb–Sr WR	71.5 ± 2.5	Age of intrusion	Pamic et al. (1988)
Ciclova (R)	Diorite	K–Ar biotite	79–74.3	Cooling below 300 °C	Soriou et al. (1986)
Bocsa (R)	Granodiorite	U–Pb zircon	79.6 ± 2.5	Age of intrusion	Nicolescu et al. (1999)
Ocna de Fier (R)	Granodiorite	U–Pb zircon	75.5 ± 1.6	Age of intrusion	Nicolescu et al. (1999)
Bradisurul de Jos (R)	Granodiorite	K–Ar biotite	67 ± 3	Cooling below 300 °C	Soriou et al. (1986)
Mehadica (R)	Granodiorite porphyry	K–Ar biotite	78 ± 3	Cooling below 300 °C	Soriou et al. (1986)
Moldova Noua (R)	Granodiorite porphyry	K–Ar biotite	69 ± 3	Cooling below 300 °C	Soriou et al. (1986)
Lapusnicul Mare (R)	Granodiorite porphyry	K–Ar biotite	71 ± 3	Cooling below 300 °C	Soriou et al. (1986)
Bolovenii Vechi (R)	Diorite porphyry	K–Ar hornblende	76 ± 4	Cooling below 300 °C	Soriou et al. (1986)
Bania (R)	Granodiorite porphyry	K–Ar biotite	78 ± 3	Cooling below 300 °C	Soriou et al. (1986)
Sopotul Vechi (R)	Quartz monzonite porphyry	K–Ar hornblende	89 ± 4	Cooling below 500 °C	Soriou et al. (1986)
Sopotul Vechi (R)	Granodiorite porphyry	K–Ar biotite	83 ± 3	Cooling below 300 °C	Soriou et al. (1986)
Herepea (R)	Andesite dyke	K–Ar	66.0 ± 0.27	Intrusion	Downes et al. (1995)
Roscani (R)	Andesite dyke	K–Ar	71.7 ± 2.7	Intrusion	Downes et al. (1995)
Rochovce (SK)	Granite	U–Pb zircon	75.6 ± 1.1	Intrusion	Poller et al. (2001)

The list is incomplete. Only ages with good sample descriptions are listed here. B, Bulgaria; C, Croatia; R, Romania; SK, Slovakia.

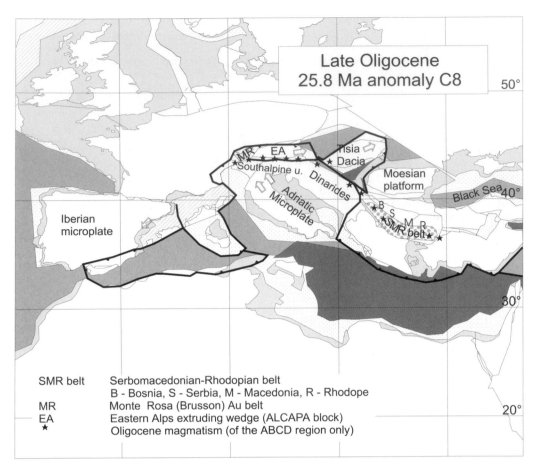

Fig. 3. Reconstruction of the Oligocene–Miocene palaeogeography and geodynamic setting in the ABCD region (modified after Neugebauer *et al.* 2001).

In contrast, in the Eastern Alps and Western Carpathians many deposits of industrial minerals (talc, magnesite), siderite and vein-type Cu–siderite were formed or were remobilized during Cretaceous metamorphism (e.g. Weber *et al.* 1997; Pohl & Belocky 1998; Ebner *et al.* 2000). No links with magmatic processes have been observed. The formation of many collisional type ore deposits in veins (e.g. Cu, As–Au) and shear zones (Au) is related to the syn-collisional extension and exhumation of metamorphic core complexes, which culminated around 80 Ma.

Oligocene–Neogene tectonic restoration and structures

During the Tertiary, two major zones of mineralization are of interest: the Serbomacedonian–

Rhodopian zone with Oligocene to Early Miocene Pb–Zn and porphyry Cu–Au deposits in the southern sectors of the ABCD orogen; and the Alpine–Carpathian belt in the north with numerous Pb–Zn and Au deposits of mainly Miocene age (Fig. 1; Table 1). These are considered to result from rather different geodynamic processes (see below).

During the Palaeogene, the Neotethys oceanic lithosphere continued to subduct northwards beneath the Cretaceous belt (e.g. Barr *et al.* 1999; Neugebauer *et al.* 2001). During the Oligocene, the Adriatic microplate rotated 30° counter-clockwise (Neugebauer *et al.* 2001 and references) and collided with the intra-Alpine microcontinent collage in the northwest, whilst subduction continued in the southeast (Fig. 3). The indentation of the Adriatic microplate resulted in oroclinal bending and formation of the curvature of the Hellenic–

Dinaric belt. A magmatic belt mimics the arcuate belt and comprises earlier, Eocene–Oligocene granitoids that are mostly exposed in northern Greece and Turkish Thracia, and mostly Late Oligocene–Miocene andesitic volcanic rocks in the Macedonian to Serbian sector (Fig. 1). It appears that a northwestward migration of volcanic activity occurred along the strike of the belt, although protolith ages are generally uncertain.

The Penninic ocean had been consumed by slab roll-back during the latest Cretaceous to Palaeogene to form a land-locked oceanic basin (e.g. Wortel & Spakman 2000; Wortmann et al. 2001 and references therein). Collision of the Cretaceous microcontinent collage with the European continent started during the Eocene in West Alpine sectors and migrated along strike to the Carpathians where collision occurred during the Neogene (Figs 1, 3 and 4). The previous southern European continental margin had been subducted beneath the Alps and subsequently exhumed as Penninic continental units. These are exposed within the Tauern window (Figs 1, 4 and 5) that is overridden by the Austroalpine nappe complex. The timing of the overthrusting of the Austroalpine units onto the Penninic units is Eocene (Liu et al. 2001 and references therein). Oblique plate collision and associated stacking of lower plate

continental units were followed by partitioning of convergence into northward thrusting along the northern leading edge of the orogen and emplacement of the entire Alpine nappe edifice onto European foreland units, and orogen-parallel strike-slip motions along wrench corridors due to general NE–SW shortening (e.g. Ratschbacher et al. 1989, 1993). The latter stage was also governed by indentation of the rigid South Alpine indenter, which formed the northern extent of the Adriatic microplate. Furthermore, emplacement of post-collisional Late Oligocene to Early Miocene intermediate and mafic, calc-alkaline plutons and dykes along southern sectors of the Eastern Alps has been interpreted as a result of the break-off of the subducted continental lithospheric slab following continent–continent collision (von Blanckenburg & Davies 1995; von Blanckenburg et al. 1998). Except for Recsk in the easternmost extension, these magmatic rocks are largely barren of mineralization. Lower plate units are exposed within the Lepontin and Tauern metamorphic core complexes, which have been metamorphosed to amphibolite facies grade through burial during collision (see Frey et al. 1999 for a review). Interlinkage of sub-horizontal shortening with strike-slip movements led to the updoming of the Tauern metamorphic core complex along an overstep of

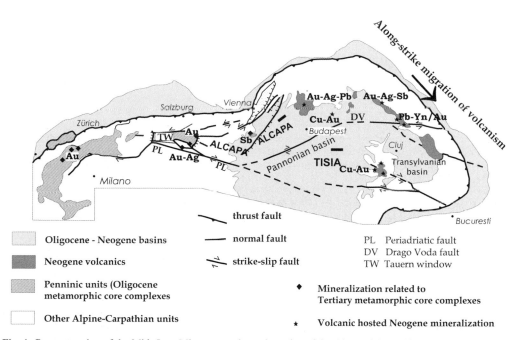

Fig. 4. Reconstruction of the Mid–Late Miocene geodynamic setting of the Alps and Carpathians with indentation, collision and extrusional processes in the west and collision and slab break-off processes in the Carpathian realm (tectonic model based on Linzer 1996 and Nemcok et al. 1998). Note that structures are shown in their present-day distribution.

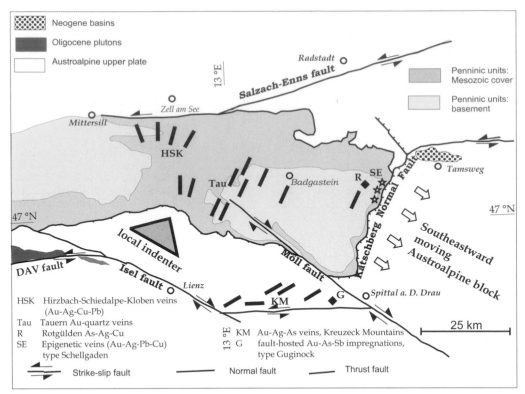

Fig. 5. Mineralization in the central Eastern Alps related to the exhumation of the Tauern metamorphic core complex in an oblique convergent system.

an orogen-parallel shear zone and the formation of low angle normal faults at the structural top of the window (e.g. Neubauer *et al.* 1999 and references therein). The Tauern window is therefore an antiformal window that is laterally confined by shear zones and detachment surfaces (Neubauer *et al.* 1999). Juxtaposition of hot rocks within the window and cold Austroalpine rocks outside the window led to a high geothermal gradient across its margins during the Cenozoic (Genser *et al.* 1996).

A combination of slab roll-back of the remnant intra-Carpathian ocean and eastward extrusion of the Alps led to the closure of a remnant oceanic basin in the Carpathian arc (e.g. Ratschbacher *et al.* 1991, 1993; Linzer 1996; Linzer *et al.* 1997; Nemcok *et al.* 1998; Fig. 3). Collision with the European foreland migrated from the Alps eastwards around the Carpathian arc (e.g. Linzer 1996). The extruding blocks were affected by variable palaeostress fields within which the maximum principal stress was mostly sub-horizontal and largely sub-parallel to the eastward and northeastward movement direction of the extruding blocks (e.g. Peresson & Decker 1997; Nemcok *et*

al. 1998; Fodor *et al.* 1999). Note that the ALCAPA block rotated *c.* 50–80° counter-clockwise and the Tisia–Dacia block *c.* 90° clockwise (e.g. Márton 1997; Patrascu *et al.* 1994; Panaiotu pers. comm.). Starting in the Mid–Late Miocene, back-arc extension affected both the Aegean region and the Pannonian basin (Gauthier *et al.* 1999; Nemcok *et al.* 1998). In the Pannonian basin, back-arc extension is associated with alkaline magmatism.

Oligocene and Miocene mineralization of the Serbomacedonian–Rhodopian belt

The Serbomacedonian–Rhodopian belt is characterized by mostly andesitic to dacitic volcanic sequences, which range in age from *c.* 35 to 19 Ma in the southeastern sectors (e.g. Frei 1995; Harkovska *et al.* 1998; Yannev *et al.* 1998; Marchev *et al.* 1998). In Turkish Thrace and Macedonia, subordinate Pliocene and even Quaternary (Thrace) volcanism also occurs (Yilmaz & Polat 1998; Harkovska *et al.* 1998). The magmatism is largely calc-alkaline, acidic, and in part highly potassic and is interpreted to represent collisional type magmatism formed partly by melting of continental crust. Oligocene and Miocene mineralization is prominent in the Dinarides and Hellenides and is mainly closely related to magmatism. Following previous work (Mitchell 1996; Serafimovski 2000), we distinguish two sub-belts with different types of deposits (Fig. 1, Table 1):

1   porphyry Cu–Mo–Au and subordinate epithermal Au deposits which are more common in the southeastern sector;
2   Pb–Zn(–Ag) breccia- and vein deposits in volcanic rocks, hydrothermal veins and fault-related breccias, the latter connected with metamorphic core complexes.

The first group comprises deposits such as Skouries and Maronia in Greece and Spahievo in Bulgaria (Serafimovski 2000; Singer & Marchev 2000; Kroll et al. 2002; Melfos et al. 2002). The age of mineralization is well constrained in only a few cases (Borovitsa–Spahievo c. 30 Ma, Singer & Marchev 2000; Skouries c. 19 Ma, Frei 1995), although geological relationships indicate in general that ages are Oligocene and Miocene.

The Pb–Zn deposits occur mainly in two settings: (1) hydrothermal veins like Madan in Bulgaria (Dimov et al. 2000; Ovtcharova et al. 2001) and in fault-related breccias at the upper margin of a metamorphic core complex as in Olimpias in Greece and, more commonly, (2) in various settings closely related to andesitic volcanism stretching from Macedonia to Bosnia (Jankovic 1997; Mitchell 1996; Serafimovski 2000). The latter type includes major ore districts such as Trepca in Kosovo, which is the premier Pb–Zn producer of the region.

Late Eocene to Miocene mineralization in the Alps and the Inner Carpathians

Many sites with late-stage hydrothermal, generally Au- and Ag-bearing, polymetallic ores have been mined in the central Eastern (Weber et al. 1997 and references therein) and Western Alps (Pettke et al. 1999, 2000 and references therein). A well studied example of orogenic mesothermal vein-type Au mineralization is the Brusson–Crodo area in the Western Alps (Figs 1 and 4). There, mesothermal Au-quartz formed in the hanging wall of the Simplon detachment fault between 31.6 and 24.5 Ma, and within the Lepontin gneiss dome at c. 10.6 Ma (Pettke et al. 1999). Ore formation post-dates metamorphism of country rocks and is related to devolatilization of calc-schists according to a model involving a scenario

of deep-later metamorphism and metamorphogenic gold mineralization proposed by Stüwe (1998). Formation temperatures of ores are between 240 and 450 °C. Similarly, several types of ore deposit and mineral occurrences are known that closely link to structures formed during exhumation of the Tauern metamorphic core complex and oblique shortening in the Eastern Alps. These include (see Fig. 5):

1   Sub-vertical NNW–SSE-trending Au–Ag–Cu–Pb–quartz–carbonate veins (type Hirzbach–Schiedalpe–Kloben, following Weber et al. 1997) in the central Tauern window;
2   sub-vertical earlier NE–SW-, later NNE–SSW-trending Au–quartz veins at the northwestern and southern terminations of the Möll valley fault (Feitzinger & Paar 1991);
3   saddle reefs and replacement ores in marbles (type Rotgülden; Horner et al. 1997);
4   Au–Ag–Pb–Cu–W veins and quartz lenses (type Schellgaden) close to the upper margins of the Tauern window (Amann et al. 1997);
5   ENE-trending Au–Ag–As veins in the Kreuzeck Mountains, which are interlinked with intermediate sub-volcanic rocks (Feitzinger et al. 1995);
6   Au–As–Sb impregnations along east-trending sinistral fault zones, type Guginock (Paar in Weber et al. 1997; Amann et al. 2002).

Further examples of Tertiary mineralization are described by Pohl & Belocky (1998) and Prochaska et al. (1995). All these ore bodies were formed after peak temperature conditions of Tertiary metamorphism and cross-cut regional structures like foliation, except for most of the Schellgaden-type. The Schellgaden-type ore bodies are part of a ductile low-angle normal fault close to the upper margins of a metamorphic core complex. The Hirzbach–Schiedalpe–Kloben lodes occur close to the northern margin of the Alps where internal structures of the Tauern window form a bend with a c. 30° opening angle. The Au–quartz lodes are widely distributed over eastern–central sectors of the Möll valley fault, which formed at c. 20 Ma (Kurz & Neubauer 1996). The lodes formed as extension gashes which show an apparent rotation from NE to NNE-trends during progressive development (Kurz & Neubauer 1996). Handler & Neubauer (2002) argued that Au-mineralization is c. 20–18 Ma by dating vein minerals such as sericite and adularia. The Rotgülden ores are confined to the upper portions of some subordinate Au–quartz lodes where these reach cover marbles on top of basement gneisses. The sites of the Rotgülden ores are, therefore, controlled by the chemical environment. The

Schellgaden ore bodies are in part epigenetic and intimately related to extension like the Au–quartz lodes. Furthermore, these quartzitic ore bodies are involved in ductile low-angle normal faults that developed close to the upper margins of the basement gneisses during exhumation of previously buried crustal units (Amann *et al.* 1997).

The lodes of the Kreuzeck Mountains post-date the intrusion of tonalitic to dacitic calc-alkaline dykes which were dated between 40 and 26 Ma with the K–Ar method (Deutsch 1984) although mineralization is largely unrelated to magmatism. These lode-type Au–Ag ore bodies follow a *c.* 50 km long east–west-trending zone along the southern margins of the Kreuzeck Mountains. The east–west-trending zone containing lodes terminates in a semi-brittle sinistral shear/fault zone where many old gold mines are known. Detailed investigations on one major site (Guginock) reveal that ore occurs as impregnations with considerable Au abundances within the fault zone (Amann *et al.* 2002). The metals of these ore bodies include Pb, Zn, Cu, Au, Ag in variable proportions. Fluid inclusion studies display maximum homogenization temperatures of *c.* 400–200 °C (Mali 1996; Pohl & Belocky 1994, 1998; Horner *et al.* 1997). Lead isotopic data constrain variable, distinct, mostly continental sources, and no inter-linkage between various hydrothermal systems existed (Köppel in Weber *et al.* 1997; Horner *et al.* 1997).

The timing of ore deposition within the Tauern window is controlled by post-metamorphic cooling and exhumation through the temperature interval of *c.* 400–200 °C, as indicated by fluid inclusions (Pohl & Belocky 1998). Further constraints are given by dating of dykes in the Kreuzeck Mountains where biotites from dykes linked with ores were dated at *c.* 20 Ma by the K–Ar method (Deutsch 1984). Handler & Neubauer (2002) reported [40]Ar/[39]Ar ages on adularia and sericite at *c.*20–16 Ma. These minerals post-date ore formation. I propose, therefore, an age of *c.* 20–18 Ma for the ore-forming event.

The Schlaining deposit includes stibnite veins within the Rechnitz metamorphic core complex (Fig. 4). Interestingly, the major veins are oriented east–west, parallel to the proposed motion direction of the extruding ALCAPA block.

In the Western Carpathians, several types of Neogene ore deposits are widespread and are intimately related to orogenic, subduction-related volcanism (Vityk *et al.* 1994; Dill 1998; Gatter *et al.* 1999; Lexa 1999; Lexa *et al.* 1999; Molnár *et al.* 1999). Initial andesitic volcanism is related to subduction of an ocean (outer flysch basin) during the Early Miocene, and final silicic volcanism is associated with back-arc extension and possibly slab detachment. Nemcok *et al.* (2000) reported a

regional control on ore veins. Major ore veins formed parallel to the movement direction during northeastward extrusion and to NW–SE extension. Ore deposits are variable and include porphyry Cu along with high-sulphidation epithermal Au, intrusion-related base metal and Au mineralization as well as epithermal base metal and Au–Ag(–Sb) veins (e.g. Gatter *et al.* 1999, Lexa 1999).

In the Eastern Carpathians and Apuseni Mountains, two major ore provinces were formed. These are the Baia Mare ore province and the Brad ore province in the Apuseni Mountains, and both are related to Neogene volcanics and regional faults. The Baia Mare ore province is associated with the Drago Voda fault, which represents the northern, sinistral, confining wrench corridor of the eastward extruding Tisia–Dacia block. Grancea *et al.* (2002) infer a pluton at depth along the fault. In detail, the area comprises a number of structures parallel to the main fault, associated with cross faults. The principal ore type is epithermal Pb–Au–Ag veins deposits (e.g. Kovacs 2001; Grancea *et al.* 2002; Marcoux *et al.* 2002). [40]Ar/[39]Ar dating of adularia and sericite proves a Neogene age for the magmatism and a younger age (10–11 Ma), by *c.* 1.5 Ma, of ore emplacement (Lang *et al.* 1994). The Brad ore province in the southern Apuseni Mountains is related to the emplacement of Neogene volcanic rocks mainly between 14.9 and 9 Ma (Pécskay *et al.* 1995; Rosu *et al.* 1997). The ore deposits are of porphyry Cu–type (Alderton *et al.* 1998; Alderton & Fallick 2000) and a related low-sulphidation epithermal type. The Au deposit Rosia Montana is related to silicic domes. Major ore deposits occur along a WNW–ESE-trending dextral strike-slip fault which is considered to represent a secondary fault zone that formed within the eastward moving Tisia–Dacia block during the Neogene (Linzer *et al.* 1997). Drew & Berger (2001) showed that magmatism and mineralization are related to a major transfer zone between subbasins within the Neogene Pannonian–Transylvanian basin system. My own unpublished structural data indicate that mineralized extensional veins trend roughly east–west, again parallel to the direction of motion of the extruding Tisia–Dacia block. The ores are unusually rich in gold. Ore mineralogical, isotopic and fluid inclusion data constrain their formation as epithermal mineralization (Alderton *et al.* 1998).

Discussion

The data presented above mainly demonstrate two facts:

1   the Late Cretaceous and Late Palaeogene–Miocene mineralization in the arcuate ABCD

mountain belt is a discontinuous process that mostly relates to particular stages of syn- and post-collisional tectonic events;

2 the type of mineralization varies strongly along strike within both the Late Cretaceous and the Oligocene–Neogene orogenic belts.

Mineralization is often related to magmatism, which calls into question the origin of the magmatism. With respect to collisional coupling of continental upper and lower plate crust, three basic situations regarding the generation of magmatism can be distinguished in the ABCD belt: (1) pre-collisional subduction-related magmatism, (2) syn-collisional slab break-off magmatism, and (3) syn-/post-collisional back arc extension. Interestingly, longer-lasting subduction-related magmatism is strikingly absent in the whole belt, although subduction of oceanic crust started in the Jurassic and continued until the Late Eocene–Oligocene. The nature of 'banatite' magmatism is uncertain, but the Late Eocene to Early Miocene magmatism along the Peridriatic fault of the Alps is interpreted to represent syn-collisional slab break-off magmatism (von Blanckenburg et al. 1998). The initial stages of Neogene volcanism in the Western and Eastern Carpathians are often interpreted as subduction-related magmatism, although collisional coupling had already occurred. Interpretation as slab break-off magmatism (Linzer 1996) seems more appropriate because the slab tear is still visible in mantle tomography (Wortel & Spakman 2000 and references therein).

For the Late Cretaceous metallogenic belt, the controls on ore deposit type and distribution by the ongoing geodynamic process are clear. In the Alpine–Inner Carpathian area, mineralization is clearly related to collision and post-collisional extensional collapse because the timing of mineralization post-dates collisional coupling of the foreland with the overriding Alpine nappe edifice. Here, early Late Cretaceous collision led to reorganization of, and changes in, plate motion directions and subduction continued with the southeastward subduction of the Penninic oceanic lithosphere. This was not possible in the east where the Cretaceous collisional belt was attached to the Moesian platform. Consequently, slab break-off appears to be a reasonable mechanism to explain the specific structural relationships as these are hosted in a post-collisional setting. The westward propagation along strike of both magmatism and mineralization supports westward propagation of the tear of the subducted slab (Fig. 6).

The Oligocene–Neogene Serbomacedonian–Rhodope metallogenic belt is also a distinct spatial and temporal belt of magmatism and mineralization which mimics the trace of the present-day Hellenic subduction zone. The magmatism and mineralization are both closely coincident with the timing and location of extension, post-dating subduction and subsequent collision of units forming units now exposed between Rhodope and Aegean region (e.g. Gauthier et al. 1999). Thus the specific type of collisional magmatism and mineralization can also be explained by slab break-off, which probably migrated from northwest to southeast as magmatism is apparantly younger in that direction. The present-day mantle lithosphere may still record this process, as suggested by seismic tomography (de Boorder et al. 1998; Wortel & Spakman 2000), where the northward dipping Mediterranean oceanic lithosphere is broken and a thermal dome is visible, underlying the Serbomacedonian–Rhodope zone. A similar process probably occurred in the Alps, as Oligocene intrusions of tonalite are explained by the same mechanism (von Blanckenburg & Davies 1995; von Blanckenburg et al. 1998), but by break-off of the southward-subducted Penninic oceanic lithosphere. This interpretation is supported by the probable westward migration of magmatism in the Serbomacedonian–Rhodopian belt which mimics the westward migration of collision and subsequent tear of the subducted slab.

Based on modelling and the tomographic work of Wortel & Spakman (2000), a generalized model is essentially based on the work of de Boorder et al. (1998) and Wortel & Spakman (2000 and references therein), for along-strike migration of magmatism and mineralization due to slab tear migration as shown in Figure 6. Van de Zedde & Wortel (2001) showed recently that slab break-off could result in an increase of transient temperature of as much as c. 500 °C and affect lithospheric levels as shallow as 35 km (see also Fig. 6). The Late Oligocene to Neogene mineralization in the Inner Carpathians and Alps can be explained in a similar way, but it requires a number of complexities. These involve lateral, orogen-parallel extrusion due to oblique convergence, collision and slab break-off processes at the front of the extrusional wedges (e.g. Linzer 1996).

The data presented above from the Neogene mineralization in and around the Tauern Window show that all the ore deposits were formed during cooling and exhumation of the Tauern metamorphic core complex, which represented an initially subducted, cool piece of mixed oceanic and continental crust. The occurrence of mostly subvertical tension gashes argues for a dominant strike-slip regime during formation. The data indicate that lodes were formed during tensional failure in the course of exhumation. Tensional failure clearly led to open pathways for both

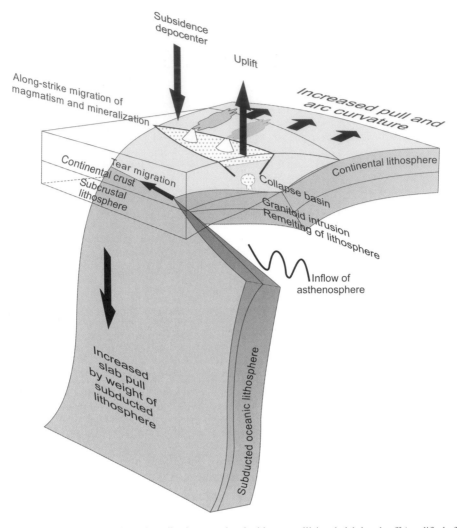

Fig. 6. Generalized model displaying mineralization associated with post-collisional slab break-off (modified after Wortel & Spakman 2000), reproduced with permission from *Science*, **290** (2000), 1910, fig. 4. Copyright 2000 American Association for the Advancement of Science.

uprising metamorphic and descending meteoric fluids within the still hot crust. Furthermore, seismic pumping of metamorphogenic fluids may have contributed to development of brittle fault systems (Sibson & Scott 1998; Pohl & Belocky 1998). The sources of metals were inferred in some cases to have their origin in continental basement rocks. In some cases, lateral secretion may have largely contributed to metal contents. In other cases, such as the As-rich Rotgülden ore deposits, the source of As is unidentified. It is assumed to have its origin in subducted black shales (Horner *et al.* 1997). As a model for the formation of these ore bodies, we propose that

they formed by initial brittle failure during exhumation of the Tauern metamorphic core complex. A pre-requisite of formation is the oblique convergence that occurred within a sinistral wrench corridor (Neubauer *et al.* 1999). All young ore bodies along the southern margins of the Kreuzeck and Goldeck mountains outside of the Tauern window may have formed during this stage. This is indicated by the orientation of palaeostress fields deduced from structural data from ore systems. Formation of sub-vertical ore-bearing tension gashes occurred during updoming of the central portions with a significant contribution from indentation of the southern central sectors of the Tauern window, as Kurz & Neubauer (1996) already noted for that area. A hydrothermal fluid system was established where waters originated from various sources, both metamorphogenic and meteoric, as indicated by isotopic data (Prochaska 1993). The Hirzalpe–Schiedalpe–Kloben type was formed during indentation and rotation of the eastern and western sectors of the Tauern window. Thus, these lodes result from tensional stresses along the northern margins of the Tauern window.

Conclusions

The data presented above allow following general conclusions to be drawn for late-stage, orogenic mineralization within the ABCD belt.

(1) Orogenic mineralization is limited to mainly two periods, Late Cretaceous and Late Palaeogene–Miocene, and thus represents a discontinuous process, which mainly relates to particular stages of syn-collisional tectonic events.

(2) The type of mineralization varies strongly along strike within both the Late Cretaceous and the Oligocene–Neogene orogenic belts, with mainly magmatic-hosted mineralization in the Carpathian, Dinaride and Balkan–Rhodopian sectors, and mainly metamorphogenic mineralization in the Alps.

(3) In the ABCD region, the principal mineralization is syn- or immediately post-collisional with respect to the country rocks.

(4) The distinction of subduction-related magmatism and slab break-off magmatism is often ambiguous. Calc-alkaline magmatism is post-collisional relative to the country rocks, arguing in favour of a genesis by slab tear.

(5) Slab break-off is a principal tectonic process which is able to explain the post-collisional origin of magmatism and mineralization in the eastern sectors of the ABCD belt, both during the Late Cretaceous, Oligocene–Neogene and the Quaternary. A strong argument is the lateral, along-strike migration of magmatism and its syn- to post-collisional nature relative to collisional coupling of upper and lower continental plates.

(6) The locations of many Oligocene and Neogene ore deposits of the Alps and Inner Carpathians are controlled by the presence of large-scale faults due to the indentation process. Many ore veins are parallel to the principal motion direction of intruding and extruding blocks.

The author gratefully acknowledges discussions with many people, including Gerhard Amann, Tudor Berza, Thomas Driesner, Fritz Ebner, Robert Handler, Chris Heinrich, Walter Kurz, Albrecht von Quadt and Ernst Willingshofer, in the course of this study. He especially wants to thank Hugo de Boorder and Jaroslav Lexa who helped much to clarify ideas and their presentation. Derek Blundell gave some further advice and polished the English. This is also highly appreciated. Work has been carried in the framework of the GEODE programme of the European Science Foundation.

References

AIELLO, E., BARTOLINI, C., BOCALETTI, M., GOCEV, P., KARAGJULEVA, J., KOSTADINOV, V. & MANETTI, P. 1977. Sedimentary features of the Srednogorie zone (Bulgaria), an Upper Cretaceous intra-arc basin. *Sedimentary Geology*, **19**, 39–68.

ALDERTON, D.H.M., THIRLWALL, M.F. & BAKER, J.A. 1998. Hydrothermal alteration associated with gold Mineralization in the southern Apuseni Mountains-Romania: preliminary Sr isotopic data. *Mineralium Deposita*, **33**, 520–523.

ALDERTON, D.M.H. & FALLICK, A.E. 2000. The nature and genesis of gold-silver-tellurium mineralization in the Metaliferi Mountains of Western Romania. *Economic Geology*, **95**, 495–516.

AMANN, G., DAXNER, G., NEUBAUER, F., PAAR, W.H. & STEYRER, H.P. 1997. Structural evolution of the Schellgaden Gold-District, eastern Tauern window, Austria – a preliminary report. *Zentralblatt für Geologie und Paläontologie*, **I**, 215–228.

AMANN, G., PAAR, W.H., NEUBAUER, F. & DAXNER, G. 2002. Auriferous arsenopyrite-pyrite and stibnite mineralization from the Sifnitz-Guginock area (Austria): indications for hydrothermal activity duringe Tertiary oblique terrane accretaion in the Eastern Alps. *In:* BLUNDELL, D.J., NEUBAUER, F. & VON QUADT, A. (eds) *The Timing and Location of Major Ore Deposits in an Evolving Orogen*. Geological Society, London, Special Publications, **204**, 103–107.

BARR, S.R., TEMPERLEY, S. & TARNEY, J. 1999. Lateral growth of the continental crust trough deep level subduction-accretion: arc-evaluation of central Greek Rhodope. *Lithos*, **46**, 69–94.

BEAUDOIN, J., TAYLOR, B.E. & SANGSTER, D.F. 1991. Silver-lead-zinc veins, metamorphic core complexes and hydrologic regimes during crustal extension. *Geology*, **19**, 1217–1220.

BERZA, T., CONSTANTINESCU, E. & VLAD, S.N. 1998. Upper Cretaceous Magmatic Series and Associated mineralisation in the Carpathian-Balkan Orogen. *Resource Geology*, **48**, 291–306.

BLEAHU, M., SOROIU, M. & CATALINA, R. 1984. On the Cretaceous tecionic–magmatic evolution of the Apuseni Mountains as revealed by the K–Ar dating. *Revue Roumaine Physique*, **29**, 367–377.

BLUNDELL, D.J. 2002. The timing and location of major ore deosits in an evolving orogen: the geodynamic context. *In:* BLUNDELL, D.J., NEUBAUER, F. & VON QUADT, A. (eds) *The Timing and Location of Major Ore Deposits in an Evolving Orogen*. Geological Society, London, Special Publications, **204**, 1–12.

BOCALETTI, M., MANETTI, P. & PECCERILLO, A. 1974. The Balkanides as an instance of Back-Arc Thrust Belt: Possible relation with the Hellenides. *Geological Society of America Bulletin*, **85**, 1077–1084.

BOCALETTI, M., MANETTI, P., PECCERILLO, A. & STANISHEVA-VASILIEVA, G. 1978. The Late Cretaceous high-potassium volcanism in eastern Srednogoriem, Bulgaria. *Geological Society of America Bulletin*, **89**, 439–447.

BOJAR, A.V., NEUBAUER, F. & FRITZ, H. 1998. Cretaceous to Cenozoic thermal evolution of the southwestern South Carpathians: evidence from fission-track themochronology. *Tectonophysics*, **297**, 229–249.

BURCHFIEL, B.C. 1980. Eastern European alpine system and the Carpathian orocline as an example of collision tectonics. *Tectonophysics*, **63**, 31–61.

BURG, J., RICOU, L., IVANOV, Z., GODFRIAUX, I., DIMOV, D. & KLAIN, L. 1993. Crustal-scale thrust complex in the Rhodope massif. Structure and kinematics. *Bulletin of the Geological Society of Greece*, **28**, 71–85.

BURG, J.P., IVANOV, Z., RICOU, L.E., DIMOR, D. & KLAIN, L. 1990. Implication of shear-sense criteria for the tectonic evolution of the Central Rhodope massif, southern Bulgaria. *Geology*, **18**, 451–454.

CHANNELL, J.E.T. & KOZUR, H. 1997. How many oceans? Meliata, Vardar, and Pindos oceans in Mesozoic Alpine paleogeography. *Geology*, **25**, 183–186.

CIOBANU, C.L., COOK, N.G. & STEIN, H. 2002. Regional setting and Re-Os age of ores at Ocna de Fier Dognecea (Romania) in the context of the banatitic magmatic and metallogenic belt. *Mineralium Deposita*, **37**, 541–567.

CIOFLICA, O., PÉCSKAY, Z., JUDE, R. & LUPULESCU, M. 1994. K-Ar ages of Alpine granitoids in the Hauzesti Drinova area (Poiana Ruscati Mountains, Romania). *Revue Roumaine de Géologie*, **8**, 3–8.

CIOFLICA, G., JUDE, R. & LUPULESCU, M. 1992. Evidence of porphyry copper mineralizations and additional remobilisation phenomena in the Lilieci-Purcariu area (Banat, Romania). *Revue Roumaine de Géologie*, **36**, 3–14.

CSONTOS, L. 1995. Tertiary tectonic evolution of the Intra-Carpathian area: a review. *Acta Vulcanologica*, **7**, 1–13.

CSONTOS, L., NAGYMAROSHI, A., HORVATH, F. & KOVAC, M. 1992. Tertiary evolution of the Intra-Carpathian arc: a model. *Tectonophysics*, **208**, 221–241.

DALLMEYER, R.D., NEUBAUER, F., HANDLER, R., FRITZ, H., MÜLLER, W., PANA, D. & PUTIS, M. 1996. Tectonothermal evolution of the internal Alps and Carpathians: Evidence from $^{40}Ar/^{39}Ar$ mineral and whole rock data. *Eclogae geologicae Helvetiae*, **89**, 203–227.

DE BOORDER, H., SPAKMAN, W., WHITE, S.H. & WORTEL, M.J.R. 1998. Late Cenozoic mineralization, orogenic collapse and s detachment in the European Alpine Belt. *Earth and Planetary Science Letters*, **164**, 569–575.

DERCOURT, J., RICOU, L.E. & VRIELINCK, B. 1993. *Atlas Tethys, paleoenvironmental maps*. Gauthier- Vilars, Gauthier- Vilars, Paris.

DEUTSCH, A. 1984. Young Alpine dykes south of the Tauern Window (Austria): A K-Ar and Sr isotopice study. *Contributions to Mineralogy and Petrology*, **85**, 45–57.

DILL, H.G. 1998. Evolution of Sb mineralization in modern fold belts: a comparison of the Sb mineralization in the Central Andes (Bolivia) and the Western Carpathians (Slovakia). *Mineralium Deposita*, **33**, 359–378.

DIMOV, D., DOBREV, S., IVANOV, Z., KOLKOVSKI, B. & SAROV, S. 2000. Structure. *Alpine Evolution and Mineralizations of the Central Rhodopes Area (South Bulgaria). ABCD-GEODE*, **2000**, 1–50.

DOWNES, H., VASELLI, O., SEGHEDI, I., INGRAM, G., REX, D., CORADOSSI, N., PÉKSAY, Z. & PINARELLI, L. 1995. Geochemistrv of late Cretaceous-early Tertiary magmatism in Poiana Rusca (Romnania). *Acta Vulcanologica*, **7**, 209–217.

DREW, L.J. & BERGER, B.R. 2001. Model of the porphyry copper/polymetallic vein kin-deposit system: Application in the Metaliferi Mountains, Romania. *In*: PIESTRZYŃSKI, L.A. (ed.) *Mineral Deposits at the Beginning of the 21st Century*. Balkema, Lisse, The Netherands, 519–522.

EBNER, F., CERNY, I., EICHHORN, R., GÖTZINGER, M.A., PAAR, W.H., PROCHASKA, W. & WEBER, L. 2000. Minerogeny of the Eastern Alps and adjoining areas. *Mitteilungen der Österreichischen Geologischen Gesellschaft*, **92**, 157–184.

EVANS, M.A. 1975. Mineralization in Geosynclines - the Alpine Enigma. *Mineralium Deposita*, **10**, 254–260.

FEITZINGER, G. & PAAR, W.H. 1991. Gangförmige Gold-Silber-Vererzungen in der Sonnblickgruppe (Hohe Tauern. *Kärnten). Archiv für Lagerstättenforschung an der Geologischen Bundesanstalt (Wien)*, **13**, 17–50.

FEITZINGER, G., PAAR, W.H., TARKIAN, M., RECHE, R., WEINZIERL, O., PROCHASKA, W. & HOLZER, H. 1995. Vein type Ag-(Au)-Pb,Zn, Cu-(W-Sn) mineralization in the Southern Kreuzeck Mountains, Carinthia Province, Austria. *Mineralogy and Petrology*, **53**, 307–332.

FODOR, L., CSONTOS, L., BADA, G., GYÖRFI, I. & BENKOVICS, L. 1999. Tertiary tectonic evolution of the Pannonian Basin system and neighbouring orogens: a new synthesis of palaeostress data. *In*: DURAND, B., JOLIVET, L., HORVÁTH, F. & SERANNE, M. (eds) *The Mediterranean Basins: Tertiary Extension within the Alpine Orogen*. Geological Society, London, Special Publications, **156**, 295–334.

FREI, R. 1995. Evolution of Mineralizing fluid in the porphyry copper system of the Skouries deposit, northeast Chalkidike (Greece): evidence from combined Pb-Sr and stable isotope data. *Economic Geology*, **90**, 746–762.

FREY, M., DESMONS, J. & NEUBAUER, F. 1999. The new metamorphic maps of the Alps: Introduction. *Schweizerische Mineralogische und Petrographische Mitteilungen*, **79**, 1–4.

FROITZHEIM, N., CONTI, P. & VAN DAALEN, M. 1997. Late Cretaceous, synorogenic, low-angle normal faulting along the Schlinig fault (Switzerland, Italy, Austria) and its significance for the tectonics of the Eastern Alps. *Tectonophysics*, **280**, 267–293.

GATTER, I., MOLNÁR, F., FÖLDESSY, J., ZELENKA, T., KISS, J. & SZEBÉNYI, G. 1999. High- and Low-Sulfidation Epithermal Mineralization of the Mátra

Mountains, Northeast Hungary. *In:* MOLNÁR, F., LEXA, J. & HEDENQUIST, J.W. (eds) *Epithermal mineralization of the Western Carpathians.* Society of Economic Geology Guidebook Series, **31**, 155–179.

GAUTHIER, P., BRUN, J.P., MORICEAU, R., SOKOUTIS, D., MARTINOD, J. & JOLIVET, L. 1999. Timing, kinematics and cause of Aegean extension: a scenario based on a comparison with simple analogue experiments. *Tectonophysics*, **315**, 31–72.

GENSER, J., VAN WEES, J.D., CLOETINGH, S. & NEUBAUER, F. 1996. Eastern Alpine tectono-metamorphic evolution: constraints from two-dimensional P–T–t modelling. *Tectonics*, **15**, 584–604.

GRANCEA, L., BAILLY, L., LEROY, J.L., BANKS, D., MARCOUX, E., MILESI, J.P., CUNEY, M., ANDRÉ, A.S., ISTVAN, D. & FABRE, C. 2002. Fluid evolution in the Baia Mare epithermal gold/polymetallic district, Inner Carpathians, Romania. *Mineralium Deposita*, **37**, 630–647.

GRECULA, P., APONYI, A. & APONYIOVA, M. 1995. *Mineral deposits of the Slovak Ore Mountains.* Geocomplex, Bratislava, **1**.

HAEUSSLER, P.J., BRADLEY, D., GOLDFARB, R., SNEE, L. & TAYLOR, C. 1995. Link between ridge subduction and gold mineralization in southern Alaska. *Geology*, **23**, 995–998.

HANDLER, R. & NEUBAUER, F. 2002. Formation of veins in the Tauern window related to continental escape in the Eastern Alps: Constraints from $^{40}Ar/^{39}Ar$ dating of white mica and adularia. *Geologische-Paläontologische Mitteleilungen Innsbruck*, **25**, 111.

HANDLER, R., VELICHKOVA, S.H., NEUBAUER, F. & IVANOV, Z. 2002. Late Cretaceous magmatic and tectonic processes in the Srednogorie zone, Bulgaria: constraints from $^{40}Ar/^{39}Ar$ age dating results. *GEODE ABCD Workshop, Sofia*, **2002**, 7.

HARKOVSKA, A., MARCHEV, P. & PECSKAY, Z. 1998. Paleogene magmatism in the Central Rhodope Area, Bulgaria - A review and new data. *Acta Vulcanologica*, **10**, 199–216.

HEDENQUIST, J.W. & LOWESTERN, J.B. 1994. The role of magmas in the formation of hydrothermal ore deposits. *Nature*, **370**, 519–527.

HEINRICH, C.A. & NEUBAUER, F. 2002. Cu-Au(-Pb-Zn-Ag) metallogeny of the Alpine-Balkan-Carpathian-Dinaride geodynamic province. *Mineralium Deposita*, **37**, 533–540.

HEINRICH, C.A., GÜNTHER, D., AUDÉDAT, A., ULRICH, T. & FRISCHKNECHT, R. 1999. Metal fractionation between magmatic brine and vapor, determined by microanalysis of fluid inclusions. *Geology*, **27**, 755–758.

HORNER, J., NEUBAUER, F., PAAR, W.H., HANSMANN, W., KOEPPEL, V. & ROBL, K. 1997. Structure, mineralogy, and Pb isotopic composition of the As-Au-Ag deposit Rotgülden, Eastern Alps (Austria): significance for formation of epigenetic ore deposits within metamorphic domes. *Mineralium Deposita*, **32**, 555–568.

JANKOVIC, S. 1997. The Carpatho-Balkanides and adjacent area: a sector of the Tethyan Eurasian metallo-genic belt. *Mineralium Deposita*, **32**, 426–433.

KARAMATA, S., KNEZEVIC, V., PECSKAY, Z. & DJORDJE-VIC, M. 1997. Magmatism and metal of the Ridanj-Krepoljin belt (eastern Serbia) and their correlation with northern and eastern analogues. *Mineralium Deposita*, **32**, 452–458.

KOUZMANOV, K. 2002. Morphology, origin and infared microthermomentry of fluid inclusions in pyrite from the Radka epithermal copper deposit, Srednogorie zone, Bulgaria. *Mineralium Deposita*, **37**, 599–613.

KOVACS, M. 2001. Subduction-related magmatism and associated metallogeny in Baia Mare Region (Romania). *Romanian Journal of Mineral Deposits*, **79**, 3–7.

KOZELJ, D.I. & JELENKOVIC, R.J. 2001. Ore forming environments of epithermal gold mineralization in the Bor metallogenetic zone, Serbia, Yugoslavia. *In:* PIESTRZYŃSKI, L.A. (ed.) *Mineral Deposits at the Beginning of the 21st Century.* Balkema, Lisse, the Netherlands, 535–538.

KRÄUTNER, H.G., VAJDEA, E. & ROMANESCU, O. 1986. K–Ar dating of the banatitic magmatites from the Southern Poiana Rusca Mts. (Rusca Montana sedimentary basin). *Dionyz Stur Institute of Geology and Geophysics*, **70-71/1**, 373–388.

KROLL, T., MÜLLER, D., SEIFERT, T., HERZIG, P.M. & SCHNEIDER, A. 2002. Petrology and geochemsitry of the shoshonite-hosted Skouries porphyry Cu-Au deposit, Chalkidike, Greece. *Mineralium Deposita*, **37**, 137–144.

KURZ, W. & NEUBAUER, F. 1996. Strain Partitioning during Formation of the Sonnblick Dome (southeastern Tauern Window, Austria). *Journal of Structural Geology*, **18**, 1327–1343.

LANG, B., EDELSTEIN, O., STEINITZ, G., KOVACS, M. & HALGA, S. 1994 Ar-Ar dating of adularia - a tool in understanding genetic relationships between volcanism and mineralization: Baia Marea Area (Gutai Mountains), northwestern Romania. *Economic Geology*, **89**, 174–180.

LANG, J.R. & BAKER, T. 2001. Intrusion-related gold systems: the present level of understanding. *Mineralium Deposita*, **36**, 477–489.

LEHMANN, B., DIETRICH, A. & WALLIANOS, A. 2000. From rocks to ore. *International Journal of Earth Sciences*, **89**, 284–294.

LEXA, J. 1999. Outline of the Alpine Geology and Metallogeny of the Carpatho-Pannonian Region. *In:* MOLNÁR, F., LEXA, J. & HEDENQUIST, J.W. (eds) *Epithermal mineralization of die Western Carpathians.* Society of Economic Geology Guidebook Series, **31**, 65–108.

LEXA, J., STOHL, J. & KONECNY, V. 1999. The Banská Stiavnica ore district: relationship between metallogenetic processes and the geological evolution of a stratovolcano. *Mineralium Deposita*, **34**, 639–665.

LINZER, H.G. 1996. Kinematics of retreating subduction along the Carpathian arc, Romania. *Geology*, **24**, 167–170.

LINZER, H.G., FRISCH, W., ZWEIGEL, P., GIRBACEA, R., HANN, H.P. & MOSER, F. 1997. Kinematic evolution of the Romanian Carpathians. *Tectonophysics*, **297**, 133–156.

LIU, Y., GENSER, J., HANDLER, R., FRIEDL, G. & NEUBAUER, F. 2001. $^{40}Ar/^{39}Ar$ muscovite ages from the Penninic/Austroalpine plate boundary, Eastern Alps. *Tectonics*, **20**, 526–547.

MALI, H. 1996. Genesis of Hg- and Sb-vein-type-mineralisations in the Austroalpine units of the Eastern Alps. *UNESCO IGCP Project No. 356, Proceedings of the Annual Meeting Sofia*, **1**, 93–98.

MARCHEV, P., DOWNES, H., THIRLWALL, M.F. & MORITZ, R. 2002. Small-scale variations of $^{87}Sr/^{86}Sr$ isotope composition of barite in the Madjarovo low-sulphidation epithermal system, SE Bulgaria: implications for sources of Sr, fluid fluxes and pathways of the ore-forming fluids. *Mineralium Deposita*, **37**, 699–677.

MARCHEV, P., ROGERS, G., CONREY, R., QUICK, J., VASELLI, O. & RAICHEVA, R. 1998. Paleogene orogenic and alkaline basic magmas in the Rhodope zone: relationships, nature of magma sources, and role of crustal contamination. *Acta Vulcanologica*, **10**, 217–232.

MARCOUX, E., GRANCEA, L., LUPULESCU, M. & MILÉSI, J.P. 2002. Lead isotope signatures of epithermal and porphyry-type ore deposits from the Romanian Carpathian Mountains. *Mineralium Deposita*, **37**, 173–184.

MÁRTON, E. 1997. Paleomagnetic aspects of plate tectonics in the Carpatho-Pannonian region. *Mineralium Deposita*, **32**, 441–445.

MELFOS, V., VAVELIDIS, M., CHRISTOFIDES, G. & SEIDEL, E. 2002. Origin and evolution of the Tertiary Porphyry copper-molybdenum deposit in the Maronia area, Thrace, Greece. *Mineralium Deposita*, **37**, 648–668.

MILSOM, J. 2001. Subduction in eastern Indonesia: how many slabs? *Tectonophysics*, **338**, 167–178.

MITCHELL, A.H.G. 1996. Distribution and genesis of some epizonal Zn-Pb and Au provinces in the Carpathian and Balkan region. *Transactions of the Institute of Mining and Metallurgy*, **105, B1**, 127–138.

MOLNÁR, F., ZELENKA, T., MÁTYAS, E., PÉCSKAY, Z., BAJNÓCZI, B., KISS, J. & HORVÁTH, I. 1999. Epithermal Mineralization of the Tokaj Mountains, Northeast Hungary: Shallow Levels of Low-sulfidation Type Systems. *In:* MOLNÁR, F., LEXA, J. & HEDENQUIST, J.W. (eds) *Epithermal mineralization of the Western Carpathians*. Society of Economic Geology Guidebook Series, **31**, 109–153.

NAFTALI, L. 2002. *Pb-Zn deposits of the Carpathian-Balkan Arc in Bulgaria, southeast Serbia and northern Greece*. Irish Geological, Irish Geological Society.

NEMCOK, M., LEXA, O. & KONECNY, P. 2000. Calculations of tectonic, magmatic and residual stress in the Stavnica Straovolcano, West Carpathians: implications for mineral precipitation paths. *Geologica Carpathica*, **51**, 19–36.

NEMCOK, M., POSPISIL, L., LEXA, J. & DONELICK, R.A. 1998. Tertiary subduction and slab break-off model of the Carpathian-Pannonian region. *Tectonophysics*, **295**, 307–340.

NEUBAUER, F., DALLMEYER, R.D., DUNKL, I. & SCHIRNIK, D. 1995. Late Cretaceous exhumation of the metamorphic Gleinalm dome, Eastern Alps: kinematics, cooling history and sedimentary response in a sinistral wrench corridor. *Tectonophysics*, **242**, 79–89.

NEUBAUER, F., GENSER, J. & HANDLER, R. 2000. The Eastern Alps: result of a two-stage collision process. *Mitteilungen der Österreichischen Geologischen Gesellschaft*, **92**, 117–134.

NEUBAUER, F., GENSER, J., KURZ, W. & WANG, X. 1999. Exhumation of the Tauern window, Eastern Alps. *Physics and Chemistry of the Earth Part A: Solid Earth and Geodesy*, **24**, 675–680.

NEUGEBAUER, J., GREINER, B. & APPEL, E. 2001. Kinematics of Alpine-West Carpathian orogen and palaeogeographic implications. *Journal of the Geological Society, London*, **158**, 97–110.

NICOLESCU, S., CORNELL, D.H. & BOJAR, A.V. 1999. Age and tectonic setting of the Ocna de Fier-Dognecea skarn deposit, SW Romania. *In:* STANLEY, C.J. *ET AL.* (eds) *Mineral Deposits: Processes to Processing*. Balkema, Rotterdam, 1279–1282.

OKAY, A.I., SATIR, M., TÜYSÜZ, O., AKYÜZ, S. & CHEN, F. 2001. The tectonics of the Strandja Massif: late Variscan and mid-Mesozoic deformation and metamorphism in the northern Aegean. *International Journal of Earth Sciences*, **90**, 217–233.

OVTCHAROVA, M., QUADT, A.V., HEINRICH, C.A., FRANK, M., ROHRMEIER, M., PEYCHEVA, I. & NEUBAUER, F. 2001. Late Alpine Extensional Stage of the Central Rhodopian Core Complex and Related Acid Magmatism (Madan Dome, Bulgaria) - Isotope and Geochronological Data. *In:* PIESTRZYŇSK, L.A. (ed.) *Mineral Deposits at the Beginning of the 21st Century*. Balkema, Lisse, the Netherlands, 551–553.

OYARZUN, R., MÁRQUEZ, A. & LILLO, J. 2001. Giant versus small porphyry copper deposits of Cenozoic age in northern Chile: adakitic versus normal calc-alkaline magmatism. *Mineralium Deposita*, **36**, 794–798.

PAMIC, J. & PÉCSKAY, Z. 1994. Geochronology of Upper Cretaceous and Tertiary igneous rocks from the Slavonija-Srijem Depression. *Nafta*, **45**, 331–339.

PAMIC, J., BELAK, M., BULLEN, T.D., LANPHERE, M.A. & MCKEE, E.H. 2000. Geochemistry and geodynamics of a Late Cretaceous bimodal volcanic association from the southern part of the Pannonian Basin in S (Northern Croatia). *Mineralogy and Petrology*, **68**, 271–296.

PATRASCU, S., BLEAHU, M., PANAIOTU, C. & PANAIOTU, C.E. 1992. The paleomagnetism of the upper Cretaceous magmatic rocks in the Banat area of the South Carpathians: tectonic implications. *Tectonophysics*, **213**, 341–352.

PATRASCU, S., PANAIOTU, C., SECLAMAN, M. & PANAIOTU, C.E. 1994. Timing of rotational motion of Apuseni Mountains (Romania): paleomagnetic data from Tertiary magmatic rocks. *Tectonophysics*, **233**, 163–176.

PÉCSKAY, Z., LEXA, J., SZKÁCS, A., BALOGH, K., SEGHEDI, I., KONECNY, V., KOVÁCS, M., MÁRTON, E., KALICIAK, M., SZÉKY-FUX, V., PÓKA, T., GYARMATI, P., EDELSTEIN, O., ROSU, E. & ZEC, B.

1995. Space and time distribution of Neogene-Quaternary volcanism in the Carpatho-Pannonian Region. *Acta Vulcanologica*, **7**, 15–28.

PERESSON, H. & DECKER, K. 1997. Far-field effects of Late Miocene subduction in the Eastern Carpathians: E-W compression and inversion of structures in the Alpine-Carpathian-Pannonian region. *Tectonics*, **16**, 38–56.

PETRASCHECK, W.E. 1986. The metallogeny of the Eastern Alps in context with the circum-Mediterranean metallogeny. *Schriftenreihe der Erdwissenschaftlichen Kommission der Österreichischen Akademie der Wissenschaften*, **8**, 127–134.

PETTKE, T., DIAMOND, L.W. & VILLA, I. 1999. Mesothermal gold veins and devolatilization in the north-western Alps: The temporal link. *Geology*, **27**, 641–644.

PETTKE, T., DIAMOND, L.W. & KRAMERS, J.D. 2000. Mesothermal gold lodes in the north-western Alps: A review of genetic constraints from radiogenic isotopes. *European Journal of Mineralogy*, **12**, 213–230.

PEYCHEVA, I., VON QUADT, A., KAMENOV, B., IVANOV, Z. & GEOGIEV, N. 2001. New Isotope Data for Upper Cretaceous Magma Emplacement in the Southern und South-Western Parts of Central Srednogorie. *Romanian Journal of Mineral Deposits*, **79**(suppl. 2), 82–83.

PLASIENKA, D. 1997. Cretaceous tectonochronology of the Central Western Carpathians, Slovakia. *Geologica Carpathica*, **48**, 99–111.

POHL, W. & BELOCKY, R. 1994. Alpidic Metamorphic Fluids and Metallogenesis in the Eastern Alps. *Mitteilungen der Österreichischen Geologischen Gesellschaft*, **86**, 141–152.

POHL, W. & BELOCKY, R. 1998. Metamorphism and metallogeny in the Eastern Alps. *Mineralium Deposita*, **34**, 614–629.

POLLER, U., UHER, P., JANÁK, M., PLASIENKA, D. & KOHÚT, M. 2001. Late Cretaceous age of the Rochovce granite, Western Carpathians, constrained by U-Pb single-zircon dating in combination with cathodoluminescence imaging. *Geologica Carpathica*, **52**, 41–47.

POPOV, P. & POPOV, K. 2000. General geologic and metallogenic features of the Panagyrishte ore region. *In:* STRASHIMIROV, S. & POPOV, P. (eds) *Geology and metallogeny of the Panagyurishte ore region (Srednogorie zone, Bulgaria). Guide to Excursions A and C.* ABCD-GEODE 2000 Workshop, Borovets Bulgaria, Publishing House St. Ivan Rilski, **2000**, 1–7.

POPOV, P. 1996. On the tectono-metallogenic evolution of the Balkan peninsula alpides. *UNESCO - IGCP project No. 356*, **1**, 5–17.

POPOV, P.N. 1987. Tectonics of the Banat-Srednogorie Rift. *Tectonophysics*, **143**, 209–216.

PROCHASKA, W. 1993. Untersuchung stabiler Isotope an alpidischen Ganglagerstätten in den Ostalpen. *Berg- und Hüttenmännische Monatshefte*, **138**, 138–144.

PROCHASKA, W., POHL, W., BELOCKY, R. & KUCHA, H. 1995. Tertiary metallogenesis in the Eastern Alps - the Waldenstein hematite deposit. *Geologische Rundschau*, **84**, 831–842.

RATSCHBACHER, L. 1986. Kinematics of Austro-Alpine cover nappes: changing translation path due to transpression. *Tectonophysics*, **125**, 335–356.

RATSCHBACHER, L., FRISCH, W., LINZER, G. & MERLE, O. 1991. Lateral Extrusion in the Eastern Alps, part 2: Structural Analysis. *Tectonics*, **10**, 257–271.

RATSCHBACHER, L., FRISCH, W., NEUBAUER, F., SCHMID, S.M. & NEUGEBAUER, F. 1989. Extension in compressional orogenic belts: The eastern Alps. *Geology*, **17**, 404–407.

RATSCHBACHER, L., LINZER, G.H., MOSER, F., STRUSIEVICZ, R.O., BEDELEAN, H., HAR, N. & MOGOS, P.A. 1993. Cretaceous to Miocene thrusting and wrenching along the central South Carpathians due to a corner effect during collision and orocline formation. *Tectonics*, **12**, 855–873.

RICOU, L.E., BURG, J.P., GODFRIAUX, I. & IVANOV, Z. 1998. Rhodope and Vardar: the metamorphic and the olistostromic paired belts related to the Cretaceous subduction under Europe. *Geodinamica Acta*, **11**, 285–309.

ROBERTSON, A.H.F. & KARAMATA, S. 1994. The role of subduction-accretion process in the tectonic evolution of the Mesozoic Tethys in Serbia. *Tectonophysics*, **234**, 73–94.

ROSU, E., PECSKAY, Z., STEFAN, A., POPESCU, G. & PANAIOTU, C.E. 1997. The evolution of Neogene volcanism in the Apuseni Mountains (Romania). Constraints from new K-Ar data. *Geologica Carpathica*, **48**, 353–359.

SANDULESCU, M. 1984. *Geotectonica Romaniei*. Edition Tehnica, Bucharest.

SCHMID, S.M., BERZA, T., DIACONESCU, N., FÜGENSCHUH, B., SCHÖNBORN, O. & KISSLING, E. 1998. Orogen-parallel extension in the Southern Carpathians. *Tectonophysics*, **15**, 1036–1064.

SERAFIMOFSKI, T. 2000. The Lece-Chalkidiki metallogenic zone: geotectonic setting and metallogenetic feature. *Geologijia*, **42**, 159–164.

SERAFIMOFSKI, T., JANKOVIC, S. & CIFLIGANEC, V. 1995. Alpine metallogeny and plate tectonics in the SW flank of the Carpatho-Balkanides. *Geologica Macedonica*, **9**, 3–14.

SIBSON, R.H. & SCOTT, J. 1998. Stress/ fault controls on the containment and release of overpressured fluids: Examples from gold-quartz vein systems in Juneau, Alaska; Victoria, Australia and Otago, New Zealand. *Ore Geology Reviews*, **13**, 293–306.

SILLITOE, R.H. 1980. The Carpathian-Balkan porphyry copper belt - A Cordilleran perspective. *In: European Copper Deposits*. Mining University, Belgrade, 28–35.

SINGER, B. & MARCHEV, P. 2000. Temporal evolution of arc magmatism and hydrothermal activity, including epithermal gold veins, Borovitsa caldera, southern Bulgaria. *Economic Geology*, **95**, 1155–1164.

SORIOU, M., CATILINA, R. & STRUTINSKI, C. 1986. K-Ar ages on some igneous rocks from the south-western end of the South Carpathians (Banat Hills). *Revue Roumaine du Physique*, **31**, 849–854.

SPENCER, J.E. & WELTY, J.W. 1986. Possible controls of base- and precious-metal mineralization associated with Tertiary detachment faults in the lower Colorado River trough, Arizona and California. *Geology*,

14, 195–198.

STAMPFLI, C.M. & MOSAR, J. 1999. The making and becoming of Apulia. *Memorie di Scienze Geologiche*, **51**, 141–154.

STRASHIMIROV, S. & POPOV, P. 2000. *Geology and metallogeny of the Panagyurishte ore region (Srednogorie zone, Bulgaria)*. ABCD-GEODE 2000 Workshop, Borovets, Guide to Excursions (A and C), Publishing House St. Ivan Rilski, Sofia.

STÜWE, K. 1998. Tectonic constraints on the timing relationships of metamorphism, fluid production and gold-bearing quartz vein emplacement. *Ore Geology Reviews*, **13**, 219–228.

ULRICH, T., GÜNTHER, D. & HEINRICH, C.A. 1999. Gold concentrations of magmatic brines and the metal budget of porphyry copper deposits. *Nature*, **399**, 676–679.

VAN DE ZEDDE, D.M.A. & WORTEL, M.J.R. 2001. Shallow slab detachment as a transient source of heat at mid lithospheric levels. *Tectonics*, **20**, 868–882.

VELICHKOVA, S., HANDLER, R. & NEUBAUER, F. 2001. Preliminary ^{40}Ar/^{39}Ar mineral ages from the Central Srednogorie Zone, Bulgaria: Implications for Cretaceous geodynamics. *Romanian Journal of Mineral Deposits*, **79/2**, 112–113.

VITYK, M.O., KROUSE, H.R. & SZAKUN, L.Z. 1994. Fluid evolution and mineral formation in the Beregovo gold-base metal deposit, Transcarpathia, Ukraine. *Economic Geology*, **89**, 547–565.

VLAD, S.N. 1997. Calcic skarns and transversal zoning in the Banat mountains. *Romania: indicators of an Andean-type setting. Mineralium Deposita*, **32**, 446–451.

VLAD, S. & BOROCS, M. 1998. Alpine metallogenesis of the Romanian Carpathians. *Romanian Journal of Mineral Deposits*, **78**, 5–20.

VON BLANCKENBURG, F. & DAVIES, J.H. 1995. Slab break-off: A model for syn-collisional magmatism and tectonics in the Alps. *Tectonics*, **14**, 120–131.

VON BLANCKENBURG, F., KAGAMI, H., DEUTSCH, A., OBERLI, F., MEIER, M., WIEDENBECK, M., BARTH, S. & FISCHER, H. 1998. The origin of Alpine plutons along the Periadriatic Lineament. *Schweizerische Mineralogische und Petrographische Mitteilungen*, **78**, 55–66.

VON COTTA, B. 1864. *Über Eruptivgesteine und Erzlagerstatten in Banat und Serbien*. Edit. V. Braunmüller, Wien, **105**.

VON QUADT, A., IVANOV, Z. & PEYCHEVA, I. 2001. The Central Srednogorie (Bulgaria) part of the Cu (Au-Mo) Belt of Europe: A review of the geochronological data and the geodynamical models in the light of the new structural and isotopic studies. *In:* PIESTRZYSKI, L.A. (ed.) *Mineral Deposits at the Beginning of the 21st Century*. Balkema, Lisse, the Netherlands, 555–558.

VON QUADT, A., PEYCHEVA, I., KAMENOV, B., FANGER, L. & HEINRICH, C.A. 2002. The Elatsite porphyry copper deposit of the Panagyurishte ore district, Srednogorie zone, Bulgaria: U-Pb zircon geochronology and isotope-geochemical investigations of ore genesis. *In:* BLUNDELL, D.J., NEUBAUER, F. & VON QUADT, A. (eds) *The Timing and Location of Major Ore Deposits in an Evolving Orogen*. Geological Society, London, Special Publications, **204**, 119–135.

WEBER, L. 1997. Handbuch der Lagerstätten der Erze, Industrieminerale und Energierohstoffe Österreichs. *Archiv für Lagerstättenforschung an der Geologischen Bundesanstalt*, **9**, 1–607.

WILLINGSHOFER, E. 2000. *Extension in collisional orogenic belts: the Late Cretaceous evolution of the Alps and Carpathians*. PhD thesis, Vrije Universiteit, Amsterdam.

WILLINGSHOFER, E., NEUBAUER, F. & CLOETINGH, S. 1999a. Significance of Gosau basins for the upper Cretaceous geodynamic history of the Alpine—Carpathian belt. *Physics and Chemistry of the Earth Part A: Solid Earth and Geodesy*, **24**, 687–695.

WILLINGSHOFER, E., VAN WEES, J.D., CLOETINGH, S.A.P.L. & NEUBAUER, F. 1999b. Thermomechanical evolution of an accretionary wedge: the Austroalpine of the Eastern Alps - indications and implications from 2D-numerical modelling. *Tectonics*, **18**, 809–826.

WILLINGSHOFER, E., ANDRIESSEN, P., CLOETINGH, S. & NEUBAUER, F. 2001. Detrital fission track thermochronology of Upper Cretaceous syn-orogenic sediments in the South Carpathians (Romania): inferences on the tectonic evolution of collisional hinterland. *Basin Research*, **13**, 379–396.

WORTEL, M.J.R. & SPAKMAN, W. 2000. Subduction and slab detachment in the Mediterranean-Carpathian region. *Science*, **290**, 1910–1917.

WORTMANN, U.G., WEISSERT, H., FUNK, H. & HAUCK, J. 2001. Alpine plate kinematics revisited: The Adria Problem. *Tectonics*, **20**, 134–147.

YANNEV, Y., INNOCENTI, F., MANETTI, P. & SERRI, G. 1998. Upper Eocene-Oligocene Collision-related Volcanism in Eastern Rhodopes (Bulgaria) - Western Thrace (Greece): Petrogenetic Affinity and Geodynamic Significance. *Acta Vulcanologica*, **10**, 199–216.

YILMAZ, Y. & POLAT, A. 1998. Geology and evolution of the Thrace volcanism, Turkey. *Acta Vulcanologica*, **10**, 293–303.

ZACHER, W. & LUPU, M. 1999. Pitfalls on the race for an ultimate Tethys model. *International Journal of Earth Sciences*, **88**, 111–115.

Auriferous arsenopyrite–pyrite and stibnite mineralization from the Siflitz–Guginock area (Austria): indications for hydrothermal activity during Tertiary oblique terrane accretion in the Eastern Alps

G. AMANN[1], W. H. PAAR[2], F. NEUBAUER[1] & G. DAXNER[1]

[1]Institut für Geologie und Paläontologie, Universität Salzburg, Hellbrunnerstr. 34, A-5020 Salzburg, Austria (e-mail: gerhard.amann@sbg.ac.at)

[2] Institut für Mineralogie, Universität Salzburg, Hellbrunnerstr. 34, A-5020 Salzburg, Austria

Abstract: Polymetamorphic schists and marbles of the Austroalpine Kreuzeck–Goldeck Complex are the host to auriferous arsenopyrite–pyrite as well as stibnite mineralization. In the Siflitz–Guginock area both types of mineralization are closely related spatially, but restricted to different lithologies. The auriferous arsenopyrite–pyrite mineralization is either disseminated or bound to quartz veins and strongly silicified fault-zones hosted in phyllites to (garnet-) micaschists. Similar disseminations within marbles are of minor importance. A stockwork-like mineralization of stibnite-filled fractures with weak metasomatic replacement is limited to marbles.

Both types of mineralization in the Kreuzeck–Goldeck Complex are intimately related to roughly east–west-trending semi-ductile to brittle strike-slip faults which formed during orogen-parallel wrenching. Semi-ductile to brittle kinematic indicators point to dextral, as well as sinistral, modes of fault movements, coeval with the formation of pyrite–arsenopyrite-bearing quartz veins, as well as the intrusion of Oligocene lamprophyre dykes. Mineralizing fluids are suggested to be derived from devolatilization of the subducted Penninic upper crust. Fluid ascent and ore precipitation is controlled by a transpressive strike-slip regime related to oblique terrane accretion during the Late Eocene to Oligocene.

Subsequent development of a Late Oligocene–Early Miocene pure shear regime with contraction trending (N)NE–(S)SW led to the development of conjugate NW–SE dextral and NE–SW sinistral brittle strike-slip faults, and overprinting by ESE–WNW oriented extension. These later events are related to the formation of auriferous mineralization from within the metamorphic core complex of the Tauern Window to the north of the Kreuzeck–Goldeck Complex, hence implying a significant change in spatial, temporal and structural control of Tertiary auriferous mineralization in the Eastern Alps.

Late tectonic gold mineralization within the Alps is a consequence of oblique continent–continent collision during the late Palaeogene, when the Penninic microcontinent collided with the Cretaceous Austroalpine nappe complex, and subsequent uplift and exhumation of Penninic continental and oceanic crust within metamorphic core complexes. Enhanced permeability along zones of structural weakness led to the formation of various styles of exclusively mesothermal gold deposits within these metamorphic core complexes, such as the Monte Rosa gold district of the Western Alps or the Tauern gold district of the Eastern Alps (Fig. 1a). Gold-rich porphyry-type deposits are not known to occur in the Alps.

^{40}Ar/^{39}Ar dating of hydrothermal muscovite from mesothermal gold lodes of the Monte Rosa gold district in the Western Alps provided ages ranging between 31.6 and 10.6 Ma (Pettke *et al.* 1999). These lodes are spatially related to post-metamorphic high K lamprophyres, providing, respectively, K–Ar whole-rock ages of 31.6 ± 1.3 Ma (Dal Piaz *et al.* 1973) and K–Ar biotite ages of 32.7 ± 1.4 Ma (Diamond & Wiedenbeck 1986). These age relationships, combined with the results of isotopic investigations (Pettke & Frei 1996; Pettke & Diamond 1997; Pettke *et al.* 1997), led to the conclusion that mineralizing hydrothermal fluids were derived from meta-morphic devolatilization of Mesozoic (Penninic) calc-schists at a deeper structural level (Pettke *et al.* 1999).

The close association between lamprophyres and mesothermal gold mineralization has been

From: BLUNDELL, D.J., NEUBAUER, F. & VON QUADT, A. (eds) 2002. *The Timing and Location of Major Ore Deposits in an Evolving Orogen*. Geological Society, London, Special Publications, **204**, 103–117.
0305-8719/02/$15.00 © The Geological Society of London 2002.

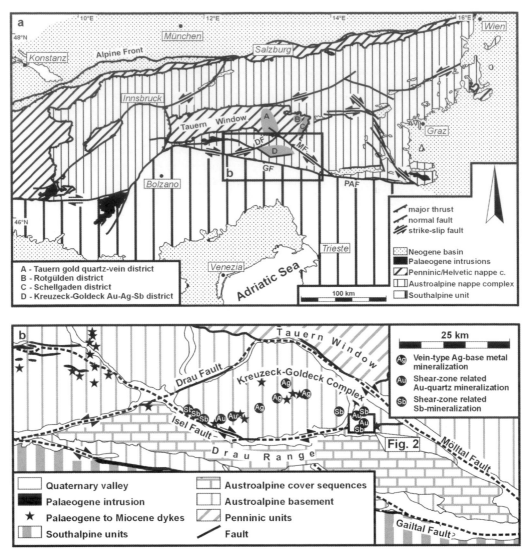

Fig. 1. (**a**) Tectonic sketch map of the Eastern Alps (modified after Neubauer *et al.* 2000). Shaded areas, labelled A, B, C, and D mark the position of different mesothermal gold deposits within the Penninic Tauern Window and the Kreuzeck–Goldeck Au–Ag–Sb district. DF, Drau Fault; MF, Mölltal Fault; PAL, Periadriatic Lineament with GF, Gailtal Fault. (**b**) Enlarged view of the Kreuzeck–Goldeck Complex to the south of the Tauern Window showing the distribution of major faults, Palaeogene to Miocene dykes and epigenetic mineralization of the Kreuzeck–Goldeck Au–Ag–Sb district (modified after Weber 1997).

known since the early twentieth century (McLennan 1915). Fundamental work published in the late 1980s to early 1990s showed, however, that there is no direct genetic link between the mesothermal gold deposits and lamprophyre emplacement, but they both share the same geodynamic regime, forming broadly simultaneously during uplift and decompression following oblique terrane collision (e.g. Wyman & Kerrich 1988, 1989; Kerrich & Wyman 1990, 1994).

According to Kerrich & Wyman (1994), some of the typical features of the gold–lamprophyre association are:

1   a close spatial and temporal relationship of gold mineralization to the emplacement of subduction-related shoshonitic lamprophyres during uplift and exhumation following continent–continent collision;

2   lamprophyres with intraplate geochemical

signatures are not known to be associated with gold deposits;

3 both shoshonitic lamprophyres and associated gold mineralization are spatially and temporally linked to zones of strike-slip displacement following oblique terrane collision;

4 mineralization and dyke emplacement closely follow peak metamorphism in the host rocks.

In contrast to the Monte Rose gold district, mesothermal gold lodes from within the Tauern Window of the Eastern Alps (Tauern gold district: Ebner et al. 2000 and references therein; Fig. 1a) are not associated with lamprophyres. Various styles of mineralization, include:

1 auriferous NNE–SSW-trending, sub-vertical, sulphide-bearing extensional quartz veins (Tauern gold–quartz veins) and shear zones in the central Tauern Window (Feitzinger & Paar 1991; Kurz et al. 1994);

2 ore impregnations within fold hinge zones, gold-bearing tension gashes and irregular, gold-bearing replacement bodies of the Rotgülden sub-district at the northeastern part of the Tauern Window (Horner et al. 1997);

3 Au–Pb–Cu–W-rich quartz layers along low-angle normal faults, sub-concordant to host rock foliation, and Au–Pb–Cu-rich, (N)NE–(S)SW-trending, sub-vertical quartz veins at the eastern border of the Tauern Window (Type II and type III ore-bodies of the Schellgaden sub-district; Amann et al. 1997).

All of these deposits post-date the peak of Late Oligocene metamorphism of the Penninic units now exposed in the Tauern Window. They are hosted by tensional structures related to rapid uplift and exhumation of the Tauern Window metamorphic core complex during NNE–SSW-oriented contraction and hydrothermal activity due to increased heat flow that is attributed to the same process (Pohl & Belocky 1994; Fügenschuh et al. 1997; Sachsenhofer 2001).

Mesothermal gold and stibnite mineralization within the Austroalpine nappe complex to the south of the Tauern Window differs significantly from the well-known deposits within the Tauern Window and, despite several publications describing them in detail (e.g. Weber 1989; Mali 1996), a comprehensive geodynamic interpretation of the former was not available until now. Analogous to the situation at the Monte Rosa gold district, this mineralization is spatially related to Oligocene lamprophyre dyke swarms. Based on new structural data from small prospects in the eastern Kreuzeck–Goldeck Complex, we discuss the possible relationships, respectively, between this me-

sothermal gold and stibnite mineralization and the tectonometamorphic evolution of the subducted Penninic microcontinent, as well as Oligocene lamprophyre dyke swarms within the Austroalpine nappe complex (Deutsch 1980). Previous age and orientation data for lamprophyre dykes in the Kreuzeck–Goldeck Complex (Müller et al. 1992), together with new structural data, will be used in this study to establish a time-frame for the succession of tectonic events and associated ore formation in the Kreuzeck–Goldeck Complex.

Geological setting

The Austroalpine nappe complex of the Eastern Alps comprises a continental basement–cover nappe pile, which received its general internal nappe structure during Cretaceous orogenic events (Neubauer et al. 2000 and references therein). Subsequent subduction and closure of the Piemontais ocean after the Early Eocene led to collision, and underplating, of the Penninic microcontinent with the combined Austroalpine–Adriatic microplate. Uplift and exhumation of Penninic units in the Tauern Window resulted from indentation of the Southalpine (Adriatic) indenter, tectonic unroofing along upper margins of Penninic sequences and the eastward extrusion of an Austroalpine tectonic wedge (Neubauer et al. 2000 and references therein).

The triangular shaped Kreuzeck–Goldeck Complex (Fig. 1) is part of the polymetamorphic Austroalpine crystalline basement, located south of the Tauern Window. Its northeastern boundary is marked by the dextral Mölltal Fault, trending NW–SE, whereas the northwestern boundary is defined by the sinistral Drautal Fault, trending NE–SW. To the south, the basement of the Kreuzeck–Goldeck Complex is overlain by the Austroalpine, Permo-Mesozoic cover sequences of the Drau Range (Fig. 1b). The dextral Gailtal Fault, trending ESE–WNW and part of the Periadriatic Lineament, separates the combined Kreuzeck–Goldeck Complex and Drau Range from the Southalpine units (Fig. 1b).

A Variscan amphibolite-facies metamorphic event in the northern Austroalpine basement of the Kreuzeck–Goldeck Complex occurred at c.340 Ma (Hoke 1990 and references therein). In contrast, the southernmost parts of the complex, immediately north of the Drau Range, reached only greenschist facies conditions during the late Variscan (c. 310 Ma; Neubauer, unpublished data).

The Variscan metamorphic overprint was superimposed by a Cretaceous metamorphic event. Amphibolite facies conditions were attained, again, only in the northernmost parts of the Kreuzeck–Goldeck Complex (Deutsch 1988;

Hoke 1990; Thöni 1999) and are indicated by white mica K–Ar cooling ages of c. 100–80 Ma (Hoke 1990). Further south, Cretaceous peak metamorphic conditions gradually decrease from greenschist facies to areas where Cretaceous metamorphic overprinting cannot be observed. K–Ar cooling ages of c. 105 Ma have been obtained for the areas affected by Cretaceous greenschist facies metamorphism (biotite coexisting with undisturbed Variscan muscovite; Brewer 1970).

Tertiary metamorphism is restricted to the Penninic units of the Tauern Window and the lowermost Austroalpine units framing the window. Ar/Ar dating of white mica from eclogites in the central southern Tauern Window provided ages for subduction related high-pressure–low-temperature metamorphism ranging between 38 and 32 Ma (Zimmermann et al. 1994; Thöni 1999; Handler et al. 2001). Only locally preserved, this metamorphic event was superimposed by a Barrovian-type metamorphic event with peak conditions of 7–9 kbar and 490–560 °C in the northeastern part of the window (Kruhl 1993) and 6.6–7.6 kbar and 490–630 °C in the southeastern part of the window (Droop 1985). An age range of 30–27 Ma was obtained for this later event by Rb–Sr dating of phengite (Inger & Cliff 1994). The onset of rapid exhumation of the Penninic units within the Tauern Window is documented by the partial resetting of white mica $^{40}Ar/^{39}Ar$ age spectra at c. 22 Ma (Liu et al. 2001).

A notable feature of the Kreuzeck–Goldeck Complex is the occurrence of undeformed, post-metamorphic lamprophyre dykes. Müller et al. (1992) distinguished two age groups, which have different geochemical characteristics and structural orientation. A first set of amphibole-bearing, shoshonitic/high-K calc-alkaline lamprophyres, together with calc-alkaline basaltic dykes, is dated at 36 ± 1 Ma by the K–Ar method. These dykes show a bimodal orientation, trending either east–west or NNW–SSE (geochemical groups 1 and 5 of Müller et al. 1992). A second set of alkaline and phlogopite-bearing shoshonitic lamprophyres, with K–Ar and Rb–Sr ages clustering at 30 ± 2 Ma, have a single orientation, trending (N)NE–(S)SW (geochemical groups 2–4 of Müller et al. 1992; Deutsch 1984). Together with similar dykes that occur near the Periadriatic Lineament, the dykes in the Kreuzeck–Goldeck Complex have been interpreted as an expression of a slab breakoff following partial subduction of the Penninic microcontinent (von Blanckenburg & Davies 1995; von Blanckenburg et al. 1998).

The metamorphic basement of the Kreuzeck–Goldeck Complex is characterized by a great number of mainly small, epigenetic mineral de-posits forming together the Kreuzeck–Goldeck Au–Ag–Sb district (Fig. 1a, b) including:

1 silver-rich, base metal-bearing veins in the central part of the Kreuzeck–Goldeck Complex, which are spatially closely related to, but post-dating, dykes of undeformed quartz porphyrite (Feitzinger et al. 1995) with K–Ar ages of 30–40 Ma (Deutsch 1984);
2 shear zone-related, mesothermal gold lodes;
3 shear zone-related antimony deposits (Weber 1989; Mali 1996; Cerny et al. 1997), which are both concentrated along the southern margin of the Kreuzeck–Goldeck Complex– the antimony deposits of the southwestern Kreuzeck–Goldeck Complex are structurally controlled by the WNW–ESE-trending Isel Fault (Weber 1989) and at least one of these deposits shows the occurrence of a nearby lamprophyric dyke hosted by the same fault system (Lahusen 1972).

We now focus on the less well-known mesothermal gold lodes and antimony occurrences of the eastern Kreuzeck–Goldeck Complex.

Mineralization in the eastern Kreuzeck–Goldeck Complex

Structurally controlled gold and stibnite mineralization is documented from several areas of the eastern Kreuzeck–Goldeck Complex. The most important locations are the abandoned Siflitz mining area to the north, and the Guginock mining area to the south of the Siflitz valley (Fig. 2). In addition to these historical mining areas, several ore showings were discovered along the Siflitz valley during fieldwork in 1995 (Fig. 2).

In the Siflitz mining area, more than 100 adits and pits are scattered over an area of 1200 × 600 m at 880–1370 m above sea level (Hiessleitner 1949). Three parallel shear zones trending ESE–WNW, each of them several metres wide, contain quartz–carbonate cataclasites within graphitic schists. Steeply plunging ore shoots, which are parallel to the shear zones, are composed of closely spaced quartz veins, up to 30 cm wide (Fig. 3a), and veinlets with abundant visible gold associated with arsenopyrite, pyrite, traces of chalcopyrite and fine-grained rutile/anatase (Paar 1995).

The stibnite and auriferous arsenopyrite–pyrite mineralization at Guginock is situated along the western flank of the Goldeck Mountains, east of Lind im Drautal, and to the south of the east–west-trending Siflitz valley, which marks the transition between Variscan amphibolite and

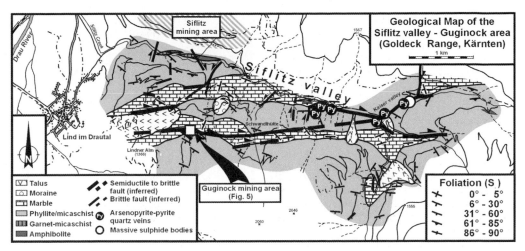

Fig. 2. New geological map of the Siflitz–Guginock area, displaying the location of the Siflitz mining area (oblique hatching), the Guginock mining area (Fig. 5) and surface outcrops of auriferous arsenopyrite–pyrite quartz veins along the Kaiser and Silflitz valleys.

greenschist facies metamorphism. The stibnite mineralization was mined here for antimony during the seventeenth century. An exploration drift was constructed in 1922 to undercut the antimony mineralization. However, instead of finding the continuation of the marble-hosted stibnite ore, an extensive zone of Au–As mineralization was encountered. The Au–As mineralization occurs in phyllites at the boundary with the marbles and was traced in another adit for more than 20 m along strike. During exploration in 1995, channel sampling across the 12 m thick mineralized phyllite yielded samples of the disseminated mineralization containing 4.4 to 7.7 ppm Au (average 6.3 ppm). Diamond core drilling (three boreholes with a total length of 260 m) along the structure proved the lateral extension of the mineralization. The best results were obtained in drill hole GU-1 (Fig. 4), where grades of between 1.7 and 6.3 ppm Au (average 3.4 ppm) were detected in a 10 m thick section.

Lithologies and structures

The area investigated (Fig. 2) is characterized by various dark, locally graphitic phyllites, which grade into micaschists towards the north. Three east–west- to ENE–WSW-trending zones of banded to massive marbles, which are in a basal stratigraphic position, occupy the cores of large anticlines trending east–west to ENE–WSW. Rare crinoids imply a Silurian–Devonian sedimentation age for the marbles (Deutsch 1988). A Late Devonian sedimentation age for the phyllites is indicated by the poorly preserved conodont fauna

within marble intercalations (Heinz 1987). Small lenses of amphibolite and garnet-micaschist are restricted to the vicinity of the Siflitz valley and its eastern extension, the Kaiser valley (Fig. 2).

Within the area, a complex sequence of deformation events is observed. The penetrative foliation is parallel to lithological boundaries. Poles of the foliation (S_1) show a well-developed great circle distribution with sub-horizontal east–west-trending fold-axes (Fig. 5a). The S–C fabrics, expressed in sections parallel to the stretching lineation (L_1), imply a top-to-WSW movement (Fig. 5b).

The penetrative foliation (S_1) is affected by small-scale tight to isoclinal F_2 folds with axes trending east–west to ENE–WSW (Fig. 5c). These folds are overturned to recumbent, face NNW, and often show sheared off overturned limbs that indicate top-NNW thrusting and folding. Locally occurring F_3 folds are open to tight, face NNE, and are characterized by a non-penetrative, sub-horizontal axial plane foliation (S_3; Fig. 5d). Both east–west to ENE–WSW-trending F_2 folds and ESE–WNW-trending F_3 folds contribute to the average north–south-oriented great circle distribution of S_1 poles. Open to tight F_3 folds are spatially connected with sub-vertical, ENE–WSW-trending semiductile to brittle strike-slip faults. One of these faults can be traced for several kilometres from Guginock, where it follows the footwall of the southernmost marble anticline, towards Schwandlhütte, where it follows the footwall of the middle marble anticline, and further towards the ENE–WSW-trending Kaiser valley (Fig. 2). Semiduc-

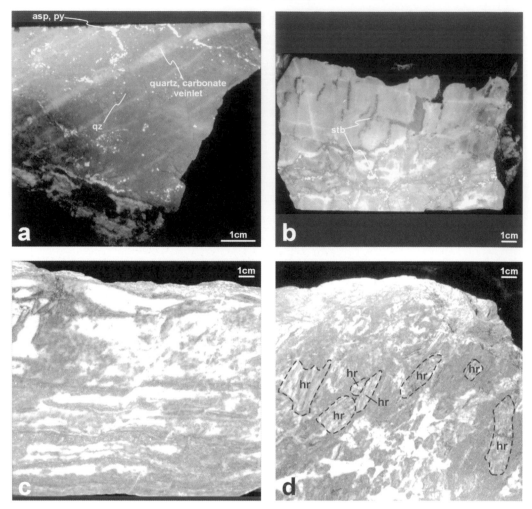

Fig. 3. (**a**) Native gold and arsenopyrite, pyrite (asp, py) bearing quartz vein (qz) cross-cut by late quartz–carbonate veinlets (Siflitz mining area). (**b**) Stockwork-like Sb mineralization of stibnite (stb) following fractures with minor metasomatic replacement within a marble (Guginock mining area). (**c**) Banded, strongly silicified ('quartzitic') auriferous arsenopyrite ore (exploration drift, Guginock mining area). (**d**) Strongly deformed 'quartzitic' ore containing auriferous arsenopyrite (grey), rounded fragments of silicified host rock (hr) and quartz gangue (white) (exploration drift, Guginock mining area).

tile shear sense indicators (S–C and ecc fabrics) within the fault zone indicate dextral movements along the fault and imply a contraction oriented approximately NW–SE to WNW–ESE (Fig. 5e; D_{4a}).

In contrast to this event, S–C and ecc fabrics along faults trending (WN)W–(ES)E imply sinistral semiductile fault movements and thus a contraction oriented approximately NE–SW (Fig. 5f; D_{4b}). Evaluation of fault data follows methods described, for example, in Angelier (1994).

Slickenside data show a succession of several stages of brittle deformation. A first stage comprising fault sets oriented WNW–ESE related to sub-horizontal contraction seems to represent late stages of dextral strike-slip movements along ENE–WSW trending faults (Fig. 5e; D_{4a}). A N(NE)–S(SW)-oriented sub-horizontal contraction led to the development of a conjugate set of (N)NW–(S)SE-trending dextral and NE–SW trending sinistral strike slip faults cutting the ENE–WSW trending faults (Fig. 5g; D_{4c}). Overprinting criteria of slickenside striations indicate a final stage of extension oriented NW–SE due to an exchange of σ_1 and σ_2 orientations (Fig. 5h; D_{4d}).

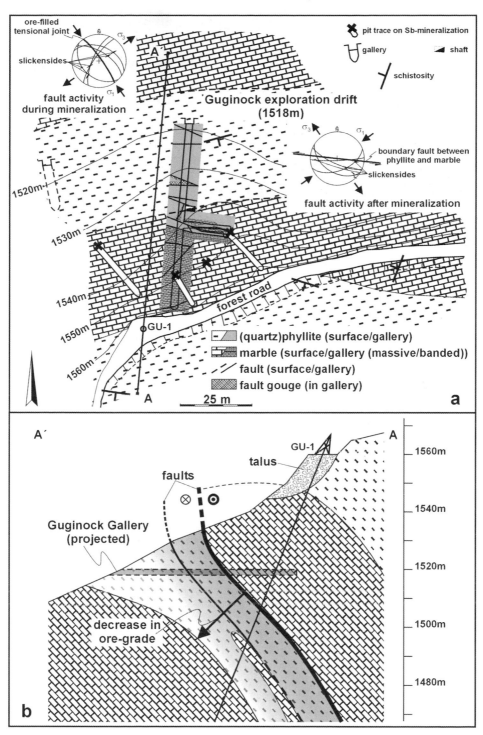

Fig. 4. (**a**) Geological map of the Guginock area, plan-view projection of the exploration drift and positions of accompanying Sb workings. A succession of brittle fault movements at the Guginock exploration drift is shown as stereographic projections of slickensides. (**b**) Schematic cross-section A–A′ along diamond core drill hole at survey point 'GU-1'.

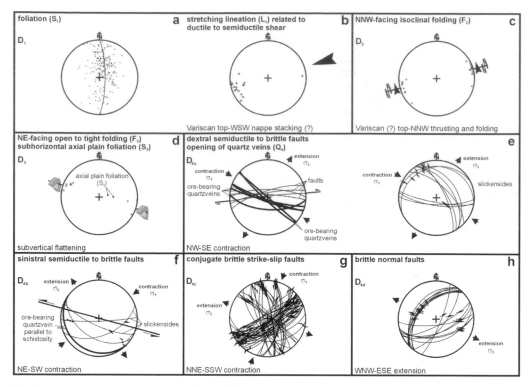

Fig. 5. Succession of deformation events in the Siflitz–Guginock area. (**a**) Poles of penetrative foliation. (**b**) Stretching lineation on penetrative foliation. (**c**) Axis of isoclinal folding. (**d**) Axis (asterisks) and axial plain foliation (dots) of open to tight folding. (**e, f**) Orientation of faults, ore-bearing quartz veins and slickensides associated with semiductile to brittle fault movements. Data were obtained, respectively, from the ore showings along the Siflitz valley and the Kaiser valley (compare Figure 2). (**g, h**) Slickensides accompanied by brittle strike-slip and normal fault movements. All plots are stereographic projection, lower hemisphere.

Guginock mining area

The NNE–SSW-trending Guginock exploration drift has its entrance within phyllites approximately 600 m to the NE of the Lindner Alm (Fig. 2), in the immediate footwall of the southernmost marble anticline. The adit exposes the contact between the phyllites and (overturned) lower limb of the marble anticline (Fig. 4). This contact is characterized by small-scale east–west-trending, north-facing recumbent folds (F₃), intense silicification of the phyllites and overprinting by an east–west-trending, sub-vertical fault gouge of a regional fault zone trending east–west. The disseminated pyrite and arsenopyrite mineralization is most strongly developed within the phyllites in the immediate vicinity of the contact zone and shows, within a few metres, a gradual decrease in intensity towards the north. In contrast, phyllites at the entrance of the gallery are not mineralized. Subordinate disseminated pyrite and arsenopyrite within the marble is restricted to a zone several

decimetres wide in the immediate vicinity of the contact zone.

Overprinting criteria on slickenside striations from the gallery indicate at least two stages of brittle deformation along the contact fault (Fig. 4a). An early stage shows oblique normal movements towards the NE to ENE. Sulphide-filled tensional joints trending NW–SE are interpreted as having formed at this stage. According to the similar orientation of the sub-horizontal extension direction (σ_3), this stage is probably related to D_{4a}. A second stage of brittle deformation, affecting the strongly silicified contact zone and thus postdating ore formation, shows sub-horizontal contraction trending NE–SW, leading to sinistral movements along the boundary fault analogous to D_{4b} indicated by surface data.

To the south of the Guginock gallery, and at a slightly higher topographic level, four pit-traces trending more or less SE–NW are situated within the marbles of the southernmost anticline. Ore samples from small dumps show stockwork-like, stibnite-filled fractures with weak metasomatic replacement (Fig. 3b). The workings within the marble are not accessible any more, hence the structural setting of the Sb mineralization remains unconstrained. Analogous stibnite deposits occurring further to the west provide better accessibility. They are arranged along splays of the WNW–ESE trending Isel Fault (Fig. 1b). Here, ore-bearing veins and breccia zones trend east–west to (S)SE–(N)NW (Weber 1989), suggesting that the formation of the stibnite mineralization was controlled by the same stress field as the c. 36 Ma lamprophyre dykes and the auriferous arsenopyrite–pyrite mineralization at Guginock (D_{4a}).

Mineralogy

The ore mineralogy exposed in the Guginock exploration drift is rather monotonous and consists predominantly of arsenopyrite, pyrite and traces of marcasite, galena, stibnite and rutile/anatase. Arsenopyrite occurs as elongated, needle-shaped crystals (10×100 μm in dimension) and is normally associated with pyrite which shows a weak chemical zonation. Small inclusions of stibnite, with an orientation parallel to former growth planes of pyrite, are frequently observed and are indications of a close genetic relationship between the Au–As and Sb mineralization. Drill cores from hole GU-1 exposed a strongly mineralized breccia zone (59 m from surface) containing clasts of sulphides (mainly pyrite) and quartz gangue and a matrix of, again, mainly pyritic ore. Strongly fragmented pyrite clasts within the breccia ore are healed by bournonite and galena which contains inclusions of Pb–Sb sulphosalts (jamesonite, boulangerite; Paar 1995). Slices of host-rock

phyllite within the breccia zone are strongly silicified, containing disseminated sulphides (mainly pyrite). Less common is a weak chloritization accompanied by the occurrence of disseminated arsenopyrite. The phyllite in the footwall of the breccia zone shows decreasing intensity of silicification and pyritization with increasing distance from the breccia zone. At the Guginock exploration drift the sulphides appear in banded 'quartzitic' ore along the strongly silicified contact between the phyllite and the marble (Fig. 3c), and as fine-grained disseminations further away from the contact. Breccia zones within the 'quartzitic' ore contain rounded fragments of silicified hostrock giving evidence for a mineralization contemporaneous with fault activity (Fig. 3d).

The gold concentrations of the Au–As ore can be assigned to invisible (refractory) gold, whereas visible gold is extremely rare and restricted to oxidized portions of the sulphides. A detailed investigation of the gold distribution in the sulphides, using an ion microprobe, showed a variation of the gold content for arsenopyrite between 32 and 160 ppm (6 grains) and 13–24 ppm Au for pyrite (3 grains).

Despite the present inaccessibility of the workings, extensive data on the mineralogy, fluid inclusions and isotope composition are available for the Guginock Sb mineralization (Table 1) which is dominated by coarse-grained stibnite. Similar occurrences elsewhere in the Kreuzeck–Goldeck Complex show that the paragenetic sequence generally starts with an auriferous arsenopyrite–association followed by a base metal paragenesis with chalcopyrite, bournonite, tetrahedrite, chalcostibnite, jamesonite, boulangerite, galena and sphalerite and is dominated by late stage stibnite (Cerny et al. 1997).

Table 1. *Chemical and isotopic characteristics of the Guginock Sb mineralization*

		Mineral	Reference
Trace elements			
Sb in galena	1000 ppm	Galena	Schroll 1954
Bi in galena	50 ppm	Galena	Schroll 1954
Fluid inclusions			
Chemistry	H_2O, $CO_2 \pm CH_4$	Quartz gangue	Mali 1996
Salinity (wt% NaCl equiv.)	22–24	Quartz gangue	Mali 1996
T_h (°C)	121–234	Quartz gangue	Mali 1996
Hydrogen & oxygen isotopes			
δD‰ (SMOW)	−44 to −63	Quartz gangue	Mali 1996
$\delta^{18}O$‰ (SMOW)	+12	Quartz gangue	Mali 1996
Sulphur isotopes			
$\delta^{34}S$ (CDT‰)	+3.0 to +3.9	Stibnite	Cerny et al. 1981
Lead isotopes			
$^{206}Pb/^{204}Pb$	18.773 ± 0.014	Stibnite	Köppel 1997
$^{207}Pb/^{204}Pb$	15.715 ± 0.016	Stibnite	Köppel 1997
$^{208}Pb/^{204}Pb$	38.365 ± 0.054	Stibnite	Köppel 1997

Arsenopyrite bearing quartz veins of the Siflitz valley and the Kaiser valley

The Kaiser valley is an eastward extension of the Siflitz valley and exposes a semiductile strike-slip fault trending ENE–WSW (Fig. 2). Outcrops of lens-shaped massive sulphide bodies, as well as arsenopyrite-bearing quartz veins several centimetres wide, accompanied by pyritization, chloritization and silicification of the phyllites, occur at several sites along this fault. The orientation of subvertical arsenopyrite-bearing quartz veins (Q_{4a}) trending (W)NW–(E)SE, arranged at the outer portions of the fault, coincides with dextral movements along the fault, indicated by S–C and ecc fabrics (Fig. 5e). The significance of the massive sulphide bodies occurring nearby is not completely resolved, but they might represent a syngenetic mineralization similar to other occurrences within the Kreuzeck–Goldeck Complex (Wallner 1981) and are not considered further here.

Similar occurrences along the Siflitz valley show a slightly different structural setting. Undeformed arsenopyrite- and locally scheelite-bearing quartz veins are arranged parallel to host rock schistosity, moderately dipping to the SE. The quartz veins are located in the vicinity of a semiductile strike-slip fault trending ESE–WNW. S–C fabrics and the orientation of the quartz veins (Q_{4b}) indicate sinistral transpressive movements due to NE contraction (D_{4b}) along the fault (Fig. 5f).

Discussion

Ductile deformation structures, like those formed during D_1 and D_2 stages, require at least greenschist facies metamorphic conditions. This is recorded for both Variscan and Cretaceous times in the Kreuzeck–Goldeck metamorphic complex. Similar structures in the western Kreuzeck–Goldeck Complex provide Variscan ages (*c.* 310 Ma; Neubauer, unpublished data). Furthermore, according to the geochronological data provided by Deutsch (1988), the Guginock and Siflitz mining areas are located outside the area affected by the Cretaceous greenschist facies metamorphic overprint. The relationship between D_3 folding and semiductile strike-slip shear zones ($D_{4a,4b}$ events) is difficult to establish. D_3 folding does, however, show a clear spatial relationship to semiductile strike-slip faults ($D_{4a,4b}$) and might be connected to local sub-vertical flattening of previously buried metamorphic rocks due to possible Cretaceous exhumation of underlying units within a transpressional strike-slip regime. Similar structures of Cretaceous age are known elsewhere in the Aus-

troalpine nappe complex (Hoke 1990; Neubauer *et al.* 2000 and references therein).

Semiductile strike-slip movements indicate both NW-directed (D_{4a}) and NE-directed contraction (D_{4b}) which are linked to the opening of ore-bearing quartz veins. Whereas extensional quartz veins trending NW–SE opened during NW-directed contraction are sub-vertical, those opened during NE-directed contraction reactivated the host-rock foliation that dips moderately to the SE, which indicates a strongly transpressive strike-slip regime. Sub-horizontal NW-directed contraction during D_{4a} coincides with both the orientation of tensional structures hosting the stibnite mineralization along the Isel Fault (east–west and (N)NW–(S)SE-trending, respectively) and the bimodal orientation (east–west and NNW–SSE-trending, respectively) of the older generation of lamprophyre dykes in the Kreuzeck–Goldeck Complex (Fig. 6a). K–Ar ages of these dykes (36 ± 1 Ma; Müller *et al.* 1992) suggest a Late Eocene to Early Oligocene time for D_{4a} deformation and hence, mineralization. Lamprophyres of this older age group show geochemical features which are typical for subduction or slab break-off related magmas (Müller *et al.* 1992; von Blanckenburg *et al.* 1998). Both NW–SE contraction during D_{4a} and NE–SW contraction during D_{4b} show a transition from semiductile to brittle deformation and the orientation of the principal axes of the stress ellipsoid remains more or less constant. Both events are related to the formation of ore-bearing quartz veins, which are only distinguishable by their different orientations. It is quite likely that D_{4a} and D_{4b} are spatial and/or temporal deviations of the same general stress field. We interpret this stress field to be related to (oblique) southeastward subduction of the Penninic microcontinent and the change from NW–SE to NE–SW contraction as a consequence of stress release due to slab break-off.

Subsequent development of conjugate brittle NE–SW-trending sinistral and NW–SE-trending dextral strike-slip faults points towards a more pure shear-like deformation regime under (N)NE–(S)SW oriented contraction (D_{4c}). Finally, ongoing uplift and exhumation of the whole area caused a shift in $\sigma_2 - \sigma_3$ orientations, due to a decrease in lithostatic pressure, leading to the development of normal faults trending NE–SW (D_{4d}). K–Ar and Rb–Sr ages of the younger, (N)NE–(S)SW-trending lamprophyric dykes (30 ± 2 Ma; Deutsch 1984; Müller *et al.* 1992) tentatively provide an age for (late) D_{4b} and/or D_{4c} (N)NE-directed contraction, or for the D_{4d} extension, respectively (Fig. 6b, c). Geochemical features of these younger dykes are transitional between subduction-related and within-plate characteristics (von

a) Late Eocene to Early Oligocene
(Oblique subduction of Penninic Microcontinent;
High heat-flow and mineralization concentrated at the KGC)

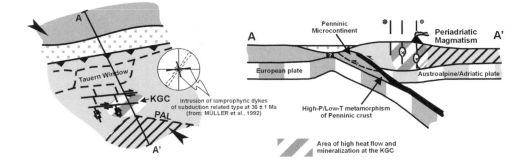

b) Oligocene
(Slab break-off, re-organization of regional stress-field
and initiation of uplift and exhumation of the Tauern Window)

c) Late Oligocene to Early Miocene
(Rapid uplift and exhumation of the Tauern Window;
High heat-flow and mineralization within the
metamorphic core complex)

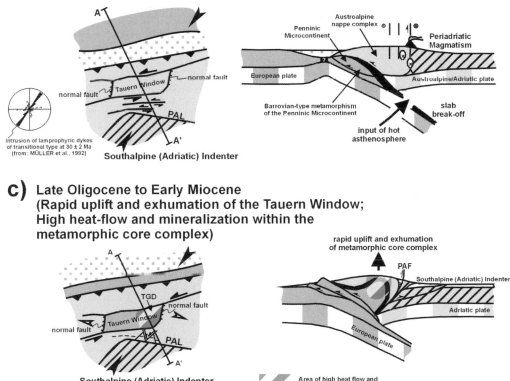

Fig. 6. Suggested development of Austroalpine crustal blocks between the Tauern Window and the approaching Southalpine indenter during the Palaeogene (strongly modified after Neubauer *et al.* 1999, 2000). KGC, Kreuzeck–Goldeck Complex; PAL, Periadriatic Lineament; TGD, Tauern Gold District.

Blanckenburg *et al.* 1998). D_{4c} and D_{4d} clearly post-date mineralization in the Kreuzeck–Goldeck Complex, but appear structurally related to the formation of mesothermal auriferous quartz veins from within the metamorphic core complex of the Tauern Window (Feitzinger & Paar 1991; Kurz *et al.* 1994; Amann *et al.* 1997; Horner *et al.* 1997).

The structural evolution of the investigated part of the Kreuzeck–Goldeck Complex reflects the large-scale geodynamic evolution of the Eastern Alps during the Tertiary which is related to Eocene accretion of the Penninic microcontinent, post-collisional indentation of the Southalpine (Adriatic) indenter and formation of associated fault systems (e.g. Ratschbacher *et al.* 1991; Frisch *et al.* 1998; Neubauer *et al.* 1999, 2000 and references therein). The Eocene dextral transpression along faults trending east–west has until now only been locally identified (e.g. Ring *et al.* 1989; Eder & Neubauer 2000). More obvious is the inversion of slip to sinistral displacement along these faults.

The suggested temperatures of gold mineralization in the Kreuzeck–Goldeck Complex (320–350 °C according to Cerny *et al.* 1997), which are well within the normal range of mesothermal gold deposits (McCuaig & Kerrich 1998), the connection of auriferous arsenopyrite–pyrite mineralization in the Siflitz–Guginock area with lateral movements within an oblique collisional regime, and their close spatial and temporal relationship to lamprophyre emplacement, show some remarkable similarities with the 'classical' mesothermal gold–lamprophyre association (e.g. Kerrich & Wyman 1994) and with orogenic gold mineralization elsewhere (Groves *et al.* 1998; Goldfarb *et al.* 2001).

Following the classification scheme of Dill (1998) the Sb mineralization in the Kreuzeck–Goldeck Complex may be specified as a Type I shear-zone hosted (sediment-hosted) mesothermal Sb deposit. Stable isotopic data and CH_4 contents of fluid inclusions from the Guginock Sb mineralization (Table 1; Mali 1996; Cerny *et al.* 1997) point towards reducing metamorphic fluids as the mineralizing agent. The elevated $\delta^{18}O$ values rule out a magmatic fluid source. According to the model proposed by Fyfe & Henley (1973), metamorphogenic mineralizing fluids generate aqueous fluids with variable amounts of CO_2, CH_4 and N_2 due to devolatilization under epidote amphibolite- and amphibolite-facies conditions (*c.* 480–580 °C; 4–5 kbar). During their ascent towards higher crustal levels these fluids take up other elements such as Au and/or Sb which are later precipitated in dilational structures.

Cretaceous metamorphism cannot have produced metamorphogenic fluids leading to ore precipitation along Oligocene structures. The only possibility to produce such fluids is due to devolatilization of subducted Permo–Mesozoic volcanosedimentary successions of Penninic units. On the other hand, lead isotopic data from similar Au–As and Sb– (As–, Au–, Pb–, Cu–) mineral deposits of the Kreuzeck–Goldeck Complex show elevated $^{207}Pb/^{204}Pb$ and $^{208}Pb/^{204}Pb$ ratios (Köppel 1997) which are typical for the entire Austroalpine nappe complex (Köppel 1983). The only exception to this general picture is the Guginock Sb mineralization which shows the most radiogenic $^{206}Pb/^{204}Pb$ and $^{207}Pb/^{204}Pb$ ratios, but relatively low $^{208}Pb/^{204}Pb$ ratios (Table 1; Köppel 1997). This suggests that different lead sources contributed to the Au–As and Sb–(As–, Au–, Pb–, Cu–) metallogenic district of the Kreuzeck–Goldeck Complex. However, many ore occurrences of the Kreuzeck–Goldeck Complex show lead isotope characteristics which indicate that rocks of the Austroalpine nappe complex was the major metal source (Köppel 1997).

Conclusion

Our new structural data show that auriferous arsenopyrite–pyrite and stibnite mineralization of the Kreuzeck–Goldeck Complex occurred during the same structural regime as the emplacement of contemporaneous Late Eocene to Early Oligocene shoshonitic/high-K calc-alkaline lamprophyres (Fig. 6a). Fluid flow in the uppermost crustal levels of the Austroalpine realm was channelled by approximately east–west-trending faults within a transpressional strike-slip regime which formed in response to oblique Late Eocene terrane accretion. The geodynamic setting of the structurally controlled auriferous arsenopyrite–pyrite and stibnite mineralization in the Kreuzeck–Goldeck Complex is significantly different from mesothermal gold deposits within the Tauern Window, which are attributed to uplift and exhumation of a metamorphic core complex during the Late Oligocene–Early Miocene (Fig. 6c). Our new results support models suggesting that epigenetic gold mineralization occurs in distinct short time intervals, closely related to the large-scale geodynamic evolution of orogens (e.g. Neubauer 2002).

We thankfully acknowledge H. Klob, Exploration manager of ARGOSY MINING Corporation (Canada), for permission to publish data on the Guginock property and to St. L. Chryssoulis, University of Western Ontario, for ion microprobe investigations. Special thanks are dedicated to Thomas Pettke and Richard Goldfarb for useful comments and specifically to John Ridley for his very detailed and constructive review, which helped a lot to improve an earlier version of this manuscript.

References

AMANN, G., DAXNER, G., NEUBAUER, F., PAAR, W.H. & STEYRER, H.P. 1997. Structural evolution of the Schellgaden Gold-District, eastern Tauern Window, Austria - a preliminary report. *Zentralblatt für Geologie und Paläontologie, Teil I*, **3/4**, 215–228.

ANGELIER, J. 1994. Fault Slip Analysis and Palaeostress Reconstruction. *In:* HANCOCK, P.L. (ed.) *Continental deformation*. Pergamon, Oxford, 53–100.

BREWER, M.S. 1970. *K-Ar age studies in the Eastern Alps - The Oberostalpindecke of Kärnten*. PhD thesis, University of Oxford.

CERNY, I., PAAR, W.H. & MALI, H. 1997. Antimon-(Arsen-, Gold-, Blei-, Kupfer-) sowie Gold-Arsen-Erzbezirk Kreuzeck-Goldeckgruppe. *In:* WEBER, L. (ed.) *Handbuch der Lagerstätten der Erze, Industrieminerale und Energierohstoffe Österreichs*. Archiv für Lagerstättenforschung der Geologischen Bundesanstalt Wien, **19**, 307–309.

CERNY, I., PAK, E. & SCHROLL, E. 1981. Schwefelisotopenzusammensetzung von Antimoniten und anderen Erzen aus Lagerstätten der Kreuzeckgruppe. *Anzeiger der mathematisch-naturwissenschaftlichen Klasse der Österreichischen Akademie der Wissenschaften*, **118**, 161–163.

DAL PIAZ, G.V., HUNZIKER, J.C. & MATNOTTI, G. 1973. Excursion to the Sesia Zone of the Schweiz. Mineralogische und Petrographische Gesellschaft, September 30th to October 3rd, 1973. *Schweizerische Mineralogisch Petrographische Mitteilungen*, **53**, 477–490.

DEUTSCH, A. 1980. Alkalibasaltische Ganggesteine aus der westlichen Goldeckgruppe (Kärnten/Österreich). *Tschermaks Mineralogisch Petrographische Mitteilungen*, **27**, 17–34.

DEUTSCH, A. 1984. Young Alpine dykes south of the Tauern Window (Austria): A K-Ar and Sr isotope study. *Contributions to Mineralogy and Petrology*, **85**, 45–57.

DEUTSCH, A. 1988. Die frühalpidische Metamorphose in der Goldeck Gruppe (Kärnten) - Nachweis anhand von Rb-Sr-Altersbestimmung und Gefügebeobachtungen. *Jahrbuch der Geologischen Bundesanstalt Wien*, **131**, 553–562.

DIAMOND, L.W. & WIEDENBECK, M. 1986. K-Ar Radiometric ages of the gold-quartz veins at Brusson, Val d'Ayas, NW Italy: Evidence of mid-Oligocene hydrothermal activity in the Northwestern Alps. *Schweizerische Mineralogisch Petrographische Mitteilungen*, **66**, 385–393.

DILL, H.G. 1998. Evolution of Sb mineralisation in modern fold belts: a comparison of the Sb mineralisation in the Central Andes (Bolivia) and the Western Carpathians (Slovakia). *Mineralium Deposita*, **33**, 359–378.

DROOP, G.T.R. 1985. Alpine metamorphism in the south-east Tauern Window, Austria: I. P-T variations in space and time. *Journal of Metamorphic Geology*, **3**, 371–402.

EBNER, F., CERNY, I., EICHHORN, R., GÖTZINGER, M., PAAR, W.H., PROCHASKA, W. & WEBER, L. 2000. Mineral Resources in the Eastern Alps and Adjoining Areas. *Mitteilungen der Österreichischen Geologischen Gesellschaft*, **92**, 157–184.

EDER, N. & NEUBAUER, F. 2000. On the edge of the extruding wedge: Neogene kinematics and geomorphology along the southern Niedere Tauern, Eastern Alps. *Eclogae geologicae Helvetiae*, **93**, 81–92.

FEITZINGER, G. & PAAR, W.H. 1991. Gangförmige Gold-Silber-Vererzungen in der Sonnblickgruppe (Hohe Tauern, Kärnten) . *Archiv für Lagerstättenforschung der Geologischen Bundesanstalt Wien*, **13**, 17–50.

FEITZINGER, G., PAAR, W.H., TARKIAN, M., RECHE, R., WEINZIERL, O., PROCHASKA, W. & HOLZER, H. 1995. Vein type Ag-(Au)-Pb, Zn, Cu-(W,Sn) mineralization in the Southern Kreuzeck Mountains, Carinthia Province, Austria. *Mineralogy and Petrology*, **53**, 307–332.

FRISCH, W., KUHLEMANN, J., DUNKL, I. & BRÜGL, A. 1998. Palinspastic reconstruction and topographic evolution of the Eastern Alps during late Tertiary tectonic extrusion. *Tectonophysics*, **297**, 1–15.

FÜGENSCHUH, B., SEWARD, D. & MANCKTELOW, N. 1997. Exhumation in a convergent orogen: The western Tauern Window. *Terra Nova*, **9**, 213–217.

FYFE, W.S. & HENLEY, R.W. 1973. Some thoughts on chemical transport processes, with particular reference to gold. *Mineral Sciences England*, **5**, 295–303.

GOLDFARB, R.J., GROVES, D.I. & GARDOLL, S. 2001. Orogenic gold and geologic time: a global synthesis. *Ore Geology Reviews*, **18**, 1–75.

GROVES, D.I., GOLDFARB, R.J., GEBRE-MARIAM, M., HAGEMANN, S.G. & ROBERT, F. 1998. Orogenic gold deposits: A proposed classification in the context of their crustal distribution and relationship to other gold deposit types. *Ore Geology Reviews*, **13**, 7–27.

HANDLER, R., KURZ, W. & BERTOLDI, C. 2001. [40]Ar/[39]Ar dating of white mica from eclogites of the Tauern Window (Eastern Alps, Austria) and the problem of excess argon in phengites. *EUG XI Journal of Conference Abstracts*, **6**, 595.

HEINZ, H. 1987. Geologie der östlichen Goldeckgruppe (Kärnten). *Jahrbuch der Geologischen Bundesanstalt Wien*, **130**, 175–203.

HIESSLEITNER, G. 1949. Die geologischen Grundlagen des Antimonbergbaues in Österreich. *Jahrbuch der Geologischen Bundesanstalt Wien*, **92**, 33–39.

HOKE, L. 1990. The Altkristallin of the Kreuzeck Mountains, SE Tauern Window, Eastern Alps - Basement Crust in a Convergent Plate Boundary Zone. *Jahrbuch der Geologischen Bundesanstalt Wien*, **133**, 5–87.

HORNER, H., NEUBAUER, F., PAAR, W.H., HANSMANN, W., KÖPPEL, V. & ROBL, K. 1997. Structure, mineralogy, and Pb isotopic composition of the As-Au-Ag deposit Rotgülden, Eastern Alps (Austria): significance for formation of epigenetic ore deposits within metamorphic domes. *Mineralium Deposita*, **32**, 555–568.

INGER, S. & CLIFF, R.A. 1994. Timing of metamorphism in the Tauern Window, Eastern Alps: Rb-Sr ages and fabric formation. *Journal of Metamorphic Geology*, **12**, 695–707.

KERRICH, R. & WYMAN, D.A. 1990. Geodynamic setting

of mesothermal gold deposits: an association with accretionary tectonic regimes. *Geology*, **18**, 882–885.

KERRICH, R. & WYMAN, D.A. 1994. The mesothermal gold-lamprophyre association: significance for an accretionary geodynamic setting, supercontinent cycles, and metallogenic processes. *Mineralogy and Petrology*, **51**, 147–172.

KÖPPEL, V. 1983. Summary of Lead Isotope Data from Ore Deposits of the Eastern and Southern Alps: Some Metallogenic and Geotectonic Implications. *In:* SCHNEIDER, H.-J. (ed.) *Mineral Deposits of the Alps and of the Alpine Epoch in Europe*. Springer, Berlin Heidelberg, 162–168.

KÖPPEL, V. 1997. Bleiisotope. *In:* WEBER, L.*In: Handbuch der Lagerstätten der Erze, Industrieminerale und Energierohstoffe Österreichs*. Archiv für Lagerstättenforschung der Geologischen Bundesanstalt Wien, **19**, 485–495.

KRUHL, J.H. 1993. The P-T-d development at the basement-cover boundary in the north-eastern Tauern Window (Eastern Alps): Alpine continental collision. *Journal of Metamorphic Geology*, **11**, 31–47.

KURZ, W., NEUBAUER, F., GENSER, J. & HORNER, H. 1994. Sequence of Tertiary Brittle Deformations in the Eastern Tauern Window (Eastern Alps, Austria). *Mitteilungen der Österreichischen Geologischen Gesellschaft*, **86**, 153–164.

LAHUSEN, L. 1972. Schicht- und zeitgebundene Antimonit-Scheelit-Vorkommen und Zinnobervererzungen in Kärnten und Osttirol / Österreich. *Mineralium Deposita*, **7**, 31–60.

LIU, Y., GENSER, J., HANDLER, R., FRIEDL, G. & NEUBAUER, F. 2001. ^{40}Ar/^{39}Ar muscovite ages from the Penninic-Austroalpine plate boundary, Eastern Alps. *Tectonics*, **20**, 526–547.

MALI, H. 1996. *Bildungsbedingungen von Quecksilber- und Antimonlagerstätten im Ostalpin (Österreich)*. Ph D Thesis, Montanuniversität Leoben.

MCCUAIG, T.C. & KERRICH, R. 1998. P-T-t-deformation-fluid characteristics of lode gold deposits: evidence from alteration systematics. *Ore Geology Reviews*, **12**, 381–453.

MCLENNAN, J.F. 1915. Quartz veins in lamprophyre intrusions. *English Mining Journal*, **99**, 11–13.

MÜLLER, D., STUMPFL, E.F. & TAYLOR, W.R. 1992. Shoshonitic and Alkaline Lamprophyres with Elevated Au and PGE Concentrations from the Kreuzeck Mountains, Eastern Alps, Austria. *Mineralogy and Petrology*, **46**, 23–42.

NEUBAUER, F. 2002. Contrasting Late Cretaceous to Neogene ore provinces in the Alpine-Balkan-Carpathian-Dinaride belt. *In:* BLUNDELL, D.J., NEUBAUER, F. & VON QUADT, A. (eds) *The Timing and Location of Major Ore Deposits in an Evolving Orogen*. Geological Society, London, Special Publications, **204**, 81–102.

NEUBAUER, F., GENSER, J. & HANDLER, R. 2000. The Eastern Alps: Result of a two-stage collision process. *Mitteilungen der Österreichischen Geologischen Gesellschaft*, **92** (1999), 117–134.

NEUBAUER, F., GENSER, J., KURZ, W. & WANG, X. 1999. Exhumation of the Tauern window, Eastern

Alps. *Physics and Chemistry of the Earth Part A: Solid Earth Geodesy*, **24**, 675–680.

PAAR, W.H. 1995. *Structurally controlled Au-As mineralization in the Kreuzeck-Goldeck complex, Carinthia*. Unpublished internal company report.

PETTKE, T. & DIAMOND, L.W. 1997. Oligocene gold quartz veins at Brusson, NW Alps: Sr isotopes trace the source of ore-bearing fluid to over a 10-km depth. *Economic Geology*, **92**, 389–406.

PETTKE, T. & FREI, R. 1996. Isotope systematics in vein gold from Brusson, Val d'Ayas (NW Italy): 1. Pb/Pb evidence for a Piemonte metaophiolite Au source. *Chemical Geology*, **127**, 111–124.

PETTKE, T., DIAMOND, L.W. & VILLA, I.M. 1999. Mesothermal gold veins and metamorphic devolatilization in the northwestern Alps: The temporal link. *Geology*, **27**, 641–644.

PETTKE, T., FREI, R., KRAMERS, J.D. & VILLA, I.M. 1997. Isotope systematics in vein gold from Brusson, Val d'Ayas (NW Italy): 2. (U + Th)/He and K/Ar in native Au and its fluid inclusions. *Chemical Geology*, **135**, 173–187.

POHL, W. & BELOCKY, R. 1994. Alpidic metamorphic fluids and metallogenesis in the Eastern Alps. *Mitteilungen der Österreichischen Geologischen Gesellschaft*, **86**, 141–152.

RATSCHBACHER, L., FRISCH, W., LINZER, G. & MERLE, O. 1991. Lateral Extrusion in the Eastern Alps, part 2: Structural Analysis. *Tectonics*, **10**, 257–271.

RING, U., RATSCHBACHER, L., FRISCH, W., BIEHLER, D. & KRALIK, M. 1989. Kinematics of the Alpine plate margin: structural styles, strain and motion along the Penninic-Austroalpine boundary in the Swiss-Austrian Alps. *Journal of the Geological Society, London*, **146**, 835–849.

SACHSENHOFER, R.F. 2001. Syn- and post-collisional heat flow in the Cenozoic Eastern Alps. *Geologische Rundschau*, **90**, 579–592.

SCHROLL, E. 1954. Ein Beitrag zur geochemischen Analyse ostalpiner Blei-Zink-Erze. *Mitteilungen der Österreichischen Mineralogischen Gesellschaft Sonderheft*, **3**, 1–85.

THÖNI, M. 1999. A review of geochronological data from the Eastern Alps. *Schweizerische Mineralogisch Petrographische Mitteilungen*, **79**, 209–230.

VON BLANCKENBURG, F. & DAVIES, J.H. 1995. Slab breakoff: A model for syncollisional magmatism and tectonics in the Alps. *Tectonics*, **14**, 120–131.

VON BLANCKENBURG, F., KAGAMI, H., DEUTSCH, A., OBERLI, F., MEIER, M., WIEDENBECK, M., BARTH, S. & FISCHER, H. 1998. The origin of Alpine plutons along the Periadriatic Lineament. *Schweizerische Mineralogisch Petrographische Mitteilungen*, **78**, 55–66.

WALLNER, P. 1981. *Integrierte Rohstoffsuche in der Kreuzeckgruppe (Kärnten/Österreich) mit besonderer Berücksichtigung der schichtgebundenen Kieslagervererzungen im Raume Strieden-Knappenstube und Politzberg*. PhD Thesis, Montanuniversität Leoben.

WEBER, L. 1989. Zur Geologie der Antimonvererzungen des Osttiroler Anteils der Kreuzeckgruppe. *Archiv für Lagerstättenforschung der geologischen Bundesanstalt Wien*, **10**, 65–74.

WEBER, L. (ed.) 1997. Handbuch der Lagerstätten, der Erze, Industrieminerale und Energierohstoffe Österreichs. *Archiv für Lagerstättenforschung der geologischen Bundesanstalt Wien*, 19.

WYMAN, D.A. & KERRICH, R. 1988. Alkaline magmatism, major structures, and gold deposits: implications for greenstone belt gold metallogeny. *Economic Geology*, **83**, 451–458.

WYMAN, D.A. & KERRICH, R. 1989. Archean shoshonitic lamprophyres associated with Superior Province gold deposits: distribution tectonic setting, noble metal abundance and significance for gold mineralization. *Economic Geology, Monograph*, **6**, 651–667.

ZIMMERMANN, R., HAMMERSCHMIDT, K. & FRANZ, G. 1994. Eocene high pressure metamorphism in the Penninic Units of the Tauern Window (Eastern Alps): Evidence from $^{40}Ar/^{39}Ar$ dating and petrological investigations. *Contributions to Mineralogy and Petrology*, **117**, 175–186.

The Elatsite porphyry copper deposit in the Panagyurishte ore district, Srednogorie zone, Bulgaria: U–Pb zircon geochronology and isotope-geochemical investigations of magmatism and ore genesis

ALBRECHT VON QUADT[1], IRENA PEYTCHEVA[2], BORISLAV KAMENOV[3], LORENZ FANGER[1], CHRISTOPH A. HEINRICH[1] & M. FRANK[1]

[1]*Isotope Geochemistry and Mineral Resources, Department of Earth Sciences, ETH-Zurich, Sonneggstr. 5, CH-8092 Zurich, Switzerland*
(e-mail: vonquadt@erdw.ethz.ch)

[2]*Central Laboratory of Mineralogy and Crystallography, Bulgarian Academy of Science, 1113 Sofia, Bulgaria*
(e-mail: peytcheva@erdw.eth.ch)

[3]*Department of Mineralogy, Petrology and Economic Geology, Sofia University St. Kliment Ohridski, 15, Bd. Tzar Osvoboditel, 1505 Sofia, Bulgaria (e-mail: kamenov@gea.uni-sofia.bg)*

Abstract: Single zircons from several porphyry dykes bracketing the time of formation of the Elatsite porphyry Cu–Au deposit (Bulgaria) were dated by high-precision U–Pb isotope analysis, using thermal ionization mass spectrometry (TIMS). On the basis of cross-cutting relationships, and the mineralogy and geochemistry of igneous and altered rocks, five dyke units are distinguished. The earliest porphyry dyke is associated with, and overprinted by, the main stage of ore-related veining and potassic alteration. U–Pb analyses of zircons yield a mean $^{206}Pb/^{238}U$ age of 92.1 ± 0.3 Ma, interpreted to reflect the time of intrusion. Zircons of the latest ore forming dyke, crosscutting the main stage veins but still associated with minor potassic alteration and veining, give an intrusion age of 91.84 ± 0.3 Ma. Thus, ore mineralization is confined by individually dated igneous events, indicating that the entire time span for the ore-forming magmatism and high temperature hydrothermal activity extended over a maximum duration of 1.1 Ma, but probably much less. Zircon analyses of a late ore dyke cutting all ore veins and hosting pyrite as the only sulphide mineral give a concordant $^{206}Pb/^{238}U$ age of 91.42 ± 0.15 Ma. Based on spatial relationships of the magnetite–bornite–chalcopyrite assemblage with coarse-grained hydrothermal biotite and K-feldspar, a Rb–Sr age of 90.55 ± 0.8 Ma is calculated using the two K-rich minerals. This age is interpreted as a closing date for the Rb–Sr system at T ≈ 300 °C consistent with published K–Ar data. Therefore the entire lifespan of the magmatic–hydrothermal system is estimated to have lasted about 1.2 Ma. Soon after, the Cretaceous complex was exposed by erosion, as shown by palaeontologically dated (Turonian; 91–88.5 Ma) sandstones containing fragments of porphyry dykes.

Geochemical discrimination ratios suggest a mixed mantle and crustal source of the Cretaceous magma. Isotope analyses of Sr, Nd and Hf confirm the conclusion that all porphyry rocks within and around the Elatsite deposit originate from an enriched mantle source at Cretaceous times, with crustal contamination indicated by moderately radiogenic Pb.

Rigorous geochemical investigation and radiometric dating are powerful tools to unravel the processes of magma-related hydrothermal ore formation, provided that they are combined with careful geological documentation of field relations. The timing of ore-forming processes requires the simultaneous use of several isotope methods to define the time span of geological events in one deposit in the broader tectonic framework, including magmatism, metamorphism and hydrothermal metal precipitation. The investigation of dyke swarms offers a considerable potential for identifying magmatic and tectonic events in calc-alkaline magmatic arcs, eroded to the plutonic–volcanic interface, where lava sequences are poorly preserved and plutonic rocks are hydrothermally altered. They provide key information through their cross-cutting relation-

From: BLUNDELL, D.J., NEUBAUER, F. & VON QUADT, A. (eds) 2002. *The Timing and Location of Major Ore Deposits in an Evolving Orogen*. Geological Society, London, Special Publications, **204**, 119–135. 0305-8719/02/$15.00 © The Geological Society of London 2002.

ships and spatial orientations, and they can be dated by robust and precise U–Pb geochronology of zircons. In addition, their geochemical characteristics can be used to define the origin of ore-forming magmas.

The Elatsite Cu–Au(–PGE) porphyry deposit of the Srednogorie zone in Bulgaria provides a good opportunity for studying the genetic relationships between magmatism, alteration and ore forming processes, as well as the lifespan of porphyry systems. The deposit is connected to multiphase intrusions (Bogdanov 1987; Popov *et al.* 2000a) of Late Cretaceous porphyry dykes into basement rocks of Palaeozoic age. The main goal of this paper is to present a precise geochronological study of different intrusion episodes, an isotope-geochemical and petrological investigation of the igneous rocks and their genetic relation to the hydrothermal mineralization, which will form a basis for geodynamic interpretations within the Cretaceous magmatic and metallogenic belt of the Carpathian–Balkan orogen.

Geological setting

The Elatsite Cu–Au porphyry deposit is the largest active mine in Bulgaria. It is located on the Etropole ridge of the Western Balkan Mountains at the northern end of the Panagyurishte ore district of the Central Srednogorie zone (Bonchev 1970; Vassilev & Staykov 1991; Fig. 1). It is considered as part of the 'Tethyan Eurasian Metallogenic Belt': an elongated sigmoidal structure of intensive Late Cretaceous to Early Tertiary (Karamata *et al.* 1997) magmatic activity and Cu mineralization. The belt can be followed from the Apuseni Mountains of Romania in the west, through the Timok area in Serbia and Srednogorie Zone in Bulgaria (Jankovic 1977, 1997).

The Elatsite deposit is associated with Late Cretaceous subvolcanic bodies and porphyry dykes (Kalaidjiev *et al.* 1984), intruded in rocks of the Berkovitsa group (Cambrian, Haydoutov *et al.* 1979; Haydoutov 1987) and granodiorites of the Vezhen pluton (Carboniferous age, 314 ± 4 Ma, Peytcheva & von Quadt, unpubl. data; Fig. 1). Greenschist facies metamorphosed rocks of the Berkovitsa group (Haydoutov *et al.*1979), together with Mid-Triassic limestones, are thrust over rocks of the Vezhen pluton (the so-called Kashana Nappe). The Berkovitsa group consists of phyllites, diabases, chlorite and actinolite schists, metamorphosed greywackes and sandstones with island arc affinity. The emplacement of the rocks of the Vezhen pluton caused contact metamorphism in the rocks of the Berkovitsa group, forming biotite–feldspar and hornblende–pyroxene hornfelses and amphibolites, and actinolite and two-mica andalusite–cordierite schists (Trashliev 1961). All these rocks (Berkovitsa group, Vezhen pluton, Elatsite dykes) host ore mineralization.

The Late Cretaceous dykes at Elatsite and around the deposit commonly occur in swarms with different orientations. Dyke frequency varies from isolated individual intrusions to closely spaced swarms separated by narrow screens of country rocks. Individual dykes are generally up to several metres wide, commonly steeply dipping, and generally have sharp chilled contacts. Intrusion mechanism is predominantly brittle, reflected by straight contacts. Data from outcrops characteristically show at least three sets of spatial orientation. The first set (striking east–west) follows the contact between the Vezhen pluton and Berkovitsa group. The thickest dykes are preferentially emplaced with this orientation. One of the largest dykes, the so-called 'big dyke', is approximately 4 km long and 100–450 m wide and dips 40–45° to the south. This dyke is probably a hypabyssal branch of a larger intrusive body not exposed at the surface. Thinner dykes of the same composition and nearly the same strike, but steeper dip, cut the Vezhen pluton. All these dykes are considered to be closely related to the main Cu–Au–PGE mineralization and are called here the 'ore-related' group. Dykes striking NE–SW are close to the orientation of the most important faults in the deposit. Usually, they are thinner (up to 5–10 m) and obviously form a second set. A set striking NNE–SSW comprises numerous thin dykes, many of them are aplitic. Dyke orientation is, however, not a reliable guide to the relative chronology.

No coeval volcanics are exposed in the vicinity of the Elatsite mine, but 6.5 km to the SSE, extrusive andesites and basaltic andesites host one of the biggest Au deposits in Europe, the Chelopech high-sulphidation epithermal gold deposit (Petrunov 1995; Andrew 1997; Moritz *et al.* 2001). Structural investigations have led some authors to consider all these rocks as one volcano-intrusive centre called the 'Elatsi–Chelopech' magmatic complex (Popov & Mutafchiev 1980; Popov *et al.*1983; Popov & Kovachev 1996; Popov & Popov 1997). The subvolcanic bodies at Elatsite and the surrounding area are assumed to be late Cretaceous in age, on the basis of their relationship to palaeontologically dated sediments and limited K–Ar isotope dating (Lilov & Chipchakova 1999). Conglomerates, sandstones and coal-bearing strata exposed 6–20 km WSW of Elatsite were determined as Turonian in age (Moev & Antonov 1978). They contain andesitic clasts, are possibly younger than the first volcanic and intrusive activity, but are intruded by subvolcanic bodies with dacite–andesite composition (Popov *et al.* 2000a). Lower Senonian (88.5–86.6 Ma)

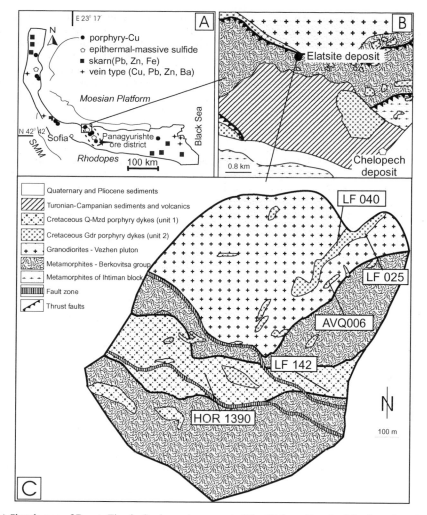

Fig. 1. (**a**) Sketch map of Banat–Timok–Srednogorie segment of the 'Tethyan Eurasian Metallogenic Belt' (modified from Jankovic 1976, 1977). (**b**) Sketch map of the northern part of the Panagyurishte ore district between Elatsite and Chelopech deposits (modified from Popov *et al.* 2000*b*). (**c**) Detailed geological map of the open pit for the Elatsite deposit; localities are shown where samples HOR 1390, LF 025 LF 040, LF 142 and AVQ006 were collected for U–Pb dating.

volcanomictic sandstones (Karagyuleva *et al.* 1974) overlie the basaltic–andesites of Chelopech. The 'big dyke' of Elatsite was determined as Turonian in age (*c.* 90 Ma) by Lilov & Chipchakova (1999), using K–Ar analyses of whole-rock samples. This age overlaps with the limited K–Ar determinations of basaltic trachyandesites at Chelopech (*c.* 91 Ma, Lilov & Chipchakova 1999). The same authors dated the later stage of magmatites in the Chelopech region and related alteration

and mineralization as Senonian, with quite a wide age span (88–65 Ma).

Petrology and geochemistry

The dykes in the Elatsite deposit and close surroundings range from almost aphyric (rare) to porphyritic in texture (more common). Based on field evidence and cross-cutting relationships between dykes and hydrothermal veins (Fanger

2001) and our petrographical observations and geochemical data, the intrusives can be divided into several units.

Quartz monzodiorite porphyries (unit 1) are the first and volumetrically most important part of the ore-related group of dykes, including the 'big dyke' of the Elatsite deposit.

Granodiorite porphyries (with minor granite porphyries and quartz syenite porphyries) form the second generation of dykes *(unit 2)* and intrude as multiple dykes into basement rocks and the first dyke generation. Their main orientation is NE–SW (striking) with steep dips. Characteristic for all rock varieties in this group are porphyritic textures with needle-like amphiboles.

K-feldspar-rich thin aplitic dykes represent the third and final generation of ore-related dykes *(unit 3)*, probably intruded after the second one as they show textural transitions along some of the dykes. These dykes are mainly orientated NNE–SSW and are not cut by ore veins and significant potassic alteration. Based on the studies of Fanger (2001) the third generation represents apophyses of the second one.

Mafic dykes represent the fourth (micro-diorites, micro-monzodiorites, diorite porphyries and their quartz-bearing varieties; *unit 4*) and a fifth (quartz diorite porphyries; *unit 5*) generation of dykes observed in the area. They are rare in the Elatsite deposit and more abundant in the surrounding area. The dykes are mostly subvertical and 0.2–3.0 m wide but usually narrower than 1 m. Aphyric varieties predominate, but porphyritic varieties also occur.

To obtain representative material for the magmatic history of the area, 96 dykes were sampled within the open pit of the Elatsite deposit and in the broad surroundings between the towns of Etropole and Zlatitsa. Based on thin section studies, whole-rock K_2O contents in line with the correlation of petrochemical indexes of differentiation, low loss-on-ignition values (LOI <3 wt.%) and only minor alteration of igneous minerals, only 24 samples were selected for further studies.

The range in SiO_2 concentrations in dykes of units 1, 2 and 3 is between 60 and 76% (Fig. 2). The rocks comprise varieties from slightly SiO_2 oversaturated (7–16% normative quartz in unit 1) to strongly SiO_2 oversaturated (16–25% normative quartz in the dykes of unit 2 and 25–37%) in the unit 3 dykes. The general calc-alkaline affinities are apparent in the AFM diagram by the

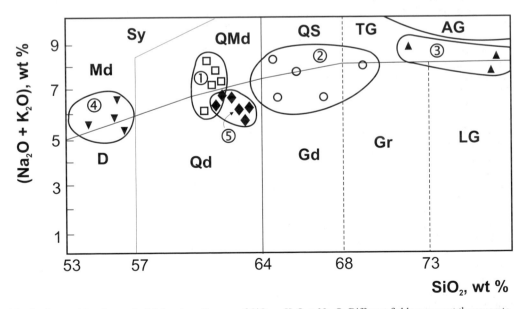

Fig. 2. Compositional spread of dykes in a diagram of SiO_2 v. $K_2O + Na_2O$. Different fields represent the separate dyke units (1 to 5). Only dykes without alteration are plotted. Abbreviations: Md, monzodiorite; D, diorite; Sy, syenite; QMd, quartz monzodiorite; Qd, quartz diorite; QS, quartz syenite; Gd, granodiorite; TG, transitional granite; Gr, granite; LG, leucogranite; AG, alkali granite. Symbol explanation: dyke units 1 (open squares), 2 (open circles) and 3 (filled triangles), units 4 (inverted filled triangles) and 5 (filled squares diamonds for the fresh samples and empty – for slightly altered samples).

rather large ranges of SiO_2 content and alumina saturation index (meta-aluminous to per-aluminous on ANK v. ACNK diagram).

Oxide and element variations in the ore-related dykes (units 1, 2 and 3) show considerable scatter on the plots, but sub-linear trends of decreasing MgO, FeO, CaO, TiO_2, Al_2O_3, P_2O_5, Sr, Ba, Zr, Y and Sc with increasing SiO_2 are revealed when strongly altered samples are omitted. These trends are broadly consistent with subtraction of plagioclase, clinopyroxene, amphibole, magnetite, zircon and apatite from the most mafic end member of the series. Ni and Sr variations, as well as MnO, are strongly influenced by the post-magmatic alterations. The Rb/Sr ratio increases with differentiation (the average values are 0.30 for unit 1, 0.50 for unit 2, and 1.40 for unit 3 dykes), which is indicative of the fractional crystallization of plagioclase in the magmatic evolution.

The SiO_2 concentration of dykes from unit 4 (53–63%) and unit 5 dykes (60–63%) are lower than those from units 1 to 3 (Fig. 2). Unit 4 dykes are slightly SiO_2 oversaturated (<5% normative quartz). Unit 5 dykes are strongly SiO_2 oversaturated (the average normative quartz is 15%). Negative correlations between SiO_2 and TiO_2, Al_2O_3, FeO, Fe_2O_3, MgO, CaO, Y, Sc, Co, Cr, Zn and positive ones with K_2O, U, Th, Rb, Pb, Ta, La, Nb are consistent with fractionation of clinopyroxene, hornblende, magnetite. The negative variations of Ti with Zr and with the ratio FeO_{tot}/MgO, and the invariably low TiO_2 concentrations at rather high Al_2O_3 are typical characteristics of arc magmas.

Spidergrams for incompatible elements, normalized to ocean-ridge granite (ORG; Pearce *et al.* 1984), show that LIL elements K, Rb, Th, Ta all increase, while Ba, Sr, Zr, Sm, Y and Yb decrease with fractionation (Fig. 3, Table 1). The LIL/HFS ratios are high, characteristic of subduction-related granitoids. The pronounced depletion of Ta and enrichment of Nb (Fig. 3) is an interesting geochemical signature for these dykes, representing not only typical arc-magma characteristics, but also the influence of a specific mantle contribution. The subduction component of the geochemical signature of the dykes increases from unit 1 to unit 3 (Th, Rb, K). The geochemical fingerprint of units 4 and 5 (Fig. 3b) are comparable with those for units 1, 2 and 3 (Fig. 3a).

Chondrite-normalized REE patterns (Fig. 4) are in accord with an enriched mantle source. Contamination in the formation of this type of parental magma cannot be excluded within the quartz monzodiorite porphyries (unit 1) and granodiorite porphyries (unit 2) patterns (Fig. 4). In the leuco-monzogranite porphyry and leuco-quartz-syenite porphyry dykes (unit 3) a REE-

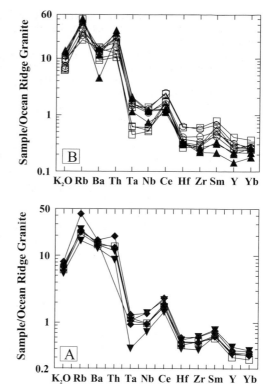

Fig. 3. Spider diagrams normalized to ocean-ridge granite reference (Pearce *et al.* 1984). (**a**) The plot contains data from dyke units 1, 2 and 3. (**b**) The plot contains data from dyke units 4 and 5. The symbols are the same as in Figure 2.

distribution typical for depleted remnant melts is noted: low sum of REEs and low amount of LREE and HREE normalized values. The progressive depletion of REEs and the decreasing Eu anomalies point to a common evolution with a gradually increasing proportion of feldspars in the fractionated assemblage.

Chondrite-normalized REE patterns of the dykes from unit 4 and 5 are similar to those of the ore-related dykes of units 1, 2 and 3 – moderately enriched in LREE relative to HREE. The negative Eu anomalies are very weak. The application of the discrimination diagrams of Mullen (1983), Pearce *et al.* (1984) and Wood *et al.* (1979) indicates a general destructive-plate boundary and calc-alkaline character of all dykes. In summary, all geochemical features are similar to those in arc-related suites and might reflect involvement of specialized mantle sources, previously modified by a subduction processes.

Table 1. *Representative geochemical analyses of the dyke units in the area of the Elatsite deposit*

Units	1		2		3		4		5	
Samples	E/8-b Mz diorite	E/15-a Mz diorite	E/43 Qz–mz diorite	E/41-g Qz–mz diorite	E/16-v Qz syenite	E/8-g Syenite	E/42-g Gabbro	E/16-b Diorite	E/46-b Qz diorite	E/19-b Grano-diorite
SiO$_2$	59.80	61.50	65.71	69.03	75.14	75.84	51.94	54.39	60.64	61.20
TiO$_2$	0.42	0.87	0.51	0.40	0.12	0.17	1.13	1.11	0.63	0.68
Al$_2$O$_3$	16.91	15.32	14.46	13.44	12.30	12.07	16.54	16.41	14.87	14.93
Fe$_2$O$_3$	2.58	2.00	1.32	0.60	0.45	0.44	3.20	2.93	1.51	1.34
FeO	2.31	4.51	4.49	4.05	1.02	0.70	6.64	5.74	6.28	4.48
MnO	0.18	0.35	0.39	0.44	0.12	0.06	0.51	0.24	0.48	0.21
MgO	2.25	3.00	2.12	1.27	0.36	1.10	4.73	3.98	3.28	2.95
CaO	5.60	4.54	2.68	2.35	1.22	0.42	5.86	6.06	4.78	5.73
Na$_2$O	4.30	3.44	4.30	3.61	3.15	3.06	2.78	3.96	3.36	3.64
K$_2$O	3.76	2.60	3.58	4.40	4.60	5.32	3.53	2.52	2.63	2.16
P$_2$O$_5$	0.31	0.25	0.25	0.17	0.10	0.09	0.31	0.33	0.24	0.31
H$_2$O$^-$	0.24	0.21	0.17	0.07	0.07	0.07	0.21	0.11	0.12	0.16
LOI	0.99	1.26	0.82	0.60	0.89	0.29	2.27	1.96	1.32	1.78
Total	99.68	100.57	100.08	100.43	99.54	99.63	100.02	99.74	100.13	99.57
Cr	23	56	77	72	47	3	65	41	108	68
Ni	9	21	45	41	16	4	19	19	58	44
Co	8.2	17.0	12.4	8.3	1.9	1.5	26.3	23.1	17.9	13.8
Sc	9.4	19.6	12.8	7.4	2.9	2.8	36.3	23.3	19.0	17.5
Y	18.1	28.5	19.5	16.6	14.3	10.2	33.6	30.2	24.0	21.6
Nb	5.1	13.7	10.9	11.4	12.1	7.4	10.2	14.0	9.4	8.5
Zr	97	206	142	115	80	74	173	209	159	162
Hf	2.6	5.4	3.7	3.3	2.9	2.9	4.4	0.50	4.1	4.1
Cu	207.3	50.5	42.1	34.6	51.5	140.8	44.8	53.2	62.3	44.7
Zn	33.1	80.2	53.5	39.3	12.3	11.8	90.5	80.7	75.0	69.8
Mo	2.7	3.1	8.7	9.7	2.3	2.8	4.3	2.8	13.3	4.0
Sr	919	374	329	231	95	94	396	361	411	397
Ba	763	784	664	634	573	226	1135	732	760	713
Rb	85	117	102	175	203	164	156	100	90	88
Ta	0.31	0.95	1.03	1.12	1.46	0.80	0.62	0.76	0.76	0.57
Pb	11.05	21.52	23.02	18.63	8.61	8.55	9.58	13.63	20.01	16.30
Th	9.91	15.25	12.78	18.69	25.18	18.71	7.24	9.43	10.38	10.06
U	3.14	4.45	4.12	7.84	5.58	4.72	1.94	2.39	4.16	4.14
La	30.73	39.22	35.95	32.70	25.11	23.45	34.37	39.15	32.50	30.27
Ce	58.74	72.06	62.26	54.35	44.82	39.29	65.96	77.52	61.26	57.86
Pr	6.55	7.97	6.61	5.69	4.40	3.88	7.89	9.05	6.96	6.47
Nd	26.44	28.16	25.08	20.15	16.33	11.90	32.76	35.74	27.93	26.61
Sm	4.70	4.75	4.57	3.74	2.97	1.90	6.64	7.06	5.55	4.96
Eu	1.46	0.96	1.15	0.92	0.61	0.24	1.73	1.92	1.39	1.14
Gd	3.52	4.07	3.61	2.92	2.34	1.61	6.41	5.17	4.51	4.47
Tb	0.56	0.66	0.64	0.46	0.40	0.27	1.04	0.98	0.67	0.67
Dy	3.10	4.39	3.54	2.64	2.21	1.46	5.98	5.29	4.20	3.82
Ho	0.71	0.88	0.72	0.62	0.52	0.31	1.23	1.14	0.84	0.81
Er	1.97	2.16	1.96	1.58	1.50	0.97	3.33	3.01	2.52	2.36
Tm	0.29	0.41	0.26	0.27	0.29	0.20	0.50	0.48	0.34	0.30
Yb	2.36	3.97	2.11	1.91	2.11	1.41	3.67	3.11	2.47	2.18
Lu	0.34	0.39	0.28	0.34	0.28	0.24	0.55	0.50	0.36	0.32

Mineralization and alteration

Mineralogical studies in the Elatsite deposit started with the beginning of exploration and became very active in the 1990's (Bogdanov 1987; Petrunov *et al.* 1992; Petrunov & Dragov 1993; Dragov & Petrunov 1996). Based on published data (Petrunov *et al.* 1992; Popov *et al.* 2000a) and pit mapping, Fanger (2001) distinguished the following sequence of mineral assemblages, named according to their most frequent and characteristic minerals and given in a succession of decreasing relative age:

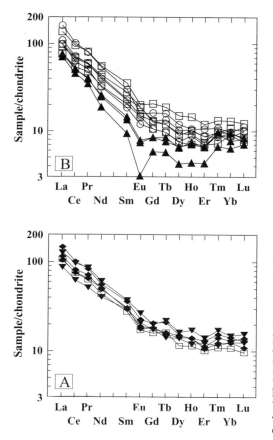

Fig. 4. Chondrite-normalized REE patterns of dykes from units 1–5. (a) Data from units 1, 2 and 3; (b) data from units 4 and 5. The symbols are the same as in Figure 2.

Pre-ore stage
  quartz + magnetite (qtz–mt)
Ore stages
  magnetite + bornite + chalcopyrite (mt–bn–cp)
  chalcopyrite + pyrite (cp–py)
  pyrite + quartz
Post-ore stages
  carbonate + zeolite + quartz (carb–zt–qtz)
  chalcocite + covellite + bornite (chalc–cov–bn)
  malachite + azurite + Fe-hydroxides

The *qtz–mt stage* is only observed in the Vezhen granodiorite. It consists of black, fine-grained veinlets. K-feldspar haloes indicate a close

relationship with potassic alteration and cross-cutting relationships with other veins and dykes document their very early formation, possibly even in pre-Mesozoic times.

The *magnetite–bornite–chalcopyrite stage* occurs locally in nests associated with rocks of pervasive potassic alteration, comprising secondary biotite (after amphibole) and K-feldspar (mostly in the matrix). A characteristic feature is the presence of PGE minerals and visible gold in the magnetite–bornite–chalcopyrite assemblage (Petrunov *et al.* 1992; Tarkian & Stribny 1999). Fluid inclusion studies in quartz of this mineralization stage show high salinities (over 50 wt% NaCl equivalent) and homogenization temperatures (T_h over 500 °C; Strashimirov & Kovachev 1992).

The *cp–py stage* is the economically most important mineralization event. It took place during and after intense stockwork veining and is associated with pervasive potassic alteration including partial replacement of plagioclase phenocrysts. It is widespread throughout the ore body and responsible for the main Cu grades. Chalcopyrite and pyrite can be disseminated in the rocks, occur as small grains in the quartz stockwork veins or in small sulphide-dominant veinlets. Molybdenite was introduced at this stage, but also precipitated abundantly in late (post-stockwork) veins. Fluid inclusion studies show a T_h between 470–450 °C and salinity around 50 wt% NaCl equivalent (Fanger 2001).

The *pyrite–quartz stage* is best developed in the marginal part of the mine, extending above and beyond the Cu cut-off contour (0.15%). Straight veins contain milky, vuggy quartz, pyrite and sometimes white to yellow calcite. Pyrite can also be found disseminated in the wallrock, in haloes of feldspar-destructive (sericite ± clay minerals) alteration overprinting potassic alteration and stockwork veining. T_h of associated fluid inclusion shows a temperature range between 300 °C and 260 °C and a low salinity (Popov *et al.* 2000*b*).

Quartz calcite–zeolite marks the end of hypogene mineralization. This assemblage occurs in fine veinlets and cavities, together with actinolite needles. Zeolites include chabasite, stilbite, laumontite, heulandite which are associated with green fluorite. Veins can be up to several centimetres thick and may contain some remobilized sulphide masses comprising chalcopyrite, bornite, chalcocite and covellite.

Secondary Cu enrichment due to weathering is restricted to the upper-most parts of the mine and then concentrated along faults. It formed minerals like malachite, azurite, cuprite, tenorite, claudeite, chalcophilite, and koettigite (Fanger 2001).

Analytical techniques

The procedures for the whole rock isotopic analysis of Sm, Nd, Rb and Sr are modified from those reported by Richard *et al.* (1976) and further detailed by von Quadt (1997). Nd isotopic ratios were normalized to $^{146}Nd/^{144}Nd = 0.7219$. Analytical reproducibility was estimated by periodical measurements of the La Jolla standard (Nd) as well as the NBS 987 (Sr). The mean of 12 runs during this work was $^{143}Nd/^{144}Nd = 0.511841 \pm 0.000007$, and 10 runs of the NBS 987 standard show an $^{87}Sr/^{86}Sr$ ratio of 0.710235 ± 0.000006. Whole-rock major oxides (Table 1) were analysed in the Chemical Laboratory of the Department of Petrology, Sofia University, as well at the ETH-Zurich. Samples with large LOI (> 3%), significant alteration, or with veins were avoided. Trace-element analyses, including REE, were carried out by Excimer Laser Ablation ICP-MS (Elan 6000) of lithium tetraborate pellets (Günther *et al.* 2001).

High-precision 'conventional' U–Pb zircon analyses were carried out on single zircon grains. Selected zircons were air-abraded to remove marginal zones with lead loss, washed in warm 4 N nitric acid and rinsed several times with distilled water and acetone in an ultrasonic bath. Dissolution and chemical extraction of U and Pb was performed following Krogh (1973), using miniaturized bombs and anion exchange columns. Blanks for the entire procedure were < 2 pg Pb and 0.5 pg U. A mixed $^{205}Pb/^{235}U$ tracer solution was used for all analyses. Both Pb and U were loaded with silica gel and phosphoric acid on single Re filaments and measured on a Finnigan MAT 262 thermal ionization mass-spectrometer using an ion counter system. The performance of the ion counter system was checked by repeated measurements of the NBS 982 standard solution. The reproducibility of the $^{207}Pb/^{206}Pb$ ratio (0.467070) is better than 0.05%. Mean age values are given at the 2 sigma level. For common lead measurements, ore minerals were dissolved in HNO_3 and the Pb was cleaned twice with a HCl–HBr process using 0.5 ml ion exchange resin. Pb was loaded with silica gel and H_3PO_4 on single Re filaments and measured with Faraday cups in static mode.

Hf isotope ratios in zircon were measured on a Nu Instruments multiple collector inductively coupled plasma mass spectrometer (MC-ICPMS; David *et al.* 2001). We checked if the high Zr concentration leads to significant matrix effects, which was not the case. During analysis, we measured the $^{176}Hf/^{177}Hf$ ratio of the JMC 475 standard of 0.282141 ± 5 (1 sigma) using the $^{179}Hf/^{177}Hf = 0.7325$ ratio for normalization (ex-

ponential law for mass correction). The Hf isotope data are summarized in Table 3. For the calculation of the ϵHf values the following present-day ratios $(^{176}Hf/^{177}Hf)_{CH} = 0.28286$ and $(^{176}Lu/^{177}Hf)_{CH} = 0.0334$ are used, and for 90 Ma a $^{176}Lu/^{177}Hf$ ratio of 0.0050 for all zircons was taken into account.

U–Pb-geochronology and Hf isotope geochemistry of the Cretaceous dykes

Isotope investigations of the Cretaceous dykes hosting the mineralization of the Elatsite deposit were made with two main purposes: to date the ore-productive magmatism, and to better define the possible source of the magmas. For the first objective, the U–Pb zircon method was used. For the second, isotope tracing by Sr–Nd–Pb in whole rocks and Hf in zircon were selected.

Zircons were separated from three rock types cropping out in the mine area: quartz monzodiorite porphyry ('big dyke', HOR 1390) representing unit 1; diorite porphyry (LF 025) and granodiorite porphyry (AVQ 006) representing unit 2 of dyke generation. Both samples of unit 2 carry needle-amphibole and crop out in the eastern part of the open pit, close to the contact between the Vezhen granite and the schists of the Berkovitsa group (Fig. 1). They cut the quartz monzodiorite of the 'big dyke' (HOR 1390) with sharp contacts. The amphibole-bearing dykes of unit 2 are intruded in several pulses and cross-cutting relations are observed between them. LF 025 is still K feldspar-altered and mineralized but belongs to the earliest members of unit 2; AVQ 006 belongs to the dykes of unit 2, post-dating the deposition of economic ore and containing disseminated pyrite.

Nine zircon grains of the monzodiorite porphyry dyke ('big dyke') were dated (Table 2). One zircon grain (2324, Table 2, Fig. 5) has a bigger proportion of inherited lead, pointing to an age of 554 Ma (Fig. 5). Two zircon points (2270, 2315) are plotting on the concordia within their uncertainties and define a mean $^{206}Pb/^{238}U$ age of 92.1 ± 0.3 Ma. We interpret that these concordant grains most closely represent the time of intrusion of the big dyke. The regression line through all zircon analyses yields a lower intercept age of 92.3 ± 1.3 Ma and an upper intercept age of 554 ± 15 Ma. Based on a variable proportion of inherited lead this calculation leads to higher intercept errors. The abraded zircons contain a wide range of U concentrations from 75.2 to 973.3 ppm, the Pb concentrations in the same grains range from 5.397 ppm to 21.49 ppm. The interpretation of the upper intercept age points to

Table 2. *U–Pb zircon isotope data for porphyry dykes of the Elatsite deposit*

Sample	Weight (mg)	*	*	U (ppm)	Pb (ppm)	$^{206}Pb/^{204}Pb$	$^{206}Pb/^{238}U$	2σ error	$^{207}Pb/^{235}U$	2σ error	$^{207}Pb/^{206}Pb$	2σ error	$^{206}Pb/^{238}U$	$^{207}Pb/^{235}U$	$^{207}Pb/^{206}Pb$	*Rho.
													Apparent ages			
HOR1390																
2283	0.0055	prism	abr.	849.5	19.48	193.3	0.01680	0.00008	0.11577	0.00208	0.04997	0.00085	107.4	111.2	193.7	0.46
2284	0.0125	prism	abr.	973.3	21.49	268.4	0.01760	0.00009	0.12108	0.00099	0.04988	0.00031	112.5	116.1	189.5	0.66
2286	0.0095	prism	abr.	673.1	12.72	233.6	0.01463	0.00007	0.09778	0.00057	0.04848	0.00016	93.6	94.7	123.1	0.81
2287	0.0205	prism	abr.	495.7	11.15	157.1	0.01537	0.00008	0.10222	0.00133	0.04821	0.00053	98.4	98.8	109.2	0.50
2270	0.0185	prism	abr.	517.9	8.351	506.1	0.01435	0.00007	0.09472	0.00080	0.04788	0.00031	91.8	91.8	93.1	0.65
2315	0.0145	prism	abr.	865.5	14.29	469.9	0.01443	0.00008	0.09521	0.00143	0.04785	0.00062	92.3	92.3	92.0	0.48
2324	0.0096	prism	abr.	75.2	5.397	128.9	0.04446	0.00026	0.34696	0.00484	0.05658	0.00068	280.5	302.4	475.5	0.52
2519	0.0072	prism	abr.	301.2	5.391	1197	0.01655	0.00008	0.11458	0.00108	0.05019	0.00039	105.8	110.1	204.0	0.58
2520	0.0049	prism	abr.	223.1	3.889	1073	0.01629	0.00008	0.11149	0.00111	0.04962	0.00044	104.2	107.3	177.2	0.54
LF 025																
2400	0.0155	prism	abr.	124.2	1.962	1271	0.01442	0.00009	0.09555	0.00084	0.04804	0.00031	92.3	92.6	101.3	0.70
2398	0.0232	prism	abr	174.4	2.811	601.4	0.01433	0.00009	0.09562	0.00105	0.04838	0.00047	91.7	92.7	118.2	0.58
2396	0.0314	prism	abr	187.6	6.667	1378	0.03211	0.00157	0.24784	0.00141	0.05580	0.00015	204.4	224.8	449.5	0.88
2401	0.0136	prism	abr	181.8	7.176	56.55	0.01435	0.00008	0.09545	0.00363	0.04823	0.00174	91.8	92.6	111.0	0.48
2397	0.0225	prism	abr	195.9	3.025	2645	0.01428	0.00013	0.09503	0.00090	0.04825	0.00061	91.4	92.1	111.5	0.96
AFQ 006																
5392	0.0070	prism.	abr.	352.6	6.018	453.8	0.01446	0.00009	0.09595	0.00134	0.04811	0.00058	92.5	93.0	104.7	0.53
5393	0.0080	prism	abr.	420.0	7.109	1158	0.01522	0.00011	0.10346	0.00107	0.04930	0.00034	97.3	99.9	162.3	0.74
5394	0.0081	prism	abr.	161.1	3.437	157.1	0.01426	0.00009	0.09602	0.00307	0.04883	0.00146	91.2	93.1	139.7	0.44
5396	0.0090	prism	abr.	203.1	3.447	411.9	0.01414	0.00014	0.09404	0.00197	0.04822	0.00087	90.5	91.2	110.1	0.52
5446	0.0041	prism	n.abr.	268.4	4.678	413.6	0.01437	0.00008	0.09618	0.00125	0.04855	0.00053	91.9	93.2	125.9	0.45
5447	0.0092	prism	n.abr	106.1	2.529	410.3	0.02013	0.00011	0.14388	0.00230	0.05183	0.00078	128.5	136.5	278.1	0.45
5448	0.0051	prism	n.abr	185.3	3.165	420.5	0.01426	0.00007	0.09613	0.00163	0.04889	0.00078	91.3	93.2	142.6	0.37
5449	0.0141	prism	n.abr	132.7	2.152	616.5	0.01424	0.00007	0.09425	0.00123	0.04799	0.00053	91.2	91.5	98.7	0.49
5452	0.0073	prism	abr.	219.9	3.819	394.9	0.01430	0.00008	0.09456	0.00236	0.04796	0.00115	91.5	91.7	97.7	0.44

*abr., abraded, n.abr, non abraded, prism., prismatic.: Rho, correlation coefficient $^{206}Pb/^{238}U$ – $^{207}Pb/^{235}U$

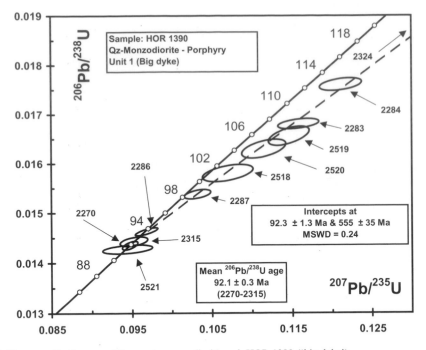

Fig. 5. U–Pb concordia diagram of the quartz monzodiorite rock HOR 1390 ('big dyke').

an input from rocks of Cambrian age. One possible source could be magmatic and sedimentary rocks of the Berkovitsa group (Haydoutov *et al.* 1979), which are Cambrian in age and have island-arc affinities. The ϵHf values of the zircons range from −2.48 to +5.6 (Table 3). Again, there is a strong correlation between the incorporation of an old lead component and low Hf ratios: the zircon analyses with the highest proportion of an old component have the lowest ϵHf value. The

concordant zircon data points show ϵHf values between +4.70 and +5.62.

Four out of five zircon analyses of the dyke LF 025 are concordant within their uncertainties and yield a mean $^{206}Pb/^{238}U$ age of 91.84 ± 0.31 Ma (Fig. 6). We have decided to take the $^{206}Pb/^{238}U$ mean age as the most likely intrusion age, and there is no reason to try an isochron age calculation. Only one zircon grain (2396) shows a minor proportion of inherited lead, pointing towards an upper intercept age of *c.* 600 Ma (Fig. 6). The U and Pb contents vary in a small range from 124 to 196 ppm and from 1.962 to 7.176 ppm, respectively. The best age estimate for the crystallization of this porphyry dyke is suggested by the calculated (mean value) $^{206}Pb/^{238}U$ age of 91.84 ± 0.31 Ma (Ludwig 1999). The concordant zircon crystals have ϵHf values between +4.56 and +7.29, except for one zircon data point with a small proportion of an old lead component, which has an ϵHf value of +2.51. The strong correlation between the incorporation of inherited lead and the ϵHf values is also shown by zircons from sample HOR 1390.

Seven out of nine zircon analyses of the post-ore granodiorite porphyry dyke (AVQ006) are concordant and yield a mean $^{206}Pb/^{238}U$ age of 91.42 ± 0.15 Ma (Ludwig 1999). Three zircon grains show a minor proportion of inherited lead.

Table 3. *Hf zircon isotope data for porphyry dykes of the Elatsite deposit*

Sample	$^{176}Hf/^{177}Hf$	2σ error	ϵHf	ϵHf $_{T90Ma}$
LF 025				
2400	0.282909	0.000004	+4.84	+6.58
2401	0.282928	0.000006	+5.52	+7.29
2397	0.282873	0.000010	+3.57	+5.34
2396	0.282794	0.000007	+0.78	+2.51
2398	0.282852	0.000006	+2.82	+4.56
HOR 1390				
2324	0.282653	0.000010	−4.21	−2.48
2270	0.282881	0.000008	+3.85	+5.62
2318	0.282802	0.000009	+1.06	+2.79
2287	0.282856	0.000010	+2.97	+4.70
2286	0.282851	0.000007	+2.79	+4.53

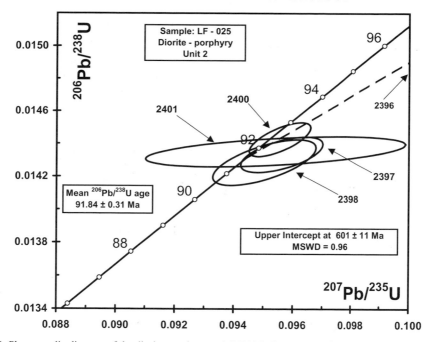

Fig. 6. U–Pb concordia diagram of the diorite porphyry rock LF 025; the mean age is calculated from the weighted average of the $^{206}Pb/^{238}U$ ratios.

Similar to those from sample HOR 1390 (Fig. 5) and LF 025 (Fig. 6), the upper intercept age of 464 Ma (Fig. 7) points to pre-Variscan crustal material. Most zircon grains are prismatic to long prismatic and have U contents ranging from 106.1 to 420 ppm and the Pb contents range from 2.152 to 7.109 ppm in the same zircons.

Isotope geochemistry of porphyries and their magma sources

The Sr isotope analyses of whole rock samples for the Elatsite deposit give initial $(^{87}Sr/^{86}Sr)_i$ ratios between 0.70492 and 0.70571 (Table 4) and the calculated ϵNd_{190} values are between −0.03 and +2.27 (Table 5). The low negative ϵNd values suggest a contribution of a mixed mantle–crustal material, and are in agreement with the low Nd model ages (Table 5). The upper intercept ages of zircon discordia lines also give an indication of a pre-Variscan crustal (Cm?) contribution.

The geochemical characteristics of the Cenomanian dykes of Elatsite deposit suggest a mixed mantle–crust source for the Late Cretaceous magmas. Isotope analyses of Sr, Nd, and Hf confirm this conclusion. All porphyritic rocks within and around the Elatsite deposit reflect an enriched mantle source in Cretaceous times,

mixed in varying proportions with crustal material. For rocks that assimilated or mixed with the mantle magma, we can infer mixed crust–mantle to crustal characteristics: characteristic $(^{87}Sr/^{86}Sr)_i$ values remain at 0.705 even after mixing of mantle and crustal components and ϵNd_{T90} values range from 0.03 to +2.27. However, ϵHf values of the concordant zircon vary between +5.4 and +6.9, indicating a possible enriched mantle source (Blichert-Toft & Albarède 1997) for the porphyry intrusions, which otherwise is more typical for MORB and OIB environments (Nowell et al. 1998; Vervoort et al. 1999). All other zircon grains with inherited lead components as well as lower ϵHf values point to typical crustal characteristics of the assimilated magma.

Isotope investigation of alteration and mineralization

K-rich minerals at Elatsite and their close relationship with mineralization provides additional constraints on the timing of cooling history of the system. We selected coarse-grained biotite and potassium feldspar, precipitated in irregular veins together with the early mt–bn–chp assemblage, for study of Rb–Sr isotope systematics (Fig. 8).

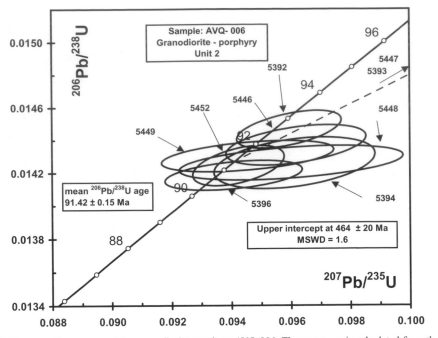

Fig. 7. U–Pb concordia diagram of the granodiorite porphyry AVQ 006. The mean age is calculated from the average of the $^{206}Pb/^{238}U$ ratios.

Table 4. *Rb–Sr isotope data for mineral and whole rock samples of the Elatsite deposit*

Sample No			Rb (ppm)	Sr (ppm)	$^{87}Rb/^{86}Sr$	$^{87}Sr/^{86}Sr$	2σ error 10xE-5	$(^{87}Sr/^{86}Sr)_{i90}$
	Minerals	Host rock/rock type						
LF-142	Bi from Bi–Fs nest	Gd of Vezhen pluton	624.8	22.08	82.72	0.81161	0.8E–5	0.70583
LF-142	Feldspar from Bi–Fs nest	Gd of Vezhen pluton	239.8	210.7	3.294	0.70942	0.5E–5	0.70521
LF-040	Chabasite	Bi–Amph dyke	53.59	2749	0.00563	0.70693	0.8E–5	0.70693
LF-040	Amphibole	Bi–Amph dyke	12.06	21.89	1.6729	0.70969	1.5E–5	0.70754
LF 025	Whole rock	Dr-porph	110.1	609.1	0.5231	0.70559	1.3E–5	0.70492
HOR 1390	Whole rock	Q–Mz–Dr-porph	95.67	714.2	0.3874	0.70576	1.3E–5	0.70526
E 8b	Whole rock	Mz–Dr-porph	81.01	936.9	0.2501	0.70543	3.0E–5	0.70511
E 43	Whole rock	Q–Mz–Dr-porph	88.64	422.6	0.6069	0.70941	1.0E–5	0.70862
AVQ 006	Whole rock	Gd-porph	89.77	975.9	0.2660	0.70611	0.6E–5	0.70577

Abbreviations: Q, quartz; Mz, monzonite; Dr, diorite; Gd, granodiorite; porph, porphyry; Bi, biotite; Fs, feldspar; Amph, amphibole

Table 5. *Sm–Nd isotope data for mineral and whole rock samples of the Elatsite deposit*

Sample No	Rock type	Sm (ppm)	Nd (ppm)	147Sm/ 144Nd	143Nd/ 144Nd	2σ error	$(\epsilon Nd)_{i90}$	T_{DM}	T_{CH}
LF 025	Dr-porph	6.32	32.95	0.1157	0.512589	1.6E–5	−0.03	0.88	0.093
HOR1390	Q–Mz–Dr-porph	3.84	20.89	0.1111	0.512704	0.6E–5	2.27	0.68	0.0
E 8g	Syenite-aplite	3.04	19.11	0.09621	0.512645	3.1E–5	1.29	0.67	0.0

Abbreviations: Q, quartz; Mz, monzonite; Dr, diorite; porph, porphyry.

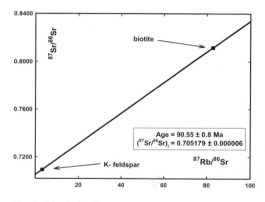

Fig. 8. Rb–Sr isochron plot for the biotite and feldspar from rock unit 2.

Biotite and feldspar define a two-point reference line, whose slope corresponds to an age of 90.55 ± 0.8 Ma with an initial $^{87}Sr/^{86}Sr$ ratio of 0.705179. This age is lower than the youngest U–Pb zircon age (91.42 Ma), but coincides with the Ar/Ar age determination on biotite from the Elatsite mine (Velichkova *et al.* 2001). The age of 90.5 Ma is interpreted as the closing time of the Rb–Sr and Ar/Ar isotope system at around 300 °C, and not as the time of mineral forming. This is an additional arguement, that the Rb–Sr, K–Ar and Ar–Ar isotope methods are not suitable for dating the high temperature ore-forming stages.

Fluid inclusion measurements in quartz of the mt–bn–chp paragenesis show a high salinity of over 50 wt% NaCl equiv and homogenization temperatures T_h of about 5007 °C (Strashimirov & Kovachev 1992). These observations point to a magmatic origin of the fluids during the high-temperature stages of ore formation at Elatsite bracket by zircon ages, about a million years before the cooling age of 90.5 Ma. The Sr initial, which is not significantly affected by this time difference, shows no notable influence of Sr from the Palaeogene host rocks during the magmatic or the ore-forming hydrothermal stage. Minerals of the post-ore phyllic assemblage (Table 4) show

slightly higher $^{87}Sr/^{86}Sr_{T90}$ ratios between 0.70520 and 0.70754. The enrichment of radiogenic ^{87}Sr in chabasite and amphibole of veins cutting a late dyke (unit 5) reflect either an evolution of the magma source or a growing influence of more radiogenic Sr added by fluid convection through basement rocks.

Lead isotope ratios are a sensitive tool to trace the evolution of the lead-bearing minerals in ore deposits and to determine the lead sources. We have analysed pyrite and chalcopyrite from the main ore stage (py–chp) and the post-ore qtz–py stage. All $^{207}Pb/^{204}Pb$ ratios are in the range 15.634–15.663 (Table 6). The high μ values ($^{238}U/^{204}Pb$) of 9.78–9.87 (Table 6) point to a Pb evolution in source magmas dominated with regard to Pb by continental crust (Stacey & Kramers 1975), and the trends of the measured $^{206}Pb/^{204}Pb$–$^{207}Pb/^{204}Pb$ ratios are best explained by relatively young upper-crustal contamination (Kramers & Tolstikhin 1997).

In Figure 9, new Pb/Pb data for pyrites and chalcopyrites of the Elatsite deposit are compared with published Pb/Pb data of galena from Elatsite, Chelopech, Medet and Assarel (Amov 1999). These data show no evidence for a significant mantle contribution of lead to the Cu deposits of the Panagyurishte ore district. Samples from the porphyry-Cu deposits of Medet and Assarel have a slightly lower $^{206}Pb/^{204}Pb$ ratio compared with Elatsite and Chelopech deposits, indicating comparatively low uranium concentrations in the Pb source of these two magmatic–hydrothermal deposits. The higher $^{207}Pb/^{204}Pb$ ratios measured in Pb-bearing minerals of Chelopech and Elatsite could be explained by assimilation of larger proportions of older crust (Kouzmanov *et al.* 2001). The involvement of granulitic lower crust is unlikely since the $^{208}Pb/^{204}Pb$ ratios of all samples show a common Pb evolution and no elevated χ values ($^{232}Th/^{204}Pb$).

Pyrite is formed at the lower-temperature stages of ore mineralization, when mixing of the primary magmatic fluids with meteoric water may be significant. The influence of the meteoric water and the infiltration trough the host rocks cannot be

Table 6. *Results of Pb–Pb isotope measurements of ore minerals*

Nr	Mineral	$^{206}Pb/^{204}Pb$	2σ error	$^{207}Pb/^{204}Pb$	2σ error	$^{208}Pb/^{204}Pb$	2σ error	μ-value
2382	cp	18.5374	0.0032	15.6382	0.0033	38.6680	0.0125	9.82
2383	cp	18.6015	0.0036	15.6342	0.0037	38.6857	0.0069	9.78
2384	py	18.5689	0.0023	15.6370	0.0019	38.6373	0.0035	9.80
2385	py	18.6184	0.0055	15.6557	0.0072	38.6691	0.0220	9.87
2386	cp	18.6370	0.0029	15.6630	0.0038	38.7167	0.0100	9.90

Abbreviations: cp, chalcopyrite; py, pyrite.

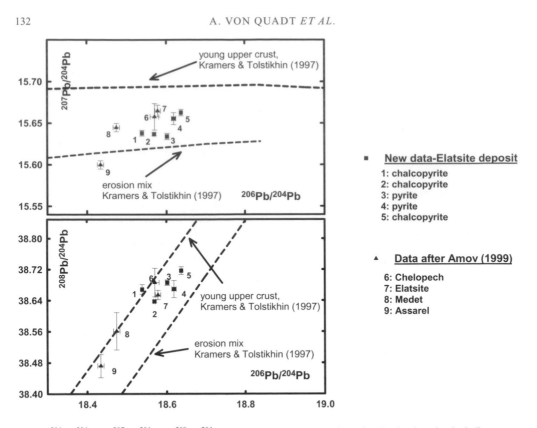

Fig. 9. $^{206}Pb/^{204}Pb$ v. $^{207}Pb/^{204}Pb$ v. $^{208}Pb/^{204}Pb$ diagram of ore minerals from the Elatsite deposits, including published Pb–Pb isotope data from Amov (1999).

quantified but is clearly not the main reason for the crustal characteristics of lead in all late Cretaceous ore deposits of the Central Srednogorie zone.

In contrast to the Pb, the few isotope data for Sr-bearing minerals (Table 4) indicate a mixed mantle–crust source for the Sr. The situation may be different in the case of other elements, such as Au, Cu or PGE, whose geochemical behaviour differs from that of Sr and Pb.

Discussion: dating of porphyry-copper mineralization

The high-precision U–Pb zircon data demonstrate that the whole time span of the ore-producing magmatism in the region of Elatsite deposit extended over a maximum duration of *c.* 0.87 Ma, starting at 92.1 ± 0.3 Ma ago with the pre- to syn-ore quartz monzodiorites of the 'big dyke' (HOR 1390, unit 1) and finishing at 91.84 ± 0.31 Ma according to $^{206}Pb/^{238}U$ age of zircons from the granodiorite porphyry dyke (LF 025, unit 2).

Considering the uncertainties of the individual age determinations, a maximum lifetime of the porphyry system of 0.87 Ma is estimated. The high temperature ore mineral assemblage (mt–bn–cp, PGEs and with potassic alteration) is confirmed by the two overlapping (within errors) dates of 92.1 ± 0.3 Ma and 91.84 ± 0.31 Ma; this means the high-temperature ore forming process including PGE mineralization is constrained to within 0.87 Ma. The latest dyke of unit 2 with an age of 91.42 ± 0.15 Ma contains only low-temperature ore minerals of the py–qtz assemblage, and fluid inclusions in quartz show a temperature range between 260 and 300 °C and a low salinity (Strashimirov & Kovachev 1992). The magmatic–hydrothermal system responsible for the main high-temperature Cu–Au deposition at Elatsite therefore collapsed after about 91.84 Ma. At about 90.5 Ma, the whole complex was cooled down to temperatures around 3007 °C based on new Rb–Sr and earlier K–Ar data. In the absence of detailed tectonic investigations, we speculate that regional uplift was the reason for termination and closure of these isotopic systems, consistent with conglomerates and sandstones formed in continental basins south of the Elatsite deposit during Turonian times (Moev & Antonov 1978). The multiphase dyke intrusions account for a rather long-living magmatic–hydrothermalsystem, even though individual events of dyke emplacement and hydrothermal mineralization were probably much shorter. Generally, at depths of about 3–5 km where typical porphyry-style deposits are formed, a short time of about 10 000 years is sufficient to cool single intrusive stocks (Cathles 1981), and considerably less in the case of dyke-shaped bodies.

This study shows the power of U–Pb zircon dating of porphyry-copper deposits, in comparison with K–Ar and ^{40}Ar/^{39}Ar isotope methods commonly used for dating ore deposits (K–Ar; Ar/ Ar;Richards *et al.* 1999; Arribas *et al.* 1995); the Rb–Sr system was used additionally by some authors (Kostitsyn 1993, 1994). The latter methods are suitable for young deposits and enable precise dating of the latest epithermal phases of mineralization based on widespread alteration minerals such as sericite, biotite, illite, adularia and alunite. A distinct advantage of ^{40}Ar/^{39}Ar dating is the high precision, especially for small differences measured in a single batch of samples (e.g. Arribas *et al.* 1995). However, neither ^{40}Ar/ ^{39}Ar nor Rb–Sr can reliably date the early high-temperature history of porphyry systems, above the blocking temperatures of the commonly dated minerals. In addition, in fluid dominated systems, an exchange of elements and isotopes is expected even at lower temperatures. For dating the ore-productive magmatism including high-temperature potassic alteration, U–Pb dating of kinetically robust igneous zircons, from magmatic rocks in unambiguous geological relationship with hydrothermal veins and alteration stages, arguably offers the most reliable dating method, which is in most cases very precise (Schaltegger 2000). Only Re–Os dating of molybdenite offers a comparably robust and precise geochronometer for the timing of porphyry-copper mineralising events (Raith & Stein 2000). Absolute timing of the entire sequence of ore-forming processes generally requires the combined use of several isotopic methods, in conjunction with careful field observation to determine the relative sequence of tectonic, magmatic and hydrothermal events.

The authors thank the geologists of Elatsite mine for their kind support during fieldwork and sampling, and W. Wittwer and Z. Cherneva for their help in sample preparation and mineral separation. Part of this work was supported by grant 2-77689-00 of the Swiss National Science Foundations SNSF and an SNSF-SCOPES grant 7BUPJ062396. The research is carried out as part of the ESF Programme 'Geodynamics and Ore Deposit Evolution'. Critical comments from reviews by Yuri A.Kostitsyn, Werner Halter, Marion Tichomirowa and Franz Neubauer have helped to improve the manuscript.

References

AMOV, B.G. 1999. Lead isotope data for ore deposits from Bulgaria and the possibility for their use in archaeometry. *Berliner Beiträge zur Archäometrie*, **16**, 5–19.

ANDREW, C. 1997. The geology and genesis of the Chelopech Au-Cu deposit, Bulgaria: Europe's largest gold resource. *In:* HARNEY, S.*In: Europe's Major Gold Deposits. Abstract volume and Programme, Newcastle, Northern Ireland.* Irish Association for Economic Geology, 68–72.

ARRIBAS, A., HEDENQUIST, J., ITAYA, T., OKADA, T., CONCEPCION, R. & GARCIA, J. 1995. Contemporaneous formation of adjacent porphyry and epithermal Cu-Au deposits over 300 ka in northern Luzon, Philippines. *Geology*, **23**, 337–340.

BLICHERT-TOFT, J. & ALBARÈDE, F. 1997. The Lu-Hf isotope geochemistry of chondrites and the evolution of the mantle-crust system. *Earth and Planetary Science Letters*, **148**, 243–258.

BOGDANOV, B. 1987. Copper deposits in Bulgaria. *Technica, Sofia*, 1–388.

BONCHEV, E. 1970. On some problems of the Srednogorie. *Review Bulgarian Geological Society*, **31.3**, 281–288.

CATHLES, L.M. 1981. Fluid flow and genesis of hydrothermal ore deposits. *Economic Geology 75th Anniversary Volume*, 424–457.

DAVID, K., FRANK, M., O'NIONS, R.K., BELSHAW, N.S. & ARDEN, J.W. 2001. The Hf isotope composition of global seawater and the evolution of Hf isotopes in the deep Pacific Ocean from Fe-Mn crusts. *Chemical Geology*, **178**, 23–42.

DRAGOV, P. & PETRUNOV, R. 1996. Elatsite porphyry copper-precious metals (Au and PGE) deposit. *In:* POPOV, P. (ed.) *Plate Tectonic Aspects of the Alpine metallogeny in the Carpato-Balkan Region.* UNESCO - IGCP Project 356, Proceedings of the annual meeting, Sofia, **1**, 171–174.

FANGER, L. 2001. *Geologie einer Kupferporphyr-Lagerstätte: Elatsite, Bulgarien.* Diploma work, ETH-Zurich.

GÜNTHER, D., VON QUADT, A., WIRZ, R., COUSIN, H. & DIETRICH, V.J. 2001. Elemental analyses using laser ablation-inductively coupled plasma-mass spectrometry (La-ICP-MS) of geological samples fused with $Li_2B_4O_7$; calibrated without matrix-matched standards. *Microchimica Acta*, **136**, 100–107.

HAYDOUTOV, I. 1987. Ophiolites and island-arc igneous rocks in the Caledonian basement of the South Carpathian-Balkan region. *In:* FLÜGEL, H., SASSI, F.P. & GRECULA, P. (eds) *Pre-Variscan and Variscan events in the Alpine-Mediterranean mountain belts.* Mineralia Slovaca, Special Monographs, 279–292.

HAYDOUTOV, I., TENCHEV, Y. & JANEV, S. 1979. Lithostratigraphic subdivision of the Diabase-Phyllitoid Complex in the Berkovitsa Balkan Mountain. *Geologica Balcanica*, **9**(3), 13–25.

JANKOVIC, S. 1977. The copper deposits and geotectonic setting of the Tethyan-Eurasian Metallogenic Belt. *Mineralium Deposita*, **12**, 37–47.

JANKOVIC, S. 1997. The Carpatho-Balkanides and adjacent area: a sector of the Tethyan Eurasian metallogenic belt. *Mineralium Deposita*, **32**, 426–433.

KALAIDJIEV, S., HADJIISKI, G. & ANGELKOV, K. 1984. Structural conditions for localization of the porphyry copper deposit Elatsite. *Review of the Bulgarian Geological Society*, **45**(2), 189–196.

KARAGYULEVA, J., KOSTADINOV, V., TZANKOV, TZ. & GOTCHEV, P. 1974. [Composition of the Panagyurishte strip east of the Topolnitza river]. *Review Geological Institute of Bulgarian Academy of Science Series Geotectonic*, **23**, 231–301. [in Bulgarian].

KARAMATA, S., KNEZEVIC, V., PECSKAY, Z. & DJORDJEVIC, M. 1997. Magmatism and metallogeny of the Ridanj-Krepoljin belt (eastern Serbia) and their correlation with northern and eastern analogues. *Mineralium Deposita*, **32**, 452–458.

KOSTITSYN, Y.A. 1993. Rb/Sr, isotope study of Muruntau deposits. Dating of the ore veins. *Geochimiya*, **9**, 1308–1319.

KOSTITSYN, Y.A. 1994. Rb/Sr isotope study of Muruntau deposit. Ore-bearing metasomatites. *Geochimiya*, **4**, 486–498.

KOUZMANOV, K., MORITZ, R., CHIARADIA, M. & RAMBOZ, C. 2001. Mineralogy, fluid inclusion study, and Re-Os dating of Mo-bearing mineralization from Vlaiko Vrah prophyry copper deposit, Panagyurishte District, Bulgaria. *EUG XI, Journal of Conference Abstracts*, **6**(1), 557.

KRAMERS, J.D. & TOLSTIKHIN, I.N. 1997. Two terrestrial lead isotope paradoxes, forward transport modelling, core formation and the history of the continental crust. *Chemical Geology*, **139**, 75–110.

KROGH, T.E. 1973. A low-contamination method for hydrothermal decomposition of zircon and extraction of U and Pb for isotopic age determination. *Geochimica et Cosmochimica Acta*, **37**, 485–494.

LILOV, P. & CHIPCHAKOVA, S. 1999. [K-Ar dating of the Upper Cretaceous magmatic rocks and hydrothermal metasomatic rocks from the Central Srednogorie]. *Geochemistry, Mineralogy and Petrology, Sofia*, , **36**, 77–91. [in Bulgarian].

LUDWIG, K.R. 1999. User's Manual for Isoplot/Ex Version 2, A Geochronological Toolkit for Microsoft Excel. *Berkeley geochronology Center Special Publication 1a*, 1–47.

MOEV, M. & ANTONOV, M. 1978. Stratigraphy of the Upper Cretaceous in the Eastern Part of Sturguel-Chelopech Strip. *Annuaire de l'Ecole Superior des Mines et Geologie (Year Book of the Higher Mining and Geological University, Sofia, Bulgaria)*, **23**(2), 7–30. [in Bulgarian with English and Russian abstract].

MORITZ, R., CHAMBEFORT, I., CHIARADIA, M., FONTIGNIE, D., PETRUNOV, R., SIMOVA, S., ARISANOV, A. & DOYCHEV, P. 2001. The Late Cretaceous high-sulfidation Au-Cu Chelopech deposit, Bulgaria: geological setting, paragenesis, fluid inclusion microthermometry of enargite, and isotopic study (Pb, Sr, S). *In:* PIESTRZYNSKY, L.A. (ed.) *Mineral deposits at the beginning of the 21st Century.* Balkema, Lisse, Netherlands, 547–550.

MULLEN, E.D. 1983. $MnO/TiO_2/P_2O_5$: a minor element discriminant for basaltic rocks of oceanic environments and its implications for petrogenesis. *Earth and Planetary Science Letters*, **62**, 53–62.

NOWELL, G.M., KEMPTON, P.D., NOBLE, S.R., FITTON, J.G., SAUNDERS, A.D., MAHONEY, J.J. & TAYLOR, R.N. 1998. High precision Hf isotope measurements of MORB and OIB by thermal ionization mass spectrometry: insights into the depleted mantle. *Chemical Geology*, **149**, 211–233.

PEARCE, J.A., HARRIS, N.B.W. & TINDLE, A.G. 1984. Trace element discrimination diagrams for the tectonic interpretation of granitic rocks. *Journal of Petrology*, **4**, 956–983.

PETRUNOV, R. 1995. Ore mineral parageneses and zoning in the Chelopech deposit. *Geochemistry, Mineralogy and Petrology*, **30**, 89–98. [in Bulgarian with English abstract].

PETRUNOV, R. & DRAGOV, P. 1993. PGE and gold in the Elacite porphyry copper deposit, Bulgaria. *In:* FENNEL HACH-ALI,*In: Current Research in Geology Applied to Mineral Deposits.* University of Granada, Granada, Spain, 543–546.

PETRUNOV, R. & DRAGOV, P. 1992. Hydrothermal PGE-mineralization in the Elatsite porphyry-copper deposit (Srednα Gora metallogenic zone, Bulgaria). *Compte rendue Academy Bulgarian Science*, **45**(4), 37–40.

POPOV, P. & KOVACHEV, V. 1996. Geology, composition and genesis of the mineralization in the Central and Southern part of the Elatsite-Chelopech ore field. *In:* POPOV, P.*In: Plate Tectonic Aspects of the Alpine Metallogeny in the Carpathian-Balkan Region.* UNESCO-IGCP Project - 356, Proceedings of the Annual meeting, Sofia, **1**, 159–170.

POPOV, P. & MUTAFCHIEV, I. 1980. [Structure of the Chelopech ore field]. *Year-book. Higher Mining and Geological Institute*, **25**(2), 25–41. [in Bulgarian].

POPOV, P. & POPOV, K. 1997. Metallogeny of Panagyurishte Ore Region. *In:* ROMIC, K. & KONDZULOVIC, R. (eds) *Ore Deposits exploration.* Belgrade, 24 April 1997, **1997**, 327–338.

POPOV, P., VLADIMIROV, V. & BAKURDZHIEV, S. 1983. [Structural model of the Chelopech ore field]. *Geologia Rudnih Mestorojdeniya*, **5**, 3–10. [in Russian].

POPOV, P., BERZA, T. & GRUBIC, A. 2000a. Upper Cretaceous Apuseni-Banat-Timok-Srednogorie (ABTS) Magmatic and Metallogenic Belt in the Carpathian-Balcan Orogen. *In:* BOGDANOV, K., STRASHIMIROV, S. & BONEV, I. (eds) ABCD-GEODE workshop, Abstract Volume, Borovets - Bulgaria, 69–70. Geodynamics and Ore Deposit Evolution of the Alpine-Balkan-Carpathian-Dinaride Province.

POPOV, P., PETRUNOV, R., KOVACHEV, V., STRASHIMIROV, S. & KANAZIRSKI, M. 2000b. Elatsite-Chelopech ore field. *In:* STRASHIMIROV, S. & KOVACHEV, B. (eds) *Geology and Metallogeny of the Panagyurishte Ore Region, Srednogorie Zone, Bulgaria.* Excursion Guide, ABCD-GEODE 2000 Workshop, Borovets, Bulgaria, **2000**, 8–18.

RAITH, J.G. & STEIN, H. 2000. Re-Os dating and sulfur

isotope composition of molybdenite from tungsten deposits in western Namaqualand, South Africa: implications for ore genesis and the timing of metamorphism. *Mineralium Deposita*, **35**, 741–753.

RICHARD, P., SHIMIZU, N. & ALLÈGRE, C.J. 1976. ^{143}Nd/^{146}Nd, a natural tracer: an application to oceanic basalts. *Earth and Planetary Science Letters*, **31**, 269–278.

RICHARDS, J.P., NOBLE, ST.R. & PRINGLE, M.S. 1999. A revised late Eocene age for porphyry Cu Magmatism in the Escondisa Area, Northern Chile. *Economic Geology*, **94**, 1231–1248.

SCHALTEGGER, U. 2000. U-Pb geochronology of the Southern Black Forest Batholith (Central Variscan Belt): timing of exhumation and granite emplacement. *International Journal of Earth Sciences*, **88**, 814–828.

STACEY, J.S. & KRAMERS, J.D. 1975. Approximation of terrestrial lead isotope evolution by a two-stage model. *Earth and Planetary Science Letters*, **26**, 207–221.

STRASHIMIROV, S. & KOVACHEV, B. 1992. Temperature of formation in copper deposits from the Srednogorie zone - data from fluid inclusion studies in minerals. *Review Bulgarian Geological Society*, **53**(2), 1–12.

TARKIAN, M. & STRIBNY, B. 1999. Platinum group elements in porphyry copper deposits: a reconnaissance study. *Mineralogy and Petrology*, **65**, 161–183.

TRASHLIEV, S. 1961. [On the genesis and age of the barite locality Kashana, District of Pirdope]. *Review of the Bulgarian Geological Society*, **32**, 245–252. [in Bulgarian].

VASSILEV, L. & STAYKOV, M. 1991. [Short Metallogeny of Bulgaria. Explanatory notes to the metallogenic map in scale 1:1 000 000]. *Review of Bulgarian Geological Society*, **52**(2), 1–55. [in Russian].

VELICHKOVA, S., HANDLER, R., NEUBAUER, F. & IVANOV, Z. 2001. Preliminary ^{40}Ar/^{39}Ar mineral ages from the Central Srednogorie Zone, Bulgaria: Implications for Cretaceous geodynamics. *ABCD – GEODE 2001 workshop Vata Bai, Romania, Abstract Volume*, **79**(Suppl. 2), 112–113.

VERVOORT, J.D., PATCHETT, P.J., BLICHERT-TOFT, J. & ALBARÉDE, F. 1999. Relationship between Lu-Hf and Sm-Nd isotopic systems in the global sedimentary system. *Earth and Planetary Science Letters*, **168**, 79–99.

VON QUADT, A. 1997. U-Pb zircon and Sr-Nd-Pb whole rock investigations from the continental deep drilling (KTB). *Geological Rundschau*, **86**, 258–271.

WOOD, D.A., JORON, J.-L. & TREUL, M. 1979. A reappraisal of the use of trace elements to classify and discriminate between magma series erupted in different tectonic settings. *Earth and Planetary Science Letters*, **45**, 326–336.

^{40}Ar/^{39}Ar geochronology of magmatism and hydrothermal activity of the Madjarovo base–precious metal ore district, eastern Rhodopes, Bulgaria

PETER MARCHEV[1] & BRAD SINGER[2]

[1]*Geological Institute, Bulgarian Academy of Sciences, Acad. G. Bonchev St., Bl.24, 1113 Sofia, Bulgaria (e-mail: pmarchev@router.geology.bas.bg)*

[2]*Department of Geology and Geophysics, University of Wisconsin-Madison, 1215 West Dayton St., Madison, WI 53706, USA*

Abstract: The Madjarovo volcanic complex and ore district comprise alteration styles from potassium silicate, advanced argillic and sericite alteration to adularia–sericite alteration/mineralization with a close and unambiguous spatial relationship to specific magmatic events. New ^{40}Ar/^{39}Ar laser fusion and incremental heating experiments on nine sanidine, biotite, adularia, K-feldspar, and alunite samples constrain the ages and time span of lavas and tephras comprising the complex and their relationship to the hydrothermal activity. These results demonstrate that high-K calc-alkaline to shoshonitic volcanic activity began *c.*32.7 Ma and terminated *c.*500 ka later with the extrusion of quartz latite lavas at 32.2 Ma. The final stage of volcanism was accompanied by intrusion of compositionally similar monzonite stocks and trachytic dykes (*c.*32.2–32.1 Ma) and associated barren advanced argillic and sericite alteration (lithocap) and adularia–sericite base/precious metal vein mineralization. A probable thermal event at *c.*12–13 Ma disturbed the ages of alunite and sericite-bearing alteration at low stratigraphic levels. However, field relations combined with a plateau age of 32.1 ± 0.2 Ma from adularia in low-sulphidation veins that cross-cut lithocap indicate that hydrothermal activity, including base- and precious-metal vein deposition, was coeval with the youngest magmatic activity.

Although most recent models of epithermal systems emphasize close spatial association among volcanic and plutonic events, alteration, and deposition of ore (Hedenquist & Lowenstern 1994; Sillitoe 1993, 1995), only rare combinations of geological events provide examples where relationships between igneous rocks, porphyry and epithermal systems (both high and low sulphidation) are exposed. This is mainly because young epithermal systems are poorly exposed, whereas older deposits are eroded or deformed (Hedenquist & Arribas 1999; Cooke & Simmons 2000). Most economic epithermal ore deposits are Tertiary and conventional K–Ar dating cannot precisely constrain the timing of magmatism, alteration, and mineralization. At the 95% confidence level the precision of ^{40}Ar/^{39}Ar age determinations of K-rich minerals is typically better than ± 0.5%, thus for the Oligocene materials of this investigation differences in age of about 200 ka may be distinguished (e.g. Marsh *et al.* 1997; Singer & Marchev 2000).

Mining at Madjarovo ore district in Bulgaria is believed to have commenced during Thracian times, with most development beginning in the 1950s. More than 10 million tons of base metal ore have been mined; another 6.5 million tons of base metal reserves and low-grade ore still exist. However, because of changes in the Bulgarian economy and a decrease in ore grade with depth, mining for base metals has ceased. Jambol Exploration Organization undertook an extensive gold exploration programme of the upper parts of the base metal veins between 1988 and 1996. Eight major veins contain reserves of *c.* 2 million tons grading 3.9 g t^{-1} Au, but a 1995 feasibility study by Euraust Mineral Developments indicated that the deposit is not economic for Au.

We chose to focus on the Madjarovo ore district because the Arda River has dissected the whole volcanic complex and ore district, exposing the close spatial relationships between porphyry monzonite stocks and dykes in the central part of the complex and potassic, barren acid–sulphate (advanced argillic) and adularia–sericite (low-sulphidation) alteration and accompanied base/precious metal mineralization. This geological setting provides an excellent opportunity to constrain the timing and

From: BLUNDELL, D.J., NEUBAUER, F. & VON QUADT, A. (eds) 2002. *The Timing and Location of Major Ore Deposits in an Evolving Orogen*. Geological Society, London, Special Publications, **204**, 137–150.

0305-8719/02/$15.00 © The Geological Society of London 2002.

duration of magmatic and hydrothermal events using $^{40}Ar/^{39}Ar$ methods. Our new age determinations support a close connection between magmatic events and hydrothermal alteration and potential contemporaneity of the contrasting fluids responsible for barren advanced argillic/sericitic alteration and adularia–sericite alteration/mineralization.

Geological setting

Regional setting

The Madjarovo ore district of southeastern Bulgaria is situated approximately 200 km SE of Sofia (Fig. 1) in the centre of the Oligocene Madjarovo volcano. The volcano itself is the easternmost volcanic structure on the Bulgarian territory of an Eocene–Oligocene continental magmatic belt that extends c. 500 km from Serbia and Macedonia to NW Turkey (Fig. 1). The magmatic belt resulted from post-Palaeocene–Eocene extension that followed Upper Cretaceous collision of the Serbo-Macedonian and Rhodope Massifs with the Pelagonian microplate (Ricou 1994). The eastern part of this belt is occupied by the Rhodope Massif, which is comprised of Precambrian to Mesozoic metamorphic rocks. Palaeogene magmatic rocks consist of calc-alkaline to shoshonitic intermediate, acid and subordinate basic volcanic rocks and their intrusive equivalents (Ivanov 1968; Innocenti et al. 1984; Harkovska et al. 1989; Del Moro et al. 1990; Christofides et al. 1998; Marchev et al.

1998a; Yanev et al. 1998), showing a distinct south to north enrichment (from Greece to Bulgaria) in K_2O and LILE. Minor alkali basalts have been described in the southeastern part of the Eastern Rhodopes (Marchev et al. 1998b). The Palaeogene magmatism was accompanied by the formation of small Cu–Mo and abundant epithermal deposits (Mavroudchiev et al. 1996; Arikas & Voudouris 1998), which form the Rhodope metallogenic province (Stoyanov 1979).

District geology

The Madjarovo volcanic complex covers 120 km^2. New geological mapping at 1:10 000 is summarized in Figure 2. Pre-Tertiary rocks, that crop out south of the Chernichevo Fault, are overlain by Upper Priabonian conglomerate, sandstone and limestone. A tephrostratigraphic marker, composed of pumice and ash-fall tuff of unknown source, named the Reseda Tuff (Ivanov & Kopp 1969), overlies these sediments and underlies the Madjarovo volcanic complex. The Arda river, exploiting a large east–west fault (Arda zone), exposes the deep stratigraphy of the volcano. Volcanism was predominantly fissure-fed and dominated by large sheet-like lava flows and subordinate epiclastic rocks that formed a shield volcano (Ivanov 1960). The unaltered mafic to intermediate volcanic rocks (Marchev et al.1989) are shoshonitic to high-K calc-alkaline rocks dominated by latites. Burchina quartz latites are

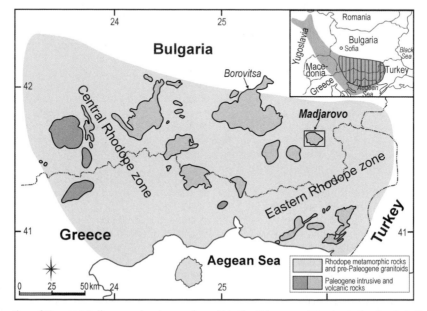

Fig. 1. Location of Eocene Madjarovo volcanic complex within the Palaeogene intrusive and volcanic belt, Rhodope Mountians. Inset shows the Palaeogene Macedonian–Rhodope–North-Aegean Volcanic Belt.

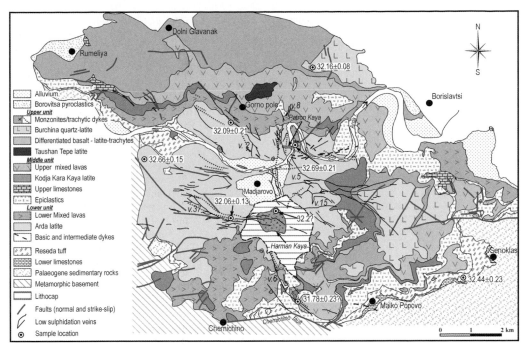

Fig. 2. Simplified geological map of the Madjarovo ore district, showing sample locations and their preferred ages. Source: unpublished mapping by P. Marchev, 1989–1995.

the youngest volcanic rocks. They are overlain by another tephrostratigraphic marker, Borovitsa rhyolite tuff, which is an outflow tuff originated from Borovitsa caldera located 20 km WNW of Madjarovo (Singer & Marchev 2000).

On the basis of petrology and Sr and O isotope compositions (Marchev & Rogers 1998), three units have been divided within the Madjarovo complex. The first two (Lower and Middle unit) start with thick (up to 150 m) latites and end with basalts. Less voluminous intercalated flows of basaltic andesites, shoshonites and andesites show petrographic and isotopic evidence for large-scale magma mixing between end-members (Raicheva et al. 2001). The third unit includes three lavas ranging from high-K high-Al basalt through latite and quartz-latite.

Early volcanic rocks are intruded by numerous monzonitic stocks and rare gabbroic and syenitic bodies, similar in composition to the Upper unit. It is presumed that the monzonite stocks coalesce at depth (Mavroudchiev 1959). The largest monzonitic body, Harman Kaya, and trachytic dykes of identical composition crop out in the central part of the complex over 40–50 km^2. A gravity minimum and magnetic anomaly beneath the centre of the complex have been interpreted as a syenitic pluton between depths of 1 and 4 km

(Iosifov et al. 1987). Intrusion of monzonite-trachyte magma was accompanied and followed by destruction of the central part of the volcano. Gergelchev (1974) postulated the existence of a caldera structure coinciding in size with the pluton, but the results of extensive drilling and our field observations did not confirm this interpretation.

Four major fault systems, striking east–west, NW–SE, north–south and NE–SW (Atanasov 1959; Velinov et al. 1977), accommodate the feeding dyke swarms for the lavas, sub-volcanic intrusions and the base/precious metal veins and alteration, suggesting that these structures controlled volcanism and subsequent mineralization. The most prominent structures are the east–west Chernichevo fault delimiting the metamorphic and volcanic rocks in the southern part of the volcano (Fig. 2), and the east–west Arda structural zone. A series of faults between them bound blocks tilted to the NNE and decreasing in altitude northward with total displacements of up to 600–700 m. The richest veins (e.g., #2, 6, 8; Fig. 2) also represent faults with displacement up to 200 m (Atanasov 1959). They show abundant evidence of brecciation and recementation, suggesting movement during ore deposition. However, preservation of the epithermal zonation in the elevated southern and down-dropped northern halves of the complex suggests that the current structure most probably developed synchronously with trachyte/monzonite intrusions and that ore deposition post-dated most deformation.

Alteration

Alteration in the Madjarovo ore district is documented by Radonova (1960), Velinov *et al.* (1977), Velinov & Aslanian (1981), Velinov & Nokov (1991) and Marchev *et al.* (1997). Current knowledge of hydrothermal alteration is summarized in McCoyd (1995) and herein. The epithermal system generated acid–sulphate and adularia–sericite alteration (Velinov & Nokov 1991) and subjacent K-silicate alteration (Marchev *et al.* 1997). Skarns were discovered by drilling near the Harman Kaya and Patron Kaya monzonite stocks (Breskovska *et al.* 1976, Figure 2). McCoyd (1995) subdivided the epithermal alteration into advanced argillic, sericitic, propylitic, quartz–sericite and adularia–sericite; these alteration zones are described below.

K silicate alteration occurs adjacent to two small monzonite stocks in the Arda zone, Patron Kaya and Kjumiurluka (the latter not shown in Fig. 2). It comprises hydrothermal biotite, K-feldspar, albite and quartz. The altered rocks contain disseminated pyrite with rare Cu-bearing (chalcopyrite) mineralization.

Propylitic alteration is developed around K-silicate and sericite alteration over *c.* 10 km^2 in the central part of the district. This alteration was originally designated as pre-ore regional propylitization (Radonova 1960). Propylitic alteration also developed surrounding fracture-controlled base/precious metal vein mineralization. The propylitic assemblage consists of all or some of the following minerals: chlorite, epidote, albite, carbonate and pyrite.

Sericite alteration consisting predominantly of muscovite+quartz accompanied by minor pyrite and dickite, this alteration affected the uppermost Harman Kaya monzonite and its wall rocks (Fig. 2). Sericite alteration forms envelopes around zones of advanced argillic alteration, together forming a lithocap (Sillitoe 1995), which probably at depth passes into K-silicate alteration. A large number of base-metal veins occur within the sericitic alteration zone and Harman Kaya monzonite intrusion but the ore grade in the veins diminishes and visible mineralization is not present. This area was determined by Breskovska *et al.* (1976) to be unsuitable for vein type mineralization.

Advanced argillic alteration comprising kaolinite, pyrophyllite, alunite, diaspore and zunyite (Velinov & Nokov 1991; McCoyd 1995), this alteration is in two east–west trending zones in the southern half of the district with alunite alteration occurring at the highest elevation (Fig. 2). The phase assemblage indicates a temperature of *c.* 270 °C, and S isotope geothermometry using an alunite–pyrite pair suggests temperatures of 300–310 °C. Alunite veins located within this alteration are up to 30 cm thick, several metres long, comprise coarse-grained crystals to 150 μm, and strike N130°E. SEM analysis (McCoyd 1995) showed that it consists of woodhousite + svanbergite + alunite. The upper part of the intrusion was silicified locally.

Quartz–sericite alteration is associated with the deeper portions of the base-metal veins, but is also found at higher levels. It is strongly fracture-controlled and forms envelopes around quartz veins. Typically, the width of the zone is a few tens of centimetres but can be up to several metres around larger veins. Quartz and sericite predominate but illite and smectite and disseminated pyrite also occur. Sericite replaces feldspar and also occurs in the groundmass, whereas quartz occurs as veinlets near the main vein deposits.

Adularia–sericite alteration is typical in the upper levels of the palaeo-volcano, it extends up to 120 m on either side of the veins (Fig. 2). It developed either as a halo around large open veins or comprises the brecciated zones. The most intensely altered rocks consist exclusively of a fine-grained aggregate of quartz and adularia. Chemical analyses indicate up to 13 wt% K$_2$O, suggesting that about 80–85% of the rock is adularia.

Mineralization

The base metal–Au mineralization at Madjarovo is located in the four major fault systems (Fig. 2). Of nearly 150 quartz–sulphide veins that have been identified, 50 veins are economically significant. Vein widths vary from half a metre to thirty metres and the largest of these are up to 3 km long (Fig. 2). Brecciation is typical of the narrowest veins whereas the larger veins are massive and internally banded.

Zoning similar to current models of epithermal deposits, with base metals precipitating below precious metals (Buchanan 1981; Berger & Eimon 1983) was suggested by Atanasov (1962). The mineralized interval persists to 1500 m below the present surface with the precious metal mineralization located in the uppermost 150–200 m. Numerous (>90) primary minerals form the deposit. The predominant sulphides are pyrite, chalcopyrite, Fe-poor sphalerite and Ag-bearing galena with minor Se, Bi, Ag, Sb, As sulfosalts and native Au and Ag (Breskovska & Tarkian 1993). Quartz is the dominant gangue mineral with lesser amounts of barite, chalcedony, jasper and calcite. The paragenesis is complex and six stages of mineralization have been identified (Atanasov 1962; Breskovska & Tarkian 1993). Nokov & Malinov (1993) report findings of molybdenite and hypogene wulfenite in some veins.

Fluid inclusion studies (Breskovska & Tarkian 1993; Nokov & Malinov 1993; McCoyd 1995) have shown that quartz associated with main stage Pb–Zn mineralization was deposited at 240–270 °C from neutral low salinity fluids (3–5 eq. wt% NaCl). Barite associated with the precious-metal mineralization was deposited from similar fluids. Although there is no direct evidence, given the estimated geothermal gradient and occurrence of hydrothermal breccias and adularia, boiling was possible (McCoyd 1995).

Geochronology

Conventional K–Ar dating at Madjarovo broadly constrained the ages of various igneous rocks between 33.5 Ma and 31.0 Ma (Lilov *et al.* 1987). Similarly, Rb–Sr mineral isochrons from the lowermost and uppermost latitic lava flows comprising Madjarovo volcano were 31.6 ± 1.2 Ma and 32.3 ± 0.6 Ma, respectively (Marchev & Rogers 1998). Arnaudova *et al.* (1991) reported a K–Ar age of 33–32 Ma for an adularia sample from the adularia–sericite alteration. A similar age (32.6 ± 1.2 Ma) was obtained by McCoyd (1995) for the vein alunite from the advanced argillic alteration. However, two sericite separates of sericite alteration and one from the quartz–sericite alteration at the southeastern end of vein #2, obtained by the same author, yielded much younger ages (13.0 ± 0.5 Ma–13.7 ± 0.6 Ma) and (12.2 ± 1.0 Ma, respectively. These data, along with the newly obtained $^{40}Ar/^{39}Ar$ data, will be discussed in a later section.

$^{40}Ar/^{39}Ar$ analytical techniques

To define more precisely the timing of igneous and hydrothermal events we undertook $^{40}Ar/^{39}Ar$ experiments on sanidine, biotite, adularia, K-feldspar and alunite from nine samples of volcanic, intrusive and hydrothermal rocks using a CO_2 laser probe. Minerals were separated from 100–250 micron sieve fractions using standard magnetic, density, and handpicking methods. Five milligrams of each mineral were irradiated at

Oregon State University for 12 or 50 hours along with 27.92 Ma sanidine from the Taylor Creek rhyolite (Duffield & Dalrymple 1990) as the neutron fluence monitor. Isotopic measurements were made of the gas extracted by either totally fusing individual crystals of < 0.1 mg, or by incrementally heating larger 2–3 mg multi-crystal aliquots using a defocused CO_2 laser beam. The isotopic composition of gas from each fusion or heating step was measured using an MAP 216 spectrometer at the University of Geneva. Six to eleven total fusion measurements were pooled together to calculate a weighted mean age and uncertainty. Incremental-heating results are generally given as weighted mean plateau ages. Analytical procedures, including mass spectrometry, procedural blanks, reactor corrections, and estimation of uncertainties are described by Singer & Marchev (2000). All uncertainties are $\pm 2\sigma$

Samples and results

The results of laser total fusion and incremental heating experiments are summarized in Table 1 and age spectra of the latter are illustrated in Figure 3. The locations of the samples and their preferred ages are shown on the geological map (Fig. 2). The significance and preferred age determined for each sample are as follows.

Reseda Tuff Sample M96-5 is from the Reseda Tuff. Although heavily zeolitized, phenocrysts of sanidine, plagioclase, biotite, sphene and zircon are unaltered and suggest that the source magma was a low-silica rhyolite. Total fusion analyses of six sanidine crystals from a pumice fragment in the upper part of the tuff yielded a weighted mean age of 32.44 ± 0.23 Ma.

Lowermost Arda latite lava flow Sample M96-8 is from the periphery of the complex (Fig. 2) where the lava overlies Reseda Tuff. It contains phenocrysts of plagioclase ($An_{77.5-51}$), augite, orthopyroxene, titanomagnetite, biotite, and minor amphibole and apatite. The weighted mean of eight total-fusion measurements of biotite yielded an age of 32.66 ± 0.15 Ma (Table 1). Sample M92-46 is from the central part of the volcano (Fig. 2) about 50 m from the Patron Kaya monzonite stock. Potassium alteration associated with the stock comprises fine-grained biotite, K-feldspar, albite and quartz pervasively replacing groundmass, plus calcite and chlorite replacement of pyroxene, leaving relict phenocrysts of plagioclase and biotite. Disseminated pyrite and rare chalcopyrite accompany these silicate minerals. Seven of eight total-fusion measurements of unaltered biotite phenocrysts yielded a weighted

Table 1. *Summary of* $^{40}Ar/^{39}Ar$ *geochronological data for laser fusion and incremental heating experiments on adularia, biotite and sanidine from Madjarovo, Bulgaria*

Sample	Mineral	Description	Total fusion		Incremental heating 'Plateau age'
			Number of fusions	Weighted mean age (Ma)	
M96-11	Adularia	Low-sulphidation alteration, vein 6	7 of 9	31.78 ± 0.23	$32.57 \pm 0.14?$
M96-11Pl	Adularia	Low-sulphidation alteration, vein 6			$31.12 \pm 0.35?$
M97-02	Adularia	Low-sulphidation alteration, vein 2	7 of 7	31.91 ± 0.40	32.09 ± 0.21
M96-4	Sanidine	Stratigraphically highest quartzlatite	11 of 11	32.16 ± 0.08	32.23 ± 0.19
4-69	Sanidine	Trachyte dyke	7 of 9	32.06 ± 0.13	
M92-46	Biotite	Stratigraphically lowest latite	7 of 8	32.69 ± 0.21	32.72 ± 0.23
M96-8	Biotite	Stratigraphically lowest latite	8 of 8	32.66 ± 0.15	
M96-5	Sandine	Reseda Tuff	6 of 6	32.44 ± 0.23	

Ages are calculated relative to 27.92 Ma Taylor Creek sanidine (Duffield & Dalrymple 1990). $\pm 2\sigma$ errors.

mean age of 32.69 ± 0.21 Ma (Table 1), whereas a ten-step incremental heating experiment gave a concordant spectrum and a plateau age of 32.72 ± 0.23 Ma that is indistinguishable from the total fusion age (Fig. 4). Neither age is indistinguishable from that of biotite M96-8, moreover the age of Arda latite and Reseda Tuff overlap, suggesting a rapid growth of Madjarovo volcano concomitant with eruption of the tuff.

Uppermost Burchina quartz latite lava flow Sample Md96-4 is from the uppermost Burchina quartz latite that contains phenocrysts of plagioclase (An $_{58.5-41.2}$), sanidine (Or$_{75.1-72.4}$), clinopyroxene; orthopyroxene rimmed by clinopyroxene, biotite, and titanomagnetite, plus zircon and apatite. Eleven total-fusion measurements of sanidine gave a weighted mean age of 32.16 ± 0.08 Ma (Table 1). The plateau age of 32.23 ± 0.19 Ma from a concordant five-step incremental heating experiment is indistinguishable from the total fusion age. Although these ages are in excellent agreement with the 32.3 ± 0.6 Ma clinopyroxene–plagioclase–biotite–sanidine Rb–Sr isochron of Marchev & Rogers (1998), our preferred total fusion age is nearly an order of magnitude more precise.

Trachytic dyke Sample 4-69 is from a trachytic dyke similar in composition and probably coeval with the Harman Kaya monzonite and spatially associated with younger mineralized veins discussed below. Sanidine phenocrysts up to 2.5 cm

are common and seven of nine total-fusion measurements yielded a weighted mean age of 32.06 ± 0.13 Ma, that is younger than the Arda latite, but indistinguishable from the Burchina quartz latite age.

Harman Kaya monzonite stock Although sample M89-150 is from the least altered part of the largest monzonite stock (Fig. 2), it is propilitized and dated K-feldspar crystals (up to 2.5 cm) are partly replaced by minor sericite and clay. The 13 step age spectrum is discordant with apparent ages increasing gradually from about 29.85 ± 0.60 Ma in low-temperature steps to the highest temperature step at 31.71 ± 0.26 Ma (Fig. 3). The oldest apparent age approaches that of the younger lava flows and the trachyte dyke. Eight total fusion ages (not shown) range from 30.51 ± 0.64 Ma to 32.24 ± 0.56 Ma with the older of these ages also similar to those of the younger lavas. Given the alteration and discordant age spectrum suggestive of minor loss of radiogenic argon, we take the older age of 32.24 ± 0.56 Ma with caution as our best estimate of time elapsed since rapid cooling of this shallow intrusion.

Adularia–sericite epithermal alteration/mineralization Samples M97-02 and M96-11 are from prominent alteration zones associated with low-sulphidation base-metal–Au veins. M97-02 is from altered latite 2 m from the northern margin of the 25 m wide vein #2 (Fig. 2). Adularia that we measured is a replacement of plagioclase and

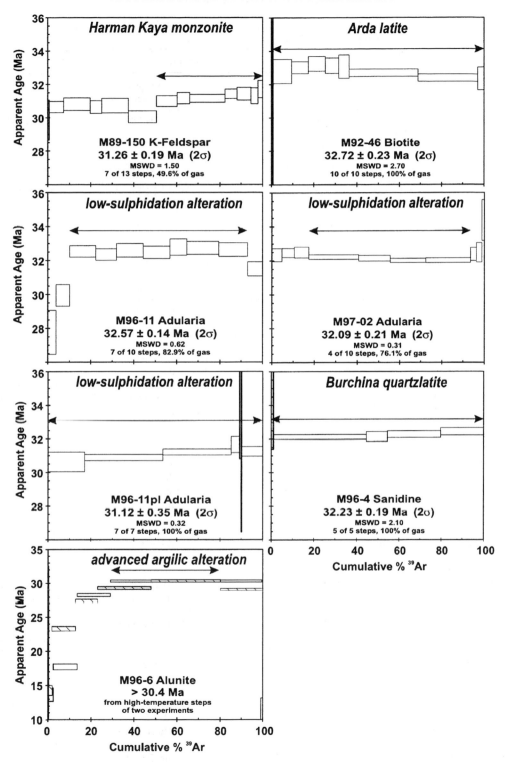

Fig. 3. Age spectra from incremental-heating experiments on sanidine, biotite, K-feldspar, adularia and alunite samples from Madjarovo.

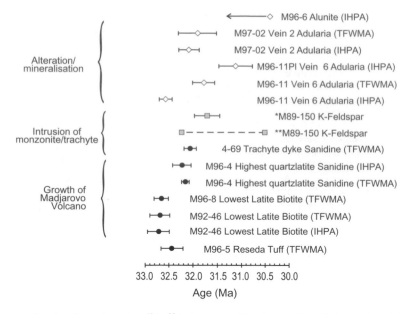

Fig. 4. Summary of preferred ages based on ^{40}Ar/^{39}Ar incremental heating and laser fusion results of this study. * K-fs age interval of monzonite based on the total fusion experiments. ** K-Fs age of the highest temperature age of incremental heating experiment. The alunite age is a minimum based on the highest age steps in the age spectra of Figure 3. Abbreviations: TFWMA, total fusion weighted mean age; IHPA, incremental heating 'plateau' age

contains minor sericite. M96-11 is from the breccia in Chatal Kaya vein #6, (Fig. 2) that comprises angular clasts of strongly adularized and silicified shoshonite lava, cemented by veinlets of quartz and ore minerals. The adularia occurs as four textural types: (1) replacement of large plagioclase phenocrysts intergrown with minor sericite (M96-11pl), (2) plagioclase microphenocrysts replaced by adularia with no sericite, (3) mixture of fine-grained quartz and adularia replacing the groundmass and (4) in veins and cavities intergrown with quartz as coarse-grained rhombs 30–300 μm long with minor sericite (M96-11). Total fusion ages of M97-02 and M96-11 are indistinguishable: 31.91 ± 0.40 Ma and 31.78 ± 0.23 Ma, respectively (Table 1). Incremental heating of sample M97-02 yielded a nearly concordant age spectrum with four steps containing 76.1% of the ^{39}Ar giving a plateau age of 32.09 ± 0.21 Ma (Fig. 3). These ages overlap that of associated trachytic dykes (32.06 ± 0.13 Ma; Table 1) and are consistent with field observations that magmatic and hydrothermal activity were channelized along the same structures. Experiments on M96-11 and M96-11pl gave discordant age spectra, the former similar to that of the Harman Kaya K-feldspar (Fig. 3), with apparent ages increasing from 27–30 Ma at lower temperature to a maximum age of 33.8 Ma, suggesting

that the sample experienced partial ^{40}Ar loss. Seven of ten steps from M96-11 defined a plateau of 32.57 ± 0.14 Ma, but this is older than the ^{40}Ar/^{39}Ar age obtained from the host trachytic dykes (32.06 ± 0.13 Ma) and the plateau age (32.09 ± 0.21 Ma) of sample M97-02 (Fig. 3). Because of this discrepancy, we believe that this age is slightly too old. We believe that the plateau age from M97-02 adularia (32.09 ± 0.21 Ma) gives the best estimate of time elapsed since formation of the adularia–sericite mineralization.

Vein alunite–advanced argillic alteration Sample M96-6 comprises coarse-grained alunite with subordinate woodhousite and svanbergite from a small vein located 200 m east of the Harman Kaya monzonite. Replicate incremental heating experiments produced similar strongly discordant age spectra with apparent ages increasing from *c.* 13–15 Ma at low temperature to maximum ages of *c.* 30.4 Ma (Fig. 4). These age spectra suggest a loss of radiogenic argon possibly due to later reheating, or weathering. Thus, even the maximum age of *c.* 30.4 Ma is probably only a minimum for the time since deposition of alunite. A sample of alunite from the same locality was measured by conventional K–Ar methods (McCoyd 1995) and gave an imprecise age of 32.6 ± 1.2 Ma (2σ).

Discussion

Thermal event at 12–13 Ma

The age of the first low-temperature step (13.65 ± 0.5) of the strongly disturbed $^{40}Ar/^{39}Ar$ age spectrum of the Madjarovo alunite coincides with the ages of several other samples from Madjarovo and neighbouring areas: (1) 13.0 ± 0.5 Ma and 13.7 ± 0.6 Ma ages of two sericite separates of sericite alteration and 12.2 ± 1.0 Ma of the sericite from the quartz–sericite vein alteration obtained by McCoyd (1995) using the K–Ar method, (2) 12 Ma Rb–Sr isochron age of a biotite–muscovite gneiss from the lower sedimentary–volcanoclastic part of the Reseda Tuff (Pljusnin et al. 1988). Thus, it appears that sericite and alunite experienced partial ^{40}Ar loss at about 12–13 Ma, recording a thermal effect that is distinct in time from the magmatic and hydrothermal activity of Madjarovo volcano. It deserves noting that earlier, Kasukeev et al. (1979) recorded an identical thermal event (12 ± 4–16 ± 5 Ma) in the mica-bearing pegmatites from Kamilski Dol, an area about 20 km east of Madjarovo, using the fission track method. Disturbed ages of adularia M96–11 also seen to reflect this thermal event. The only important difference between adularia samples M97-02 and M96-11 is their position with respect to metamorphic basement. Sample M97-02, is located approximately 500–600 m above the metamorphic basement, whereas sample M96-11 is less than 50 m above the metamorphic basement. All the other samples showing disturbed $^{40}Ar/^{39}Ar$ age spectra or younger ages (e.g. McCoyd's sericites and Pliyusnin et al.'s biotite–muscovite gneiss from the Reseda Tuff) are taken either close to or directly from the metamorphic basement, suggesting that the ^{40}Ar loss is not consistent with supergene processes.

The closure temperatures for argon diffusion in muscovite and sericite depend mainly on the temperature and to a lesser extend on the cooling rate, being higher for rapid cooling and lower for slow cooling (Dodson 1973). Snee et al. (1988) estimated a muscovite argon closure temperature of about 325 °C under conditions of rapid cooling and short reheating and a temperature of about 270 °C during slow cooling or extended reheating. The biotite $^{40}Ar/^{39}Ar$ ages indicate that there was no heating above 300 °C. The low-temperature release can be facilitated by the fine-grained size of the sericite. However, although this may be applicable to the sericite in the quartz–sericite alteration it is not thought to apply to the coarser-grained muscovite in the metamorphic rocks. In a study of fluid disturbed ages in metamorphic rocks Miller et al. (1991) state that a convecting fluid at

approximately 275 °C may have caused the loss of radiogenic argon from samples of muscovite with the amount lost dependant upon the length of time the mineral remained at this temperature.

Our preferred explanation for the age difference between disturbed and non-disturbed rocks at Madjarovo is a partial resetting of K–Ar isotope system in the muscovite and alunite that occurred 12–13 Ma ago as the result of endogenous processes.

Duration of magmatic and hydrothermal activity

Our $^{40}Ar/^{39}Ar$ data constrain the duration of magmatic activity and the timing of principal magmatic/hydrothermal events in the complex. Volcanism began shortly after emplacement of the Reseda tuff at 32.44 ± 0.23 Ma. The biotite single-crystal ages of 32.66 ± 0.14 Ma and 32.69 ± 0.15 Ma (Table 1) from the lowermost latite lava combined with the plateau age of 32.72 ± 0.12 Ma (Fig. 3) are indistinguishable at the 95% confidence level from the underlying Reseda Tuff, implying that the two events were closely spaced in time. The ages of 32.23 ± 0.19 Ma and 32.16 ± 0.08 Ma (Fig. 3, Table 1) from the uppermost lavas agree with the 32.16 ± 0.15 Ma age of the Borovitsa pyroclastic flow that covers these lavas (Singer & Marchev 2000). The last volcanism was accompanied or followed by the intrusion of the Harman Kaya stock and trachytic dykes c.32.2?– 32.06 ± 0.13 Ma. Association of the stock with this volcanism is supported by their similar Sr and Nd isotope ratios (Marchev et al. 2002).

Thus, magmatic activity took place over a period of not more than 900 ka, most probably c.500 ka which falls well within the lifespan of well-documented composite volcanoes in Japan, the Cascades, and Southern Andes that were active for 80 ka to 900 ka (e.g., Wohletz & Heiken 1992; Singer et al. 1997).

Hydrothermal activity began with intrusion of the monzonite stock (c.32.2? Ma) and trachytic dykes (32.06 ± 0.13 Ma) and created the advanced argillic (>30.4 Ma) and adularia–sericite alteration/mineralization (32.09 ± 0.21), respectively. The loss of Ar in the alunite did not allow dating of the advanced argillic alteration. However, two types of alteration spatially overlap in the central part of the volcano with cross-cutting relationships, suggesting that adularia–sericite postdates the lithocap formation. The emplacement of trachytic dykes at 32.06 ± 0.13 Ma and subsequent adularia alteration at 32.09 ± 0.21 Ma indicate that the duration of both the advanced argillic, and adularia–sericite systems was less than 250 ka.

This includes time for the monzonite stock to cool, solidify, and develop advanced argillic and sericite alteration. Numerical modelling (Cathles et al. 1997; Marsh et al. 1997) indicates that a shallow stock of this size would cool below the closure temperature of dateable minerals in less than 100 ka.

The Madjarovo epithermal system

Alteration and epithermal deposits traditionally have been divided into acid sulphate or high-sulphidation and adularia–sericite or low-sulphidation (Heald et al. 1987; Hedenquist et al. 2000 and references therein). Hedenquist et al. (2000, 2001), partly based on the studies of John et al. (1999) and John (2001), add another type, dividing the low-sulphidation epithermal deposits into intermediate-sulphidation and proper low-sulphidation deposits. According to these authors, the intermediate-sulphidation deposits are typically hosted in arc-related andesites and dacites. Mineralization is Ag and base-metal rich with Mn carbonates and barite as gangue and massive to comb textured quartz. Sericite is a common alteration and gangue mineral but adularia is rare. Salinity of fluid inclusions is from 3–5 to 10–20 wt% NaCl. Typical low-sulphidation deposits formed in extensional settings hosted by rhyolitic–dacitic or more alkaline rocks. They are Au-rich with the sulphides (pyrite, pyrrotite, arseno-pyrite, and high-Fe sphalerite) recording a low-sulphidation state. Veins show crustiform textures with dominated chalcedony, whereas adularia and illite are common gangue and alteration minerals. The salinity of the fluids is generally <1 wt% NaCl.

Based on the high Ag content and barite, sulphide mineralization (including the Fe-poor sphalerite), zonation of the deposit, and salinity of the fluids, the epithermal mineralization at Madjarovo should be an intermediate-sulphidation type. However, Madjarovo is distinguished in having extensive adularia–sericite alteration, a small amount of carbonate, extensional structures, and shoshonitic volcanism, which are more typical of low-sulphidation systems.

Comparison with other high-, low-, and intermediate-sulphidation systems

Although Heald et al. (1987) argue that the adularia–sericite and acid–sulphate environments are mutually exclusive, a number of ore districts preserve both. These deposits originated in different arc settings: ocean island, e.g. the Plio-Pleistocene Baguio volcano, Northern Luzon, Philippines (Aoki et al. 1993) and Tavua Caldera, Fiji (Setterfield et al. 1992), continental arcs, e.g. Comstock Lode (Vikre 1989); Eocene Mount Skukum Au deposit, Yukon Territory, Canada (Love et al. 1998) and Oligocene Chala deposit, Bulgaria (Singer & Marchev 2000). Most of these (Comstock Lode, Baguio and Chala) fall in the intermediate-sulphidation type of Hedenquist et al. (2001). The intermediate-sulphidation Victoria deposit, high-sulphidation Lepanto and Far Southeast porphyry Cu–Au deposits in Luzon, Phillipines are additional examples of a close spatial and temporal association between these different types of hydrothermal systems (Arribas et al. 1995; Hedenquist et al. 1998, 2001). Precise K–Ar dating of K-silicate to sericite alteration in Lepanto–Far Southeast revealed synchronicity of these processes that occurred over less than 100 ka (Hedenquist et al. 1998, 2001).

High-precision K–Ar and $^{40}Ar/^{39}Ar$ dating suggests that age differences between the acid-sulphate and adularia–sericite alteration types range from 300 to 800 ka and possibly up to 1.6 Ma (Setterfield et al. 1992; Aoki et al. 1993; Love et al. 1998; Singer & Marchev 2000). The temporal separation at Madjarovo was less than 250 ka and probably less than 100–200 ka. In this respect Madjarovo resembles the Far Southeast–Lepanto–Victoria system, where the intermediate-sulphidation Victoria deposit formed c. 150 ka after the Far Southeast porphyry and the Lepanto high-sulphidation deposit (Hedenquist et al. 2001).

Changes in alteration and mineralization styles in several epithermal systems were accompanied by shifts from intermediate to silicic magma compositions (Bonham 1986; Sillitoe 1989; McEwan & Rice 1991; Singer & Marchev 2000).

It appears that changes in alteration style at Madjarovo were not related to different magma compositions. The composition and age of monzonite stocks and trachytic dykes are identical, reflecting shallow emplacement of a single batch of magma and its associated hydrothermal system. However, a short time gap between intrusion of trachytic dykes and emplacement of the low-sulphidation mineralization could be enough for a change in the composition of the source magma towards more acid composition (Fig. 5).

The contributions from cooling magmas to low-sulphidation epithermal systems may be restricted to the heat needed to drive hydrothermal circulation. Low-sulphidation hydrothermal systems typically form distal to intrusions and may apparently persist up to 1.5 Ma after volcanism (Silberman 1985; Heald et al. 1987; Conrad et al. 1993; Hedenquist & Lowenstern 1994). However, based on precise $^{40}Ar/^{39}Ar$ measurements, Conrad & McKee (1996), Henry et al. (1997) and Singer &

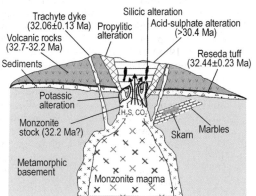

A   **32.7 to 32.1 Ma**

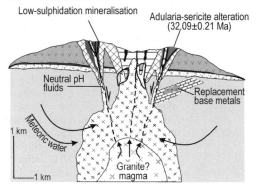

B   **32.1 to 32.0 Ma**

Fig. 5. Model for the evolution of Madjarovo volcanic complex, associated alteration and mineralization.

Marchev (2000) showed that high-grade adularia–sericite deposits at Sleeper and Round Mountain, Nevada and Chala, Bulgaria, were generated 100 ka to 300 ka after emplacement of rhyolitic magmas. Madjarovo deposits are similar, having formed rapidly after trachytic magma intrusion.

A close spatial and temporal relationship between rhyolitic dykes and adularia–sericite alteration/mineralization in the Spahievo Ore District, Bulgaria was interpreted by Singer & Marchev (2000) as reflecting a genetic connection. Similarly, the association of monzonite/trachyte magma and low-sulphidation vein mineralization at Madjarovo is interpreted to favour a direct genetic relationship. However, comparison of the Sr isotope compositions of the monzonite intrusion and gangue barite from the largest veins (Marchev *et al.* 2002) make the Harman Kaya intrusion an unlikely source for the low-sulphidation epithermal system. Thus, we propose that the fluids forming the low-sulphidation mineralization at Madjarovo were related to a magma that solidified

at a deeper level, perhaps 3–4 km depth, and that the fluids were additionally contaminated by radiogenic Sr from Rhodope basement rocks which host the intrusions.

Model

Our model for the evolution of the Madjarovo volcanic complex, its alteration and mineralization, is outlined in Figure 5 and summarized here:

1   a shield volcano grew between 32.7 and 32.2 Ma;
2   magmatism continued 32.2–32.1 Ma with intrusion of monzonitic stocks and trachytic dykes, as apophyses of a larger pluton located at depths of 2–4 km (Fig. 5a);
3   intrusion of the latter magma was followed by a period of rapid erosion and down-cutting that removed 300–500 m from the south half of the complex; the palaeosurface encroached the top of the Harman Kay monzonitic stock;
4   the deeper crystallizing volume of the pluton exolved a S- and Cl-rich fluid that separated to form a low-saline acid vapour and saline brine (e.g. Hedenquist & Lowenstern 1994; Shinohara & Hedenquist 1997); acid vapour ascended and reacted with the top of the monzonitic stocks and their host lavas, thereby forming barren lithocap zones; brines also condensed to form K-alteration and minor porphyry mineralization;
5   progressive cooling and fractionation of the pluton produced more silicic (granitic or granosyenitic) magma evinced as the quartz latite lavas; separation, neutralization, and dilution of this mainly magmatic fluid gave rise to the low-sulphidation mineralization; the time span separating intrusion of the monzonite-trachyte magma, formation of lithocap, and the low-sulphidation system was most probably not more than 200 ka.

This work was funded by Swiss NSF Grant 7BUPJ048659 for Bulgarian–Swiss cooperative research. We appreciate the comments and helpful reviews by R. Nakov and M. Economou. The comments and constructive advice of A. von Quadt are greatly acknowledged. We thank Danko Jelev for assistance in the field and Raya Raicheva for help with the figures.

References

AOKI, M., COMSTI, E.C., LAZO, F.B. & MATSUHISA, Y. 1993. Advanced argillic alteration and geochemistry of alunite in an evolving hydrothermal system at Baguio, Northern Luzon, Philippines. *Resource Geology*, **43**, 155–164.

ARIKAS, K. & VOUDOURIS, P. 1998. Hydrothermal alterations and mineralizations of magmatic rocks in the southeastern Rhodope massif. *In:* CHRISTO-FIDES, G., MARCHEV, P. & SERRI, G. (eds) *Teriary magmatism of the Rhodopian region*. Acta Vulcanologica, **10**, 353–365.

ARNAUDOVA, R., VELINOV, I., GOROVA, M., ARNAUDOV, V., BONEV, I., BATANDJIEV, I. & MARCHEV, P. 1991. Geochemistry of quartz-adularia metasomatites from the Madjarovo Tertiary polymetal Au-Ag deposits, the East Rhodopes. *In:* MRNA, F. (ed.) *Exploration Geochemistry 1990*. Proceedings of the 3rd International Joint Symposium IAGC and AGC, Prague, 13–16.

ARRIBAS, A. JR, HEDENQUIST, J.W., ITAYA, T., OKADA, T., CONCEPCIÓN, R.A. & GARCIA, J.S. JR 1995. Contemporaneous formation of adjacent porphyry and epithermal Cu-Au deposits over 300 ka in northern Luson, Philippines. *Geology*, **23**, 337–340.

ATANASOV, A. 1959. The structure of the lead zinc ore deposit of Madjarovo. *Annuaire de l'Université de Sofia, Géologie*, **52**(2), 313–348. [in Bulgarian with English Summary].

ATANASOV, A. 1962. Mineralization stages, primary zoning and genesis of the complex ore-deposit of Madjarovo. *Annuaire de l'Univversité de Sofia, Géologie*, **55**(2), 229–267. [in Bulgarian with English Summary].

BERGER, B.R. & EIMON, P. 1983. Conceptual models of epithermal precious metal deposits. *In:* SHANKS, W.C. III (ed.) *Volume on unconventional mineral deposits*. Society of Mining Engineers, American Institute of Mining Engineering, 191–205.

BONHAM, H.F. JR 1986. Models for volcanic-hosted epithermal precious metal deposits; a review. *In: Proceedings Symposium 5: Volcanism, Hydrothermal systems and Related Mineralization*. International Volcanological Congress, Auckland, 13–17.

BRESKOVSKA, V., ILIEV, Z., MAVROUDCHIEV, B., VAPTZAROV, I., VELINOV, I. & NOZHAROV, P. 1976. The Madjarovo ore field. *Geochemistry, Mineralogy and Petrology*, **5**, 23–57. [in Bulgarian with English abstract].

BRESKOVSKA, V. & TARKIAN, M. 1993. Mineralogy and fluid inclusion study of polymetallic veins in the Madjarovo ore field, Eastern Rhodope, Bulgaria. *Mineralogy and Petrology*, **49**, 103–118.

BUCHANAN, L.J. 1981. Precious metal deposits associated with volcanic environments in the southwest. *In:* DICKSON, W.R. & PAYNE, W.D. (eds) *Relations of Tectonics to Ore Deposits in the Southwest Cordillera*. Arizona Geological Society Digest, **14**, 237–262.

CATHLES, L.M., ERENDI, H.J. & BARRIE, T. 1997. How long can a hydrothermal system be sustained by a single intrusive event? *Economic Geology*, **92**, 766–771.

CHRISTOFIDES, G., SOLDATOS, T., ELEFTERIADIS, G. & KORONEOS, A. 1998. Chemical and isotopic evidence for source contamination and crustal assimilation in the Hellenic Rhodope plutonic rocks. *In:* CHRISTOFIDES, G., MARCHEV, P. & SERRI, G. (eds) *Tertiary magmatism of the Rhodopian region*. Acta Vulcanologica, **10**, 305–318.

CONRAD, J.E., McKEE, E.H., RYTUBA, J.T., NASH, J.T. & UTTERBACK, W.C. 1993. Geochronology of the Sleeper deposit, Humboldt County, Nevada: Epithermal gold-silver mineralization following emplacement of a silicic flow-dome complex. *Economic Geology*, **88**, 317–327.

CONRAD, J.E. & McKEE, E.H. 1996. High-precision $^{40}Ar/^{39}Ar$ ages of rhyolitic host rocks and mineralized veins at the Sleeper Deposit, Humbolt County, Nevada. *In:* COYNER, A.R. & FAHEY, P.L. (eds) *Geology and ore Deposits of the American Cordilera*. Geological Society of Nevada Symposium Proceedings, Reno/Sparks, Nevada, April 1995, 257–262.

COOKE, D.R. & SIMMONS, S.F. 2000. Characteristics and Genesis of Epithermal Gold Deposits. *Society of Economic Geologists, Reviews in Economic Geology*, **13**, 221–244.

DEL MORO, A., KYRIAKOPOULOS, K., PEZZINO, A., ATZORI, P. & LO GUIDICE, A. 1990. The metamorphic complex associated to the Kavala plutonites: An Rb–Sr geochronological, petrological and structural study. 2nd Hellenic–Bulgarian Symposiun, Thessaloniki. *Geologica Rhodopica*, **2**, 143–156.

DODSON, M.H. 1973. Closure temperature in cooling geochronological and petrological systems. *Contributions to Mineralogy and Petrology*, **40**, 259–274.

DUFFIELD, W. & DALRYMPLE, G.B. 1990. The Taylor Creek Rhyolite of New Mexico: a rapidly emplaced field of lava domes and flows. *Bulletin of Volcanology*, **52**, 475–487.

GERGELCHEV, V.N. 1974. Main features and formation stages of the Madjarovo cauldron subsidence and structural conditions for its ore-bearing properties. *Bulletin of the Geological Institute Series of Metal and Non-metal Mineral Deposits*, **23**, 5–29.

HARKOVSKA, A., YANEV, Y. & MARCHEV, P. 1989. General features of the Paleogene orogenic magmatism in Bulgaria. *Geologica Balcanica*, **19**(1), 37–72.

HEALD, P., FOLEY, N.K. & HAYBA, D.O. 1987. Comparative anatomy of volcanic-hosted epithermal deposits: acid-sulphate and adularia-sericite types. *Economic Geology*, **82**, 1–26.

HEDENQUIST, J.W. & ARRIBAS, A. JR 1999. I. Hydrothermal processes in intrusion-related systems. II. Characteristics, examples and origin of epithermal gold deposits. *In:* MOLNAR, F., LEXA, J. HEDENQUIST, J.W.) (eds) *Epithermal Mineralization of the Western Carpathians*. Society of Economic Geologists, Guidebook Series, **31**, 13–61.

HEDENQUIST, J.W., ARRIBAS, A. JR & REYNOLDS, T.J. 1998. Evolution of an intrusion-centered hydrothermal system, Far Southeast-Lepanto porphyry and epithermal Cu-Mo deposit, Philippines. *Economic Geology*, **93**, 373–404.

HEDENQUIST, J.W., ARRIBAS, A. JR & URIEN-GONZALES, E. 2000. Exploration for epithermal gold deposits. *Society of Economic Geologists, Reviews in Economic Geology*, **13**, 245–277.

HEDENQUIST, J.W., CLAVERIA, R.J. & VILLAFUERTE, G.P. 2001. Types of sulfide-rich epithermal deposits and their affiliation to porfiric systems: Lepanto-

Victoria-Far Southeast deposits, Philippines, as examples. *In: Pro Explo 2001, Lima, Volume Proceedings, CD.*

HEDENQUIST, J.W. & LOWENSTERN, J.B. 1994. The role of magmas in the formation of hydrothermal ore deposits. *Nature*, **370**, 519–527.

HENRY, C.D., ELSON, H.B., McINTOSH, W.C., HEIZLER, M.T. & CASTOR, S.B. 1997. Brief duration of hydrothermal activity at Round Mountain, Nevada, determined from ^{40}Ar/^{39}Ar geochronology. *Economic Geology*, **92**, 807–826.

INNOCENTI, F., KOLIOS, N., MANETTI, P., MAZZUOLI, R., PECCERILLO, A., RITA, F. & VILLARI, L. 1984. Evolution and geodynamic significance of the Tertiary orogenic volcanism in Northeastern Greece. *Bulletin Volcanologique*, **47**, 25–37.

IOSIFOV, D., TSVETKOVA, D., PCHELAROV, V., REVYAKIN, P., GERGELCHEV, V. & TSVETKOV, A. 1987. Deep structure of Avren-Madzarovo ore zone. *Review of the Bulgarian Geological Society*, **48**(2), 73–86. [in Bulgarian with English abstract].

IVANOV, R. 1960. [Magmatism in the East Rhodopean depression: I. Geology]. *Travaux sur la Géolologie de Bulgarie, Série Géochemie et des cîtes métallliféres et non-métalliféres*, **1**, 311–387. [in Bulgarian with German abstract].

IVANOV, R. 1968. Zonal Arrangement of Rock Series with respect to Deep-Seated Masses. *Repr 23rd International Geological Congress, Prague*, **1**, 43–56.

IVANOV, R. & KOPP, K.-O. 1969. Das Alttertiar Thrakiens und der Ostrhodope. *Geologica et Palaeontologica*, **3**, 123–151.

JOHN, D.A. 2001. Miocene and Early Pliocene epithermal gold-silver deposits in the Northern Great Basin: Characteristics, distribution, and relationship to magmatism. *Economic Geology*, **96**, 1827–1854.

JOHN, D.A., GARSIDE, L.J. & WALLACE, A.R. 1999. Magmatic and tectonic setting of late Cenozoic epithermal gold-silver deposits in Northern Nevada, with an emphasis on the Rah Rah and Virginia ranges and the Northern rift. *Geological Society of Nevada, Special Publications*, **29**, 65–158.

KASUKEEV, N., AMOV, B., KASUKEEVA, M., TANEVA, T. & IGNATOVA, R. 1979. Thack-method studies on the geological age of mica-bearing pegmatites from the Eastern Rhodope Mountains. *Geochimistry, Mineralogy and Petrology*, **10**, 3–10.

LILOV, P., YANEV, Y. & MARCHEV, P. 1987. K/Ar dating of the Eastern Rhodope Paleogene volcanism. *Geologica Balcanica*, **17**(6), 49–58.

LOVE, D.A., CLARK, A.N., HODGSON, C.J., MORTENSEN, J.K., ARCHIBALD, D.A. & FARRAR, E. 1998. The timing of adularia-sericite-type mineralization and alunite-kaolinite-type alteration, Mount Skukun epithermal gold deposit, Yukon Territory, Canada. ^{40}Ar/^{39}Ar and U-Pb geochronology. *Economic Geology*, **93**, 437–462.

MARCHEV, P., DOWNES, H., THIRLWALL, M.F. & MORITZ, R. 2002. Small-scale variations of ^{87}Sr/^{86}Sr isotope composition of barite in the Mafjarovo low-sulphidation epithermal system, SE Bulgaria: implications for sources of Sr, fluid fluxes and pathways of the ore-forming fluids. *Mineralium*

Deposita, **37**, 669–677.

MARCHEV, P., ILIEV, Z. & NOKOV, S. 1989. Oligocene volcano Madjarovo. *Guide-book of Scientific excursion*, **E-2**, 97–103.

MARCHEV, P., NOKOV, S., McCOYD, R. & JELEV, D. 1997. Alteration processes and mineralizations in the Madjarovo Ore Field - a brief review and new data. *Geochimistry, Mineralogy and Petrology*, **32**, 47–58.

MARCHEV, P. & ROGERS, G. 1998. New Rb-Sr data on the bottom and top lava flow of the Madjarovo volcano: Inferences for the age and genesis of the lavas. *Geochimistry, Mineralogy and Petrology*, **34**, 91–96.

MARCHEV, P., ROGERS, G., CONREY, R., QUICK, J., VASELLI, O. & RAICHEVA, R. 1998a. Paleogene orogenic and alkaline basic magmas in the Rhodope zone: relationships, nature of magma sources, and role of crustal contamination. *In:* CHRISTOFIDES, G., MARCHEV, P. & SERRI, G. (eds) *Tertiary magmatism of the Rhodopian region.* Acta Vulcanologica, **10**, 217–232.

MARCHEV, P., VASELLI, O., DOWNES, H., PINARELLI, L., INGRAM, G., ROGERS, G. & RAICHEVA, R. 1998b. Petrology and geochemistry of alkaline basalts and lamprophyres: implications for the chemical composition of the upper mantle beneath the Eastern Rhodopes (Bulgaria). *In:* CHRISTOFIDES, G., MARCHEV, P. & SERRI, G. Acta Vulcanologica, **10**, 233–242.Tertiary magmatism of the Rhodopian region.

MARSH, T.M., EINAUDI, M.T. & McWILLIAMS, M. 1997. ^{40}Ar/^{39}Ar geochronology of Cu-Au and Au-Ag mineralization in the Potrerilos district, Chile. *Economic Geology*, **92**, 784–806.

MAVROUDCHIEV, B. 1959. Upper Oligocene intrusions from the Madjarovo ore district. *Annuaire de l'Université de Sofia, Géologie*, **52**(2), 251–300. [in Bulgarian with English Summary].

MAVROUDCHIEV, B., BOYANOV, I., JOSSIFOV, D., BRESCOVSKA, V., DIMITROV, R. & GERGELCHEV, V. 1996. Late Alpine Metallogeny of the Eastern Rhodope Collision-collapse units and Continental-rift Structures. *In:* POPOV, P. (ed.) *Plate tectonic aspects of the Alpine metallogeny in the Carpatho-Balkan region.* Sofia, UNESCO, IGCP Project N356, **1**, 125–135.

McCOYD, R. 1995. *Isotopic and geochemical studies of the epithermal-mesothermal Pb-Zn deposits of S.E. Bulgaria.* PhD thesis, University of Aberdeen.

McEWAN, C.J.A. & RICE, C.M. 1991. Trace-element geochemistry and alteration facies associated with epithermal gold-silver mineralization in an evolving volcanic centre, Rosita Hills, Colorado, U.S.A. *Transactions of the Institution of Mining and Metallurgy*, **100**, B19–B27.

MILLER, W., FALLICK, A., LEAKE, B., MACINTYRE, R. & JENKIN, G. 1991. Fluid disturbed hornblende K-Ar ages from Dalradian rock of Connemara, Western Ireland. *Journal of the Geological Society, London*, **148**, 985–992.

NOKOV, S. & MALINOV, O. 1993. Quartz-adularia metasomatites and molybdenium mineralization from the Madjarovo ore field (SE Bulgaria). *Review of the Bulgarian Geological Society,* **54**(1), 1–12.

PLJUSNIN, G.S., MARCHEV, P. & ANTIPIN, V.S. 1988. Rb-Sr age and genesis of the shoshonite-latite Eastern Rhodopes series. *Doklady Akademii Nauk SSSR,* **303**, 719–724.

RADONOVA, T. 1960. [Studies on the mineral composition and alterations of Madjarovo base-metal ore deposit in the Eastern Rhodopes]. *Travaux sur la Géolologie de Bulgarie, Série Géochemie et des cîtes métalllifères et non-métallifères,* **1**, 115–197. [in Bulgarian with German abstract].

RAICHEVA, R., MARCHEV, P. & VASELLI, O. 2001. Mixed and mingled lavas at Lower Oligocene Madjarovo and Zvezdel volcanoes, Eastern Rhodopes (Bulgaria). ABCD-GEODE 2001 Workshop Vata Bai, Romania. *Romanian Journal of Mineral Deposits,* **79**(suppl. 2), 88–89.

RICOU, L-E. 1994. Tethys reconstructed: plates, continental fragments and their boundaries since 260 Ma from Central America to Southeastern Asia. *Geodinamica Acta,* **7**, 169–218.

SETTERFIELD, T.H., MUSSETT, A. & OGLETHORPE, R.D.J. 1992. Magmatism and associated hydrothermal activity during the evolution of the Tavua Caldera: $^{40}Ar/^{39}Ar$ dating of the volcanic, intrusive, and hydrothermal events. *Economic Geology,* **87**, 1130–1140.

SHINOHARA, H. & HEDENQUIST, J.W. 1997. Constraints on magma degassing beneath the Far Southeast porphyry Cu-Au deposit, Philippines. *Journal of Petrology,* **38**, 1741–1752.

SILBERMAN, M.L. 1985. Geochronology of hydrothermal alteration and mineralization: Tertiary epithermal precious-metal deposits in the Great Basin. *U.S. Geological Survey Bulletin,* **1646**, 55–70.

SILLITOE, R.H. 1989. Gold deposits in western Pacific island arcs: The magmatic connection. *In:* KEAYS, R.R., RAMSAY, W.R.H. & GROVES, D.I. (eds) *The Geology of Gold Deposits. The perspective in 1988.* Economic Geology Monographs, **6**, 274–291.

SILLITOE, R.H. 1993. Intrusion-related gold deposits. *In:* FOSTER, R.P. (eds) *Gold metallogeny and exploration.* Chapman and Hall, London, 168–208.

SILLITOE, R.H. 1995. Exploration of porphyry copper lithocaps. *In:* MAUK, J.L. & ST GEORGE, J.D. (eds) *PACRIM Congress 1995.* Australasian Institute of Mining and Metallurgy, Publication Series, **9/95**, 527–532.

SINGER, B.S., DUNGAN, M.A. *ET AL.* 1997. Volcanism and erosion during the past 930 thousand years at the Tatara–San Pedro complex, Chilean Andes. *Geological Society of America Bulletin,* **109**, 127–142.

SINGER, B. & MARCHEV, P. 2000. Temporal evolution of arc magmatism and hydrothermal activity, including epithermal gold veins, Borovitsa caldera, southern Bulgaria. *Economic Geology,* **95**, 1155–1164.

SNEE, L.W., SUTTER, J.F. & KELLY, W.C. 1988. Thermochronology of economic mineral deposits: Dating the stages of mineralization at Panasqueira, by high-precision $^{40}Ar/^{39}Ar$ age spectrum techniques on muscovite. *Economic Geology,* **83**, 335–354.

STOYANOV, R. 1979. *[Metallogeny of the Rhodope Central Massif].* Nedra, Moscow [in Russian].

VELINOV, I. & ASLANIAN, S. 1981. Vein-type alunite from Bulgaria. *Comptes rendus de l'Académie bulgare des Sciences,* **34**, 1417–1419.

VELINOV, I., BATANDJIEV, I., TCHOLAKOV, P. & BLAZHEV, B. 1977. [New data about the relationships between structure forming and postmagmatic processes in Madjarovo ore field]. *Comptes rendus de l'Académie bulgare des Sciences,* **30**, 1749–1752. [in Russian].

VELINOV, I. & NOKOV, S. 1991. Main types and metallogenic significance of the Madjarovo hydrothermally altered Oligocene volcanics. *Comptes rendus de l'Académie bulgare des Sciences,* **44**(9), 65–68.

VIKRE, P.G. 1989. Fluid-mineral relations in the Comstok Lode. *Economic Geology,* **84**, 1574–1613.

WOHLETZ, K. & HEIKEN, G. 1992. *Volcanology and geothermal energy.* University of California Press, Berkeley.

YANEV, Y., INNOCENTI, F., MANETTI, P. & SERRI, G. 1998. Upper Eocene-Oligocene collision-related volcanism in Eastern Rhodopes (Bulgaria)-Western Thrace (Greece): Petrogenetic affinity and geodynamic significance. *In:* CHRISTOFIDES, G., MARCHEV, P. & SERRI, G. (eds) *Tertiary magmatism of the Rhodopian region.* Acta Vulcanologica, **10**, 279–292.

Multiple generations of extensional detachments in the Rhodope Mountains (northern Greece): evidence of episodic exhumation of high-pressure rocks

ALEXANDER KROHE[1] & EVRIPIDIS MPOSKOS[2]

[1]*University of Muenster, Institute for Mineralogy, Laboratory for Geochronology, Corrensstrasse 24, D-48149 Muenster, Germany (e-mail: krohe@nwz.uni-muenster.de)*
[2]*National Technical University of Athens, Department of Mining and Metallurgical Engeneering, Section of Geological Sciences, 9, Heroon Polytechniou Str., Gr-15780 Zografou (Athens), Greece*

Abstract: An integrated structural and petrological study shows that exhumation of high-P rocks of the Rhodope occurred in several pulses. The structurally uppermost Kimi Complex recording an Early Cretaceous high-P metamorphism was exhumed between about 65 and >42 Ma. The Sidironero, Kardamos and Kechros Complexes that record Early Tertiary high-P metamorphism (at least 19 kbar at 700 °C) were exhumed at >42–30 Ma. Exhumation occurred with isothermal decompression. Strain episodes depict thrusting of medium/high grade above lower grade high-P rocks, syn-thrusting extension and post-thrusting extension. Polyphase extension created several generations of detachment zones, which, in sum, excise about 20 km of material within the crustal profile. This reduction in crustal thickness is consistent with a reduction of the present crustal thickness from more than 40 km to less than 30 km in the eastern Rhodope. We mapped the post-thrusting Xanthi low-angle detachment system over 100 km, from its break up zone above the Sidironero Complex (Central Rhodope) into the eastern Rhodope. This detachment shows an overall ESE-dip with a ramp and flat geometry cutting across the earlier thrust structures. The Kimi Complex is the hanging wall of all syn- and post-thrusting extensional systems. On top of the Kimi Complex, marine basins were formed from the Lutetian (*c.* 48–43 Ma) through the Oligocene, during extension. Successively, at *c.* 26 to 8 Ma, the Thasos/Pangeon metamorphic core complexes were exhumed. In these times representing the early stages of Aegean back-arc extension, the Strymon and Thasos detachment systems caused crustal thinning in the western Rhodope. Renewed heating of the lithosphere associated with magmatism and exhumation of hot middle crust from beneath the Sidironero Complex occurred. We focus on the geometry, timing and kinematics of extension and contraction structures related to the >42–30 Ma interval and how these exhumed high-P rocks. We interpret high-P rocks exhumed in this interval as a window of the Apulian plate beneath the earlier (in the Cretaceous) accreted Kimi Complex.

Exhumation of continental material that has been earlier buried to depths corresponding to 20 kbar and more is associated with reduction of at least 60 km of material from above the high pressure (high-P) rocks. Various tectonic mechanisms have been suggested, describing different mass movement paths that result in exhumation of high-P rocks.

(i) '*Late-/post-collisional extension*'. The buoyant thick continental crust and/or the hot lithospheric mantle cause buoyancy forces and surface uplift; the uplifted thick continental domain exerts a force on the surrounding and tends to extend ('collapse'). Surface uplift may also be caused by convective removal of the cold and heavy lithospheric mantle (e.g. Platt *et al.*1998).

(ii) '*Back-arc extension*'. Drag on retreating subduction boundaries initiates profound extension tectonics in continental areas above the subducting plate (e.g. Royden 1993).

(iii) '*Upward extrusion*'. Detachment and ascent of tectonic wedges from a continually subducting continental plate occurs on concurrently active deeply rooted thrusts and higher level normal faults. Ascent is essentially controlled by buoyancy forces of the subducted material during ongoing collision tectonics (e.g. Chemenda *et al.* 1995).

Such processes may explain the close association of extensional and contractional structures reported in several high-P terrains. However, each mechanism demands substantially distinct driving

From: BLUNDELL, D.J., NEUBAUER, F. & VON QUADT, A. (eds) 2002. *The Timing and Location of Major Ore Deposits in an Evolving Orogen*. Geological Society, London, Special Publications, **204**, 151–178. 0305-8719/02/$15.00 © The Geological Society of London 2002.

forces and mechanical properties of the lithospheric wedge. Thus, detailed structural, petrological and geochronological investigations of high-P metamorphic terrains are a key for understanding the kinematics of large-scale mass movements associated with exhumation of deep rocks.

In this article we present results of our integrated structural and petrological research on Alpine high-P rocks in northern Greece (Rhodope Domain, Fig. 1). We will address the spatial and temporal distribution of compression and extension tectonics, the role of both in the exhumation history of high-P rocks and the magnitude of displacements on the respective detachment/shear zone systems.

Geological context and scope of work

Structural setting

Exhumation mechanisms of high-P rocks have been intensely debated throughout the Mediterranean Alpine Belt that extends from southern Spain over Corsica, the western and eastern Alps, southern Bulgaria/northern Greece, the Aegean into Turkey (Fig. 1). From the Late Oligocene to Recent, in southern Spain, Corsica and the Aegean Sea, subduction of remnants of Triassic/Jurassic oceanic lithosphere has triggered back-arc extension overprinting the structures related to collision (Royden 1993). Importantly, high-P rocks were exhumed both within such sites of back-arc extension (western Alps and Rhodope in southern Bulgaria/northern Greece, Fig. 1), and also outside such sites. In either case, exhumation is linked

with extensional structures. The Greek territory is assembled by largely different tectono-metamorphic complexes from this suture zone.

The Rhodope Domain (Fig. 1) comprises Variscan continental crust, Mesozoic metasediments and remnants of oceanic crust (Burg *et al.*1996; Ricou *et al.*1998; Barr *et al.*1999). In places, grades of Alpine metamorphism exceed 19 kbar at *c.* 700°C (Liati & Seidel 1996). Recently, Ultra high-P metamorphism has been detected, indicated, amongst other things, by the presence of diamond inclusions in garnets of pelitic gneisses (Mposkos *et al.*2001). Probably Early Cretaceous Ultrahigh-P and Early Tertiary high-P stages occurred (Wawrzenitz & Mposkos 1997; Liati & Gebauer 1999). This domain reflects Cretaceous and Tertiary episodes of crustal thickening and extension (Kilias *et al.*1997; Mposkos 1989; Dinter *et al.*1995; Burg *et al.* 1996; Barr *et al.*1999, and below).

The Rhodope Domain is located to the north of the Aegean basin: a major back-arc structure in the Mediterranean, sited behind the Dinaric–Hellenic thrust belt (Royden 1993; Fig. 1). From the central Rhodope (Bulgarian part) to the Aegean Sea, and the Turkish boundary (eastern Rhodope) the crustal thickness has been reduced from above 50 km to below 30 km, as estimated from gravimetric data (Makris 1985).

Previous ideas

Early ideas considered the Rhodope Domain as a pre-Alpine (Variscan or older) 'stable block' between the Dinaric–Hellenic and the Carpathian

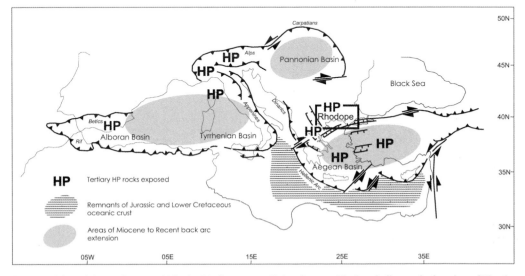

Fig. 1. Position of the study area within the Mediterranean Alpine Orogen. The box indicates the location of Figs 2, 3 and 7.

thrust belt ('Zwischengebirge' of Kober 1928). Since the discovery of Alpine high-P metamorphism, the tectonic setting has been discussed controversially.

(i) According to Burg *et al.* (1996) and Ricou *et al.* (1998) the Rhodope Domain consists of superimposed 'mixed' (continental and ophiolitic) and continental units ('Drama continental fragment') that both accreted/collided with Europe during the Cretaceous, along the 'Axios (Vardar) suture zone' situated to the east of the Pelagonian Zone (Fig. 2, insert map).

(ii) According to Dinter (1998), the western Rhodope Domain is composed of parts of the Apulian continental plate (his 'Rhodope Metamorphic Core Complex') exposed as a tectonic window beneath an Early Tertiary accretionary complex.

The nature, timing and kinematics of tectonic movements causing exhumation of the Rhodope Domain have also been the subject of controversial discussion: Burg *et al.*(1995) and Ricou *et al.*(1998) interpret the dominant structures as to reflect Cretaceous collision and upward extrusion

Fig. 2. New tectonic subdivision of the Rhodope Domain and location of major detachment systems. Locations of profiles (Fig. 12) are also indicated.

Table 1. *Geochronological data*

	Section I		Section II		
	Vertiskos Complex	Kimi Complex	Kerdilion Complex	Upper Sidironero Complex	Lower Sidironero Complex
K–Ar Hbl	93 ± 3–123 ± 3			37 ± 1–41 ± 1 (amphibolite) 38 ± 1 (Kentavros Metagranitoid)	45 ± 1–47 ± 1 (amphibolite) 57 ± 1–95 ± 1 ('eclogitogen. amphib.')
K–Ar Ms	'Variscan'		48 ± 3–56 ± 3		37 ± 1
K–Ar Bi	88 ± 3–115 ± 4		32 ± 3–39 ± 3	35 ± 1	
Rb–Sr whole rock					
Rb–Sr white mica	'Variscan' (orthogneiss, pegmatoids)	65 ± 3 (pegmatoid)	51 ± 1 (Thasos Upper Unit)		
Rb–Sr Bi			39 ± 1 (Thasos Upper Unit)		
Sm–Nd Grt		119 ± 2 (grt–pyroxenite)			
U–Pb Zr				294 ± 8 (orthogneiss) 42 ± 1 (eclogite) 40 ± 1 (migmatite)	
U–Pb Mo					
U–Pb Ttn					

of high-*P* rocks. This was followed by Oligocene/ Miocene extension. In Dinter's (1998) view, Apulian rocks were exhumed by Miocene extension in the Aegean back-arc, following Early Tertiary thrusting and thickening.

Scope of work

We have carried out structural and microstructural investigations of the high-*P* rocks of the Greek part of the Rhodope Mountains (Fig. 2) in order to elucidate the mass movements associated with the exhumation of high-*P* rocks. Our investigations particularly focus on the geometry, kinematics and timing of multiple generations of normal detachment systems. The Rhodope Mountains were selected as a study area, because they preserve the tectono-metamorphic imprint of a long-lasting geodynamic history including Late Oligocene–Miocene back-arc extension (Dinter *et al.*1995; Wawrzenitz & Krohe 1998), and exhumation of high-*P* rocks prior to Aegean back-arc extension (Burg *et al.*1995). The Rhodope Mountains expose the structural setting of high-*P* rocks over large distances, contrasting to the central Aegean Sea (Fig. 1), where high-*P* rocks occur isolated on the Cyclades' islands (Vandenberg & Lister 1996).

We newly introduce large-scale discrete low-angle detachment systems, which we mapped using the 1:50 000 scale geological maps of IGME (Institute of Geological and Mining Exploration of Greece), in the following referred to as the Xanthi, West Kardamos, East Kardamos and Kechros normal detachment fault systems (Fig. 2). We give detailed descriptions of these systems and particularly address the following questions:

1 what are the temporal and spatial relationships of these extensional structures to contractional structures?

2 how do the various geochronological data constrain absolute ages of inferred deformation increments?

3 how are the inferred deformation increments related to the *P–T* path of the tectonic complexes?

General tectonic subdivision

In the eastern Rhodope Mountains, the high-*P* metamorphic terrain is overlain by the supracrustal eastern Circum Rhodope Zone (eastern CRZ; Fig. 2), comprising various ophiolitic sequences. K–Ar hornblende ages and apatite fission track ages both ranging between 150 and 160 Ma (review in Bigazzi *et al.*1989; Hatzipanagiotou *et al.*1994), indicate cooling at shallow crustal levels no later than in the late Jurassic. Associated phyllites, marbles, albite gneisses and greenschists were weakly metamorphosed to temperatures below *c.*400 °C (chlorite zone), indicated by the mineral assemblages Qtz–Ab–Phe–Kfs–Stp (albite gneisses) and Qtz–Ab–Chl–Czo–Act (greenschists; abbreviations after Kretz 1983 and Bucher & Frey 1994). These remain undated.

We subdivide the underlying high-*P* metamorphic terrain into three crustal sections separated by the large scale detachment faults. A compilation of published geochronological data (Table 1; see below for references) suggests that each crustal section has a distinct geochronological record (Table 1) and that geochronological

	Section II (continued)			Section III	
	Kardamos Complex	Kechros Complex	Post-Tectonic Plutons	Pangeon & Thasos Complexes, Intermediate Unit	Thasos Complex, Lower Unit
K–Ar Hbl	42 ± 1		30 ± 1 (Xanthi Pluton) 29 ± 1–33 ± 1 (Vrondou Pluton)	22 ± 1–23 ± 1 (Symvolon Metapluton)	
K–Ar Ms					
K–Ar Bi	39 ± 1		29 ± 1 (Xanthi Pluton) 30 ± 1–32 ± 1 (Leptokaria Pluton) 30 ± 1 (Vrondou Pluton)	15 ± 1	
Rb–Sr whole rock				26 ± 2	18 ± 1
Rb–Sr white mica		334 ± 5 (pegmatoid) 37 ± 1 (gneiss)		23 ± 1	18 ± 1 (gneiss) 'Variscan' (pegmatoid)
Rb–Sr Bi				15 ± 1	12 ± 1–15 ± 1
Sm–Nd Grt					'Variscan' (pegmatoid)
U–Pb Zr		334 ± 3 (orthogneiss)			
U–Pb Mo U–Pb Ttn				'Variscan' (orthogneiss) 23 ± 1 (Symbolon Pluton)	18 ± 1–25 ± 1

ages decrease downward, implying three major episodes of exhumation/cooling of structurally deeper sections. From the top to the bottom, the three crustal sections are as follows.

Section I was exhumed before 48 Ma. It consists of the high grade Kimi, Vertiskos and Kotili–Melivia–Complexes (Fig. 2). This section comprises migmatitic gneisses, intercalated marbles and boudins of eclogite–amphibolites, metaperidotites and also leucocratic high-P/high-T rocks (Burg et al.1995). This section broadly coincides with the mixed ophiolitic/continental unit of Burg et al.(1996).

Section II is interpreted to have been exhumed between 50 and 30 Ma. It consists of the Kerdilion, Sidironero, Kardamos and Kechros Complexes (Fig. 2) representing separate windows of Section II in the west, central, and east Rhodope respectively, beneath Section I. Predominant are partly migmatitic para- and orthogneisses, marbles and amphibolites containing intercalations of ultramafic rocks and relics of eclogites in amphibolites (Sidironero and Kechros Complex). Within Section II, high-P mineral assemblages indicate different maximum pressures and temperatures ranging from low to high grade. Section II incorporates Variscan crust. Variscan protolith ages of granitoid gneisses and metapegmatites are indicated by U–Pb SHRIMP data of single zircons (294 ± 8 Ma) in the Sidironero Complex (Liati & Gebauer 1999; Table 1), by conventional U–Pb zircon data in the Kechros Complex (334 ± 5 Ma, Peitcheva & von Quadt 1995), and by a Rb–Sr muscovite age of 334 ± 3 Ma in the Kechros Complex (Mposkos & Wawrzenitz 1995, cf. Table 1).

Section III is interpreted to have been exhumed between 26 and 10 Ma and comprises the Thasos/ Pangeon Metamorphic Core Complex in the western Rhodope (Fig. 2; Wawrzenitz & Krohe 1998). This section occupies the lowermost tectonic position, underneath the Kerdilion and the Falakron/ Sidironero Complex and consists of orthogneisses, metapelites, huge marble complexes and mafic rocks. Characteristic are a Late Oligocene/Miocene medium-P low to medium grade metamorphism and mineral relics of earlier high-P metamorphism. Also, this section incorporates Variscan crust as is revealed by U–Pb zircon and monazite dates from orthogneisses, R–Sr magmatic muscovite dates and Sm–Nd magmatic garnet data from a metapegmatite on Thasos Island (Table 1; Wawrzenitz 1997).

CRZ and Section I of the high-P metamorphic domain are overlain by Tertiary marine deposits (Fig. 3). Two major pulses of sedimentation and basin formation occurred. The first lasted from the Priabonian, locally Lutetian (i.e. from 42–50 Ma after Harland et al.1990), through the Oligocene and mainly created the eastern and central Rhodopian basins. The second lasted through the Early/ Mid-Miocene and largely created the western Rhodopian basins (Fig. 3). These two pulses temporally coincide with the c. 50–30 Ma and the 26–12 Ma exhumation episodes of Sections I and II, respectively (Table 1, Fig. 3).

Major low-angle detachment systems that separate Sections I, II and III from each other were formed during different episodes, respectively: In the 26–10 Ma interval, the Strymon and Thasos low-angle extensional detachment systems caused exhumation of the Thasos/Pangeon metamorphic

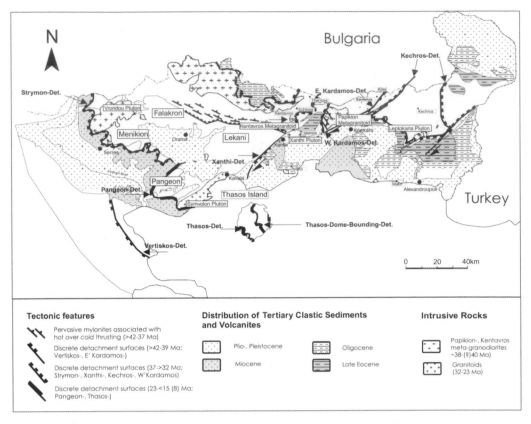

Fig. 3. Distribution of granitoid rocks and Tertiary basins in the Rhodope domain. Modified according to the geological maps of IGME.

core complex (Section III, western Rhodope, Figure 2), concurrent with detachments on the Cyclades' islands (Sokoutis *et al.* 1993; Dinter *et al.* 1995; Dinter 1998; Wawrzenitz 1997; Wawrzenitz & Krohe 1998). Section II was already exhumed in the Eocene and Oligocene. In this article, we describe for the first time in detail the Eocene and Oligocene detachment sytems of the Greek Rhodope Mountains: the Xanthi, western Kardamaos, eastern Kardamos and Kechros detachment sytems.These systems separate Section II from a hanging-wall unit consisting of the eastern CRZ and Section I (Fig. 2) and correlate with crustal scale detachments known from the Bulgarian Rhodope (cf. Burg *et al.* 1996).

Hanging wall complex: Section (I) of the high-*P* metamorphic domain

The Kimi Complex is a high-*T*/high-*P* metamorphic complex generally situated in the hanging wall of all Eocene and Oligocene detachment fault systems in the eastern Rhodope. Geothermobaro-

metric studies on garnet–spinel pyroxenites (high-*P* mineral assemblage is Grt–Cpx(Sp–Ol–Hbl), former eclogites (Grt–Cpx(Jd25)– Qtz–Rt) and metaperidotites (Grt–Cpx–Opx–Ol–Spl) estimate 15.5 kbar and 770 °C (Mposkos & Perdikatsis 1989; Wawrzenitz & Mposkos 1997; Mposkos & Krohe 2000). All these rocks are boudins in migmatitic gneisses that contain phengitic muscovite (up to 6.6 Si atoms p.f.u calculated for 22 O atoms; Table 2, analyses 1 and 2). The *P–T* path is associated with cooling during decompression (Fig. 4; cf. Mposkos & Krohe 2000 for details on the *P–T* path).

A Sm–Nd garnet–clinopyroxene–whole rock age from the garnet–spinel pyroxenite yielded *c.*119 ± 2 Ma and is interpreted as crystallization age under high-*P* conditions (Wawrzenitz & Mposkos 1997). An eclogite strongly deformed at granulite facies conditions also shows a U–Pb zircon SHRIMP age of *c.*119 Ma (Gebauer & Liati 1997). A R–Sr magmatic mica age of 65 ± 3 Ma in a metamorphic pegmatite crosscutting the foliation of the migmatites (Mposkos & Wawrzenitz 1995) is interpreted as the crystallization age of

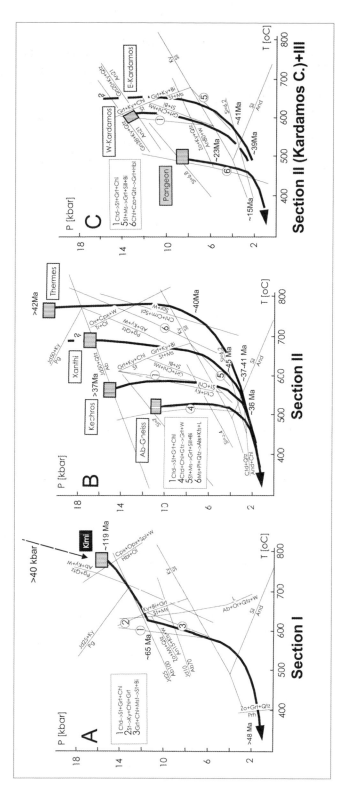

Fig. 4. *P–T* exhumation paths of the different tectonic complexes of the Rhodope Domain (modified after Mposkos & Krohe 2000). The petrogenetic grid is based on appropriate mineral parageneses and mineral compositions analysed in partially amphibolitized eclogites, metapelites, gneisses and ultramafic rocks. Reaction curves Ctd → St + Grt + Chl, St → Ky + Chl + Grt, Grt + Chl + Ms → St + Bi + Gr, Ctd → Ky → St + Chl, are from Vuichard & Ballèvre (1988), St + Ms → Grt + Sill + Bi is from Powell & Holland (1990), Ab + Or + Qtz + W → L, is from Johannes (1985), Cpx + Opx + Spl → Hbl + Ol is from Jenkins (1981), Chl + Cm → Spl → Spr + W is from Ackermand *et al.* (1975), Pg + Qtz → Ab + Al$_2$SiO$_5$ + H$_2$O is from Chatterjee (1972), Ms$_{ss}$ + Ab + Qtz → Kf$_{ss}$ + Als + L is from Thompson & Thompson (1976), Pg → Jd$_{50}$ + Ky + W is from Holland (1979). Isopleths Si 7, Si 6.8, 6.4 and Si 6.2 are from Massonne & Szpurka (1997). The triple point for the aluminosilicates is from Bohlen *et al.* (1991). The reaction curve Chl + Czo + Qtz → Grt + Hbl is calculated with the THERMO-CALC program. The remaining reaction curves were calculated with the GEO-CALC program (Brown *et al.* 1988) and the database of Berman (1988), using analysed compositions of coexisting mineral phases. Abbreviations of mineral names after Kretz (1983) and Bucher & Frey (1994).

mica and hence as a minimum age for the migmatitic stage (Mposkos & Wawrzenitz 1995).

In the eastern Rhodope, the overall thickness of the hanging wall unit (Kimi Complex and CRZ) does not exceed a few hundred metres. Both are transgressively overlain by Lutetian/Priabonian (48–42 Ma) conglomerates (IGME unpublished maps, 1999). Thus, Late Eocene–Oligocene clastic basins are situated almost immediately above Section II that was still at great depths and exposed to high temperatures at the time of sedimentation. This means that extension on these normal detachment fault systems cut out several tens of km of crustal material between the crustal Sections I and II.

Sidironero Complex (Section II):

Bounding detachment system

In the central Rhodope, the Xanthi Detachment System separates the Kimi Complex from the underlaying Sidironero Complex (Section II, Fig. 5a, b). This system consists of a discrete detachment surface underlain by a 0.5–2 m thick cataclastic zone (Fig. 6a). It trends NE–SW and dips 25–45° SE (Fig. 2) and truncates at intermediate to high angle the older mylonitic foliation, the lithological layering, and the metamorphic profile of the underlying Sidironero Complex. Microfabrics suggestive of dynamic recrystallization of calcite in the cataclastic zone indicate that translation on the detachment occurred above c. 200–250 °C. Deformation continuing at lower temperatures is indicated by cataclastic calcite. In quartz–feldspar rocks, cataclastic flow created an interconnecting system of narrowly spaced shear fractures and microbreccias defining a weak macroscopic cataclastic foliation (Fig. 6b). Locally, particularly around the Kentavros Granodiorite (Fig. 3), limited plastic elongation and rotation recrystallization of quartz are present (Fig. 6c).

Lineation trends show variations probably caused by later folding and reactivation of the detachment during the early Miocene. To the west of Xanthi, in the Falakron marble, striations plunging SSW at low angle on SE dipping planes (Fig. 7) are defined by elongated quartz segregations. There, recrystallized calcite grains oblique to the shear plane indicate top down-dip movement to the SSW or dextral oblique normal faulting. To

Fig. 5. Xanthi detachment fault surface near the town of Xanthi. (**a**) The detachment surface behind Xanthi town – viewed looking to the west – is bounded to low grade Falakron Series marble (lower part of Section II) forming the large mountain range in the background. Relics of the upper plate (black arrows) containing eclogite amphibolites of the upper plate (attributed to the Kimi Complex) are transgressively overlain by Eocene to Oligocene sediments. The basin (left side of photograph) consists of Miocene to Pliocene deposits that in turn transgressively overly the Eocene to Olicogene sediments, suggesting Neogene reactivation of segments of the detachment surface. The Oligocene Xanthi granodiorite (mountains in the foreground) intruded into the detachment surface. White arrows show where the detachment surface emerges from the eastern margin of the granodiorite. (**b**) Closer view on the detachment surface (D) bounded to weak Falakron Series marble north of Xanthi town. White broken line indicates a corrugation axis; D' is the part of the detachment surface in the corrugation synform. White arrows indicate eclogite amphibolites of the upper plate Kimi Complex preserved in a corrugation synform.

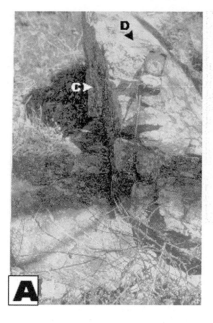

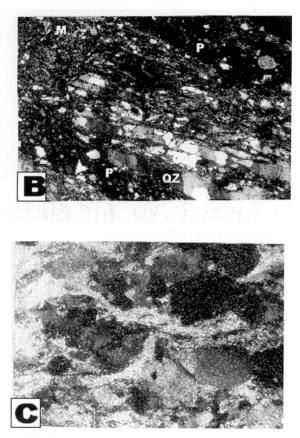

Fig. 6. Xanthi-detachment, microfabrics. (**a**) Outcrop of the Xanthi detachment surface D (3 km south of Kenntavros). C denotes foliated cataclastic zone beneath the detachment surface. Arrow points into the direction of movements of the hanging wall defined by long axis orientation of fractured fragments and secondary ('Riedel') shear zones cross-cutting the detachment surface in the lower part of the photograph. Scale: ? ? (**b**) Foliated cataclasites immediately beneath the Xanthi Detachment Surface (Echinos area) indicate totally brittle deformation. Narrow-spaced cataclastic shear fractures, microbreccia (M) and probably pseudotachylites (P) cut across quartz fabrics showing characteristics of dynamic recrystallization and static annealing (QZ). Note the straight boundaries between P and M domains (arrows). Scale: long side of photo is 6 mm. (**c**) Low temperature shear zones beneath the Xanthi Detachment Surface (area around the Kentavros Granodiorite); quartz shows plastic elongation and local rotation recrystallization; plagioclase behaves rigidly. σ-clasts of plagioclase show top to the right sense of shear indicating eastward normal faulting in this area. Scale: long side of photo is 6 mm.

the east of Xanthi, a lineation trend is poorly defined, striation/lineation trends, the configuration of shear fractures and shear zones show (Fig. 6a) top down-dip movement predominantly to the SSE, locally to the east (Fig. 7).

Previously, the Kimi Complex was considered as a part of the upper Sidironero Complex, as both are typified by a high-T–high-P metamorphism (summarized as upper tectonic unit; e.g. Mposkos & Liati 1993). However, we have separated the two complexes (Mposkos & Krohe 2000) because:

1   the Kimi Complex was earlier (between 65 and 48 Ma) exhumed than the Sidironero Complex indicated by the oldest transgressive

conglomerates (Lutetian) and the Rb–Sr muscovite age of the pegmatites;

2   the Kimi Complex and the Tertiary sediments are separated from the Sidironero–, Kardamos and Kechros Complexes by this detachment system (Fig. 2);

3   the P–T–t path of the Kimi Complex suggests an episode of cooling during exhumation/decompression (Mposkos & Krohe 2000), contrasting with that of the underlying Sidironero Complex.

The structurally higher levels of the Sidironero Complex record the highest P–T conditions during high-P metamorphism in Section II. In the

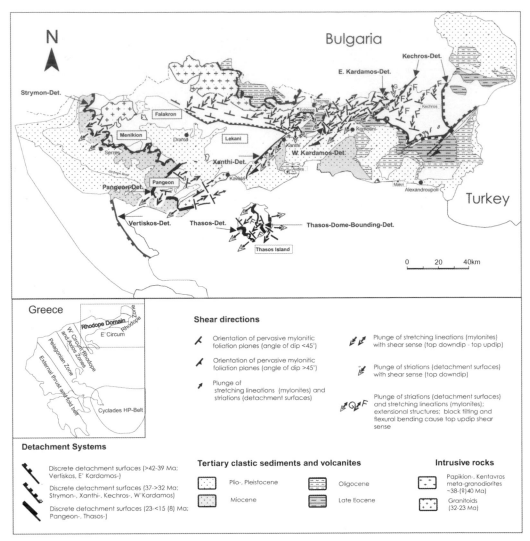

Fig. 7. Most important detachment surfaces and major kinematic data of mylonites formed in the 48–30 Ma and 26–12 Ma intervals. Each kinematic indicator symbol summarises 5–20 measurements. Shear senses were determined in the field and in oriented thin sections.

Thermes area (Fig. 2), migmatic orthogneisses and pelitic gneisses enclose lenses of kyanite eclogites indicating minimum peak pressures of c.19 kbar and at least 700 °C (Liati & Seidel 1996, 'Thermes'-curve, Fig. 4b). Crystallization of migmatites was within the stability field of silli-manite + K-feldspar (Mposkos & Liati 1993). This suggests near isothermal decompression from the maximum depth (>19kbar) to depths corre-sponding to c. 6–7 kbar (Mposkos & Liati 1993; Mposkos 1998, Fig. 4b). Crystallization of musco-vite in migmatitic pegmatites, showing the mineral assemblage Ms + Bi + Kfs + Pl + Qtz, and the

low Si content of white mica (Si = 6.2, Table 2, analysis 3) constrain subsequent cooling into the stability field of Ms+Qtz below 6 kbar (using the phengite barometer of Massonne & Szpurka 1997, 'Thermes' area in Fig. 4b). In migmatites, kyanite was preserved through these stages probably as a result of rapid decompression followed by fast cooling.

In amphibolites underneath these migmatite series ('Xanthi area', Fig. 2), hornblende ($K_{0.26}Na_{0.3}Ca_{1.78}Mg_{2.58}Fe_{1.16}Ti_{0.14}Al_{2.61}Si_{6.49}$ ca-tions to 23 O) formed after the eclogite facies metamorphism and produced An-rich plagioclase

Table 2. *Chemical Composition of white K-micas in gneisses and pegmatites from various tectonic complexes of the Rhodope Domain.*

Rock type	Kimi Complex		Sidironero Complex				Kardamos Complex				Kechros Complex		Pangeon Complex		
	Migmatitic gneiss		Pegmatite	Orthogneiss			Orthogneiss				Orthogneiss		Orthogneiss		
Sample No.	1	2	3	4c	5rim	6re	7c	8r	9c	10r	11c	12r	13c	14r	15re
SiO_2	48.86	47.83	46.64	51.71	46.89	47.61	49.74	47.00	48.33	46.38	48.35	47.88	51.41	47.32	49.26
TiO_2	0.73	1.10	0.79	0.28	0.56	0.70	0.96	1.03	0.80	0.77	0.66	0.80	0.22	1.05	0.94
Al_2O_3	27.11	27.72	33.45	23.99	31.52	30.41	26.51	30.52	27.73	30.63	27.96	28.95	26.12	30.95	28.09
FeO_{tot}	4.76	4.00	1.96	5.04	4.00	4.62	3.70	4.32	5.00	4.75	4.96	5.01	3.06	3.02	3.23
MnO	0.01	0.03	–	–	–	–	–	–	–	–	0.14	0.11	0.21	0.13	-
MgO	2.36	2.52	1.05	2.76	0.59	0.77	2.96	1.65	2.23	1.57	1.99	1.73	3.30	1.51	2.53
CaO	–	0.02	–	–	–	0.19	0.21	–	0.13	0.33	0.10	–	–	–	–
Na_2O	0.16	0.10	0.47	–	0.84	0.38	–	0.84	–	–	–	0.25	-	0.50	0.27
K_2O	10.88	11.04	10.64	11.40	10.26	10.48	11.41	10.50	11.56	11.35	11.52	11.26	11.70	10.77	10.95
Total	94.87	94.36	95.01	95.18	94.66	95.16	95.49	95.86	95.78	95.78	95.68	95.99	96.02	95.25	95.27
Cations based on 22 oxygen atoms															
Si	6.598	6.489	6.228	7.010	6.344	6.420	6.690	6.279	6.507	6.227	6.520	6.423	6.852	6.346	6.601
Ti	0.074	0.112	0.079	0.028	0.057	0.071	0.097	0.103	0.081	0.078	0.066	0.080	0.022	0.107	0.094
Al	4.314	4.423	5.265	3.833	5.027	4.833	4.203	4.806	4.400	4.846	4.443	4.576	4.104	4.892	4.436
Fe^{3+}	0.425	0.423	0.186	0.120	0.178	0.228	0.204	0.421	0.399	0.501	0.378	0.425	0.158	0.229	0.227
Fe^{2+}	0.112	0.031	0.033	0.452	0.274	0.293	0.212	0.062	0.164	0.033	0.181	0.136	0.184	0.110	0.135
Mn	0.001	0.003	–	–	–	–	–	–	–	–	0.016	0.012	0.024	0.015	–
Mg	0.475	0.509	0.209	0.558	0.119	0.155	0.593	0.328	0.448	0.313	0.399	0.345	0.656	0.302	0.504
Ca	–	0.003	–	–	–	0.028	0.030	–	0.018	0.048	0.014	–	–	–	–
Na	0.042	0.026	0.122	–	0.220	0.099	–	0.218	–	–	–	0.064	–	0.131	0.071
K	1.874	1.910	1.813	1.972	1.771	1.803	1.958	1.790	1.986	1.944	1.981	1926	1.990	1.843	1.873

Fe^{3+}. Fe^{2+} are calculated for 8 tetrahedral and 4 octahedral cations. Porphyroclast: c,core; r, rim; re, recrystalized.

(An 67%) and clinopyroxene $(Ca_{0.90}Mg_{0.74}Fe_{0.31}Al_{0.09}Si_{1.95}O_6)$ (Fig. 8b). According to the experimental results of Spear (1981), the reaction hornblende $\rightarrow$ clinopyroxene + plagioclase occurred above 700 °C in accordance with decompression at high temperatures (Fig. 4b, 'Xanthi'-curve).

The structurally lower levels of the Sidironero Complex consisting of alternating metapelites and orthogneisses (Albite Gneiss Series), plus marble layers and thick marble of the Falakron Series (Fig. 2), record significantly lower P–T conditions of 11 kbar and 500 °C (Fig. 4b; 'Ab-Gneiss'). In metapelites temperatures were within the stability field of the mineral assemblage Grt + Ctd + Chl. Minimum pressures are confined by the intersection between the Si = 7.00 isopleth of Massonne & Szpurka (1997) corresponding to the analysed maximum phengite component in albite gneisses (Table 2, analysis 4c) and the reaction curve Ctd + Chl + Qtz $\rightarrow$ Grt + H_2O calculated from compositions of Grt core $(Grs_{0.24}Prp_{0.05}Alm_{0.69}Sps_{0.15})$, Ctd $(Mg_{0.26}Fe_{1.69}Al_{4.03}Si_{1.99}$ cations to 12 O) and Chl $(Mg_{4.58}Fe_{4.48}Al_{5.48}Si_{5.26}$ cations to 28O) in metapelites (Fig. 4b). Garnets from metapelites show decreasing grossular composition and increasing Mg/Mg + Fe ration from the core to the rim (Grt rim $Grs_{0.06}Prp_{0.17}Alm_{0.75}Sps_{0.01}$, Grt) indicating a temperature increase during (the earlier stages of) decompression. Generally in this work, the phengite barometer from

Massonne & Szpurka (1997) is applied to white K-micas associated with the critical mineral assemblage Phen + Bi + Kfs + Qtz.

Strain associated with decompression

Within the Sidironero Complex, mineral assemblages of high-P metamorphism are essentially relics preserved in boudins of (ultra) mafic rocks, or as relict grains in quartzo-feldspatic mylonites. Thus, the pervasive foliation planes, lineation trends, grain scale fabrics, and shear sense indicators postdate peak pressures and therefore potentially depict the kinematics and deformation mechanisms related to exhumation of high-P metamorphic rocks. Typical mineral assemblages reveal downward-decreasing P–T conditions through the complex.

At the structurally higher levels of the Sidironero Complex, migmatites and, particularly the strained rims of amphibolitized eclogite boudins, amphibolites are deformed with microfabrics typical of high-T mylonites. Hornblende and the coexisting post-eclogitic clinopyroxene of amphibolites of the Xanthi area (see above) show characteristics of dynamic recrystallization (Fig. 8a, b) inferring deformation at high temperatures during decompression. S–C fabrics defined by elongated hornblende grains asymmetrically aligned to the compositional layering indicate top-to-SW movements during high-T shearing (Fig.

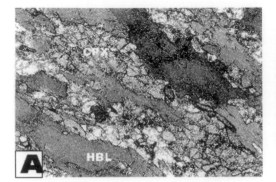

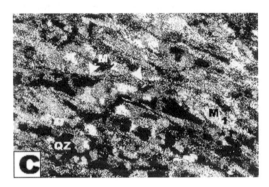

Fig. 8. (a) Amphibolites (Upper Sidironero Complex; 5 km NE Xanthi). During decompression hornblende (HBL) reacted producing An-rich plagioclase and clinopyroxene (CPX). Scale: long side of photo is 2.3 mm. (b) Amphibolite (Upper Sidironero Complex, 5 km NE of Xanthi) deformed during decompression at high temperatures. Diopside (CPX) was formed syntectonically at the expense of hornblende; hornblende crystals and diopside crystals in Cpx aggregates show weak shape preferred orientation and S–C fabrics. Shear sense is top right (SW directed overthrusting in the outcrop). Scale: long side of photo is 6 mm. (c) S–C fabric in pelitic gneiss (Albite Gneiss Series, Lower Sidironero Complex, Xanthi area). Shearing is top to the left corresponding to top to S–W overthrusting. Large phengitic white mica fishes (M1) in S-domains recrystallize to fine grained less phengitic white mica (M2) in C-domains, indicating deformation during decompression. Quartz ribbons show dynamic recrystallization. Scale: long side of photo is 6 mm.

8b). Through cooling, strain localized into biotite, white mica and quartz rich domains; top to SE sense of shear at that stage is indicated by secondary shear zones.

In the Albite Gneiss Series, in the structurally lower part of the Sidironero Complex, pre-mylonitic phengitic white mica grains from orthogneisses and metapelites show decreasing Si content from core to rim (Table 2, analyses 4c, 5r). The mica grains recrystallized during deformation have a low Si content (Si = 6.4–6.3, Table 2, analysis 6re) consistent with deformation/recrystallization (Fig. 8c) during decompression to 5–6 kbar (Fig. 4b). In amphibolites, syn-deformational hornblende oriented with long axes parallel to the lineation, show compositional zoning from tremolitic in the core ($Na_{0.18}Ca_{1.70}Mg_{3.85}Mn_{0.03}Fe_{1.18}Al_{0.76}$ $Si_{7.49}$ cations to 23 O) to tschermakitic hornblende in the rim ($Na_{0.38}Ca_{1.65}Mg_{2.70}Mn_{0.03}Fe_{1.50}Al_{2.14}$ $Si_{6.52}$ cations to 23 O) in accordance with increasing temperature during decompression. Thus, importantly, deformation took place at increasing temperature during decompression (albeit at generally lower temperatures than in the higher

Sidironero Complex) and continued during successive cooling (Fig. 4b). Within the lower Sidironero Complex, shear sense indicators such as S–C relationships and fishes of relic phengitic mica also consistently indicate top to SW (updip) shearing.

We interpret formation of mylonites to be associated with thrusting in the deeper crust (Fig. 7; cf. Dinter et al. 1995; Burg et al.1996; Kilias et al. 1997, 'Nestos thrust' zone), and thus to have formed before formation of the Xanthi detachment system. Consistent with this are (i) shear senses indicating updip SW movements, (ii) downward-decreasing maximum pressures and temperatures during high-P metamorphism (previous to mylonite formation) within the Sidironero Complex and (iii) downward decreasing temperatures during the deformation/decompression stage, indicating thrusting of hotter on top of cooler crustal sections and/or cooling from below within the Sidironero Complex. The Xanthi-detachment system cross-cutting this mylonitic foliation brought the Kimi Complex into contact with the high and low grade parts of the Sidironero Complex. This system was clearly formed after juxtaposition of high grade upon low grade units and after emplacement of both at a shallow depth.

Kardamos Complex (Section II)

Bounding detachment systems

Detachment surfaces bounding the Kardamos Complex show structural variations.

(i) The eastern Kardamos detachment surface that separates the structurally higher (eastern) parts of the Kardamos Complex from the Kimi Complex (Section I) is underlain by a shear zone a few hundred metres thick (Fig. 9) formed at greenschist facies conditions. This detachment surface trends at low angle to the orientation of the foliation of such shear zones that shape a gently NE-plunging dome structure (see below, Fig. 7).

(ii) The western Kardamos detachment that bounds the western Kardamos Complex (Section II, Fig. 7) is, like the Xanthi detachment system, a discrete detachment surface underlain by a 0.5– 2 m thick cataclastic zone. This detachment surface and the cataclastic zone trend approximately north–south, dip at low angle to the west and cut across the mylonitic foliation, the lithological succession and metamorphic profile of the footwall Kardamos Complex. Thus, the western part

exposes the deeper levels of the Kardamos Complex. Locally, the detachment surface is tightly corrugated about east–west-trending axes.

P–T history: general

The structurally higher parts of the Kardamos Complex, beneath the eastern Kardamos detachment system, comprises migmatites, meta-blastic gneisses, metapelites and metabasites that record upper amphibolite facies metamorphism. In metapelites, this is indicated by the mineral assemblages Grt–Ky–Bi (Fig. 4c). Retromorphic staurolite and sillimanite coexist with muscovite and biotite. According to the petrogenetic grid of Vuichard & Ballevre (1988), coexisting Grt + St + Bi and St + Bi + Als and the absence of chlorite in the metapelites indicates minimum temperatures of 600–650 °C during this stage of decompression (Fig. 4c). In metabasites, synmylonitic mineral assemblages include plagioclase (An_{20-25}) and hornblende.

The western Kardamos Complex exposes structurally lower parts beneath the cross-cutting western Kardamos detachment. This part is composed of a sequence of ortho- and paragneisses, metapelites (albite gneisses), metabasites and intercalated marbles, which are overlain by the 'Papikion Mt. Metagranodiorite' (Figs 2 and 3). Gneisses locally preserve the high-*P* mineral assemblage Grt Ky

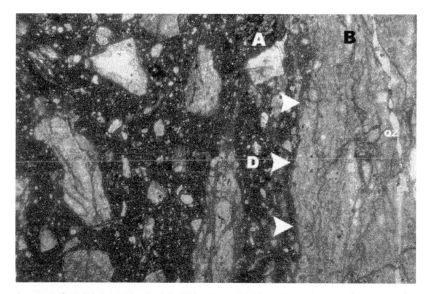

Fig. 9. Detachment surface (D) separating brecciated marbles of the upper plate Kimi Complex (A) from plastically flowing marble of the eastern Kardamos Complex (B), 10 km NE of Komotini. The marble of the Kardamos Complex contains quartz veins (qz) that were also plastically deformed and boudinaged subparallel to the extension of the detachment surface. Boudinage asymmetry indicates dextral sense of shear (dextral oblique normal faulting in the outcrop). Scale: (long side of photo is 9.4 mm.

Zo–Chl–Pl (An$_{19-25}$). The reaction Grt + Ky + Qtz → An calculated for the analysed mineral compositions (Fig. 4c) yielded 13 kbar for garnet core and 12 kbar for garnet rim compositions, assuming 600 °C (Mposkos & Krohe 2000), indicating garnet growth during decompression.

The retromorphic mineral assemblages are Grt–Ms–Bi–Ab–Pl$_{(15-20)}$ ± Kfs–Qtz in pelitic gneisses and Grt–Hbl–Bi–Czo–Ab–Pl$_{(14-21)}$– Qtz in metabasites characterizing the $P–T$ conditions during decompression. Albite (An$_{0-4}$) has a rim of coexisting oligoclase (An$_{14-21}$). Staurolite that replaced garnet or kyanite coexisted with chlorite, constraining the decompression path within the garnet–staurolite–chlorite stability field and limiting the peak temperature between 580 °C and 620 °C (Fig. 4c). This suggests that decompression was near isothermal but at lower temperatures than in the structurally higher part, and thus downward-decreasing temperatures within the Kardamos Complex (Fig. 4c).

Strain partitioning associated with decompression

The Kardamos Complex also experienced shearing during decompression. Mineral assemblages recording maximum $P–T$ conditions predate shearing and are only locally preserved. Directions of shear planes and lineations in mylonitic gneisses vary over the Kardamos Complex due to a complex tectonic history (Fig. 7). Detachment surfaces bounding the Kardamos Complex show structural variation.

Structurally higher part. Only in the uppermost parts, within shear zones close to the eastern Kardamos detachment surface, are pre-mylonitic phengitic white micas preserved. In granodioritic orthogneisses immediately beneath the detachment surface, the maximum phengite component is Si = 6.5 atoms p.f.u. (Fig. 10, sample K31, Table 2, analysis 9c). At *c.*300 m beneath the detachment surface it increases to Si = 6.7 atoms p.f.u. (Fig. 10, sample K33, Table 2, analysis 7c). This is interpreted as representing an increase in pressure from 8 to 10 kbar (at temperatures of 600 °C) over that short distance and is in accordance with higher level extension. Pre-mylonitic phengites tend to have accommodated to lower pressures during deformation as is indicated by strongly varying Si contents decreasing from the core to the rim (Fig.10, Table 2), thus corroborating shearing during decompression. In shear zones close to the detachment surface, deformation continued through low temperatures (Fig. 11a, b), as is suggested by brittle behaviour of feldspar.

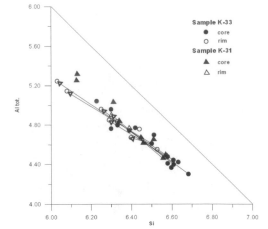

Fig. 10. Compositional variations of white K-micas in granodioritic orthogneiss mylonites from the eastern Kardamos carapace shear zone, in terms of Al$_{tot}$ v. Si. Deviation of the plotted analyses from the ideal muscovite–celadonite composition indicates that part of Al is substituted by Fe^{3+}. Sample K-31 is from immediately beneath the eastern Kardamos detachment surface, and sample K-33 from 300 m beneath (K31) the eastern Kardamos detachment surface. The range of Si values indicates the presence of phengites and less phengitic muscovites in each sample. Arrows show core to rim compositions. In pre-mylonitic white K-micas and in recrystallized flakes celadonite component tends to decrease from the core to the rim. Close to the detachment surface (sample K-31) cores of the pre-mylonitic white micas show lower maximum seladonite components.

This is consistent with extension-related juxtaposition of the cold Kimi Complex above the upper Kardamos Complex.

In mylonites at distances of *c.* 500 m and more from the eastern Kardamos detachment surface, white K-mica is frequently intergrown with biotite and has low phengite components. This suggests decomposition of an older, pre-mylonitic, phengitic mica during deformation. Thus, in this part, the record of peak pressures has been almost entirely erased during amphibolite-facies metamorphism. The formation of St and Sil (coexisting with Ms and Bi) was coeval with shearing, indicating syn-deformational temperatures above 600–650 °C (see above, Fig 4c). Characteristic in metabasites are recrystallization of plagioclase and hornblende that form elongated aggregates subparallel to the stretching lineation. Deformation was outlasted by annealing, producing grain growth and straight grain boundaries (Fig. 11c).

In all parts of the eastern Kardamos Complex, the shear zone has a normal component. In the

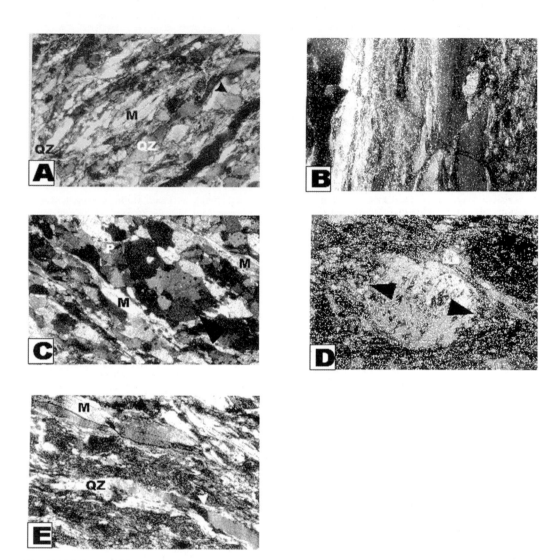

Fig. 11. (a) Quartz feldspar LT-mylonite in the carapace shear zone, a few metres beneath the eastern Kardamos detachment surface; phengitic white mica (M) is preserved; feldspar behaves by brittle fracturing (arrow) and plastically stretched quartz (Qz) shows a low degree of recovery/recrystallization. Scale: long side of photo is 2.3 mm. (b) Low temperature plastic strain of quartz in the carapace shear zone, a few centimetres beneath the eastern Kardamos detachment surface; extreme elongation of grains and low degree of recovery/rotation recrystallization. Scale: long side of photo is 9.4 mm. (c) Quartz feldspar high-*T* mylonite in the carapace shear zone, *c.* 300 m beneath the eastern Kardamos detachment surface; relict phengitic white mica; feldspar shows dynamic recrystallization and grain growth at low differential stresses and high temperatures (annealing). Scale: long side of photo is 2.3 mm. (d) Albite gneiss mylonite in the structurally deepest (western) Kardamos Complex. Albite overgrew an internal foliation defined by biotite and quartz aligned at high angle to the external foliation. Asymmetric strain shadows in albite clast (arrows) indicate top-to-the-left sense of shear (SW-directed in the outcrop). Scale: long side of photo is 6 mm. (e) Eastern Rhodope. Low temperature mylonite a few metres beneath the detachment surface separating the Kechros from the Kimi Complex. Low temperature plastic strain of quartz (QZ) is indicated by strong elongation of host grains and low degree of recovery/rotation recrystallization. Top left sense of shear is indicated by prism subgrain walls oblique to the elongations of quartz ribbons (arrow) and phengitic white mica (M) forming fishes consistent with SSW-ward translation of the hanging wall. Scale: long side of photo is 2.3 mm.

southern limb of the dome the foliation trends NE–SW, dips moderately SE and bears low angle SW to SSW plunging lineations. Shear sense is top to the SW indicating dextral oblique normal faulting (Fig. 7). Further to the east, toward Komotini (Fig. 7), the plunge of lineations changes to south and ESE; top-to-the-south and - ESE movements indicate normal faulting. In the northeastern Kardamos Complex (in the NE-plunging hinge of the antiform) the mylonitic foliation swings to a north–south trend, dips at intermediate angle to the east and bears east to ENE plunging lineations. Shear senses indicate top down-dip movements to the east and ENE, respectively (Fig. 7).

Structurally Lower Part. The retromorphic mineral assemblages are Grt–Ms–Bi–Ab–$Pl_{(15-20)}$ ± Kfs–Qtz (pelitic gneisses) and Grt–Hbl–Bi–Czo–Ab–$Pl_{(14-21)}$– Qtz (metabasites), which are syn-mylonitic. During non-coaxial strain in the mylonites of these lower parts albite–oligoclase progressively overgrew an internal foliation defined by biotite and quartz (metapelites) or hornblende (metabasites). This internal foliation is aligned at c.30–80° to the external foliation and rotates into the direction of the external foliation at the rim of the grain, in accordance with a SW sense of shear (Fig. 11d).

Beneath the cross-cutting western Kardamos detachment surface, the mylonitic foliation shows changing orientations. In the core of the Kardamos dome, both the dip of the foliation and the plunge of the lineation are to the NE. Toward the outer part of the dome (SW Kardamos Complex; Fig. 7), the dip of the mylonitic foliation successively changes to SE and the plunge of the lineation changes to (S)SW. Yet, sense of shear invariably indicates top-to-the-SW movements (cf. Burg *et al.* 1993, 1996; Ricou *et al.* 1998, Fig. 7). We interpret this geometric pattern of the shear zones as follows.

Shear senses indicating updip SW movements in the core of the dome suggest deeper level thrusting in the structurally lowermost exposed metamorphic sequences. Consistent with this are upward increasing syn-deformational temperatures indicating cooling from below and/or thrusting of hotter on top of cooler crustal sections. Variations of the orientation of shear planes from the inner (lower) toward the outer Kardamos Complex suggest a continuous change from contractional to (dextral oblique) transtensional and high level extension (Fig. 7). This high level extension is the dominant structure of the eastern Kardamos Complex.

The gently west dipping western Kardamos detachment cross-cuts all the shear zones and developed after cooling of the western Kardamos complex. Hence, this detachment surface is interpreted as both genetically unrelated to the mylonites and younger than the eastern Kardamos detachment system.

Kechros Complex (Section II): metamorphism and deformation

In the eastern Rhodope, a discrete detachment (Kechros-detachment) that is underlain by a shear zone only about 50 m thick separates the Kechros Complex from the Kimi Complex (Section I) and the CRZ. This shear zone was formed at lower greenschist facies and is oriented at low angle to the detachment surface. Both, detachment and shear zone, are flat lying and shape an open antiform structure (Fig. 7).

The Kechros Complex representing the footwall unit of this detachment system largely consists of orthogneisses with intercalations of metapelites, ultramafic rocks and eclogites. In eclogites, the relict high-*P* mineral assemblage Grt + Omp(Jd_{35-55}) ± Ky + Tr + Hbl + Czo + Qtz + Rt + Phe indicates minimum *P*–*T* conditions of 15 kbar (corresponding to a depth of about 53 km) at c. 550 °C (Fig. 4b, Mposkos & Perdikatsis 1989; Liati & Mposkos 1990). Kyanite occurs as inclusions in garnet cores, associated with clinozoisite and quartz, being formed during the prograde path in an earlier stage of the high-*P* event. In gneisses, the maximum phengite component of white micas (Si = 7 atoms p.f.u.) reflects this high-*P* event (Mposkos, 1989). Maximum *P*–*T* conditions and the *P*–*T* paths are close to those of the Albite Gneiss Series of the Sidironero Complex (Fig. 4b).

Upper greenschist/lower amphibolite-facies metamorphism at about 4–6 kbar overprints the high-*P* metamorphism. This stage is indicated by staurolite formation by chloritoid consuming reactions (e.g. Ctd + Ky → St + Chl; Ctd + Phen → St + Ms + Chl and Ctd + Ms → St + Bi), inferring nearly isothermal decompression (Fig. 4b; Mposkos 1989; Mposkos & Liati 1993).

The Kechros Complex was pervasively deformed during earlier decompression, prior to development of the detachment. This is indicated by widespread post-deformational growth of St and Chl in thick mylonites. Recrystallized white mica grains aligned within the mylonitic foliation are less phengitic than pre-mylonitic white micas. Microprobe analyses of these micas from Orthogneisses (cf. Mposkos 1989, Table 2, Sample R 9A core Si = 6.51, rim Si = 6.44 and Table 2 analyses 11c, 12r) suggest that shearing occurred during decompression to pressures of c.8–6 kbar.

Foliation planes dip at low angle approximately to the north; generally top-to-the-south and -SSW shear senses occur (cf. Burg *et al.* 1996, Fig. 7).

In the younger shear zone directly beneath the detachment surface, deformation continuing at lower greenschist facies is indicated by green, low-Ti biotite replacing garnet and muscovite and, in metabasites, by actinolite and chlorite replacing hornblende. Feldspar behaved brittly (Fig. 11e). Chloritoid is replaced by andalusite and chlorite. Sense of shear is top to the SW (Fig. 11e) consistent with the orientations and shear senses of the thick shear zones (cf. Burg *et al.* 1996, Fig. 6).

Importantly, the upper greenschist facies Kechros Complex lies directly beneath the Kimi Complex. This contrasts with central Rhodope, where the Kimi Complex is situated directly above the medium-high grade (high-*T*/high-*P*) rocks of the upper Sidironero Complex that in turn overlies the upper greenschist facies albite gneisses of the lower Sidironero Complex. This suggests that the Kechros detachment system cuts out the medium–high grade (high-*T*/high-*P*) rocks of the upper Sidironero Complex.

Geochronological record of Eocene–Oligocene metamorphism

Upper and lower parts of the Sidironero Complex differ in geochronological records: In the structurally uppermost part (Thermes area, Fig. 2) of the Sidironero Complex, a U–Pb SHRIMP zircon age from partly amphibolitized eclogites gives 42 ± 1 Ma (Table 1; Liati & Gebauer 1999). These zircons are interpreted as having crystallized at eclogite facies, and thus dating the high-*P* stage (Liati & Gebauer 1999) but may rather reflect a minimum age for high-*P* metamorphism (see discussion). A U–Pb SHRIMP zircon age from an adjacent migmatite of 40 ± 1 Ma is inferred to reflect the migmatization age (Table 1; Liati & Gebauer 1999). In this part, both K–Ar hornblende ages (Liati 1986) and R–Sr-whole rock white mica ages (Kyriakopulos pers. comm. 1999) range

between 37 ± 1 and 41 ± 1 Ma, setting a minimum age for mylonitization.

In structurally deeper levels (but still above the Albite Gneiss Series; northern Xanthi area, Figs, 2 and 3), a quartz vein yielded a slightly older U–Pb SHRIMP zircon age of 45 ± 1 Ma; this age is interpreted as reflecting quartz vein formation during dehydration at low temperatures (*c.*300 °C) along the prograde *P–T* path (Table 1; Gebauer & Liati 1997). However, in the structurally deeper levels, K–Ar-hornblende dates of amphibolites without relics of eclogites are significantly older, yielding 45 ± 1 to 47 ± 1 Ma (beneath the marble horizon, northern Xanthi, Table 1, Liati 1986, see discussion). K–Ar biotite and white mica ages of 36 ± 1 and 37 ± 1 Ma indicate cooling below 350 °C and 300 °C, respectively (Table 1; Liati 1986) and are thus a maximum age of frictional slip on the Xanthi detachment surface. This late stage is also reflected by a zircon date from a pegmatite cross-cutting the mylonitic foliation of 35 Ma ± 1 (Gebauer & Liati 1997).

In the Kechros Complex a mica sieve fraction of large (>500 m) (largely pre-mylonitic) phengitic white mica grains from a mylonitic orthogneiss yielded a R–Sr age of 37 ± 1 Ma (Table 1, Wawrzenitz & Mposkos 1997). In contrast to the upper Sidironero Complex, *P–T* conditions of the Kechros Complex did not exceed the stability field of Ms + Qtz and the suggested 550 °C closure temperatures of the R–Sr white mica systems. Thus, Wawrzenitz & Mposkos (1997) interpreted this date as a minimum age of high-*P* metamorphism in the Kechros Complex. Moreover, *c.* 37 Ma is a maximum age of mylonitization and juxtaposition of the Kechros Complex next to the Kimi Complex (Wawrzenitz & Mposkos 1997).

As no geochronological data from the Kardamos Complex existed, samples of white mica and biotite from an orthogneiss of the eastern Kardamos Complex were analysed for K–Ar age determination by Geochron Labs (Massachusetts). K–Ar ages of 42 ± 1 and 39 ± 1 Ma, were found, respectively (Table 3), indicating cooling below 350–300 °C, significantly before the upper Sidironero and Kechros Complexes. Thus, according to these ages, the eastern Kardamos detachment

Table 3. *K–Ar data of muscovite and biotite from the eastern Kardamos Complex*

	$^{40}Ar/^{40}K$	Avge. ^{40}Ar [ppm]	Ave. K [%]	^{40}K [ppm]	Age [Ma]
Muscovite (80–200 μm)	0.002473	0.2234	7.570	9.030	42.1 ± 1
Biotite (80–200 μm)	0.002316	0.2042	7.394	8.820	39.4 ± 1

Age = $1/\lambda_\beta + (\lambda_0 + \lambda'_0)$ [ln [$\lambda_\beta + (\lambda_0 + \lambda'_0)/(\lambda_0 + \lambda'_0) \times {}^{40}Ar/^{40}K + 1$]
Constants: $\lambda_\beta = 4.962 \times 10^{-10}$ year; $(\lambda_0 + \lambda'_0) = 0.581 \times 10^{-10}$ year; $^{40}Ar/^{40}K = 1.1193 \times 10^{-6} g/g$

cooled prior to the footwall units of Xanthi detachment.

Granitoids associated with detachments

Most granitoid intrusions of the Rhodope Mountains are spatially linked to detachment fault systems (Fig. 3). The ages and intrusion depths of these granitoids provide further constraints on depths and time intervals over which the discrete, brittle detachments were active.

The highly sheared Kentavros granodiorite, about 10 km to the north of Xanthi (Fig. 2) is a syn-detachment intrusion into the upper Sidironero Complex, immediately beneath the Kimi Complex and the Xanthi detachment surface. K–Ar hornblende ages of 38 ± 1 Ma (Table 1, Liati 1986) match the R–Sr white mica–whole rock ages of the Sidironero Complex, indicating concurrent cooling of the pluton and enclosing metamorphic rocks. Hornblende geobarometry (Schmidt 1992) yielded estimated pressures of 5.1–5.4 kbar (Al_{tot} in Hbl is 1.70–1.76 atoms p.f.u. calculated for 23 O), corresponding to intrusion depths of 14–15 km. This constrains the depth of formation of the detachment in this area, after 38 Ma.

The Xanthi granodiorite is a late or post-detachment intrusion (Fig. 3). Mineral assemblages produced by contact metamorphism (Liati 1986), indicate an intrusion depth at less than 10 km (corresponding to 2.5–3kbar). K–Ar hornblende and biotite ages of 30 ± 1 and 28 ± 1 Ma, respectively (Table 1), are probably close to the intrusion age because the pluton cooled rapidly in a cool environment as is indicated by contact metamorphism of adjacent Eocene sediments. These ages also provide a minimum time constraint for slip on the detachment surface.

To the north of Alexandroupoli (eastern Rhodope) late–post-kinematic granodiorites producing contact metamorphism in the Kechros and Kimi Complex (Fig. 3, Leptokaria granite) have R–Sr biotite whole-rock ages of 30 ± 1 to 32 ± 1 Ma (Table 1; Del Moro et al. 1988). Along with the mineral ages of the gneisses of the Kechros Complex, this indicates that slip on the detachment in the Kechros and Xanthi areas occurred in the same time interval, between 37 and 30–32 Ma.

No geochronological data exist from the Papikion meta-granodiorite, which separates the higher and lower Kardamos Complex (Fig. 3). Intrusion depth of the granodiorite estimated by hornblende geobarometry (Schmidt 1992) is 14–19 km (4.9–6.7 kbar; Al_{tot} in Hbl is 1.67–2.05 atoms p.f.u.); mafic enclaves have significantly higher crystallization depths of 23–29 km, corresponding to

8.4–10.5 kbar (Al_{tot} in Hbl is 2.40–2.85 atoms p.f.u.).

Unroofing the Thasos/Pangeon Complex (Section III; 26–10 Ma interval)

The Thasos/Pangeon Complex mylonites in the western Rhodope (Dinter et al. 1995; Sokoutis et al. 1993; Wawrzenitz & Krohe 1998; Fig. 2) were formed during exhumation. They show predominant SW–NE-trending lineations with mainly top-to-SW- and locally top-to-NE-directed shear senses (Fig. 7). In metabasites the mineral assemblage Hbl–Ab(An_{01})– Pl(An_{32})– Grt–Czo–Chl + Qtz–Rt–Ttn indicates epidote–amphibolite facies. In orthogneiss mylonites the Si content in white K-micas decreases from the pre-mylonitic to the recrystallized white micas from c.6.85 to 6.35 atoms p.f.u. (Pangeon Mountains, Table 2, analyses 13c, 14r). Pressures of 9 kbar were attained at temperature of c. 490 °C. This is constrained by the intersection of the isopleth $Si = 6.8$ of Massonne & Szpurka (1997) and the reaction curve $Chl + Czo + Qtz \rightarrow Grt + Hbl$ calculated with the Thermocalc program, using the analysed mineral compositions Grt ($Grs_{0.38}Alm_{0.52}Prp_{0.06}$ $Sps_{0.04}$), Hbl ($K_{0.19}$ $Na_{0.43}$ $Ca_{2.00}$ $Mg_{1.92}$ $Mn_{0.04}$ $Fe^{2+}_{2.14}$ $Fe^{3+}_{0.36}$ $Ti_{0.03}$ $Al_{2.07}$ $Si_{6.44}$ cations for 23O), Chl ($Mg_{7.61}Fe_{2.43}Al_{4.59}Si_{5.48}$ cations for 28 O) and Czo (Ps = 20%).

However, pervasive shearing and exhumation of the Thasos/Pangeon Complex occurred during the Miocene. R–Sr dates of white mica and biotite between 23 ± 2 and 15 ± 1 Ma in the Pangeon mountains (Table 1, Del Moro et al. 1990) and on Thasos Island reflecting cooling and constrain in part the age of deformation (Wawrzenitz 1997, Wawrzenitz & Krohe 1998). Calc-alkaline granitoid intrusions interpreted as syn-deformational (Symvolon granodiorite, Fig. 3) yielded U–Pb titanite and $^{40}Ar/^{39}Ar$ hornblende dates of 22 ± 1 to 23 ± 1 Ma (Table 1; Dinter et al. 1995). On Thasos Island the metamorphic grade increases downward. Also the R–Sr dates of white mica and biotite decrease downward from 23 ± 1 and 15 ± 1 Ma to 18 ± 1 and 12 ± 1 Ma, respectively (Table 1). The later interval is interpreted as extensional exhumation of a deeper crustal level on a second detachment system ('dome bounding detachment' in Figs 2 and 3; Wawrzenitz & Krohe 1998). According to Wawrzenitz (1997) and Wawrzenitz & Krohe (1998), the Thasos metamorphic core complex was formed during early Aegean back-arc extension. So far, no accurate geochronological data exist to constrain the age of high-P metamorphism of the Pangeon Mts.

From $c.15$ Ma until 8 Ma, low angle normal (Strymon and Thasos) detachments associated with mylonite zones were active (Dinter & Royden 1993; Wawrzenitz & Krohe 1998). Early/Mid-Miocene basins developed on the hanging walls of these younger detachment faults. The basins are interpreted to represent supra-detachment basins of the Strymon and Thasos detachment systems (Dinter et al.1995). The extent of such basins increases toward the western Rhodope (Figs 2 and 3). The earlier exhumed metamorphic complexes (including Sidironero Complex etc.) are in the hanging-wall position of these detachments (Fig. 12a, b).

Discussion: kinematic and dynamic model

Geochronological constraints on the tectonic evolution (<42 and 30 Ma interval)

In summary, Section II (Sidironero, Kardamos and Kechros Complexes) is a part of a high-P crustal domain that, within the 42–30 Ma interval, experienced strong pervasive deformation during decompression at temperatures of 500–700 °C and during cooling. This history contrasts with those of the over- and underlying complexes, which show older and younger metamorphic histories, respectively. At its top and bottom, this crustal section is bounded by detachment systems formed in the <42–30 Ma and 23–8 Ma intervals, respectively (Fig. 12a–d). Exhumation of the high-P rocks was concurrent with transgression in the Lutetian or Priabonian and basin formation continued to the Late Eocene–Oligocene, particularly in the eastern Rhodope. We interpret this sedimentation as related to movements on detachment systems bounding this crustal section at its top.

According to Liati & Gebauer (1999), the sequence of U–Pb SHRIMP zircon ages from quartz vein, to eclogite and migmatite of the Sidironero Complex brackets the subduction–exhumation history of the entire Rhodope high-P terrain within a short period of time between 45 and 35 Ma. However, the Rhodope high-P terrain does not show a homogeneous P-T and exhumation history. Even among different parts of the Sidironero Complex, P–T histories and deformation mechanisms vary. Such variations seem to correlate with differences of geochronological data that show a broadly downward-increase in age. This sets important constraints on the reconstruction of differential mass movements in the <45–30 Ma time span.

The uppermost migmatitic part of the Sidironero Complex (Thermes area) had already exhumed into the middle crust (stability field of sillimanite) at $c.40$ Ma. This is shown by the 40 ± 1 Ma U–Pb zircon SHRIMP date constraining the age of migmatization. Pegmatoids associated with migmatites contain muscovites that show low phengite components. In deeper parts of the higher Sidironero Complex, above the marble horizon north of Xanthi (Fig. 2), R–Sr muscovite ages and K–Ar hornblende ages, which both range between 37 ± 1 and 41 ± 1 Ma (Table 1), reflect a minimum age for recrystallization during mylonitization associated with thrusting of this part over the Albite Gneiss Series. In this part, mylonitization was at high temperatures and succeeded migmatization and syn-migmatic deformation.

In the lower Sidironero Complex, beneath the marble horizon north of Xanthi (Fig. 2), K–Ar ages of hornblende from amphibolite mylonites yielding 45 ± 1 Ma (Liati 1986) are interpreted as dating (re)crystallization of hornblende during mylonitization of the lower Sidronero Complex, after high-P metamorphism. Mylonite formation is interpreted to continue through cooling below the supposed closure temperatures of the K–Ar hornblende system of 550 °C. The $c.$ 45 Ma hornblende data are also a minimum age for high-P metamorphism in the lower Sidironero Complex as mylonitization developed after high-P metamorphism (registered by the Si contents of recrystallized white mica). Downward increasing ages might reflect cooling of the footwall due to thrusting on top of cold Falakron Series (Fig. 2; Dinter 1998). The abrupt increase in hornblende ages toward the lower Sidironero Complex also suggests that thrusting continued along discrete faults at temperatures lower than closure of these isotope systems.

Frequently, excess Ar is observed in hornblende of high-P metamorphic complexes. However, when excluding samples that preserve distinct textural remnants of the eclogite stage ('eclogitogenic amphibolites' of Liati 1986), there is a clear correlation of the higher ages with the lower part (beneath the marble horizon north of Xanthi) and of the lower ages with the high-grade higher part of the Sidironero Complex (Table 1). In the case of excess argon, a scatter of various values throughout the Sidironero Complex would rather be expected.

The 45 ± 1 Ma U–Pb age of the quartz vein from the lower Sidironero Complex (Liati & Gebauer 1999) approximates the K–Ar hornblende ages within this part. Thus, alternatively, this age may be related to metamorphic reactions and zircon crystallization associated with deformation and hydration during decompression prior to cooling. So far, the age of high-P metamorphism in this part is poorly constrained.

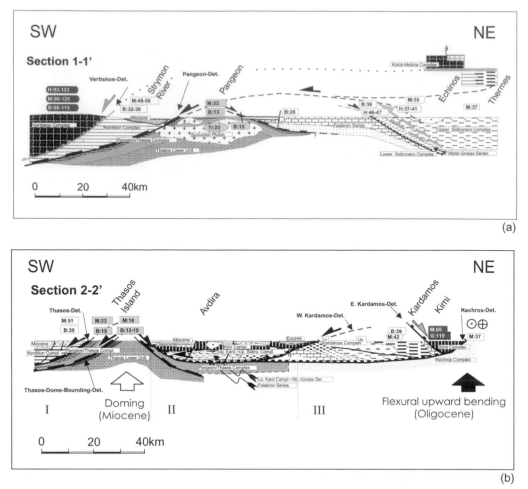

Fig. 12. Profiles through the Rhodope domain. Vertical scale is two times horizontal scale See Figure 2 for location of profiles, Figures 2 and 3 for unit fillings and Table 1 for ages.

Kardamos Complex. PCooling of the (structurally higher) amphibolite facies of eastern Kardamos Complex below 350–300 °C at *c.*41–39 Ma is interpreted to result from its emplacement beneath the cold Kimi Complex on the extensional eastern Kardamos detachment fault. Importantly, these K–Ar data also show that extensional unroofing in the higher Kardamos Complex was contemporaneous with contractional deformation in the lower Sidironero Complex. Besides, cooling, shearing in low-*T*-mylonites, and frictional slip on detachment systems in the higher Kardamos Complex roughly occurred within the same time span as the high-*T* deformation in the upper Sidironero Complex. Presently, no geochronological data exist from the underlying upper greenschist facies western Kardamos Complex.

Kechros Complex. A R–Sr white mica date of

37 ± 1 Ma is interpreted to reflect a maximum age of mylonitization (Wawrzenitz & Mposkos 1997). In the analysed sample, syn-mylonitic white micas indicate pressures (6–8 kbar) higher than those indicated by the pegmatite muscovites crystallized at 41 Ma in the upper Sidironero Complex. This has the following implications:

1   mylonitization of the Kechros Complex is younger (<37 Ma) than that of the Sidironero Complex;
2   at *c.* 37 Ma, the Kechros Complex was at greater depth than the Sidironero Complex;
3   high-*P* metamorphism of the Kechros Complex may postdate high-*P* metamorphism in the Sidironero Complex, i.e. the Kechros Complex was part of a deeper crustal segment during exhumation of the Sidironero Complex.

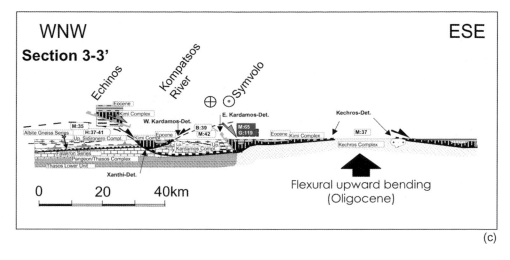

(c)

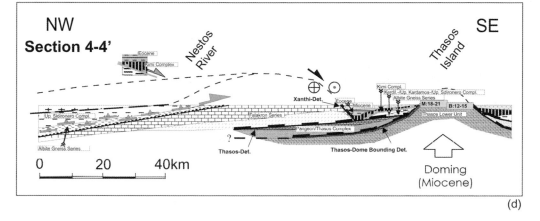

(d)

Fig. 12. (continued)

Two-stage extension tectonics: kinematics of syn- and post-thrusting extension (<42–30 Ma interval)

Thrusting within the Sidironero Complex occurred during decompression. This means continuous removal of material from the top of (i.e. exhumation of) the Sidironero Complex in close association, or even simultaneously, with thrusting. The main removal mechanism was extension. This is evident from preservation of a metamorphic hanging wall consisting of pre-Late Eocene metamorphic rocks (Kimi Complex) and transgressive Late Eocene to Oligocene marine (eastern Rhodope) sediments. Migmatites of the footwall Sidironero Complex also have Late Eocene/Early Oligocene ages (Fig. 12a–d). Thus, several tens of kilometres of crustal material have been removed along the contact of the Kimi and Sidironero Complex, which is indicative for normal faulting.

Yet, kinematic reconstruction requires juxtaposition of these two complexes on two successive detachment systems formed in two successive extension stages for reasons summarized as follows.

Lutetian marine sedimentation (c.48 Ma) occurs almost directly upon migmatites that were at the time of deposition hotter than 700 °C. However, transgression of marine sediments upon the Kimi Complex predated formation of the Xanthi detachment surface about 10 Ma (north of Xanthi).

The Xanthi detachment is associated with cataclastic deformation and clearly developed after the migmatic stage of the Sidironero Complex. Thus, the high grade high-P upper Sidironero Complex must have been emplaced in the middle crust by exhumation episodes older than the Xanthi detachment.

We suggest that early extension (c.42–30 Ma), coeval with thrusting of hot over cold metamorphic rocks, excised a crustal section between the Sidironero and Kimi Complex. However, at the contact between these two complexes, in the Xanthi area, no mylonites related to such earlier, syn-metamorphic extension are recognized. Therefore, these structures have probably been cut out by the Xanthi detachment (Fig. 13). In our structural interpretation (Fig. 13), the eastern Kardamos system that is associated with shearing represents an older detachment system that was excised by the Xanthi detachment system between the Kimi and upper Sidironero Complexes. The eastern Kardamos detachment system is clearly older than the western Kardamos detachment (situated east of Xanthi–Echinos, see Fig. 13, see above).

The structure of the western Kardamos detachment surface is similar to that of the Xanthi detachment (Xanthi area) that ramps through the thrust structures, from the upper, high grade to the lower, low grade Sidironero Complex (Fig. 13). We interpret the western Kardamos detachment fault as a satellite fault rooting into the Xanthi master detachment fault. Both excise the Kardamos Complex between the Kimi Complex and the deeper Sidironero Complex (or even deeper crustal sections, Figure 13). In this interpretation (i) the Kardamos Complex forms part of the hanging wall of the Xanthi master detachment; (ii) the deeper Kardamos Complex with hot rocks thrust over cold is equivalent to the lower and upper Sidironero Complex (symbolized in Figure 12 by indentation of the specific signatures); (iii) the higher Kardamos Complex contains syn-thrusting extensional structures between Kimi and upper Sidironero Complex cut out by the Xanthi detachment (Fig. 13).

Xanthi Detachment System: large-scale extensional ramp–flat structure

According to geochronological data, the detachment juxtaposing the Kimi Complex above the Kechros Complex (eastern Rhodope), between 37 and 32 Ma, was coeval with formation of the Xanthi detachment (Central Rhodope). However, contrasting with the Xanthi detachment surface, this detachment developed at greater depth with shearing, being associated with cooling. We consider this detachment as the eastern continuation of the Xanthi detachment, which dips roughly to the SE or ESE from the brittle upper crust into the middle crust. This connected detachment system thus extends over a length of about 100 km showing a ramp and flat geometry (Fig. 12c). The extensional ramp structure of the Xanthi detachment is directly observed in the central Rhodope.

It is important to note that in the eastern Rhodope, the Kimi Complex was transported above the low grade high-P rocks of Kechros Complex by detachment systems. This contrasts with the central Rhodope, where the high grade high-P upper Sidironero Complex is inserted between the Kimi Complex and the low grade high-P lower Sidironero Complex. In our structural reconstruction (Fig. 12b, c, d) the Xanthi connected detachment system excised the hot high-P rocks in the eastern Rhodope. Translation on this detachment system removed about 20 km of crustal thickness, the estimated thickness of the Kardamos and Sidironero Complexes, including Albite Gneiss and Falakron Series, from above the Kechros Complex. This is in coincidence with the reduction of the present crustal thickness from the central to the eastern Rhodope from about 50 km to below 30 km, interpreted from gravimetric data (Makris 1985).

Overall structural relationships and tectonic history of the Rhodope Domain

Based on the geochronological data and structural relationships already outlined, the tectono-metamorphic evolution of the Rhodope can be summarized as follows (cf. Fig. 14).

(1) Section I, i.e. the Kimi Complex, and probably also parts of the Vertiskos Complex and the Melivia Kotili Complex (Fig. 2) occupying the structurally high positions, show a pre-Eocene metamorphic history. They are bounded at their base by Late Eocene or Oligocene detachments (Fig. 14). In the Kimi Complex, high-P metamorphism is Early Cretaceous. These complexes represent a lithospheric fragment having continental crustal components and mantle fragments that accreted the European continental margin in the (Early) Cretaceous (cf. Burg et al.1996).

(2) The retrograde P-T path of the Kimi Complex, between c. 119 and c.65 Ma is characterized by decompression to depths larger than 30 km (medium-P stage), formation of migmatites and metamorphic pegmatites, cooling, and continuous hydration of former high-P rocks and granulites. Hydration and exhumation at that time has been linked to subduction of oceanic material beneath the European plate margin that included the Kimi Complex (Mposkos & Krohe 2000). A final stage of exhumation occurred after 65 Ma, but before the middle Eocene.

(3) The Sidironero, Kechros, Kardamos and Kerdilion Complexes generally occur beneath either Late Eocene (Kardamos and Vertiskos detachments, Fig. 2) or Oligocene detachments (Fig. 14). Geochronological data essentially indi-

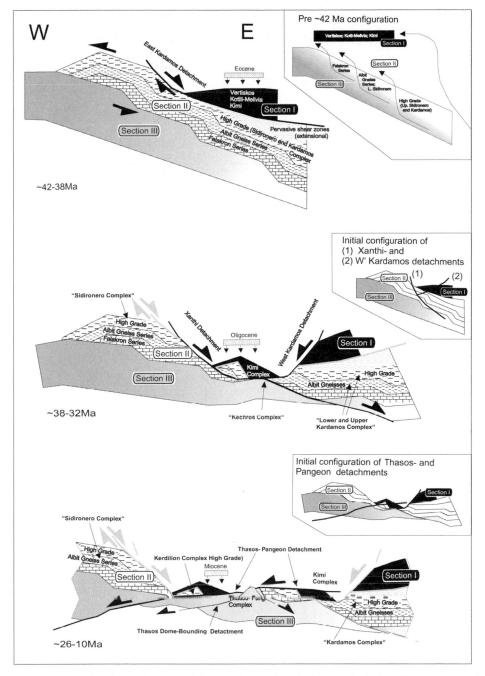

Fig. 13. Sketch illustrating the emplacement of the tectonic complexes in the central Rhodope. Two stage extension tectonics emplaced the Kimi Complex and the Eocene sediments on top of the high grade Sidironero Complex:*C*. 42–38 Ma: thrust emplacement of high grade high-*P* rocks (Section II) was accompanied by high level extension associated with a broad shear zone on top of the migmatic rocks (eastern Kardamos detachment system), exhumation of high grade high-*P* rocks and transgression of basins on the upper plate (Section I; Kimi Complex). Approx. 38–32 Ma: post-contractional extension ('out of sequence') low angle normal faults cut through the earlier formed thrust and extensional structures and emplaced Eocene sediments immediately above high grade high-*P* rocks migmatized at the time of sedimentation. Approx. 26–10 Ma: Miocene metamorphic core complexes were formed between 26 and 10 Ma, substantially after the exhumation stage of high-*P* rocks.

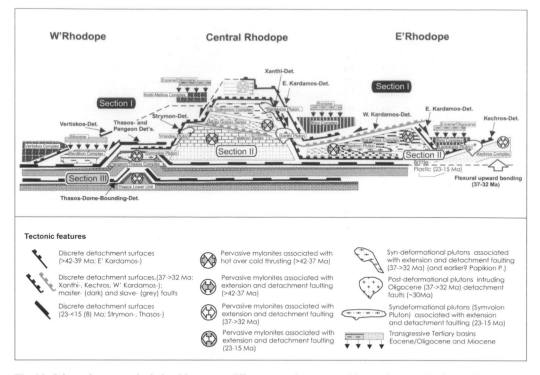

Fig. 14. Schematic structural relationships among different tectonic metamorphic complexes and basins, and the geometry of various generation of extensional detachment systems. See Figures 2 and 3 for explanation of the unit fillings.

cate Early Tertiary metamorphic histories. High-*P* metamorphism indicates maximum burial depths of 53 km (15 kbar) in eastern Rhodope and over 68 km (min. *P* 19 kbar) in central Rhodope (Thermes area) at temperatures of *c.* 550 °C and *c.* 700 °C respectively. Nearly isothermal decompression from the maximum depth up to depths of <20 km infers rapid exhumation.

(4) In the Late Eocene/Early Oligocene, two episodes of extension created two sets of detachment systems.

(5) The older is clearly coeval with deeper level thrusting of hot over cold high-*P* rocks (eastern Kardamos detachment system, between 42 and 39 Ma) separating the Kimi Complex from the eastern Kardamos Complex. This detachment is probably linked with the detachment system separating the Vertiskos Complex from the Kerdilion Complex (Fig. 14).

(6) The younger detachment is the Xanthi (– Kechros) detachment system (37–30 Ma) that dismembered the thrust structures. This detachment separates the Melivia Kotili and Kimi Complexes from the Sidironero, western Kardamos and Kechros Complexes (Fig. 14). It has a low overall dip angle, shows an ESE dipping ramp-flat structure, cross-cutting into deeper crustal levels to the

east, and extends over about 100 km (Fig. 12, a–d). Syn- and post-thrusting detachment systems are both associated with exhumation of high-*P* rocks.

(7) The crustal thickness excised by the Xanthi connected detachment system is larger in the eastern Rhodope than in the central Rhodope. In the eastern Rhodope, this system emplaced the upper plate Kimi Complex upon a structurally much lower level of the previous thrust structure (Kechros Complex, Figure 14). Hence, in the eastern Rhodope, crustal thinning, inferred from gravimetric data, occurred in the Oligocene on the eastern dipping Xanthi–Kechros normal detachment system (Fig. 14).

(8) Ages of detachment systems are constrained from structural relationships and geochronological data of sheared metamorphic rocks and granitoids (Fig. 14). In the Lutetian/Priabonian and Oligocene, simultaneous with detachment formation, several pulses of sedimentation occurred. Transgression of basins linked to both detachment systems is generally upon Vertiskos, Melivia–Kotili, and Kimi Complexes, forming the upper plate (Fig. 14).

(9) After exhumation of high-*P* rocks in the Sidironero/Kardamos and Kechros Complex, be-

tween *c.* 23 and *c.* 8 Ma, the structurally lower-most Pangeon/Thasos metamorphic core Complex was exhumed along the Strymon and Thasos detachment (Figs 13 and 14). This extension is interpreted to reflect early Aegean back-arc extension (Lister & Forster 1996; Dinter & Royden 1993; Wawrzenitz & Krohe 1998). Apparently, this stage was associated with lithospheric heating, magmatism and pervasive plastic flow of the middle crust, underneath the Sidironero Complex (Figs 12a, b, 13 and 15). Probably this extension reactivated upper crustal segments of the earlier Oligocene detachment surfaces (Fig. 14) in the northern Strymon Valley and area west of Xanthi (Fig. 12a, b).

(10) In the western Rhodope, crustal thinning inferred from gravimetric data (Makris 1985), essentially occurred in the 26–12 Ma interval on the Strymon–Thasos detachment systems that also created the Miocene supra-detachment basins (Figs 12 and 13). In the eastern Rhodope, the structurally deep-seated footwall Kechros Complex escaped Miocene heating due to earlier exhumation and flexural upward bending beneath the Xanthi–Kechros detachment system (Fig. 14).

Late Eocene to Miocene detachment formation is linked with transgression of marine clastic sediments and limestones, with calc alkaline magmatism that lasted from the Lutetian until at least *c.* 23 Ma ago.

We interpret Sidironero, Kechros, and Kardamos Complexes (Section I) that were exhumed in the <42–30 Ma interval as slivers of the extended Apulian continental margin (Variscan basement, sedimentary cover) and of oceanic material. These collided in the Early Tertiary with the European plate margin. During this collision the Kimi and Vertiskos Complexes were probably already part of the European plate. The Kimi and Vertiskos Complexes substantially differ in *P–T* histories and geochronological record from the Sidironero, Kechros, and Kardamos Complexes (Section II); geochronological data of these compelexes show no indications of an Early Cretaceous metamorphism, but the age of high-*P* metamorphism is poorly constrained.

Correlation with the Bulgarian Rhodope

The tectonic complexes in the Greek Rhodope, presented in this article, correlate with the tectonic units of the Bulgarian Rhodope, which earlier subdivisions did not allow. The Kimi Complex correlates with the Krumovica Unit, and the Kotili–Melivia Complex probably with the Madan Unit of the Bulgarian Rhodope (Burg *et al.* 1996; Ricou *et al.* 1998; Z. Ivanov pers. comm. 1999). Upper Sidironero, Kardamos and Kechros complexes correspond to the Central Rhodopian Gneiss Dome, Kecebir Dome and Biela Recka Unit in the Bulgarian Rhodope, respectively. Xanthi, western and eastern Kardamos detachment systems, and also the detachment system bounding the Kechros Complex, deeply extend into Bulgaria, where they represent prominent structures (cf. Burg *et al.*1996; Ricou *et al.* 1998; Z. Ivanov pers. comm. 1999).

Conclusions

Our data show first that the predominant NE–SW trend of stretching lineations in mylonites and top-to-SW senses of shear were formed during principally different events and thus do not in general reflect progressive thrusting (cf. Burg *et al.*1993; Burg *et al.*1996; Ricou *et al.* 1998), secondly that several phases of extension from the Late Eocene to the Miocene had a major impact on the metamorphic fabrics and on the large-scale structures of the Rhodope Domain (Dinter *et al.* 1995; Wawrzenitz & Krohe 1998, this work) and thirdly that a large proportion of tectonic contacts juxtaposing various superimposed metamorphic complexes are extensional. In the individual tectonic complexes, the lineation trends of the mylonites do in fact depict spatial partitioning of flow within the crust and a succession of compression and extension events. Syn-thrusting extension was followed by post-thrusting extension. All these different episodes contributed to exhumation of high-*P* rocks and are genetically linked with formation of basins on the hanging wall (downthrown) blocks of detachments.

Our kinematic/geodynamic model sketches out that:

1 most parts of the west, central and east Rhodope domain in Greece expose high-*P* rocks that were exhumed in the Eocene/Oligocene beneath a thin lid of material that had been exhumed earlier;

2 except for this thin lid, the dominant pervasive structures and tectonic boundaries reflect Early Tertiary tectonic processes (compression and extension);

3 low-angle detachments had already formed as early as the Late Eocene, thus preceding Aegean back-arc extension;

4 such older detachment systems connect to a principal large-scale tectonic boundary that caused exhumation of a high-*P* terrain in their footwall, about 10–20 Ma before the beginning of Late Oligocene/Miocene back-arc extension;

5 the structurally deeper part is interpreted as showing Early Tertiary high-*P* metamorphism and was a part of the Apulian plate; its *P–T*

history and geochronological record differ from the higher part, reflecting the Cretaceous accretion/collision history.

We are very greatly indebted to Periklis Papadopulos (ITME, Xanthi) for a spirited exchange of information and for giving us a preview of the 1:50 000 scale maps all over the study area, which he is presently working on. We are much obliged to Zivko Ivanov, Dimo Dimov and Kalina Shipkova (Sofia, Bulgaria) for their kind hospitality and the long time they devoted to introduce us to the Bulgarian Rhodope. We sincerely thank Nicole Wawrzenitz for inspiring discussions and a lively exchange of ideas. Detailed criticisms by Jean-Pierre Burg helped to improve style and content of the paper substantially. We also appreciate the constructive reviews from Mark Anderson, Simon Cluthbert and Mark Krabbendam. The project has been financially supported by an EU-Marie Curie Fellowship given to A.K. (Grant number ERFBMBICT972586), which is gratefully acknowledged.

References

ACKERMAND, D., SEIFERT, F. & SCHREYER, W. 1975. Instability of saphirine at high pressures. *Contributions to Mineralogy and Petrology*, **50**, 79–92.

BARR, S.R., TEMPERLEY, S. & TARNEY, J. 1999. Lateral growth of the continental crust through deep level subduction-accretion: a re-evaluation of central Greek Rhodope. *Lithos*, **46**, 69–94.

BERMAN, R.G. 1988. Internally consistent thermodynamic data for minerals in the system Na$_2$O-K$_2$O-CaO-MgO-FeO-Fe$_2$O$_3$-Al$_2$O$_3$-SiO$_2$-TiO$_2$-H$_2$O-CO$_2$. *Journal of Petrology*, **29**, 445–522.

BIGAZZI, G., DEL MORO, A., INNOCENTI, F., KYRIAKOPOULOS, K., MANNETTI, P., PAPADOPOULOS, P., NORELLITI, P. & MAGGANAS, A. 1989. The magmatic intrusive complex of Petrota, West Thrace: Age and geodynamic significance. *Geologica Rhodopica*, **1**, 290–297.

BOHLEN, S.R., MONTANA, A. & KERRICK, D.M. 1991. Precise determinations of the equilibria kyanite-sillimanite and kyanite-andalusite and a revised triple point for Al$_2$SiO$_5$ polymorphs. *American Mineralogist*, **76**, 677–680.

BROWN, T.H., BERMAN, R.G. & PERKINS, E.H. 1988. GEO-CALC: Software for calculation and display of pressure-temperature-composition phase diagrams using an IBM or compatible personal computer. *Computer & Geoscience*, **14**, 279–289.

BUCHER, K. & FREY, M. 1994. *Petrogenesis of metamorphic rocks*. Springer Verlag, Berlin.

BURG, J.P., IVANOV, Z., RICOU, E., DIMOV, D. & KLAIN, L. 1993. Implications of shear sense criteria for the tectonic evolution of the Central Rhodope Massif, Southern Bulgari. *Geology*, **18**, 451–454.

BURG, J.P., GODFRIAUX, I. & RICOU, E. 1995. Extension of the Rhodope thrust units in the Vertiskos.kerdilion Massifs. *Comptes Rendues de l'Academie des Sciences, Paris*, **320**, 889–896.

BURG, J.P., RICOU, E., IVANOV, Z., GODFRIAUX, I.,

DIMOV, D. & KLAIN, L. 1996. Syn-metamorphic nappe complex in the Rhodope Massif: Structure and Kinematics. *Terra Nova*, **8**, 6–15.

CHATTERJEE, N.D. 1972. The upper stability of the assemblage paragonite+quartz and its natural occurrences. *Contributions to Mineralogy and Petrology*, **34**, 288–303.

CHEMENDA, A.I., MATTAUER, M., MALAVIEILLE, J. & BOKUN, A.N. 1995. A mechanism for syn-collisional rock exhumation and associated normal faulting: Results from physical modelling. *Earth and Planetary Science Letters*, **132**, 225–232.

DEL MORO, A., INNOCENTI, F., KYRIAKOPULOOS, K., MANNETTI, P. & PAPADOPOULOS, P. 1988. Tertiary granitoids from Thrace (Northern Greece): Sr isotopic and petrochemical data. *Neues Jahrbuch Mineralogie, Abhandlungen*, **159**, 113–135.

DEL MORO, A., KYRIAKOPULOOS, K., PEZZINO, A., ATZORI, P. & LO GIUDICE, A. 1990. The metamorphic complex associated to the Kavala plutonites: An Rb–Sr geochronological, petrological and structural study. *Geologica Rhodopica Thessaloniki*, **2**, 143–156.

DINTER, D.A. 1998. Late Cenozoic extension of the Alpine collision orogen, northeastern Greece: Origin of the north Aegean basin. *Geological Society of America Bulletin*, **110**, 1208–1230.

DINTER, D.A. & ROYDEN, L. 1993. Late Cenozoic Extension in Northeastern Greece - Strymon Valley Detachment System and Rhodope Metramorphic Core Complex. *Geology*, **21**, 45–48.

DINTER, D.A., MACFARLANE, A., HAMES, W., ISACHSEN, C., BOWERING, S. & ROYDEN, L. 1995. U-Pb and ^{40}Ar/^{39}Ar geochronology of the Symvolon granodiorite: implication for the thermal and structrural evolution of the Rhodope metamorphic core complex, northern Greece. *Tectonics*, **14**, 886–908.

GEBAUER, D. & LIATI, A. 1997. Geochronological evidence for Mesozoic rifting and oceanisation followed by Eocene subduction in the Rhodope complex (northern Greece). *Terra Nova Abstract Supplement*, **7**, 10–11.

HARLAND, W.B., ARMSTRONG, R.L., COX, A., CRAIG, V.L.E., SMITH, A.G. & SMITH, D.G. 1990. *A Geologic Time Scale 1989*. Cambridge University Press, New York.

HATZIPANAGIOTOU, K., PE-PIPER, G. & PYRGIOTI, S.L. 1994. Subophiolitic amphibole sole from the Dafnospilia–Kedros area, western Thessaly, Greece. *Neues Jahrbuch Mineralogie, Monatshefte*, **9**, 391–402.

HOLLAND, T.J.B. 1979. Experimental determination of the reaction: Paragonite = jadeite+kyanite+H$_2$O, with applications to eclogites and blueschists. *Contributions to Mineralogy and Petrology*, **68**, 293–301.

JENKINS, D.M. 1981. Experimental phase relations of hydrous peridotites modelled in the system H2O-CaO-MgO-Al$_2$O$_3$-SiO$_2$. *Contributions to Mineralogy and Petrology*, **77**, 166–176.

JOHANNES, W. 1985. The significance of experimental studies for the formation of migmatites. *In:* ASHWORTH (ed.) *Migmatites*. Blackie, New York, 36–85.

KILIAS, A., FALALAKIS, G. & MOUNDRAKIS, D. 1997.

Alpine tectonometamorphic history of the Serbomacedonian metamorphic rocks: Implication for the Tertiary unroofing of the Serbomacedonian-Rhodope metamorphic complexes (Macedonia, Greece). *Mineral Wealth*, **105**, 9–27.

KOBER, L. 1928. *Der Bau der Erde*. Borntraeger, Berlin.

KRETZ, R. 1983. Symbols for rock-forming minerals. *American Mineralogist*, **68**, 277–279.

LIATI, A. 1986. *Regional metamorphism and overprinting contact metamorphism of the Rhodope zone, near Xanthi N.Greece: Petrology, Geochemistry, Geochronology*. PhD thesis, Techn.Univ. Braunschweig.

LIATI, A. & MPOSKOS, E. 1990. Evolution of the eclogites in the Rhodope zone of northern Greece. *Lithos*, **25**, 89–99.

LIATI, A. & SEIDEL, E. 1996. Metamorphic evolution and geochemistry of kyanite eclogites in central Rhodope, northern Greece. *Contributions to Mineralogy and Petrology*, **123**, 293–307.

LIATI, A. & GEBAUER, D. 1999. Constraining the prograde and retrograde P-T-t path of Eocene HP rocks by SHRIMP dating of different zircon domains: inferred rates of heating, burial, cooling and exhumation for central Rhodope, northern Greece. *Contributions to Mineralogy and Petrology*, **135**, 340–354.

LISTER, G.S. & FORSTER, M.A. (EDS) 1996. *Inside the Aegean Core Complexes*. Australian Crustal Research Center, Technical Publications, **45**.

MAKRIS, J. 1985. Geophysics and geodynamic implications for the evolution of the Hellenides. *In:* STANLEY, D.S. & WEZEL, F.C. (eds) *Geological evolution of the Mediterranean Basin*. Springer Verlag, New York, 231–248.

MASSONNE, H.J. & SZPURKA, Z. 1997. Thermodynamic properties of white micas on the basis of high-pressure experiments in the systems K_2O-MgO-Al_2O_3-SiO_2-H_2O and K_2O-FeO-Al_2O_3-SiO_2-H_2O. *Lithos*, 229–250.

MPOSKOS, E. 1989. High-Pressure metamorphism in gneisses and pelitic schists in the East Rhodope Zone (N.Greece). *Mineralogy and Petrology*, **41**, 25–39.

MPOSKOS, E. 1998. Cretaceous and Tertiary tectonometamorphic events in Rhodope zone (Greece). Petrological and geochronological evidences. *Bulletin of the Geological Society of Greece*, **32/3**, 59–67.

MPOSKOS, E. & LIATI, A. 1993. Metamorphic evolution of metapelites in the high-Pressure terrane of the Rhodope zone, Northern Greece. *Canadian Mineralogist*, **31**, 401–424.

MPOSKOS, E. & PERDIKATSIS, V. 1989. Eclogite-amphibolites in East Rhodope Massif. *Geologica Rhodopica Sofia, Bulgaria*, **1**, 160–168.

MPOSKOS, E. & KROHE, A. 2000. Petrological and structural evolution of continental high pressure (HP) metamorphic rocks in the Alpine Rhodope Domain (N.Greece). *In:* PANAYIDES, I., XENOPONTOS, C. & MALPAS, J. (eds) *Proceedings of the 3rd International Conf. on the Geology of the Eastern Mediterranean (Nicosia, Cyprus)*. Geological Survey, Nicosia, Cyprus, 221–232.

MPOSKOS, E., KOSTOPOULOS, D. & KROHE, A. 2001.

Ultra-high pressure metamorphism from the Rhodope Metamorphic Province, Northeastern Greece: A prelimary report on a new discovery. *Journal of Conference Abstracts (EUG XI)*, **5**(2), 341.

MPOSKOS, E. & WAWRZENITZ, N. 1995. Metapegmatites and pegmatites bracketing the time of highP-metamorphism in polymetamorphic rocks of the E-Rhodope, N.Greece: Petrological and geochronological constraints. *In:* PAPANIKOLAOU, D., (ed.) *Proceedings of the 15th Congress of the Cespatho-Balkan Geological Association*. Geological Society of Greece, Special Publications, **4/2**, 602–608.

PEITCHEVA, I. & VON QUADT, A. 1995. U-Pb zircon dating of metagranites from Byala Reka region in the east Rhodopes. Bulgaria. *In:* PAPANIKOLAOU, D., (ed.) *Proceedings of the 15th Congress of the Cespatho-Balkan Geological Association*. Geological Society of Greece, Special Publications, **4/2**, 637–642.

PLATT, J.P., SOTO, J.I., WHITEHOUSE, M.J., HURFORD, A.J. & KELLEY, S.P. 1998. Thermal evolution, rate of exhumation, and tectonic significance of metamorphic rocks from the floor of the Alboran extensional basin, western Mediterranean. *Tectonics*, **17**, 671–689.

POWELL, R. & HOLLAND, T. 1990. Calculated Mineral equilibria in the pelite system, KFMASH (K_2O-FeO-MgO-Al_2O_3-SiO_2-H_2O). *American Mineralogist*, **75**, 367–380.

RICOU, L.-E., BURG, J.-P., GODFRIAUX, I. & IVANOV, Z. 1998. Rhodope and Vardar: the metamorphic and olistostromic paired belts related to Cretaceous subduction under Europe. *Geodinamica Acta*, **11**, 285–309.

ROYDEN, L.H. 1993. Evolution of retreating subduction boundaries formed during continental collision. *Tectonics*, **12**, 629–638.

SOKOUTIS, D., BRUN, J.-P., VAN DEN DRIESSCHE, J. & PAVLIDES, S. 1993. A major Oligo-Miocene detachment in the southern Rhodope controlling north Aegean extension. *Journal of the Geological Society, London*, **150**, 243–246.

SCHMIDT, W.M. 1992. Amphibole composition in a tonalite as a function of pressure: an experimental calibration of the Al-in hornblende barometer. *Contributions to Mineralogy and Petrology*, **110**, 304–310.

SPEAR, F.S. 1981. An experimental study of hornblende stability and compositional variability in amphibolite. *American Journal of Science*, **281**, 697–734.

THOMPSON, J. JR & THOMPSON, A.B. 1976. A model system for mineral facies in pelitic schists. *Contributions to Mineralogy and Petrology*, **58**, 243–277.

VANDENBERG, L.C. & LISTER, G.S. 1996. Structural analysis of basement tectonites from the Aegean metamorphic core complex of Ios,Cyclades, Greece. *Journal of Structural Geology*, **18**, 1437–1454.

VUICHARD, J.P. & BALLEVRE, M. 1988. Garnet-chloritoid equilibria in eclogitic pelitic rocks from the Sezia zone (Western Alps): their bearing on phase relations in high pressure metapelites. *Journal of Metamorphic Geology*, **6**, 135–157.

WAWRZENITZ, N. 1997. *Mikrostrukturell unterstützte*

Datierung von Deformationsinkrementen in Myloniten: Dauer der Exhumierung und Aufdomung des metamorphen Kernkomplexes der Insel Thasos (Süd-Rhodope, Nordgriechenland). Doctoral Thesis, Univ. Erlangen-Nürnberg, Erlangen, Germany.

WAWRZENITZ, N. & KROHE, A. 1998. Exhumation and doming of the Thasos metamorphic core complex (S. Rhodope, Greece): Structural and geochronological constraints. *Tectonophysics*, **285**, 301–332.

WAWRZENITZ, N. & MPOSKOS, E. 1997. First evidence for Lower Cretaceous high-P/high-T metamorphism in the Eastern Rhodope, North Aegean Region, North-East-Greece. *European Journal of Mineralogy*, **9**, 659–664.

The relationship between ore deposits and oblique tectonics: the SW Iberian Variscan Belt

FERNANDO TORNOS[1], CÉSAR CASQUET[2], JORGE M. R. S. RELVAS[3], FERNANDO J. A. S. BARRIGA[3] & REINALDO SÁEZ[4]

[1]*Instituto Geológico y Minero de España. Azafranal 48. 37001 Salamanca, Spain*
(e-mail: f.tornos@igme.es)
[2]*Dpt. Petrología. Fac. Geologia. Universidad Complutense Madrid. 28040 Madrid, Spain*
[3]*CREMINER/Dpt Geologia, Fac. Ciências, Universidade Lisboa, Edifício C2, Piso 5, 1749-016 Lisboa, Portugal*
[4]*Dpt. Geología. Fac. Ciencias. Universidad Huelva, Spain*

Abstract: The Ossa Morena and South Portuguese Zones of the Variscan Belt of Iberia are interpreted to represent continental fragments that collided during the Variscan orogeny. Oblique northward subduction of an oceanic realm beneath the Ossa Morena Zone and subsequent collision induced thrusting and left-lateral transcurrent motion of crustal blocks and formation of a variety of ore deposits in both terranes. Most of the mineralization is related to dilational openings within thrusts and shear zones, extensional faults and pull-apart basins. A discontinuous diachronous vertical section from exhalative to deep mesozonal hydrothermal systems of Variscan age can be inferred. Volcanic-hosted massive sulphides are formed in third order pull-apart basins, but deeper related extensional structures are the *loci* for epithermal Hg, fluorite and Pb–Zn vein systems, Cu–Ni magmatic mineralization and iron-rich calcic skarns. Dilational regions along major shear zones also host mesozonal gold-bearing quartz veins. The overall Variscan mineralization pattern is inferred to be representative of an oblique collisional, (transpressional) geodynamic regime.

A critical factor in the evolution of major metallogenic provinces and world-class ore deposits is the existence of crustal discontinuities that allow the shallow emplacement of magmas or facilitate large-scale and long-lived convective hydrothermal circulation. Both processes are directly related to the formation of mineral deposits since a heat source and the circulation of large amounts of hydrothermal fluids are prerequisites in ore forming processes. Geochemical traps leading to specific ore precipitation are more local features that can be found in most upper crustal environments.

Global geodynamic models have shown that ore deposits were often formed in orogen-related extensional environments, usually back arc settings, or were related to late orogenic extensional collapse and subsequent rift-related stages (e.g. Mitchell & Garson 1981). However, less attention has been paid to the role of extensional structures related to oblique collision during the final stages of continental convergence. Here, regional strike-slip structures usually host extensional zones (tensional fractures, transfer zones, pull apart basins, etc.) that can be the *loci* of major hydrothermal

systems. Such cases have been described in several ore belts, including those hosting porphyry copper deposits (Davidson & Mpodozis 1991; Mpodozis *et al.* 1994; Tomlinson 1994; Reutter *et al.* 1996; Unrug *et al.* 1999), epithermal systems (Tosdal & Nutt 1998), mesothermal gold veins (Sanderson *et al.* 1991; Patey & Wilton 1993; Miller *et al.* 1994; Kerrich 1997) or shale-hosted base metal deposits (O'Brien & Davies 1986).

The SW margin of the Variscan Belt of Iberia shares many of the features of transpressional orogenic belts (e.g. Ribeiro 1981; Sanderson *et al.* 1991). Despite its rather well known geodynamic evolution (see Quesada *et al.* 1991; Eguíluz *et al.* 2000 and references therein), little is known about the influence of oblique tectonics on the formation of the abundant ore deposits of the area. In this review we present an overview of the more significant ore deposits, which show an important structural control that is considered as critical to their formation. The model includes different types of mineralization formed at different crustal levels as well the world class deposits of the Iberian Pyrite Belt.

From: BLUNDELL, D.J., NEUBAUER, F. & VON QUADT, A. (eds) 2002. *The Timing and Location of Major Ore Deposits in an Evolving Orogen*. Geological Society, London, Special Publications, **204**, 179–198.
0305-8719/02/$15.00 © The Geological Society of London 2002.

Geodynamic setting and evolution of the SW Iberian Variscan Belt

The SW Iberian Variscan Belt is part of the Iberian (or Hesperian) Massif, i.e. the pre-Alpine basement of the Iberian Peninsula, which crops out in the southwestern Iberian Peninsula. The Iberian Variscan Belt has long been subdivided into zones parallel to the belt axis, mainly on the basis of stratigraphic and palaeogeographic considerations (Julivert *et al.* 1974; inset in Fig. 1). The SW Iberian Variscan Belt comprises the Ossa Morena Zone (OMZ) and the South Portuguese Zone (SPZ) (Figs 1 and 2). Recent tectonic models indicate that during the Cadomian orogeny (late Proterozoic–early Cambrian) the Ossa Morena Zone was accreted to the southern margin of the Central Iberian Zone, which is considered to be an autochthonous terrane (Abalos *et al.* 1991; Abalos & Eguiluz 1992, Ochsner 1993), shown in Figure 3. The South Portuguese Zone was in turn accreted to the Ossa Morena Zone during the Variscan orogeny (Silva *et al.* 1990).

The geodynamic evolution of this part of the Iberian Massif and the location of its precise boundaries has been the subject of much debate in the recent past. The cryptic Cadomian suture is tentatively located somewhere to the north or within the Badajoz–Córdoba Shear Zone, a first order, 10–20 km wide, mylonitic corridor resulting from strong left-lateral shearing during Variscan deformation (Burg *et al.* 1981; Abalos *et al.* 1991, Azor *et al.* 1993). The Variscan suture is defined by the Pulo de Lobo Terrane and the Beja Acebuches ophiolite (Silva *et al.* 1990; Quesada *et al.* 1994). The Ossa Morena Zone is a complex domain consisting of a Proterozoic–lower Cambrian basement and a Palaeozoic cover, reworked during the Variscan orogeny. The basement shows evidence of penetrative deformation, metamorphism and ensialic magmatism, both metaluminous and peraluminous, that on the basis of geochronology can be attributed to the Cadomian orogeny (e.g. Galindo *et al.* 1988; Ochsner 1993). The lithological record consists of a lower sequence of gneisses, black shales, sandstones and amphibo-

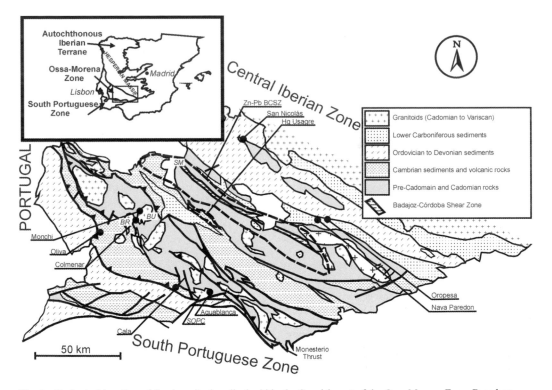

Fig. 1. Geological location of the deposits described within the Spanish part of the Ossa Morena Zone. Based on Locutura *et al.* (1990). SOPC, Santa Olalla Plutonic Complex; BR, Brovales Pluton; BU, Burguillos Pluton; SM, Santa Marta Stock. The inset shows the general zone division of the Variscan Belt of Iberia. The Pulo de Lobo Group and the Beja–Acebuches Ophiolite are located between the Ossa Morena Zone and the South Portuguese Zone. The Iberian Pyrite Belt is within the South Portuguese Zone.

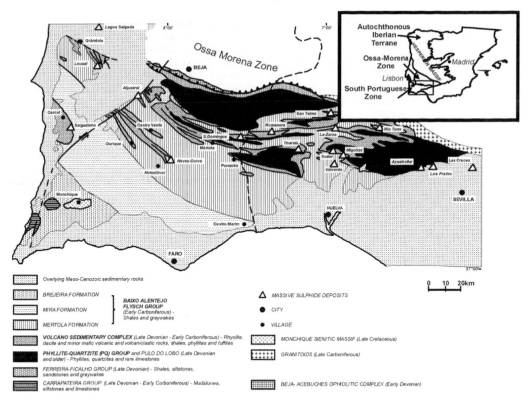

Fig. 2. Geological sketch of the Iberian Pyrite Belt showing the location of the major ore deposits. Modified from Carvalho *et al.* (1999).

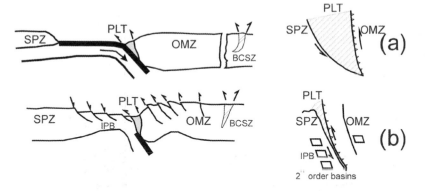

Fig. 3. Interpretative cross section and map of the SW margin of the Variscan Belt of Iberia during Variscan times. Modified from Silva *et al.* (1990) and Quesada (1992). (**a**) Plate convergence, subduction and minor oblique collision and partial obduction during the late Devonian (pre-Fammenian);(**b**) Generalized oblique collision, lateral escape of the South Portuguese Zone and formation of second order pull-apart basins and volcanic alignments in the Iberian Pyrite Belt and Ossa Morena Zone during the late Devonian–Lower Carboniferous. Not to scale. SPZ, South Portuguese Zone; IPB, Iberian Pyrite Belt; OMZ, Ossa Morena Zone; BCSZ, Badajoz–Córdoba Shear Zone; PLT, Pulo de Lobo Terrane (including the Beja–Acebuches Ophiolite).

lites (see Quesada *et al.* 1991; Eguíluz *et al.* 2000 for a complete description) which is overlain by a complex syn-orogenic sequence that includes calc-alkaline volcanic and plutonic rocks, limestones and siliciclastic sediments formed in an Andean-type orogen (Sánchez Carretero *et al.* 1990). After collision and extensional collapse of the orogen, rifting begun and a shelf sequence was deposited unconformably above the Cadomian basement in Early to Mid-Cambrian times. This stage reached a climax during the mid–late Cambrian (Liñán & Quesada 1990) when intraplate alkaline magmatism, mainly transitional-alkaline, set in at 510–500 Ma (Galindo *et al.* 1988; Ochsner 1993). During the Ordovician, Silurian and Early Devonian, the region was apparently a quiet shallow water passive continental margin (Robardet & Gutierrez Marco 1990). The Variscan deformation in the OMZ resulted in widespread plutonism synchronous with thrust-style deformation and low grade metamorphism. A key feature of the Ossa Morena Zone is the existence of long-lived crustal-scale faults of WNW–ESE trend that have conditioned the paleogeography at least since early Cambrian times (Liñán & Quesada, 1990).

Modern terrane analysis (Quesada 1998) envisages that the South Portuguese Zone consists of two exotic terranes that were sequentially accreted to the Ossa Morena Zone. The northern Pulo de Lobo Group (Fig. 2) is an oceanic terrane consisting of sediments of an accretionary prism and an ophiolite, the Beja–Acebuches Ophiolite, derived from the partial obduction of a pre-Famennian oceanic basin (Silva *et al.* 1990). The southern terrane shows continental features and is made of the Iberian Pyrite Belt and the SW Portugal Domains (Oliveira 1983, 1990). The Pulo do Lobo Group consists of several Devonian metamorphic detrital formations, some of them including scarce metavolcanics. The Beja–Acebuches Ophiolite consists of a strip of rocks, 100 km long and about 1.5 km thick, showing the characteristics and organization of a typical ophiolite, i.e. metagabbros (mafic granulites, flaser gabbros and amphibolites), serpentinites, metabasalts and cherts (Munhá *et al.* 1986; Quesada *et al.* 1994). Both the Pulo de Lobo Group and the ophiolite were affected by a low pressure/high temperature regional metamorphism, ranging from lower greenschist to granulite facies. The Iberian Pyrite Belt constitutes the central domain of the SPZ. It is characterized by the presence of a subvolcanic to volcanic, bimodal magmatic suite of late Famennian to early late Visean age. Both the magmatism and the massive sulphide formation took place in an intracontinental rift or pull-apart basin environment. Late and post-Variscan magmatism is also recorded in the form of small plutonic bodies (e.g. Simancas 1983; Schütz *et al.* 1987; de la Rosa 1992; Stein *et al.* 1996; Thiéblemont *et al.* 1998). Conditions of metamorphism range from prehnite–pumpellyite to lower greenschist facies, the latter restricted to local shear zones. The geology, geotectonic setting and palaeogeographic evolution of the Iberian Pyrite Belt are described in more detail later. The SW Portugal Domain is composed of anchimetamorphic monotonous sedimentary rocks, deposited in a shallow marine environment (Oliveira 1983). No magmatic rocks have been reported here.

Oblique north-verging Variscan subduction of oceanic crust beneath the OMZ took place from the mid–late Devonian to the Visean (Dallmeyer *et al.* 1993) with the situation merging into the subduction of the SPZ under the OMZ during the late Devonian to middle Westphalian interval. Further continental collision produced the amalgamation of the Pulo de Lobo and South Portuguese terranes to the southern margin of the OMZ and the lateral escape eastwards of the South Portuguese Zone, with the formation of large left-lateral strike-slip faults and shear bands (Fig. 2).

The Ossa Morena Zone

Variscan geological features

Initial Variscan transpressional deformation within the OMZ was largely accommodated along discrete mylonitic shear zones (thrusts and, mostly left-lateral, steep NW–SE strike-slip faults). The resulting pattern is one of domains formed by the stacking of imbricate southward directed thrusts that involve a Cadomian foliated basement and the cover sequences (e.g. the Monesterio thrust, Fig. 1; Apalategui *et al.* 1990). Thrust sheets can display large-scale recumbent folding (e.g. Vauchez 1975) at the leading edge of the thrusts. In turn, these thrust domains are bounded by steep strike-slip faults striking NW–SE, which define the limits of several narrow longitudinal tectonic domains (Apalategui *et al.* 1990).

Upon exhumation, progressive deformation was accommodated by ductile–brittle and brittle faults of NW–SE, north–south and NE–SW trend that partially overprinted older tectonic lineaments. A complex pattern of transpressional late Variscan faults has been recognized by Sanderson *et al.* (1991) in the southern margin of the Central Iberian Zone near the boundary with the Badajoz–Córdoba Shear Zone (Fig. 1). Here a domino or bookshelf fault zone consisting of north–south and NE–SW trending strike-slip right-lateral faults rotated toward a NW–SE orientation due to continued left-lateral motion along the shear zone.

An ensialic magmatic arc of I-type high-K calc-alkaline plutonic rocks developed during Variscan evolution. Available crystallization ages are still few, but range between 352 ± 4 Ma and 332 ± 3 Ma (Dallmeyer *et al.* 1995; Casquet *et al.* 1998; Pin *et al.* 1999; Montero *et al.* 2000). During subduction, small intra-montane continental basins filled with terrigenous sediments and coal seams formed in pull-apart basins in northern OMZ. On the other hand, relics of an older marine platform of Tournaisian age are found as small sedimentary–volcanic inliers discordant with the metamorphic basement (Gabaldón *et al.* 1985).

A wide variety of ore deposits of different ages and significance has been long recognized in the OMZ (up to 20 ore deposits and several hundreds of ore showings; Locutura *et al.* 1990, and references therein). Most of them are of Variscan age, formed in relation to the active magmatic arc.

Mineralization related to shear zones (mesozonal mineralization)

Variscan deformation in the deeper structural levels led to the formation of large-scale imbricate thrusting and recumbent folding with a SW vergence. One of the more remarkable structures is the NW–SE-trending Monesterio Thrust which separates two major tectonic domains (Fig. 1) (Apalategui *et al.* 1990). Rocks within the thrust zones often show significant gold enrichment, particularly when they cross-cut late Proterozoic black shales, amphibolites, quartzites, volcaniclastic rocks and subvolcanic dacite–rhyolites of the Cadomian syn-orogenic sequence. Some of these rocks can contain significant gold and copper anomalies that are interpreted as an older (pre-Cadomian) palaeoplacer or exhalative mineralization (Locutura *et al.* 1990).

The mineralized zones are found in steep NW–SE to WNW–ESE and, to a lesser extent, north–south- and NE–SW-trending brittle–ductile faults and extensional zones, being especially intense at their intersections (Canales & Matas 1992). The major structures are coincident with the regional shear zones and probably represent dilational features within them. The NE–SW-trending structures are interpreted as extensional fractures whilst the less common north–south-trending shear bands and faults are probably synchronous antithetic faults to the main shear zones. Mineralization at tectonic intersections is probably the consequence of major focused fluid flow. The Variscan gold mineralization is hosted by irregular halos of pervasive structurally controlled sericitization and chloritization in the deformed rocks. The more altered zones show an intense texturally destructive silicification with local tourmalinization, albitization and K feldspar growth. The mineralization occurs disseminated in these rocks or in the cross-cutting quartz veins (Canales & Matas 1992). The sulphide proportion is always low and consists of arsenopyrite, pyrrhotite, pyrite and chalcopyrite. Concentrations of magnetite can be locally found (Canales, pers. comm.). Dramatic changes of fluid pressure are to be expected at these places. Hydraulic failure induced fluid separation and concomitant precipitation of metals transported by weak acids or bases (As, Au). Very few works have dealt with this style of mineralization, but preliminary studies show that they have a great economic potential (Canales & Matas 1992). They share many features with the mesothermal gold-bearing sulphide-poor quartz veins, typically formed in metamorphic belts (Kerrich & Wyman 1990; Hodgson 1993). The relationship with brittle–ductile structures and the style of hydrothermal alteration and assemblage indicate that they correspond to deposits formed at intermediate to shallow depths (6–12 km; GebreMariam *et al.* 1995).

Epizonal Ni–Cu magmatic mineralization related to strike-slip faults

The recently discovered Aguablanca Ni–Cu deposit (Fig. 1) is here interpreted as an example of magmatic mineralization in which strike-slip faulting resulting from oblique tectonics have played a critical role in its genesis (Tornos *et al.* 2001). The ore body forms two sub-vertical magmatic pipes, which have been drilled down to 600 m depth. Here, the (semi)-massive sulphides, mainly pyrrhotite, chalcopyrite and pentlandite, cement and support a magmatic breccia of heterolithic fragments of fine-grained pyroxenitic and peridotitic cumulates, gabbros–gabbronorites and minor host rocks (Fig. 4e). This breccia is included in an epizonal (0.5–1 kbar) diorite–gabbro intrusion, the Aguablanca Stock, which is adjacent to the regional-scale NW–SE Cherneca Fault (Fig. 5). This stock is part of a larger composite, mainly tonalitic, intrusive massif, the Santa Olalla Plutonic Complex dated at 332 Ma (Montero *et al.* 2000). The Aguablanca Stock and the magmatic mineralization have a superimposed pervasive hydrothermal alteration with the magmatic assemblage retrograding to amphibole, chlorite, carbonates, sericite and a wide variety of sulphides. These late minerals host minor amounts of platinum group minerals (see Ortega *et al.* 2001).

The intrusion and mineralization are interpreted as having formed in a two-stage process, probably related to the development of extensional fractures along the Cherneca Fault (Fig. 5). The first event was the early intrusion, probably fault controlled, of a primitive mafic magma and development of a magmatic chamber in the upper crust. Assimilation and contamination with the host, late Proterozoic pyrite-rich black slates, took place. This led to igneous differentiation and the formation of an immiscible sulphide phase, which sank along with the pyroxenitic cumulates. A further reactivation

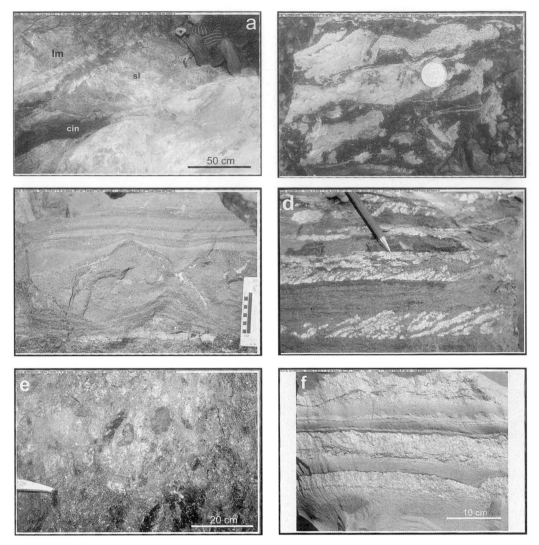

Fig. 4. Details of some of the mineralization referred to in the text. (**a**) Cambrian limestones replaced by silica and cinnabar (cin: massive cinnabar; lm: limestone; sl: silicified limestone. Location: Usagre. (**b**) Hydrothermal breccia filled by sphalerite-rich lodes located in tensional gashes (sph: sphalerite; cc: calcite; Q: quartz). Location: Llera. (**c**) Compositional banding resulting from replacement of a mylonitic rock (probably an igneous protolith) by plagioclase + magnetite + actinolite–tremolite and sulphides, enclosing magnetite + plagioclase fragments. Location: Colmenar Mine, Jerez de los Caballeros. (**d**) Tension gashes in a magnetite and actinolite–tremolite skarn filled by late pyrite and chalcopyrite. Location: Cala Mine, Cala. (**e**) Magmatic breccia with fragments of pyroxenite, gabbro and gabbronorite supported by sulphides. Location: Aguablanca. (**f**) Sedimentary banding in exhalative massive sulphides defined by alternating layers of coarse grained pyrite + siderite interbedded with turbidite-like layers of more fine-grained sulphides. Location: Tharsis Mine, Iberian Pyrite Belt.

of the structure promoted the rise and final emplacement of the unconsolidated magmas at depths of about 3 km. Partially consolidated cumulates and the sulphide magma were subsequently injected forcefully into open NE–SW extensional fractures, leading to the formation of the mineralized breccia near the northern margin

of the Aguablanca Stock and a few metres away from the Cherneca Fault. This fault and minor satellite extensional structures thus probably played an important role in driving magma ascent and controlling the location of the ore. S–C composite foliations within the fault, the sense of rotation of the regional cleavage near the Stock

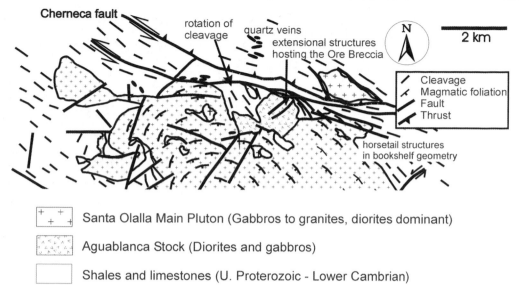

Fig. 5. Synthetical geological sketch of the Aguablanca Stock showing the relationship of the intrusion and the magmatic mineralization to the Cherneca Fault and related structures. Modified from Tornos *et al.* (2001).

and the bookshelf-like geometry of antithetic faults are consistent with a dominant left-lateral strike-slip sense of movement. The NE–SW-trending breccia pipe is compatible with this kinematic scenario.

Epizonal iron-rich calcic skarns adjacent to granitoids and shear zones

A significant type of mineralization in the Ossa Morena Zone is represented by the iron-rich calcic skarns (Casquet & Tornos 1991). They form several ore bodies with a total of 180 Mt of ore originally in place with high iron grades, significant Cu credits and locally high Au, Co, U and REE contents. The skarns are usually developed at the contact of Variscan plutons of gabbroic to monzogranitic composition and limestones, calc-silicate hornfelses and volcanic rocks of Lower Cambrian age. They show a rather typical calcic assemblage with prograde hedenbergite– salite and minor grandite (± wollastonite) zones replaced by a retrograde skarn with actinolite–hornblende and minor epidote, feldspars, quartz and calcite. Minor early magnesial skarns are locally found. The ore assemblage includes magnetite and low amounts of pyrite, chalcopyrite and other sulphides (Fig. 4c). A special case is found at the Monchi Mine (Fig. 1) where an unusual B-rich mineral assemblage with abundant vonsenite

and axinite as well as significant amounts of löllingite, cobaltite, monazite, allanite and uraninite is found in close association with the skarn (Arribas 1962).

The exact origin of the skarn-related ores remains controversial. The geological setting suggests that iron might have been introduced from nearby calc-alkaline magmas during the early, high-temperature stage of the skarn development (Casquet & Tornos 1991). However, the same carbonate and volcanic rocks that host the skarns include earlier stratabound (sub)-exhalative and breccia-like magnetite deposits genetically related to the Cambrian volcanism (Dupont 1979). The relationships between skarn-type and the earlier magnetite mineralization is not obvious, although in some cases it seems likely that some skarn magnetite might come from the remobilization of the older magnetite. Systematic radiogenic isotope geochemistry of magnetites and host rocks suggests that any isotopic heritage is masked by the magmatic/hydrothermal remobilization (Galindo *et al.* 1995; Cuervo *et al.* 1996; Darbyshire *et al.* 1998).

A key feature in the formation of these deposits is the relationship with major syn-magmatic north–south- and NW–SE-trending faults which seem to have played a significant role as shear zones, channelling fluid flow and skarn location. This relationship is particularly obvious at the El Colmenar ore body (Fig. 6a). Here, different types of vein and massive skarn are found in the exo-contact of the epizonal Variscan Brovales pluton (340 ± 7 Ma; Montero *et al.* 2000) with lower Cambrian marbles and siliciclastic sediments. The contact is defined by a brittle–ductile to brittle nearly north–south right-lateral fault zone (Sanabria 2001). The fault is a conjugate to the major NW–SE strike-slip faults and has undoubtedly played a major role in driving magma emplacement and focusing fluid flow and skarn formation. The role of faults in controlling skarn formation is evident again in the Cala Mine, the most important magnetite mine in the region, which also shows an important structural control (Fig. 6b). The ore body is found at the contact of the Lower Cambrian marbles with the small tonalitic Cala Stock (Casquet & Velasco 1978; Velasco & Amigó 1981). The magnetite ore is massive and occurs as two WNW–ENE-trending bands, about 10 m thick, dipping steeply northwards. It replaced an earlier, mostly garnet, skarn along fault planes. The sulphide minerals, mostly chalcopyrite and pyrite, postdate the magnetite

and precipitated in extensional fractures (Fig. 4d). Detailed mapping suggests that the Cala Stock was intruded along a major left-lateral fault that actually separates the Herrerías and Cumbres–Hinojales tectonic units. The intrusion probably took place in a dilational zone within the fault, probably a pull-apart structure. The stock is considered to be an offshoot of the nearby and larger Santa Olalla Plutonic Complex, that was emplaced at a depth equivalent to 1.5 kbar (Casquet 1980). Fluid flow was channelled along the contacts with the host rocks; skarn replacement with precipitation of ores probably took place later, during late extensional movement.

A remarkable example is represented by the huge massive skarns found at the northern contact of the Santa Olalla Plutonic Complex. These skarns are almost barren and consist for the most part of garnet with only a local hydrothermal retrograde alteration (Casquet, 1980). The skarns here are older than the main intrusive units of the Santa Olalla Complex but are accompanied by small precursor basic intrusives and larger granitic

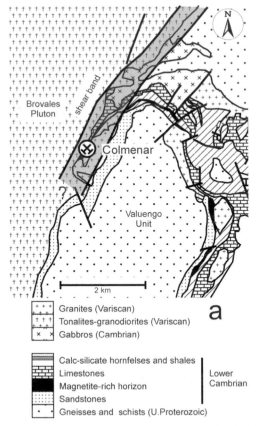

Fig. 6. Tectonic setting of the Colmenar (**a**) and Cala (**b**) iron-rich calcic skarns.

bodies that crop out along the aforementioned Cherneca Fault. Again the picture emerges here of a relatively long-lived hydrothermal and intrusive activity that is fault controlled.

Volcanic-associated massive sulphides

Along the Badajoz–Córdoba Shear Zone (Fig. 1) there are several Carboniferous syn-orogenic pull-apart basins with WNW–ESE trend (Gabaldón et al. 1985). However, volcanic-associated massive sulphides are only found in the Benajarafe–Matachel basin (Fig. 7). It is discontinuously exposed basin about 20 km long that was originally part of a larger but tectonically dismembered and partially eroded basin containing Tournaisian to Visean marine sediments. Its boundaries are defined by WNW–ESE-trending left-lateral strike-slip and normal NNW–SSE-trending faults. The basin was filled with shallow marine siliciclastic, carbonate and volcaniclastic sediments with local terrestrial equivalents (Quesada 1983; Gabaldón et al. 1985). Equivalent basins formed in back arc settings are the preferred place for the formation of stratiform volcanic-hosted massive sulphides (e.g. Kuroko, Bathurst, Urals; Ohmoto 1996; Barrie & Hannington 1999; Herrington 2000). Despite this preferential setting, massive sulphides are very scarce in the Ossa Morena Zone.

The mineralization is hosted by bimodal volcanic rocks. They consist of rhyolitic domes unconformably covered by an up to 200 m thick sequence of variegated epiclastic rocks including mass flows and shales that are lateral to submarine andesitic flows with some Cu-bearing sulphide disseminations (Late Tournaisian to Lower Visean; Gabaldón et al. 1985; Sánchez Carretero et al. 1990). These epiclastic rocks are interpreted as syn-tectonic deposits formed near a fault scarp. A

small, gold-bearing polymetallic (Zn–Pb–Cu) ore-body (Nava–Paredón) is located at the contact between the rhyolites and the epiclastic rocks (Tornos et al. 2000). The mineralization occurs within two different structures and at different stratigraphic levels, inside the uppermost brecciated rhyolites (hyaloclastites) and the epiclastites, but always near the contact between the two of them. The massive sulphides consist of sphalerite, galena and pyrite with minor amounts of chalcopyrite and traces of Ag-bearing tetrahedrite, pyrrhotite, bismuthinite, boulangerite and electrum. The mineralization is thought to have formed by the replacement of the rhyolites and the epiclastic sediments near their contact. The genetic model involves the upwelling circulation of deep fluids through a poorly defined feeder. This fluid carried the metals, probably derived from the leaching of the rhyolites and underlying basement. Near the regional aquifers they probably mixed with sulphur-bearing stratabound waters confined to the major lithological contacts, precipitating sulphides by fluid mixing (Tornos et al. 2000). Related to this mineralization there are some stratabound hematite-rich layers that are likely exhalative equivalents into an oxidized shallow marine basin.

The morphology of the basin and the orientation of the volcanic alignments can be easily interpreted as directly related to left-lateral transpressional tectonics. The faults controlling the massive sulphide formation probably correspond to synchronous tensional structures. In such a system, hydrothermal activity was probably triggered by the combination of high heat flow and the development of large hydrothermal convection cells.

Pb–Zn– (Ba) veins

The more historically significant mineralization of the Ossa Morena Zone is the abundant Pb–Zn ore located in veins within or near the Badajoz–Cordoba Shear Zone (Figs 1 and 8). Hundreds of veins were mined during the late nineteenth and early twentieth centuries, representing at that time the leading lead ore field in Europe. The mineralization occurs in short (<500 m), thick lensoidal quartz–carbonate-bearing veins. In a few cases the veins are hosted by shales and schists. However, they are abundant in the brittle rocks of the Badajoz–Córdoba Shear Zone, including ultramylonites, gneisses, quartzites and amphibolites. Moreover, the veins are especially abundant close to some minor epizonal granodioritic stocks intruded along this shear zone (e.g. Santa Marta Stock; Fig. 8). The internal structure of the veins is very uniform, consisting of open space fillings

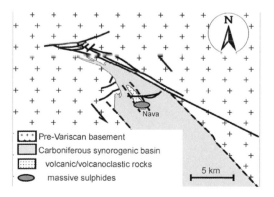

Pre-Variscan basement
Carboniferous synorogenic basin
volcanic/volcanoclastic rocks
massive sulphides

Nava

5 km

Fig. 7. Geological sketch of a sector of the Benajarafe–Matachel basin showing the structural control on the Nava Paredón massive sulphide ore body.

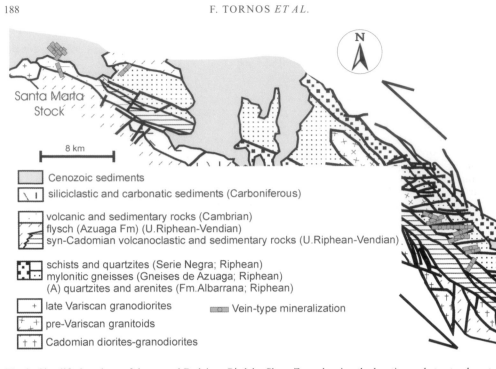

Fig. 8. Simplified geology of the central Badajoz–Córdoba Shear Zone showing the location and structural control of the late-Variscan to post-Variscan Pb–Zn veins. Based onIGME (1994).

and abundant polyphase hydrothermal breccias with fragments of altered host rocks supported by quartz and carbonates with a cockade texture (Fig. 4b). The related hydrothermal alteration is very subtle, only consisting of minor sericitization, feldspathization and silicification. The sulphide assemblage includes galena, sphalerite and chalcopyrite with minor pyrite, linneite, Ag-rich tetrahedrite, pyrargirite and pyrrhotite. Barite and fluorite are very scarce. Fluid inclusion work (Chacón et al. 1981) shows that the hydrothermal fluids were low salinity waters with the quartz precipitating at temperatures near 200 °C. Lead isotope composition suggests that the metals were scavenged from the host rocks during late Variscan to early Alpine times (Tornos et al. 1998). These veins share many features with the adularia–sericite type of epithermal systems, with the ore probably precipitated by boiling of the fluids. The dominant trends are N70–80°E, N110–120°E and N0–45°E. The cross-cutting relationships with the regional foliation, the brittle nature of the veins and the low temperature of formation indicate that the veins formed at shallow depths (<1 km) and that they post-date the dominant tectonic fabric of the Badajoz–Cordoba Shear Zone. The morphology and orientation of most of the veins (ENE–WSW and north–south to NE–SW) are consistent with

their formation along brittle extensional structures (tension planes) during the waning left-lateral movement of the shear zone (Late Carboniferous?). The WNW–ESE-trending veins seem to have formed along minor pull-apart structures within the shear zone. These structures also locally channelled the intrusion of some granitoids, such as the Santa Marta Stock. These intrusions were able to generate local high temperature thermal domes that increased the fluid flow and favoured mineralization.

Epithermal Hg

The Usagre area (central Ossa Morena Zone, Figure 1) is characterized by the presence of a thick sequence of Lower Cambrian black limestones with some slate intercalations. These rocks are intruded by hypabyssal felsic and mafic dykes of late Variscan age that are related to the major WNW ESE faults. Hosted by the limestones, but always closely associated with these WNW–ESE trending structures, there are Cu, Pb, Sb, barite and Hg prospects as well as an old Hg mine (Usagre). The sulphide mineralization is replacive on the limestones and always associated with a pervasive silicification of the limestones that also post-dates the dykes. The ore assemblage consists of cinnabar, pyrite, Hg-rich sphalerite, chalcopyrite, Hg-bearing tetrahedrite, galena and local gold with ankerite, barite and quartz (Fig. 4a). The ore-forming event took place at temperatures between 250 and 320 °C and the fluids were saline brines with 19–24 wt% NaCl equivalent (Tornos & Locutura 1989). The cross-cutting relationship with the dykes indicates that this mineralization is the result of a late Variscan extensional event and took place in an epizonal environment broadly equivalent to that of the epithermal deposits. Again, magmas and heat and fluid flow were channelled along regional faults, leading to replacements after limestones and open space filling. Ore precipitation took place by fluid-carbonate interaction and/or mixing with surficial, oxidized waters.

Other mineralization in the Ossa Morena Zone

Many other ore deposits within the Ossa Morena Zone have a strong structural control. Beside the widespread Fe, Cu, Zn–Pb, W–Bi and barite veins, we also describe here some Sn-bearing replacement deposits.

A tin-rich replacive mineralization has been recently discovered in Oropesa, in the easternmost Ossa Morena Zone (Fig. 1). The arsenopyrite–cassiterite-bearing ore is hosted by pervasively silicified slates, feldspathic sandstones and limestones of late Carboniferous age. The mineralization is located in a small pull-apart basin similar to that hosting the massive sulphides of Nava Paredon, the alteration and mineralization being controlled by north–south normal sub-vertical faults. Nearby Ordovician quartzites and shales near the edge of the basin host north–south-trending Sn-rich quartz veins which are interpreted as the feeder zones of the replacive mineralization and connected with an hypothetical underlying leucogranite, in a model similar to that of Patterson et al. (1981) for the Renison Bell deposit (Locutura et al. 1990). Vein-type deposits are widespread across the Ossa Morena Zone. N120–140°E-trending veins probably remobilizing older stratabound ores occur in late Palaeozoic siliciclastic rocks while tensional NNW–SSE quartz veins, up to 6 km long, contain quartz, barite, siderite, hematite, magnetite, pyrite, tetrahedrite, sphalerite, galena and chalcopyrite. Furthermore, the few granitic intrusions with perigranitic W– (Cu–Bi–Au)-bearing quartz veins (San Nicolás, Oliva) were probably intruded along extensional zones. However, little is known about the geological setting of these late Variscan intrusions.

The Iberian Pyrite Belt

Stratigraphic record

The sedimentary record of the IPB includes rocks of Devonian to Carboniferous age, that host the well know giant massive sulphide deposits (Fig. 2). The most useful regional stratigraphic nomenclature is that of Schermerhorn & Stanton (1969) and Schermerhorn (1971), which has been adopted with minor modifications by many authors (e.g. Oliveira 1983, 1990; Barriga 1990; Sáez et al. 1996; Leistel et al. 1998). It comprises three main units: from bottom to top, the Phyllite and Quartzite Group (PQ Group), the Volcanic–Sedimentary Complex (VS Complex) and the Culm Group (or Baixo Alentejo Flysch Group). The PQ Group is largely homogeneous throughout the IPB. It includes a monotonous detrital sequence over 1000 m thick composed of shales and sandstones (quartzwacke and quartzarenite). Available data suggest a shallow depositional environment for the PQ Group, probably in a storm-dominated platform (Moreno & Sáez 1990). These features suggest an epicontinental sea environment for the whole Iberian Pyrite Belt during late Devonian times.

The homogeneous features of the PQ Group change abruptly towards the top of the unit. Here there is an increase of the sand/lutite ratio and different exotic facies are found including fan-deltas, near-shore bars, mega-debris flows and some limestone lenses (Moreno et al. 1996). This uppermost sequence has been dated as late Famennian. These facies are indicative of collapse and fragmentation of the stable continental platform near the Devonian–Carboniferous boundary (Moreno et al. 1996). This collapse resulted in the development of several sub-basins, which constituted the realm where deposition of the VS Complex took place. The VS Complex consists of felsic (dacite–rhyolite) and mafic (basalt) magmatic rocks interbedded with a thick sedimentary sequence. They show complex intrusive and inter-fingered relationships. Massive rocks include sills, domes and cryptodomes of felsic composition with less abundant mafic flows and dykes. Sub-volcanic rocks, both felsic and mafic, are locally ubiquitous, forming the bulk of the stratigraphic column at some localities (Boulter 1993; Mitjavila et al. 1997). The VS Complex also includes: (a) detrital rocks including thick volcanic-derived epiclastics, with grain size ranging from conglomerate to fine arenites and some pumice-rich mass flows; (b) grey and black shales, sometimes directly associated to the massive sulphide deposits, (c) chemical sedimentary rocks including massive sulphides, Mn–Fe-rich chert and jasper and scarce limestone lenses. Some of the volcaniclastic rocks are perhaps related to a subaerial explosive activity (Schermerhorn 1971; Routhier *et al.* 1978; Quesada 1996) Age constraints based on scarce palaeontological and absolute age determinations place the VS Complex between late Famennian and Middle Visean times.

Magmatic rocks in the IPB have been the object of much research in the last three decades (e.g. Hamet & Delcey 1971; Routhier *et al.* 1978; Munhá 1983; Mitjavila *et al.* 1997; Thiéblemont *et al.* 1998). Nowadays it is widely accepted that the magmatism of the IPB is bimodal, with only some scarce andesites (Munhá, 1983; Mitjavila *et al.* 1997). Despite the association, both in space and time, there is no lithological transition between felsic and mafic rocks as they originated and evolved separately.

The Culm Group is composed of the Basal Shaly Formation and the Flysch Sequence (Schermerhorn 1971; Moreno 1993). It comprises a thick succession of shales, litharenites and rare conglomerates of late Carboniferous age, which overlies the VS Complex (Oliveira 1983). It represents the infill of rapidly subsiding first-order basins, mostly by turbidite sediments coming from at least two source areas, the IPB and the OMZ. Palaeocurrent analysis points to two different directions of sediment supply, ENE–SSW and NNE–SSW (Moreno 1993). It is worth noting that equivalent trends have been proposed for alignments of volcanic foci and of alternating belts of massive sulphide and Mn-rich mineralization (Routhier *et al.* 1978).

General features of the massive sulphides

The hydrothermal activity and associated massive sulphide formation in the IPB relates to local, focused and long-lasting hydrothermal systems. The importance of the ore bodies justifies a somewhat extended description. Massive sulphide-related hydrothermal alteration zoning varies from one deposit to another in the IPB, from essentially stratabound, indicating fluid flow paths controlled by permeability contrasts (e.g. Neves Corvo: Relvas *et al.* 1997, 2001; Tharsis: Tornos *et al.* 1998), to cone-shaped (e.g. Rio Tinto: García Palomero 1980; Aljustrel: Barriga & Fyfe 1988; Valverde: Toscano *et al.* 1993). Ore-related hydrothermal alteration is usually zoned, with a chlorite and/or quartz-rich core that grades outwards into sericite- and paragonite-rich zones. The overall alteration/mineralization patterns in the typical VHMS deposits of the IPB are clearly of intermediate pH, low to high temperature (70–400 °C) and low-sulphidation type. The unusually low abundance of volcanics in the overall footwall

succession suggests that most of the giant IPB deposits should perhaps be best described by hybrid genetic models, in between typical volcanic-associated massive sulphides and sedimentary–exhalative deposits (Sáez *et al.* 1999; Relvas 2000).

The Neves Corvo deposit represents a remarkable exception within the Iberian Pyrite Belt. Its extraordinary ore geochemistry constitutes a unique feature not only in the IPB but also for VHMS deposits in general. The extremely high-grade copper and tin ores of Neves Corvo call for multi-sourced metal contributions. Isotopic data for this deposit, both stable and radiogenic, are consistent with the incorporation of fluid contributions of magmatic and/or deep metamorphic origin in the seawater-dominated ore-generating system (Relvas 2000). Such a deep contribution has also been proposed for other deposits in the northern Iberian Pyrite Belt (Aguas Teñidas Este, San Telmo) by Sánchez-España *et al.* (2000).

Except for the Neves Corvo deposit, the published oxygen and hydrogen isotope signatures of the IPB ore fluids are remarkably constant throughout the whole province, clearly indicating the largely predominant or even exclusive involvement of modified seawater (including connate pore fluids) in the hydrothermal systems ($0‰ < \delta^{18}O < 6‰$; $-15‰ < \delta D < 10‰$; Barriga & Kerrich 1984; Munhá *et al.* 1986). A general consensus is emerging that the relative homogeneity of most IPB deposits in relation to their base metal ratios and radiogenic isotopic signatures points to similar histories of metal supply, source geochemistry, extraction regimes and metal transport for the ore-forming systems, regardless of the varied depositional environments and ore-precipitating mechanisms already recognized in this province (Barriga & Fyfe 1998; Tornos *et al.* 1998; Velasco *et al.* 1998; Sáez *et al.* 1996, 1999; Relvas 2000; Relvas *et al.* 2001). Significant involvement of juvenile sources for the ore-forming metals can be ruled out on the basis of the strontium, neodymium, lead and osmium isotopic signatures of the IPB ores; metals in these deposits are dominantly of crustal derivation (Marcoux 1998; Mathur *et al.* 1999; Nieto *et al.* 2000; Relvas *et al.* 2001)

It has been pointed out that some IPB deposits do not in any way fit a genetic model of accumulation on the sea floor, given the absence of sulphide oxidation and sedimentary dilution of the ores, and the frequent presence of pre-ore cap rocks (cherts, Barriga & Fyfe 1988; igneous and sedimentary rocks, Tornos 2000) with extensive hydrothermal alteration. Currently favoured models include: (1) mineral deposits related to ore fluids accumulated in brine pools (Tharsis, Tornos *et al.* 1998; Tornos & Spiro 1999); (2) mineral deposits formed via replacement of black shales (Almodóvar *et al.* 1998); (3) massive sulphide deposits formed by replacement of coherent rhyolites/rhyodacites at relatively shallow depths (<700 m) and with no relationship to shales (e.g. part of the ore bodies of the Rio Tinto camp); (4) ore bodies formed by replacement of thick pumice mass flows near the contact with overlying shales or massive volcanics (e.g. San Platón, Concepción; Tornos 2000). Some of the major IPB deposits are transitional (e.g. Neves Corvo; Relvas *et al.* 2000).

Tectonic constraints and palaeogeographic evolution of the Iberian Pyrite Belt

It is widely accepted that much of the ore geochemical characteristics and physical differences amongst VHMS deposits should be ascribed to relevant aspects of either intra-plate or subduction-related magmatism and palaeogeographic evolution of the various geotectonic settings where VHMS metallogenesis takes place (Franklin *et al.* 1981; Large 1992; Lydon 1996; Barrie & Hannington 1999; Fig. 9). In the IPB, a total of massive ores in excess of 2 billion tonnes was deposited (Schermerhorn 1970; Leistel *et al.* 1998). The geological and palaeogeographical environment required for deposition of probably

the world's largest concentration of volcanogenic massive sulphide ores should represent the paradigm of perfect conditions for VHMS generation.

As previously emphasized (Sáez *et al.* 1996; Leistel *et al.* 1998; Carvalho *et al.* 1999), the most recent geodynamic interpretations, taking into account the magmatic regime, sedimentological record and palaeogeographical evolution in this province, favour a regional extensional environment (intra-continental rifting or development of pull-apart basins) for the IPB magmatism and ore-generation (Silva *et al.* 1990; Sáez *et al.* 1996; Mitjavila *et al.* 1997; Quesada 1998).

The characteristics of Hercynian magmatism in the IPB involved crustal melting at relatively low pressures, which required extensional tectonics and high heat flow conditions. A forearc extensional setting induced by lateral escape in the marginal units of the exotic South Portuguese Zone has been envisaged as a consequence of the northward, oblique collision between the latter and the autochthonous Iberian terrane during the early stages of the Variscan orogeny (Silva *et al.* 1990; Quesada 1998). This tectonic setting is distinct from the subduction-related arc and back-arc geodynamic environments postulated for most other equivalent metallogenic provinces, and makes it difficult to compare some of the characteristics of the IPB deposits with their modern counterparts. Nevertheless, these differences could

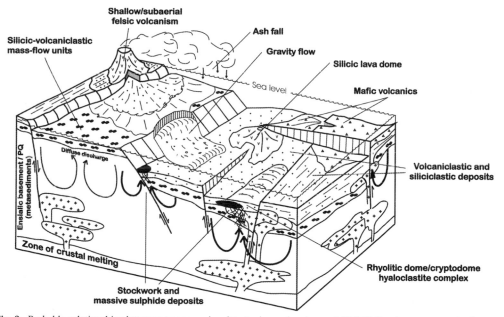

Fig. 9. Probable relationships between transpressional tectonics, volcanism and VHMS-forming processes in the Iberian Pyrite Belt during the deposition of the Volcano- Sedimentary Complex and formation of massive sulphide deposits.

actually represent the key to understanding what
in fact may have made this part of the world so
favourable for hosting about 5 Ma of intensely
productive hydrothermal activity around the end
of Devonian times (Pereira *et al.* 1996; Oliveira *et
al.* 1997; Nesbitt *et al.* 1999; Mathur *et al.* 1999;
Nieto *et al.* 2000; Relvas *et al.* 2001).

Deep crustal fault structures, originated in
response to the oblique transpressional tectonic
events, are thought to have favoured the rise of
superheated, water-depleted and poorly fractionated felsic magmas to fairly shallow crustal
levels. These magmas could have been generated
either by crustal anatexis or by partial melting of
quartz-feldspathic basement rocks, induced by
conductive heat loss of mafic magmas formed at
the base of the crust (Munhá 1983;Mitjavila *et al.*
1997; Thiéblemont *et al.* 1998). The shallow
setting and the water-undersaturated nature of
these acidic magmatic systems make it difficult to
account for the exsolution of significant amounts
of magmatic fluids, consistent with the large
predominance of modified seawater and connate
waters in the fluid budget of the majority of IPB
convective hydrothermal systems. Possible significant inputs of magmatic fluid (and metal) contributions in particular ore-forming systems (e.g.
Neves Corvo, Relvas *et al.* 2001; see also
Sánchez-España *et al.* 2000), if any, should have
been related with deeper, independent felsic intrusive bodies.

A highly segmented ensialic rift environment,
where fault-bounded, strongly compartmentalized
basins with variable subsidence on a variety of
scales, seems to be a likely scenario for the IPB
and its ore-forming processes. In the IPB, this
palaeogeography is further indicated by the general absence of lateral stratigraphic correlations in
the volcano–sedimentary complex, which reflect
very sharp facies variations, both laterally and
vertically (see Oliveira 1983, 1990). Variations in
water depth within individual basins (e.g. Rio
Tinto, Badham 1982) or between separate, com-
partmented basins, and variable sedimentary depositional rates related to differential subsidence
are to be expected in such an environment. Basin
instability resulting from strong syn-volcanic tectonic activity should have been responsible for
important soft-sediment mass displacements,
which may account for the abundance of syn-
sedimentary, gravity-driven transport features
documented in both the host volcaniclastic and
sedimentary packages, and in some massive ores
as well (Silva *et al.* 1990; Tornos *et al.* 1998;
Relvas 2000; Fig. 4f). In the IPB, compelling
evidence in favour of this tectonic interpretation
arises from the clear, regional-scale orientation of
the alignments of volcanic foci (Rambaud 1978),

from the lateral variation of many sedimentary
features (Oliveira 1990; Moreno 1993; Moreno *et
al.* 1996), from the major structural lineaments
(Ribeiro & Silva 1983; Silva *et al.* 1990; Quesada
1998), and from the distribution of both the
massive sulphide deposits (Sáez *et al.* 1996;
Leistel *et al.* 1998) and the low-temperature Mn-
bearing mineralization (Routhier *et al.* 1978).
These widely recognized features led Quesada
(1996) to propose a sub-division of the IPB into
different domains, according to their distinct
relationships between volcanism, sedimentary facies and mineralization. Arguably, these sectors
would represent regional-scale, second-order basins bounded by major east–west structures, within which third-order basins directly related to
massive sulphide deposition might have been
controlled by structural alignments with predominant east–west and NE–SW orientation.

Apart from massive sulphides, abundant vein
type hydrothermal mineralization occurs in the
Iberian Pyrite Belt, related to late- and post-
Variscan strike-slip tectonics (Sáez *et al.* 1988).
Ore assemblages include fluorite, F-(Pb–Zn), Sb–
(As–Cu), Pb–Zn– (Ag–barite), As–Cu–Bi–Co–
Ni and Sn–W–As.

Conclusions

There is an evident spatial and genetic relationship
between the more significant Variscan mineralization and extensional discontinuities in the Ossa
Morena and South Portuguese Zones of the Variscan Belt of Iberia. Dilational regions within
faults, including thrusts and strike-slip faults, have
played a significant role in magma intrusion and
channelling of heat and fluid flow. Because this
tectonic pattern is the consequence of oblique
subduction and subsequent oblique collision of the
South Portuguese Zone under the Ossa Morena
Zone, we infer that the recognized pattern of
abundant mineralization—and probably the size of
the deposits as well—might be representative of
this geodynamic scenario. In such a transpressional setting, different styles of ore deposits were
formed, involving abnormally high heat flow
evolved from fault-driven magmatic activity and
fluid flow along open fractures. Open structures
channelled magmas to shallow depths and promoted the flow of large batches of hydrothermal
fluids. The depth and timing of the extensional
structures seem to have played a major role in the
style of mineralization on both sides of the suture
(Table 1). Early deep structures are the host to
mesothermal gold deposits (6–12 km depth) while
younger brittle faults are host to epithermal Zn–
Pb veins and Sn- and Hg-rich replacements
(<1 km depth) as well as volcanic associated

Table 1. *Major types of mineralization related to transpressional structures in SW Iberia*

Structural level	Mineralization	Tectonic location	Local geological control	Examples
(Sub)superficial	Massive sulphides (SPZ)	Pull-apart basins NW–SE and N–S?	Relationship with syn-sedimentary faults of likely E–W and NW–SE trend, locally defining third order basins or volcanic alignments	Massive sulphides Iberian Pyrite Belt
	Massive sulphides (OMZ)	Pull-apart basins NW–SE	E–W syn-sedimentary extensional structures with related hydrothermal activity associated with volcanic rocks	Nava Paredon
Epizonal	Hg epithermal	WNW–ESE strike-slip faults	Local pull-apart structures in massive limestones	Usagre
	Zn–Pb veins	Regional left-lateral WNW–ESE trending shear zone	Tensional gashes (ENE–WSW, WNW–ESE and N–S to NE–SW) in the brittle rocks	Badajoz Cordoba Shear Zone
	Sn replacement	Pull-apart basins with WNW–ESE trend	Subvolcanic felsic intrusions in basins with hydrothermal alteration in reactive and permeable rocks.	Oropesa
	Ni–Cu magmatic	Pull-apart structures related to WNW–ESE trending left-lateral strike-slip faults	Extensional structures with NE–SW trend	Aguablanca
	Fe– (Cu) calcic skarns and replacements	NNW–SSE and WNW–ESE shear bands	Intersection of regional shear bands with limestones or previously mineralized rocks. Local intrusion of tonalites in pull-apart structures	Cala, Colmenar
Mesozonal	Au-bearing sulphide-poor veins	Major shear bands with WNW–ESE trend	Extensional zones in major structures; NE–SW extensional zones; N–S antithetic? faults	Chocolatero

massive sulphides. Mineralization at intermediate structural levels includes magmatic Cu–Ni ore bodies and calcic skarns, formed at depths between 1.5 and 3 km (Table 1).

In the Iberian Pyrite Belt, the situation is rather different. The mineralization is restricted to (sub-) surficial levels. Here, collision and contemporaneous lateral escape promoted the opening of crustal-scale fractures with intrusion of deep-seated magmas and a concomitant thermal anomaly. Second- and third-order pull-apart sedimentary basins were also formed, the locations of the largest massive sulphide deposits. However, many of the original features have been masked by Variscan and Alpine deformation. The influence of transpressional tectonics is not only restricted to SW Iberia. In the Central Iberian Zone, major strike-slip structures seem to control mesothermal gold mineralization (Sanderson *et al.* 1991; González Clavijo *et al.* 1993) as well as the intrusion of Sn–W-bearing granitoids.

This work is a synthesis and review of studies carried out by the authors in the framework of the GEODE project. Specific projects include DGES projects numbered PB96-0135 (F.T.), PB98-0960 (R.S.) and AMB-0918-CO2-01 (C.C., F.T.) (Spain) and SULFIBER (2/2.1/CTA/420/94) Praxis XXI, Ministry of Science and Technology (F.B., J.R.) (Portugal). We would like to acknowledge A. Canales, F. Velasco and E. Pascual for their help in the study of SW Iberian metallogenesis. Franz Neubauer, Henry Fritz and an anonymous reviewer helped to improve the original manuscript.

References

ABALOS, B. & EGUILUZ, L. 1992. Evolución geodinámica de la zona de cizalla dúctil de Badajoz-Córdoba durante el Proterozoico superior-Cámbrico inferior. *In:* GUTIERREZ-MARCO, J.C., SAAVEDRA, J. & RÁBANO, I. (eds) *Paleozoico Inferior de Ibero-América*. Publicaciones UNEX, 577–592.

ABALOS, B., GIL IBARGUCHI, J.I. & EGUÍLUZ, L. 1991. Cadomian subduction/collision and Variscan transpression in the Badajoz–Córdoba Shear Belt (SW Spain). *Tectonophysics*, **199**, 51–72.

ALMODÓVAR, G.R., SÁEZ, R., PONS, J., MAESTRE, A., TOSCANO, M. & PASCUAL, E. 1998. Geology and genesis of the Aznalcóllar massive sulphide deposits, Iberian Pyrite Belt, Spain. *Mineralium Deposita*, **33**, 111–136.

APALATEGUI, O., EGUÍLUZ, L. & QUESADA, C. 1990. Ossa Morena Zone: Structure. *In:* MARTINEZ, E. & DALLMEYER, R.D. (eds) *Pre-Mesozoic Geology of Iberia*. Springer Verlag, Berlin, 280–291.

ARRIBAS, A. 1962. Mineralogia y metalogenia de los yacimientos españoles de uranio: Burguillos del Cerro (Badajoz). *Estudios Geológicos*, **28**, 173–192.

AZOR, A., GONZÁLEZ LODEIRO, F. & SIMANCAS, J.F. 1993. Cadomian subduction/collision and Variscan transpression in the Badajoz Cordoba Shear Belt, southwest Spain: a discussion on the age of the main tectonometamorphic events. *Tectonophysics*, **217**, 343–346.

BADHAM, N.P.D. 1982. Further data on the formation of ores at Rio Tinto, Spain. *Transactions of the Institute of Mining and Metallurgy*, **91**, B26–B32.

BARRIE, C.T. & HANNINGTON, M.D. 1999. Classification of volcanic-associated massive sulfide deposits based on host rock composition. *In:* BARRIE, C.Y. & HANNINGTON, M.D. (eds) *Volcanic associated massive sulfide deposits: Processes and examples in modern and ancient settings*. Reviews in Economic Geology, **8**, 1–11.

BARRIGA, F.J.A.S. 1990. Metallogenesis in the Iberian Pyrite Belt. *In:* MARTINEZ, E. & DALLMEYER, R.D. (eds) *Pre-Mesozoic Geology of Iberia*. Springer-Verlag, Berlin, 369–379.

BARRIGA, F.J.A.S. & FYFE, W.S. 1988. Giant pyritic base-metal deposits: the example of Feitais, Aljustrel, Portugal. *Chemical Geology*, **69**, 331–343.

BARRIGA, F.J.A.S. & FYFE, W.S. 1998. Multi-phase water-rhyolite interaction and ore fluid generation at Aljustrel, Portugal. *Mineralium Deposita*, **33**, 188–207.

BARRIGA, F.J.A.S. & KERRICH, R. 1984. Extreme ^{18}O-enriched volcanics and ^{18}O-evolved marine water, Aljustrel, Iberian Pyrite Belt: Transition from high to low Rayleigh number convective regimes. *Geochimica et Cosmochimica Acta*, **48**, 1021–1031.

BOULTER, C.A. 1993. Comparison of Rio Tinto, Spain, and Guaymas Basin, Gulf of California: An explanation of a supergiant massive sulfide deposits in an ancient sill-sediment complex. *Geology*, **21**, 801–804.

BURG, J.P., IGLESIAS, M., LAURENT, P., MATTE, P. & RIBEIRO, A. 1981. Variscan intracontinental deformation: the Coimbra-Córdoba shear zone (SW Iberian Peninsula). *Tectonophysics*, **78**, 161–177.

CANALES, A.M. & MATAS, J. 1992. Anomalías de Au en el flanco Sur de l Anticlinorio de Olivenza Monesterio. *In:* RÁBANO, I. & GUTIERREZ MARCO, J.C. (eds) *Libro de Resúmenes*. VIII Reunión de Ossa Morena, 53–69.

CARVALHO, D., BARRIGA, F.J.A.S. & MUNHÁ, J. 1999. Bimodal-siliciclastic systems B the case of the Iberian Pyrite Belt. *In:* BARRIE, C.T. & HANNINGTON, M.D. (eds) *Volcanic-associated massive sulfide deposits: processes and examples in modern and ancient settings*. Reviews in Economic Geology, **8**, 375–408.

CASQUET, C. 1980. *Fenómenos de endomorfismo, metamorfismo y metasomatismo en los mármoles de la Rivera de Cala (Sierra Morena)*. PhD Thesis, Universidad Complutense, Madrid.

CASQUET, C., GALINDO, C., DARBYSHIRE, D.P.F., NOBLE, S.R. & TORNOS, F. 1998. Fc-U-REE mineralisation at Mina Monchi, Burguillos del Cerro, SW Spain: age and isotope (U-Pb, Rb-Sr and Sm-Nd) constraints on the evolution of the ores. *GAC-MAC-APGGQ, Quebec 98 Conference, Abstract Volume*, **23**, A28.

CASQUET, C. & TORNOS, F. 1991. Influence of depth and igneous geochemistry on ore development in skarns: The Hercynian Belt of the Iberian Peninsula. *In:* AUGUSTHITIS, C. (ed.) *Skarns, their petrology and metallogeny.* Augusthitis, Athens, 555–591.

CASQUET, C. & VELASCO, F. 1978. Contribución a la geología de los skarns cálcicos en torno a Santa Olalla de Cala (Huelva-Badajoz). *Estudios Geológicos*, **34**, 399–405.

CHACÓN, J., HERRERO, J.M. & VELASCO, F. 1981. Las mineralizaciones de Zn-Pb de Llera (Badajoz). *I Congreso Español Geología*, **2**, 447–455.

CUERVO, S., TORNOS, F., SPIRO, B. & CASQUET, C. 1996. El origen de los fluidos hidrotermales en el skarn férrico de Colmenar-Santa Bárbara (Zona de Ossa Morena). *Geogaceta*, **20**, 1499–1500.

DALLMEYER, R.R., FONSECA, P.E., QUESADA, C. & RIBEIRO, A. 1993. $^{40}Ar/^{39}Ar$ mineral age constraints for the tectonothermal evolution of a Variascan suture in southwest Iberia. *Tectonophysics*, **222**, 177–194.

DALLMEYER, R.D., GARCÍA-CASQUERO, J.L. & QUESADA, C. 1995. $^{40}Ar/^{39}Ar$ mineral age constraints on the emplacement of the Burguillos del Cerro Igneous Complex (Ossa Morena Zone, SW Iberia). *Boletín Geológico Minero*, **106**, 203–214.

DARBYSHIRE, D.P.F., TORNOS, F., GALINDO, C. & CASQUET, C. 1998. Sm–Nd and Rb–Sr constraints on the age and origin of magnetite mineralisation in the Jerez de los Caballeros iron district of Extremadura, SW Spain. *ICOG-9, Chinese Science Bulletin*, **43**(suppl), 28.

DAVIDSON, J. & MPODOZIS, C. 1991. Regional geologic setting of epithermal gold deposits, Chile. *Economic Geology*, **86**, 1174–1186.

DE LA ROSA, J. 1992. *Petrología de las rocas básicas y granitoides del batolito de la Sierra Norte de Sevilla.* PhD Thesis, Universidad Sevilla, **1992**.

DUPONT, R. 1979. *Cadre geologique et metallogenese des gisements de fer du sud de la province de Badajoz (Sierra Morena Occidentale-Espagne).* PhD Thesis, Institute de National Polytechnique de Lorraine.

EGUÍLUZ, L., GIL IBARGUCHI, J.I., ABALOS, B. & APRAIZ, A. 2000. Superposed Hercynian and Cadomian orogenic cycles in the Ossa Morena Zone and related areas of the Iberian Massif. *Geological Society of America Bulletin*, **112**, 1398–1413.

FRANKLIN, J.M., SANGSTER, D.M. & LYDON, J.W. 1981. Volcanic associated massive sulfide deposits. *In:* SKINNER, B. (eds) *Economic Geology, 75 Anniversary Volume.* Society of Economic Geologists, **75**, 485–627.

GABALDÓN, V., GARROTE, A. & QUESADA, C. 1985. El Carbonífero Inferior del Norte de la Zona de Ossa Morena (SW de España). *In: X Carboniferous International Geological Congress, IGME, Madrid,* 173–185.

GALINDO, C., CASQUET, C., PORTUGAL FERREIRA, M. & MACEDO, C.A.R. 1988. Geocronología del Complejo Plutónico Táliga-Barcarrota (CPTB).(Badajoz, España). *In:* BEA, F. (eds) *Geología de los granitoides y rocas asociadas del Macizo Hespérico.* Editorial Rueda, Madrid, 385–392.

GALINDO, C., DARBYSHIRE, F., TORNOS, F., CASQUET, C. & CUERVO, S. 1995. Sm-Nd geochemistry and

dating of magnetites: A case study from a Fe district in the SW of Spain. *In:* PASAVA, J., KRIBEK, B. & ZAK, K. (eds) *Mineral Deposits: From their origin to environmental impacts.* Balkema, Rotterdam, 41–43.

GARCÍA PALOMERO, F. 1980. *Caracteres geológicos y relaciones morfológicas y genéticas de los yacimientos del anticlinal de Riotinto.* PhD Thesis, Instituto de Estudios Onubenses, Huelva.

GEBREMARIAM, M., HAGEMANN, S.G. & GROVES, D.I. 1995. A classification scheme for epigenetic Archean lode gold deposits. *Mineralium Deposita*, **30**, 408–410.

GONZÁLEZ CLAVIJO, E., DÍEZ BALDA, M.A. & ALVAREZ, F. 1993. Structural study of a semiductile strike-slip system in the Central Iberian Zone (Variscan Fold Belt, Spain): structural controls on gold deposits. *Geologische Rundschau*, **82**, 448–460.

HAMET, J. & DELCEY, R. 1971. Age, synchronisme et affiliation des roches rhyolitiques de la province pyrito-cuprifère du Baixo Alentejo (Portugal): mesures isotopiques par le métode $^{87}Rb/^{87}Sr$. *Comptes Rendus de l'Academie des Sciences, Paris*, **272D**, 2143–2146.

HERRINGTON, R. 2000. Southern Urals. *In:* LARGE, R. & BLUNDELL, D. (eds) *Database on Global VMS districts.* CODES-GEODE CD-ROM, 85–104.

HODGSON, C.J. 1993. Mesothermal lode gold deposits. *In:* KIRKHAM, R.V., SINCLAIR, W.D., THORPE, R.I. & DUKE, J.M. (eds) *Mineral Deposit Modeling.* Geological Society of Canada Special Papers, **40**, 635–678.

IGME 1994. *Mapa Metalogenético de España escala 1/200000, hoja No.167-168, Cheles-Villafranca de los Baros.* Instituto Geológico y Minero de España, Madrid.

JULIVERT, M., FONTBOTÉ, J.M., RIBEIRO, A. & CONDE, L. 1974. *Mapa tectónico de la Península Ibérica y Baleares.* Instituto Geológico Minero de España, Madrid.

KERRICH, R. 1997. Late-accretion lode gold deposits through time; transpressional controls on mineralisation in external supercontinent cycles. *Abstracts with Programs. Geological Society of America*, **29-6**, 14.

KERRICH, R. & WYMAN, D. 1990. The geodynamic setting of mesothermal gold deposits: an association with accretionary tectonic regimes. *Geology*, **18**, 882–985.

LEISTEL, J.M., MARCOUX, E., THIÉBLEMOND, D., QUESADA, C., SÁNCHEZ, A., ALMODÓVAR, G.R., PASCUAL, E. & SÁEZ, R. 1998. The volcanic-hosted massive sulphide deposits of the Iberian Pyrite Belt. *Mineralium Deposita*, **33**, 2–30.

LARGE, R.R. 1992. Australian volcanic-hosted massive sulfide deposits: Features, styles and genetic models. *Economic Geology*, **87**, 471–510.

LIÑAN, E. & QUESADA, C. 1990. Ossa Morena Zone: Stratigraphy, rift phase (Cambrian). *In:* MARTINEZ GARCÍA, E. & DALLMEYER, R.D. (eds) *Pre-Mesozoic Geology of Iberia.* Springer Verlag, Berlin, 259–266.

LOCUTURA, J., TORNOS, F., FLORIDO, P. & BAEZA, L.

1990. Ossa Morena Zone: Metallogeny. *In:* MARTI-NEZ GARCÍA, E. & DALLMEYER, R.D. (eds) *Pre-Mesozoic Geology of Iberia.* Springer Verlag, Berlin, 321–332.

LYDON, J.W. 1996. Characteristics of volcanogenic massive sulphide deposits, interpretations in terms of hydrothermal convection systems and magmatic-hydrothermal systems. *Boletin Geologico Minero,* **107**, 215–264.

MARCOUX, E. 1998. Lead isotope systematics of the giant massive sulphide deposits in the Iberian Pyrite Belt. *Mineralium Deposita,* **33**, 45–58.

MATHUR, R., RUIZ, J. & TORNOS, F. 1999. Age and sources of the ore at Tharsis and Rio Tinto, Iberian Pyrite Belt, from Re-Os isotopes. *Mineralium Deposita,* **34**, 790–793.

MILLER, L.D., GOLDFARB, R.J., GEHRELS, G.E. & SNEE, L.W. 1994. Genetic links among fluid cycling, vein formation, regional deformation, and plutonism in the Juneau gold belt, southeastern Alaska. *Geology,* **22**, 203–206.

MITCHELL, A.H.G. & GARSON, M.S. 1981. *Mineral deposits and global tectonic setting.* Academic Press, London.

MITJAVILA, J., MARTI, J. & SORIANO, C. 1997. Magmatic evolution and tectonic setting of the Iberian Pyrite Belt volcanism. *Journal of Petrology,* **38**, 727–755.

MONTERO, P., SALMAN, K., BEA, F., AZOR, A., EXPÓSITO, I., LODEIRO, F., MARTINEZ-POYATOS, D. & SIMANCAS, F. 2000. New data on the geochronology of the Ossa-Morena Zone, Iberian Massif. Variscan-Appalachian dynamics: the building of the Upper Paleozoic basement. *Basement Tectonics,* **15**, 136–138.

MORENO, C. 1993. Postvolcanic Paleozoic of the Iberian pyrite belt: an example of basin morphologic control on sediment distribution in a turbidite basin. *Journal of Sedimentary Petrology,* **63**, 1118–1128.

MORENO, C. & SÁEZ, R. 1990. Sedimentación marina somera en el devónico del Anticlinorio de Puebla de Guzmán, Faja Pirítica Ibérica. *Geogaceta,* **8**, 62–64.

MORENO, C., SIERRA, S. & SÁEZ, R. 1996. Evidence for catastrophism at the Famennian-Dinantian boundary in the Iberian Pyrite Belt. *In:* STROGEN, P., SOMERVILLE, I.D. & JONES, G.L. (eds) *Recent advances in Lower Carboniferous Geology.* Geological Society, London, Special Publications, **107**, 153–162.

MPODOZIS, C., TOMLINSON, A.J. & CORNEJO, P.C. 1994. Acerca del control estructural de intrusivos Eocenos y porfidos cupriferos en la region de Potrerillos, El Salvador. *In: Actas VII Congreso Geologico Chileno,* **2**, 1596–1600.

MUNHÁ, J. 1983. Hercynian magmatism in the Iberian Pyrite Belt. *Memorias Servicio Geologico Portugal,* **29**, 39–81.

MUNHÁ, J., BARRIGA, F.J.A.S. & KERRICH, R. 1986. High [18]O ore-forming fluids in volcanic-hosted base metal massive sulfide deposits: geologic, ^{18}O/^{16}O, and D/H evidence from the Iberian pyrite belt, Crandon, Wiscosin, and Blue Hill, Maine. *Economic Geology,* **81**, 530–552.

NESBITT, R.W., PASCUAL, E., FANNING, C.M., TOSCANO,

M., SÁEZ, R. & ALMODÓVAR, G.R. 1999. First zircon U-Pb age from the Iberian Pyrite Belt, Spain. *Journal of the Geological Society, London,* **156**, 7–10.

NIETO, J.M., ALMODÓVAR, G.R., PASCUAL, E., SÁEZ, R. & JAGOUTZ, E. 2000. Estudio isotópico con el sistema Re-Os de las mineralizaciones de sulfuros de la Faja Pirítica Ibérica. *Geogaceta,* **27**, 181–184.

O'BRIEN, G.W. & DAVIES, H. 1986. Transpressional strike-slip faulting in the Mount Isa Inlier. *BMR Research Newsletter,* **4**, 1–2.

OCHSNER, A. 1993. *U-Pb geochronology of the Upper Proterozoic-Lower Paleozoic geodynamic evolution in the Ossa Morena Zone (SW Iberia): constraints on the timing of the Cadomian orogeny.* PhD Thesis, University Zurich.

OHMOTO, H. 1996. Formation of volcanogenic massive sulfide deposits: The Kuroko perspective. *Ore Geology Reviews,* **10**, 135–177.

OLIVEIRA, J.T. 1983. The marine Carboniferous of South Portugal: a stratigraphic and sedimentological approach. *In:* LEMOS DE SOUSA, M.J. & OLIVEIRA, J.T. (eds) *The Carboniferous of Portugal.* Memórias dos Serviços Geológicos de Portugal, **29**, 3–37.

OLIVEIRA, J.T. 1990. South Portuguese Zone. *In:* MARTINEZ GARCÍA, E. & DALLMEYER, R.D. (eds) *Pre-Mesozoic Geology of Iberia.* Springer Verlag, Berlin, 333–346.

OLIVEIRA, J.T., PACHECO, N., CARVALHO, P. & FERREIRA, A. 1997. The Neves Corvo Mine and the Paleozoic Geology of Southwest Portugal. *In: Geology and VMS deposits of the Iberian Pyrite Belt.* SEG Fieldbook Series, **27**, 21–71.

ORTEGA, L., LUNAR, R., GARCÍA PALOMERO, F., MORENO, T. & PRICHARD, H.M. 2001. Removilización de minerales del grupo del platino en el yacimiento de NiCuEGP de Aguablanca (Badajoz). *Boletín Sociedad Española Mineralogia,* **24**, 175–176.

PATTERSON, D.J., OHMOTO, H. & SOLOMON, M. 1981. Geological setting and genesis of cassiteritesulfide mineralisation at Renison Bell, Western Tasmania. *Economic Geology,* **76**, 393–438.

PATEY, K.S. & WILTON, D.H.C. 1993. The Deer Cove Deposit, Baie Verte Peninsula, Newfoundland, a Paleozoic mesothermal lode-gold occurrence in the Northern Appalachians. *Canadian Journal of Earth Sciences,* **30**, 1532–1546.

PEREIRA, Z., SAEZ, R., PONS, J.M., OLIVEIRA, J.T. & MORENO, C. 1996. Edad Devonica (Struniense) de las mineralizaciones de Aznalcóllar (Faja Pirítica Ibérica) en base a palinologia. *Geogaceta,* **20**, 1609–1612.

PIN, C., PAQUETTE, J.L. & FONSECA, P. 1999. 350 Ma (U-Pb zircon) igneous emplacement age and Sr-Nd isotopic study of the Beja gabbroic complex (S Portugal). *In: XV Reunión de Geología del Oeste Peninsular, Extended Abstracts, Diputación de Badajoz,* 219–222.

QUESADA, C. 1983. El carbonífero de Sierra Morena. *In:* MARTINEZ, C. (eds) *El carbonífero de Sierra Morena.* Instituto Geológico Minero España, Madrid, 243–278.

QUESADA, C. 1992. Evolución tectónica del Macizo Ibérico (una historia de crecimiento por acrecencia sucesiva de terrenos durante el Proterozoico Superior y Paleozoico. *In:* GUTIERREZ MARCO, J.C., SAAVEDRA, J. & RÁBANO, I. (eds) *Paleozoico Inferior de Ibero-América.* Universidad de Extremadura, 173–190.

QUESADA, C. 1996. Estructura del sector español de la Faja Pirítica: implicaciones para la exploración de yacimientos. *Boletin Geológico Minero,* **107**, 65–78.

QUESADA, C. 1998. A reappraisal of the structure of the Spanish segment of the Iberian Pyrite Belt. *Mineralium Deposita,* **33**, 31–44.

QUESADA, C., BELLIDO, F., DALLMEYER, R.D., GIL IBARGUCHI, I., OLIVEIRA, T.J., PÉREZ ESTAÚN, A. & RIBEIRO, A. 1991. Terranes within the Iberian Massif: Correlations with West Africa sequences. *In:* DALLMEYER, R.D. & LECORCHÉ, P.P. (eds) *The West African orogens and circum-Atlantic correlations.* Springer Verlag, Berlin, 136–153.

QUESADA, C., FONSECA, P.E., MUNHA, J., OLIVEIRA, J.T. & RIBEIRO, A. 1994. The Beja-Acebuches ophiolite (Southern Iberia Variscan fold belt): Geological characterization and geodynamic significance. *Boletín Geológico Minero,* **105**, 3–49.

RAMBAUD, F. 1978. Distribución de focos volcánicos y yacimientos en la banda pirítica de Huelva. *Boletín Geológico Minero,* **99**, 223–233.

RELVAS, J.M.R.S. 2000. *Geology and metallogenesis at the Neves Corvo deposit, Portugal.* PhD Thesis, University of Lisbon.

RELVAS, J.M.R.S., BARRIGA, F.J.A.S., FERREIRA, A. & NOIVA, P.C. 2000. Ore Geology, Hydrothermal Alteration and Replacement Ore-Forming Mechanisms in the Corvo Orebody, Neves Corvo, Portugal. *In:* GEMMELL, J.B. & PONGRATZ, J. (eds) *Volcanic environments and massive sulfide deposits.* Program and Abstracts, CODES, 167–168.

RELVAS, J.M.R.S., BARRIGA, F.J.A.S., FERREIRA, A., NOIVA, P.C. & CARVALHO, P. 1997. *Footwall alteration and stringer ores, Corvo orebody, Neves Corvo, Portugal.* Neves Corvo, Field Conference Abstracts, Society of Economic Geologists, Lisbon.

RELVAS, J.M.R.S., TASSINARI, C.C.G., MUNHÁ, J. & BARRIGA, F.J.A.S. 2001. Multi-sourced ore-forming fluids in the Neves Corvo VHMS deposit of the Iberian Pyrite Belt (Portugal): Strontium, Neodymium and Lead isotope evidence. *Mineralium Deposita,* **36**, 416–427.

REUTTER, K.J., SCHEUBER, E. & CHONG, G. 1996. The Precordilleran fault system of Chuquicamata, northern Chile; evidence for reversals along arc-parallel strike-slip faults. *Tectonophysics,* **259**, 213–228.

RIBEIRO, A. 1981. A geotraverse through the Variscan Fold Belt in Portugal. *Geologie en Mijnbow,* **60**, 41–44.

RIBEIRO, A. & SILVA, J.B. 1983. Structure of the south Portuguese zone. *In:* LEMOS DE SOUSA, M.J. & OLIVEIRA, J.T. (eds) *The Carboniferous of Portugal.* Memórias dos Serviços Geológicos Portugueses, **29**, 83–890.

ROBARDET, M. & GUTIERREZ MARCO, J.C. 1990. Ossa Morena Zone: Stratigraphy, Passive Margin Phase (Ordovician- Silurian-Devonian). *In:* MARTINEZ, E. & DALLMEYER, R.D. (eds) *Pre-Mesozoic Geology*

of Iberia. Springer Verlag, Berlin, 267–272.

ROUTHIER, P., AYE, F., BOYER, C., LÉCOLLE, M., MOLIÉRE, P., PICOT, P. & ROGER, G. 1978. La ceinture sud-ibérique à amas sulfurés dans sa partie espagnole médiane. Tableau géologique et metallogénique. Synthèse sur le type amas sulfurés volcano-sédimentaires. *In:* 26th *International Geological Congress, Paris.* Mémorie BRGM, 000–001.

SÁEZ, R., ALMODÓVAR, G.R. & PASCUAL, E. 1996. Geological constraints on massive sulfide genesis in the Iberian Pyrite Belt. *Ore Geology Reviews,* **11**, 429–451.

SÁEZ, R., PASCUAL, F., TOSCANO, M. & ALMODÓVAR, J.R. 1999. The Iberian type of volcano-sedimentary massive sulfide deposits. *Mineralium Deposita,* **34**, 549–570.

SÁEZ, R., REQUENA, A., FERNÁNDEZCALIANI, J.C. & ALMODÓVAR, G.R. 1988. Control estructural de las mineralizaciones de Sn-W-As del bajo Corumbel, La Palma del Condado, Huelva. *Studia Geologica Salmanticensia, Special Volume,* **4**, 189–204.

SANABRIA, R. 2001. *Caracterización y evolución de las mineralizaciones ferríferas y de los skarns del Coto Minero San Guillermo, Jerez de los Caballeros (Badajoz).* Masters Thesis, Universidad Complutense, Madrid, **97**.

SÁNCHEZ CARRETERO, R., EGUÍLUZ, L., PASCUAL, E. & CARRACEDO, M. 1990. Ossa Morena Zone: Igneous rocks. *In:* MARTINEZ, E. & DALLMEYER, R.D. (eds) *Pre-Mesozoic Geology of Iberia.* Springer Verlag, Berlin, 292–313.

SÁNCHEZ-ESPAÑA, J., VELASCO, F. & YUSTA, I. 2000. Hydrothermal alteration of felsic volcanic rocks associated with massive sulphide deposition in the northern Iberian Pyrite Belt (SW Spain). *Applied Geochemistry,* **15**, 1265–1290.

SANDERSON, D.J., ROBERTS, S., McGOWAN, A. & GUMIEL, P. 1991. Hercynian transpressional tectonics at the southern margin of the Central Iberian Zone, west Spain. *Journal of the Geological Society, London,* **148**, 893–898.

SCHERMERHORN, L.J.G. 1970. The deposition of volcanics and pyrite in the Iberian Pyrite Belt. *Mineralium Deposita,* **5**, 273–279.

SCHERMERHORN, L.J.G. 1971. An outline stratigraphy of the Iberian Pyrite Belt. *Boletín Geológico Minero,* **82**, 239–268.

SCHERMERHORN, L.J.G. & STANTON, W.I. 1969. Folded overthrusts at Aljustrel (South Portugal). *Geological Magazine,* **106**, 130–141.

SCHÜTZ, W., EBNETH, J. & MEYER, K.D. 1987. Trondjemites, tonalites and diorites in the South Portuguese Zone and their relations to the volcanites and mineral deposits of the Iberian Pyrite Belt. *Geologische Rundschau,* **76**, 201–212.

SILVA, J.B., OLIVEIRA, J.T. & RIBEIRO, A. 1990. Structural outline of the South Portuguese Zone. *In:* MARTINEZ, E. & DALLMEYER, R.D. (eds) *Pre-Mesozoic Geology of Iberia.* Springer Verlag, Berlin, 348–362.

SIMANCAS, F. 1983. *Geología de la extremidad oriental de la Zona Sudportuguesa.* PhD Thesis, University of Granada.

STEIN, G., THIÉBLEMONT, D. & LEISTEL, J.M. 1996.

Relations volcanisme/plutonisme dans la Ceinture Piriteuse Ibérique, secteur de Campofrío, Espagne. *Comptes Rendus de l'Academie des Sciences Paris,* **322**, 1021–1028.

THIÉBLEMONT, D., PASCUAL, E. & STEIN, G. 1998. Magmatism in the Iberian Pyrite Belt: petrological constraints in a metallogenic model. *Mineralium Deposita,* **33**, 98–110.

TOMLINSON, A.J. 1994. Relaciones entre el porfido cuprifero y la falla inversa de mina de Portrerillos; un caso de intrusion sintectonicas. *In: Actas VII Congreso Geologico Chileno,* **2**, 1629–1633.

TORNOS, F. 2000. Styles of mineralisation and mechanisms of ore deposition in massive sulfides of the Iberian Pyrite Belt. *In:* GEMMELL, J.B. & PONGRATZ, J. (eds) *Volcanic environments and massive sulfide deposits.* Program and Abstracts. CODES, 211–212.

TORNOS, F. & LOCUTURA, J. 1989. Mineralizaciones epitermales de Hg en Ossa Morena (Usagre, Badajoz). *Boletín Sociedad Española Mineralogía,* **12**, 363–374.

TORNOS, F. & SPIRO, B. 1999. The genesis of shale-hosted massive sulphides in the Iberian Pyrite Belt. *In:* STANLEY, C.J. *ET AL.* (eds) *Mineral Deposits: Processes to Processing.* Balkema, Rotterdam, 605–608.

TORNOS, F., BAEZA, L. & GALINDO, C. 2000. The Nava-Paredon orebody (SW Spain): a replacement massive sulfide in a shallow subaqueous setting. *In:* GEMMELL, J.B. & PONGRATZ, J. (eds) *Volcanic environments and massive sulfide deposits.* Program and Abstracts, CODES, 215–216.

TORNOS, F., CASQUET, C., GALINDO, C., VELASCO, F. & CANALES, A. 2001. A new style of Ni-Cu mineralisation related to magmatic breccia pipes in a transpressional magmatic arc, Aguablanca, Spain.

Mineralium Deposita, **36**, 700–706.

TORNOS, F., GONZÁLEZ CLAVIJO, E. & SPIRO, B.F. 1998. The Filon Norte orebody (Tharsis, Iberian Pyrite Belt): a proximal low-temperature shale-hosted massive sulphide in a thin-skinned tectonic belt. *Mineralium Deposita,* **33**, 150–169.

TOSCANO, M., RUIZ DE ALMODOVAR, G., PASCUAL, E. & SÁEZ, R. 1993. Hydrothermal alteration related to the Masa Valverde massive sulphide deposit, Iberian Pyrite Belt, Spain. *In:* FENOLL, P., TORRES, J. & GERVILLA, F. (eds) *Current Research in geology applied to ore deposits.* AND CITY, 389–392.

TOSDAL, R.M. & NUTT, C.J. 1998. Localization of sedimentary-rock-hosted Au deposits of the Carlin Trend, Nevada, along an Eocene accommodation zone. Abstracts with Programs. *Geological Society of America,* **30**, 372.

UNRUG, R., HARANCZYK, C. & CHOCYK, J.M. 1999. Easternmost Avalonian and Armorican-Cadomian terranes of Central Europe and Caledonian-Variscan evolution of the polydeformed Krakow mobile belt; geological constraints. *Tectonophysics,* **302**, 133–157.

VAUCHEZ, A. 1975. Tectoniques tangeantielles superposées dans le segment hercynien Sud-Ibérique: Les nappes et plis couchés de la region dÁlconchel-Fregenal de la Sierra (Badajoz). *Boletín Geologico Minero,* **86**, 573–580.

VELASCO, F. & AMIGÓ, J.M. 1981. Mineralogy and origin of the skarn from Cala (Huelva, Spain). *Economic Geology,* **76**, 719–727.

VELASCO, F., SÁNCHEZ-ESPAÑA, J., BOYCE, A.J., FALLICK, A.E., SÁEZ, R. & ALMODOVAR, G.R. 1998. A new sulphur isotopic study of some Iberian Pyrite Belt deposits: evidence of a textural control on sulphur isotope composition. *Mineralium Deposita,* **34**, 4–18.

Permo-Mesozoic multiple fluid flow and ore deposits in Sardinia: a comparison with post-Variscan mineralization of Western Europe

M. BONI[1], P. MUCHEZ[2] & J. SCHNEIDER[3]

[1]Dipartimento di Geofisica e Vulcanologia, Univ. di Napoli Federico II, Via Mezzocannone 8,
I-80134 Napoli, Italy (e-mail: boni@unina.it)
[2]Fysico-chemische Geologie, K.U.Leuven, Celestijnenlaan 200C, B-3001 Heverlee, Belgium
[3]Institut für Geowissenschaften und Lithosphärenforschung, JLU Giessen, Senckenbergstr. 3,
D-35390 Giessen, Germany

Abstract: The post-Variscan hydrothermal activity and mineralization in Sardinia (Italy) is reviewed in the framework of the geological and metallogenic evolution of Western Europe. The deposits can be grouped into (a) skarn, (b) high- to low-temperature veins and (c) low-temperature palaeokarst. The structural, stratigraphical and geochemical data are discussed. The results suggest three hydrologically, spatially, and possibly temporally, distinct fluid systems. *System 1* (precipitating skarn and high-temperature veins) is characterized by magmatic and/or (?) magmatically heated, meteoric fluids of low-salinity. The source of metals was in the Variscan magmatites, or in the Palaeozoic/Precambrian basement. *System 2* (low-temperature veins and palaeokarst) is represented by highly saline, Ca-rich (formation or modified meteoric) fluids. Sources of the metals were the pre-Variscan ores and carbonate rocks. *System 3* is characterized by low-temperature, low-salinity fluids of meteoric origin.

The hydrothermal deposits related to Systems 1 and 2 can be framed in a 'crustal-scale hydrothermal palaeofield', characterizing most of the post-orogenic mineralization in Variscan regions of Western and Southern Europe, allowing for local age differences of each single ore district and background effects. The suggested timing for the hydrothermal events in Sardinia is: (1) Mid-Permian (270 Ma), (2) Triassic–Jurassic. It is suggested that the Mesozoic events were related to the onset of Tethys spreading.

The post-Variscan geological evolution of Europe is characterized by numerous magmatic and hydrothermal mineralizing events ranging in age from the end of orogenic compression to the onset of Tethys spreading. Their actual timing and peculiar characteristics are distinctly different from one area to another throughout Europe, but some general features appear to be common to most of them.

The main subject of this paper is the multiple hydrothermal activity and mineralization in Sardinia (Italy). This review is based on the extensive structural, stratigraphical, geochemical and microthermometric data available for this area. The reconstructed evolution of post-Variscan ore deposition in Sardinia is compared to related neighbouring areas, and discussed within the framework of the post-Variscan geological and metallogenic evolution of Western Europe. Therefore, a summary of the most relevant characteristics of the late and post-orogenic hydrothermal activity in Germany, France, Spain, England, Belgium and Switzerland is also presented. From these data a broad overview can be gained of the major fluid flow and mineralization events occurring in Europe in the time span between Permian and Mesozoic.

Following the conceptual approach of 'crustal scale hydrothermal palaeofields', proposed by Groves et al. (1998) for Archaean orogenic gold deposits, and recently rejuvenated by Bouchot et al. (2000) for late Variscan ores of the French Massif Central, we have tried to interpret the post-Variscan deposits of Sardinia in the much broader frame of a European perspective.

Geological setting and ore deposits

Sardinia belongs to the Gondwana-derived Iberian–Armorican microplate assemblage (e.g. Crowley et al. 2000). Along with the neighbouring Corsica, it forms a small segment of the southern flank of the Variscan orogen (Arthaud & Matte 1977), linked to the 'Armorica' fold zone by a narrow suture (Carmignani et al. 1994). The Sardinian Palaeozoic basement generally shows strong tectono-stratigraphical and metallogenic

From: BLUNDELL, D.J., NEUBAUER, F. & VON QUADT, A. (eds) 2002. *The Timing and Location of Major Ore Deposits in an Evolving Orogen*. Geological Society, London, Special Publications, **204**, 199–211.
0305-8719/02/$15.00 © The Geological Society of London 2002.

analogies with other areas of the European Variscan belt.

In the Iglesiente–Sulcis area (Fig. 1), considered as the external zone of the 'Sardinian' Variscides, the geology is largely dominated by Palaeozoic lithotypes, of sedimentary as well of igneous origin. Cambrian sedimentary rocks are prevailing: part of the succession, namely the Gonnesa Group carbonates, host the largest Mississippi Valley Type (MVT) deposits in Italy, exploited until a few years ago. The metals for the stratabound ores have originated from a crustal source (Bechstädt & Boni 1994). The lead belongs to the same isotopic province as that occurring in the mineral

deposits of the southern Alps, Austro-Alpine nappes, southern France and Spain (Arribas & Tosdal 1994).

The incomplete Palaeozoic succession, spanning in age from the early Cambrian to the Devonian and Carboniferous, underwent at least two compressional phases of deformation and one extensional period, followed by late Variscan granite intrusions and basement uplift. Strong ductile deformation is widespread in most lithologies, with the exception of early diagenetic dolomites and coarse siliciclastics that generally show rather brittle deformation. The 'Southern Nappe Zone' in southeastern Sardinia (Carmignani *et al.* 1994), a mainly siliciclastic

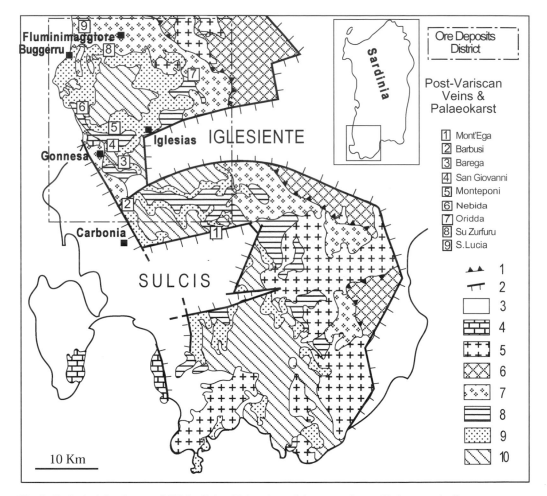

Fig. 1. Geological sketch map of SW Sardinia with locations of the economic post-Variscan ore bodies. Abbreviations: 1, overthrust; 2, normal fault; 3, Cenozoic; 4, Mesozoic; 5, Variscan granites; 6, Palaeozoic (allochthonous units); 7, Ordovician to Devonian succession; 8, Iglesias Group (Middle Cambrian–Lower Ordovician); 9, Gonnesa Group (Lower Cambrian); 10, Nebida Group (Lower Cambrian) (modified from Bechstädt & Boni 1994).

succession of Upper Cambrian to Carboniferous age, shows higher metamorphic grade (greenschist facies), also followed by the emplacement of granite batholiths.

During the late Palaeozoic and Mesozoic (Proto-Tethys to early Alpine Tethys stages), the Sardinia island (part of Sardinia–Corsica Massif *sensu* Ziegler 1988) evolved as part of the southern margin of the European plate, where repeated syn-sedimentary tectonic activity, such as strike-slip movements combined with sedimentation in discontinuous pull-apart basins, have been detected over the entire time span (Rau 1990; Vai 1991). Shear-zone tectonics, possibly related to late Variscan basement uplift, have also been described (Carosi *et al.* 1992).

Widespread calc-alkaline granitoid bodies are attributed both to syn- and post-collisional Variscan stages, the leucogranites marking the end of the sequence (from 330 to 290 Ma, late Carboniferous to Permian; Del Moro *et al.* 1975; Guasparri *et al.* 1984; Secchi *et al.* 1991; Boni *et al.* 1999). The emplacement of plutonic bodies was related to several transtensional phases, which took place during the late Variscan evolution of the Palaeozoic basement (Secchi & D'Antonio 1996).

The post-Variscan sedimentary record, though poor, starts with a clear trend towards crustal attenuation and fragmentation during the late Carboniferous to Permian, possibly related to a large transcurrent mega-shear zone (Cassinis *et al.* 1999). This resulted in the development of continental basins containing coeval magmatic products of still calc-alkaline affinity. Porphyry stocks, ignimbrite flows, rhyodacitic lavas, and basaltic dykes occur throughout the island within a poorly defined age interval between 280 and 250 Ma (Lombardi *et al.* 1974; Beccaluva *et al.* 1981; Edel *et al.* 1981; Cozzupoli *et al.* 1984; Atzori & Traversa 1986; Vaccaro *et al.* 1991). The large spread of age data may result either from widespread hydrothermal alteration of the magmatites that disturbed the isotopic systems, from imprecise and/or inaccurate dating techniques, or a combination of both.

The Mesozoic successions, though incomplete and sporadic, point to further crustal thinning (a prelude to Alpidic rifting), as demonstrated by Middle to Upper Triassic marine sediments. The magmatic counterparts of this extensional period were identified as sub-alkaline and alkaline dykes (Atzori & Traversa 1986) with a mantle Sr signature, occurring mostly in the northern part of the island, and dated by Vaccaro *et al.* (1991) at about 230 Ma, using Ar/Ar on biotites. The condensed Triassic sequences extend upwards into thicker Jurassic and Cretaceous carbonates (mostly in northern and eastern Sardinia).

The sedimentary and magmatic setting in Sardinia may be interpreted in the broader evolutionary frame of a stepwise transition from Variscan orogenic regimes (Permo-Carboniferous) to post-orogenic, Mesozoic (Triassic–Jurassic) extensional conditions.

As reported by Boni *et al.* (1992, 1999, 2000*a*), several distinct hydrothermal systems were probably active in SW Sardinia, not only related to post-kinematic magmatism (skarn and retrograde contact metamorphism), but also to other periods of post-orogenic crustal extension during the Permian and Mesozoic. This resulted in a variety of hydrothermal products, including economic ore deposits.

Post-Variscan ore deposits

The main formation of the previously called 'Permo–Triassic' ores (Boni 1985) and related phenomena, such as brecciation, dolomitization, and silicification, occurred within a time span between the emplacement of leucogranite intrusions and the deposition of the Early Tertiary coal-bearing lagoonal carbonates (Sulcis Basin), which are not affected by hydrothermal circulation. Most probably, their deposition was restricted to the lower part of the Mesozoic (Boni *et al.* 1992). The orebodies are mainly hosted by carbonate rocks of the Lower Cambrian Gonnesa Group in the Iglesiente region (Fig. 1). Some veins, however, extending a few kilometres in length, occur in Cambro–Ordovician clastic lithologies of the Arburese allochthonous unit (Montevecchio–Ingurtosu Zn–Pb>Cu vein field; Salvadori & Zuffardi 1973), and in volcaniclastic sequences of the southeastern 'External Nappes' (Pb–Ag 'Filone Argentifero del Sarrabus', Ba–F veins at Silius; Masi *et al.* 1975; Natale 1969). Neither of the localities mentioned above are shown on the map of Figure 1.

On the basis of geological, petrological and geochemical characteristics, the deposits can be grouped into (a) skarn ores, (b) high- to low-temperature veins, followed by pervasive dolomitization and (c) low-temperature palaeokarst ores (Boni *et al.* 1992, 2000*a, b*). Most of the skarn occurrences in SW Sardinia are associated with different kinds of leucogranites (Boni *et al.* 1999) that have been pervasively altered by several pulses of hydrothermal fluids. Retrograde boiling, sudden pressure and rapid temperature changes due to fault activity related to granite intrusion at shallow depths were invoked to explain the alteration phenomena (Guasparri *et al.* 1984; Secchi *et al.* 1991). The mineralized skarns are generally calcic exoskarns of Zn–Pb(–Cu) type. They may occur near the contacts to the intrusive bodies and at distance, as well as in distal fissures within carbonate rocks. The concentrations are bound to stratigraphic contacts between different lithologies and to fault zones. The Lower Palaeozoic stratabound ores were also involved to various degrees in the contact-metamorphic processes. Some of the mineralized veins occurring in Sardinia, even if not directly connected to any nearby exposed intrusive body, may be derived from high temperature hydrothermal processes similar to those generally occurring in the latest stages of skarn evolution (Boni *et al.* 1992). In fact, these veins may be considered as distal equivalents of a fully evolved skarn suite, thus also potentially related to leucogranitic intrusions. The only time constraint is given by the observation that some veins occur along the same fractures where late Variscan porphyry dykes had been previously emplaced (e.g. S. Lucia mine, Buggerru; Bakos & Valera 1972).

Low-temperature veins are more common in Sardinia. With few exceptions, they are hosted by Lower Cambrian carbonate rocks. These ores consist chiefly of barite and argentiferous galena with up to 0.8% Ag in galena concentrates. Calcite, quartz and Fe-rich dolomite are the principal gangue minerals, whereas fluorite and sphalerite form subordinate vein fillings observed in a few occurrences. The veins are aligned parallel to late to post-Variscan fault systems (Arthaud & Matte 1975, 1977). Common strike directions are: north–south, N30°E, and its conjugate N30°W, N60–70°E, and east–west (Valera 1967).

Comparable mineral associations, often spatially related to nearby veins, have been observed in extensive networks of palaeokarst cavities in the SW of the island, which deeply penetrate the repeatedly uplifted Cambrian carbonate blocks (Boni 1985; Boni & Amstutz 1982). This ore type occurs mainly as a cement of polygenetic collapse breccias and rarely replaces both cement and matrix of internal sediments. Palaeokarst mineralization ('Ricchi Argento' orebodies) is often accompanied by silicification as well as Fe-dolomitization ('yellow dolomite'), mainly occurring below old peneplanation surfaces and following horizontal geometries. The palaeokarst ores show concretionary textures and collapse breccias, cemented by several generations of calcite, quartz, barite and Ag-rich galena. Silver is contained in several sulphosalts (mostly freibergite) in the galena. The similar mineralogical and geochemical characteristics of the low-temperature veins and of mineralized palaeokarst, suggest a genetic link between the two types of occurrences (Boni 1986).

Epigenetic replacive dolomitization (saddle dolomite), frequently characterized by zebra textures ('geodic dolomite'), affected Cambrian limestones and early diagenetic dolomites over large areas of the Iglesiente–Sulcis former mining district. This late dolomitization crops out over an area of more than 500 km^2 and reaches a thickness of more than 600 m. It has been considered to be also of post-Variscan age, because of cross-cutting relationships with sedimentary and tectonic structures (Boni *et al.* 2000*a*, *b*). The 'geodic dolomite' clearly predates the above-mentioned vein-type and palaeokarst Ba and Pb–Ag ores in carbonate rocks (Bechstädt & Boni 1994).

Geochemical data

Isotopes

Isotopic data are mainly available from mineralization in the southwestern part of Sardinia, where a coherent geological and ore deposit model could be established (Bechstädt & Boni 1994; Boni *et al.* 1992, 1996).

Strontium isotopes. $^{87}Sr/^{86}Sr$ ratios of barites and calcites from veins and cavity fillings related to the palaeokarst form two distinct groups. In the skarn and in vein deposits near to granites, the values range from 0.7110 to 0.7140, whereas the low-temperature vein deposits display ratios between 0.7094 and 0.7115 (Cortecci *et al.* 1989; Boni *et al.* 1992). Sr isotopic signatures of hydrothermal dolomites (0.7088–0.7093) and associated calcites (0.7101) are between the fields of low-temperature veins (Boni *et al.* 1992), Cambrian carbonates (0.7088–0.7090, Boni *et al.* 1996), and stratabound barites (0.7086–0.7090, Boni *et al.* 1996) (Fig. 2).

Lead isotopes. Pb-isotope ratios of galenas from skarn and vein deposits near granites and from low temperature vein and palaeokarst deposits, form two distinct groups. Group (a) is characterized by $^{206}Pb/^{204}Pb = 18.01–18.29$ and $^{208}Pb/^{204}Pb = 38.17–38.47$, whereas group (b) yields values of $^{206}Pb/^{204}Pb = 17.86–18.07$ and $^{208}Pb/^{204}Pb = 37.95–38.20$, respectively (Swainbank *et al.* 1982; Ludwig *et al.* 1989; Boni & Köppel 1985). Pb-ratios of galenas from group (a) are similar to the values measured in feldspars of Variscan granites (Fig. 2).

Pb isotope ratios of hydrothermal dolomites and internal, unmineralized sediments in the palaeokarst are similar to those of the galena lead of low-temperature veins. The values of group (b) are near to the isotopic compositions of the Cambrian stratabound ore-lead ($^{206}Pb/^{204}Pb =$

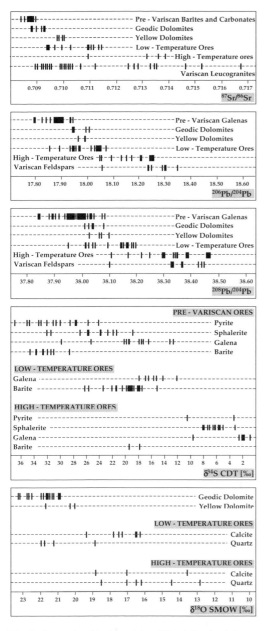

Fig. 2. Synopsis of isotope data of Cambrian-hosted stratabound ores and post-Variscan high- and low-temperature mineralization as well as hydrothermal dolomites from SW Sardinia (modified from Boni *et al.* 1992).

$17.80-17.95,^{208}Pb/^{204}Pb = 37.84-38.05$, Boni & Köppel 1985). The Pb isotopic signatures of the post-Variscan high- and low-temperature ore types indicate mixtures of Cambrian ore lead and

Variscan lead derived either from Lower Palaeozoic clastic sediments and/or from Variscan granites. The high temperature ores (b) are dominated by this 'Variscan' component, whereas the low temperature ores (a) and the hydrothermal dolomites are greatly influenced by the 'Cambrian' component (Boni *et al.* 1992; Fig. 2).

Sulphur isotopes. The $\delta^{34}S$ values of sulphides from skarn and high temperature veins (e.g. Mont'Ega, Montevecchio etc.) range between +1 and +11‰ CDT (Jensen & Dessau 1966; Cortecci *et al.* 1989; Boni *et al.* 1992); those from low-temperature veins and palaeokarst ores range from +12 to +18‰ CDT. Only a few data are available for barites occurring in upper parts of the high-temperature veins (Fig. 2). Their $\delta^{34}S$ values are isotopically indistinguishable from those measured in the low-temperature deposits (between +15.3 and +26.4‰ CDT, Cortecci *et al.* 1989; Boni *et al.* 1992). Since the $\delta^{34}S$ values of Cambrian galenas partly overlap with those of Variscan barites, Cortecci *et al.* (1989) suggested that the latter may have originated from the oxidation of Cambrian-hosted stratabound sulphides.

Oxygen and carbon isotopes. Oxygen and carbon isotope data on the gangue minerals have been published by De Vivo *et al.* (1987), Fontboté & Gorzawski (1987), Cortecci *et al.* (1989) and Boni *et al.* (1988, 1992, 2000*a*) (Fig. 2). It should be emphasized that no specific or comprehensive study on oxygen isotopes of the post-Variscan ores as a category is available. Generally, the $\delta^{13}C$ values are in the same range as those of the hosting Cambrian limestones and early diagenetic dolomites (between −2 and +1‰ PDB).

In high-temperature vein quartz, the measured $\delta^{18}O$ values range from +14.6‰ to +18.5‰ (SMOW). One quartz sample from a skarn deposit yielded $\delta^{18}O$ values of about +13‰. This compares well with the values of +13.0‰ to +17.2‰ reported for quartz from the 'Silver Vein' of Sarrabus in SE Sardinia, the origin of which was also related to post-Variscan, higher temperature hydrothermal activities (Masi *et al.* 1975). Quartz from low temperature veins and palaeokarst, on the contrary, shows higher $\delta^{18}O$ values, i.e. between +18.8‰ and +22‰ (SMOW).

The $\delta^{18}O$ values of calcites vary widely between +13.6 and +20.7 ‰ SMOW (De Vivo *et al.* 1987;Boni & Iannace 1990), the low temperature calcites being the heaviest ones (Fig. 2). $\delta^{18}O$ values of hydrothermal dolomite vary from +20.2‰ to +23.4‰ (Fontboté & Gorzawsky 1987; Boni *et al.* 1992, 2000*a*). These values are different from those of early diagenetic Cambrian dolomites (+23.5 to +27.5‰, Boni *et al.* 1988).

Their relative uniformity suggests a rapid and effective fluid circulation process in terms of temperature and/or of water/rock ratio (Boni & Iannace 1990). Oxygen isotope compositions of post-Variscan barites range between +8.2 and +12.5‰ (Cortecci *et al.* 1989; Boni *et al.* 1992). Cortecci *et al.* (1989) interpreted these values as a result of oxidation of Cambrian sulphides in the presence of moderately warm water.

Fluid inclusion microthermometry

Fluid inclusion studies of post-Variscan ore and gangue minerals in SW Sardinia have been published by Valera (1974), Boni (1986), De Vivo *et al.* (1987), Cortecci *et al.* (1989), and Boni *et al.* (1990, 1992, 2000a).

The textural and structural features of the leucogranite intrusions in SW Sardinia, with their skarn assemblage, indicate that the entire suite was emplaced at shallow level in the crust as a consequence of the lower water content of the original magmas. Their crystallization is thought to have occurred at pressures of <1 kbar, corresponding to depths of less than 3.5 km (Guasparri *et al.* 1984). Subsequent retrograde metamorphism paired with hydrothermal circulation replaced early segregated skarn minerals by epidote and chlorites and deposited a wide range of ore minerals. Fluid inclusions characteristics of epidote, quartz, fluorite, and calcite in retrograde skarn assemblages and high-temperature vein deposits reveal an early $H_2O-NaCl$ type fluid (fluid 1) with a maximum T_h of 350 °C and a salinity of less than 10 wt% NaCl eq. (Boni *et al.* 1990) (see Fig. 4). Values higher than 200 °C, however, have been measured only in the skarn.

A second $H_2O-NaCl-CaCl_2$ fluid (fluid 2) with higher salinities (above 20 wt% NaCl eq.) and lower T_h (<140 °C) was responsible for most of the low-temperature vein and palaeokarst ores (Boni 1986) (Fig. 3). These homogenization (T_h) values represent minimum temperatures of trapped fluid. However, independent geological constraints in SW Sardinia support also that the pressure correction in the case of the low-temperature veins and palaeokarst fillings is very small (Boni *et al.* 1992). A similar $H_2O-NaCl-CaCl_2$ fluid also caused the precipitation of the widespread hydrothermal dolomites in the same area (Boni *et al.* 2000a). Their homogenization temperatures shows a gradient with values decreasing from east to west. Higher temperatures (mean of around 100 °C) have been found in the eastern part of the Iglesiente area, whereas the lowermost temperatures (mean of about 85 °C) were determined for occurrences along the western coast. Salinities are also high (>23.3 wt% NaCl eq.). Inclusions in

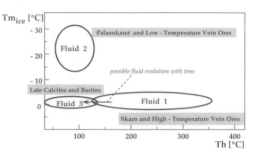

Fig. 3. Schematic diagram of homogenization versus last ice-melting temperatures from primary inclusions in epidote, quartz, fluorite, calcite and dolomite in skarn and high temperature veins, low temperature veins and palaeokarst ores and in late calcite and barite veins (modified from Boni *et al.* 1992).

calcites and barites, possibly belonging to the latest stages of low-temperature vein and palaeokarst deposits, contain a third $H_2O-NaCl$ type fluid (fluid 3), which is characterized by salinities between 0 and 1 wt. % NaCl eq. and T_h values between 70 and 130 °C (Boni *et al.* 1992).

Fluid origin during the mineralization stages

A combination of fluid inclusion and isotopic data should in principle allow the calculation of the oxygen isotopic composition of the mineralizing fluids. However, as already discussed in detail in Boni *et al.* (1992), the inhomogeneity of sampling policy for Sardinia (we have considered, at least partly, a mixed data set produced by different laboratories), the ample spread in our own measured values, and the uncertainities concerning the fractionation factors among minerals and waters, impose a more cautious approach. In fact, if we calculate the inferred absolute O-isotope values of the parent fluids for Sardinian minerals using several published equations (Clayton *et al.* 1972 or Bottinga & Javoy 1973 for quartz, O'Neil *et al.* 1969 for calcite, Northrop & Clayton 1966; Matthews & Katz 1977 and Land 1983 for dolomite), we obtain ranges of isotope values compatible with all kind of possible waters. However, Boni *et al.* (1992) showed that the sole and most consistent indication derived from the whole set of data, is that the O-isotope values of the fluid depositing the high temperature veins were on average higher than those related to low temperature mineralization, as well as to hydrothermal dolomites. The heavier values were interpreted, by comparing them with quartz from a skarn sample, as derived from a prevailing magmatic–metamorphic source. The inferred isotope composition of the fluid responsible for the main stage of low temperature mineralization can be considered compatible with either formation, or modified meteoric waters. A clear indication of meteoric waters being introduced in the system (as already suggested by De Vivo *et al.* 1987 and Boni & Iannace 1990) can be deduced from the data set measured in the latest (non mineralizing) mineral stages of the low temperature event.

Evolutionary scenario

Based on comparative isotope investigations on ores and probable metal sources, combined with the peculiar physico-chemical characteristics of the fluids, post-Variscan hydrothermal activity and mineralization in Sardinia has been assigned to three hydrologically, spatially, and possibly temporally distinct fluid systems.

System 1 (skarn and high-temperature vein deposits, 150–350 °C) was active either within the metamorphic aureoles of leucogranites or in close proximity to them. It is characterized by partly magmatic and (?)magmatically heated, meteoric fluids of relatively low-salinity. Pb and Sr isotopes of related minerals bear the imprint of a radiogenic 'Variscan' component, associated either with the late Variscan magmatites, or with the low-grade metamorphic Palaeozoic, or pre-Palaeozoic basement.

System 2 (low-temperature veins and palaeokarst, 70–150 °C) is represented by highly saline, Ca-rich fluids (meteoric modified or formation waters) that circulated over a much larger area, always independent from known intrusive bodies. The Sr, Pb, and S isotopic signatures of minerals precipitated from system 2 fluids are dominated by values similar to Cambrian stratabound ores and carbonate host rocks, with only a minor contribution of a radiogenic 'Variscan' source.

System 3 is characterized by low-temperature (70– <130 °C), low-salinity fluids, possibly of meteoric origin. Pb, Sr, and S isotope signatures of its precipitates are again similar to those of the Cambrian host rocks. This system, which may still be active today, is spatially superimposed on products of systems 1 and 2. The economic mineralization potential of system 3 in Sardinia was almost nil.

The existing data do not allow the timing and duration of hydrothermal activity in Sardinia to be precisely constrained. In particular, it is not known whether the corresponding fluid circulations are the expression of only one long-lasting tectono-thermal event over the time span between the end of the Variscan orogeny and the onset of the Alpine cycle, or are the result of temporally and genetically distinct thermal pulses. However, the reappraisal of the geochemical data on post-Variscan mineralization, combined with the reconstructed geological evolution of Sardinia, enables the reconstruction of a broad evolutionary scenario of hydrothermal circulation and base metal mineralization.

One early and important episode of hydrothermal activity is recorded by the minerals from veins and granite alteration assemblages in SW Sardinia, some of which have been recently investigated by the $^{40}Ar/^{39}Ar$ method (Boni *et al.* 1999). The effects of hydrothermal overprinting on magmatic biotites and feldspars as well as the age of newly formed Ba silicates was studied. The $^{40}Ar/^{39}Ar$ data of this first episode correspond to an age of about 270 Ma, which postdates the youngest late Variscan granitoid intrusions by at least 30 Ma. At that time the shallow intrusives had no residual heat to drive hydrothermal mineralizing systems. This same age has been reported for several calc-alkaline magmatic dykes throughout Sardinia (Vaccaro *et al.* 1991) and records an Early to Mid-Permian period of crustal extension and wrench faulting (Arthaud & Matte 1977). This Permian hydrothermal phase may also have produced, besides Ba-silicates, the high-temperature/low-salinity veins occurring near Variscan granites in SW Sardinia, which are characterized by isotopically radiogenic ore and gangue minerals. The structural pattern related to magmatic intrusions was possibly reactivated due to extensive wrenching, thus allowing a deeply reaching circulation of hydrothermal fluids, strongly interacting with basement rocks and acquiring a radiogenic character.

The other prominent hydrothermal stage(s) can only be narrowed to the time interval between the late Permian and the late Mesozoic. No unambiguous age determination of this (or these) younger phase(s) is available. The $^{40}Ar/^{39}Ar$ results discussed above demonstrate a very complex superposition of several hydrothermal alteration events. However, an age of 230 Ma has been determined in a few magmatic dykes by Vaccaro *et al.* (1991), and interpreted as evidence for early Alpidic rifting in Sardinia. The low-temperature/high-salinity veins and palaeokarst deposits, containing isotopically unradiogenic mineral phases, may be related to this rifting phase. Similar fluids, but indicative of a much larger fluid flow controlled by Variscan foliation and cleavage planes, were responsible for pervasive hydrothermal dolomitization, replacing extensive areas of Lower Palaeozoic carbonates in Sardinia (Boni *et al.* 2000a, b). However, only a relative age of this epigenetic dolomitization may be inferred by cross-cutting relationships to younger Pb–Ag–Ba veins.

Sardinia and the European perspective

The Permian and Mesozoic geotectonic evolution of Europe is characterized by the fragmentation and break-up of the Pangea supercontinent, which had been consolidated by collision of Gondwana and Laurasia in the course of the Variscan orogeny. Crustal extension started already in the Permian and resulted in a renewed global plate reorganization from the early Mesozoic onwards (e.g. Dietz & Holden 1970; Ziegler 1988, 1993). The significance of these processes for the generation and circulation of hydrothermal mineralizing fluid systems in Europe and North America has been discussed by several workers (e.g. Mitchell & Halliday 1976; Chauris 1978; Halliday & Mitchell 1984).

Early late to post-Variscan mineralization in the Variscan Massifs of western and southwestern Europe (Fig. 4a) is characterized by relatively high-temperature, low-salinity $H_2O-NaCl$ fluids (e.g. Boiron *et al.* 1996; Tornos *et al.* 2000). The oxygen isotopic composition of these fluids indicates a surface-derived origin (probably meteoric water), which partly equilibrated with basement rocks (Wilkinson *et al.* 1995). Related mineralization is interpreted to reflect extensive downward fluid penetration in the crust enhanced by brittle deformation (Wilkinson *et al.* 1995). Also, the high-temperature, low-salinity fluids (system 1) in Sardinia are regarded as deeply circulating fluids, strongly interacting with basement rocks. This interaction is clearly indicated by high Pb and Sr isotope ratios of hydrothermal minerals and possibly also by the ^{18}O-enriched nature of the mineralizing fluids. Although a spatial relationship is often present between these late to post-Variscan mineralization phases and granitic intrusions, a genetic link between both cannot be demonstrated. These magmatic rocks provided the migration pathways of the fluids, since they were preferentially emplaced in structurally active zones (Boiron *et al.* 1996).

Precipitation from highly saline brines is a common feature of post-Variscan mineralization throughout Europe (Darimont 1983; Behr & Gerler 1987; Charef & Sheppard 1988; Alderton & Harmon 1991; Behr *et al.* 1993; Muchez *et al.* 1994; Canals *et al.* 1999; Gleeson *et al.* 2000). In Western and Central Europe the origin of the highly saline fluids has been interpreted as relating to the expulsion of formation waters from Permian (and Triassic?) intracontinental basins (Behr & Gerler 1987) and of Palaeozoic brines from basement rocks (Heijlen *et al.* 2000, 2001). Many economic mineral deposits throughout Europe reflect an important Early to Mid-Permian hydrothermal event (Fig. 4a) that has also been proposed for Sardinia (Boni *et al.* 1999, see above). Isotopic ages of about 270 Ma have been determined for uranium deposits both in the Erzgebirge (Germany) and in the French Massif Central (U–Pb ages of pitchblende, Lancelot *et al.* 1984; Leroy & Holliger 1984; Pagel 1990; Respaut *et al.*

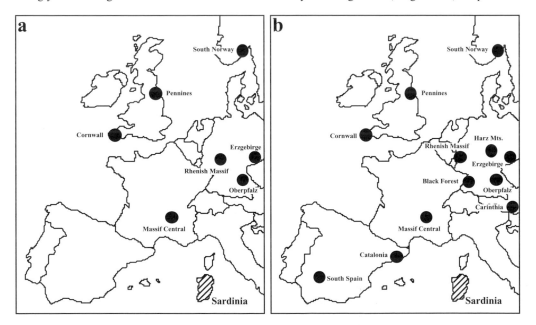

Fig. 4. Sketch map of important localities for European hydrothermal ore deposits related (**a**) to Early–Mid-Permian (260–270 Ma) and (**b**) Triassic–Jurassic (230–170 Ma) crustal extension.

1991; Förster & Haack 1995). The same age has been determined also for polymetallic veins and hydrothermal kaolinite deposits in Cornwall, England (Halliday 1980; Bray & Spooner 1983; Chen et al. 1993). Moreover, many other economic mineral veins in Germany, south Norway, northern England and the Massif Central have the same age (K–Ar, Rb–Sr and Pb/Pb ages of hydrothermal illites and feldspars, Ineson & Mitchell 1972; Ihlen et al. 1979; Ineson et al. 1979; Bonhomme et al. 1987; Zheng et al. 1991; Lippolt et al. 1985; Schneider 2000; Schneider & Haack 2000; Sm–Nd ages of fluorites, Chesley et al. 1991). These mineral occurrences reflect an early post-orogenic period of pronounced crustal extension, related to dextral wrenching and plate reorganization in the Laurasia–Gondwana realm, characterized by intense rifting and hydrothermal activity (Arthaud & Matte 1977; Dziedzic 1986; Bonin 1988; Gast 1988; Ziegler 1988).

However, Permian–Mesozoic brines also penetrated the basement (Mullis & Stalder 1987; Möller et al. 1997) and caused the formation of even more important economic deposits all over Europe. Isotopic ages of mineralization precipitated from these fluids (illites and feldspars: K–Ar and Rb–Sr; fluorites: Sm–Nd; pitchblendes: U–Pb; hematite: $(U + Th)/^4He$) mainly fall into the Triassic and Jurassic (Fig. 4b), depending on their original position within the European intraplate geometry (Joseph et al. 1973; Halliday 1980; Bonhomme et al. 1983, 1987; Halliday & Mitchell 1984; Schmitt et al. 1984; Brockamp & Zuther 1985; Carl & Dill 1985; Canals & Cardellach 1993; Haack & Lauterjung 1993; Scrivener et al. 1994; Förster & Haack 1995; Wernicke & Lippolt 1997a, b; Subias et al. 1999; Meyer et al. 2000; Schneider et al. 1999, 2001). On the base of cement stratigraphy, Zeeh & Bechstädt (1994) suggested that the famous Alpine Bleiberg-type Pb–Zn deposits are also related to extensional tectonics spanning from Late Triassic to the Jurassic.

Jurassic fluid systems have been generally interpreted as an expression of continent-wide rifting and concomitant regional fluid circulation, that reflect major tectonic disturbances and enhanced heat flow preceding the opening of the North Atlantic ocean. However, the mid- to late Triassic and early Jurassic events seem to have been related rather to the evolving Tethys and Central Atlantic margins (Bouladon & de Graciansky 1985; Schneider 2000). The low-temperature veins and palaeokarst fillings in Sardinia (System 2) typically precipitated from highly saline H_2O–NaCl–$CaCl_2$ fluids during the Mesozoic. Taking into account the geographical position of the island, a genetic relationship with the evolving Tethys margin seems most logical. In addition,

isotopic data further indicate a rock-dominated hydrothermal system, comparable to that proposed for the post-Variscan Zn–Pb deposits in Belgium (Heijlen et al. 2000, 2001).

A massive contribution of Sr-rich, concentrated sedimentary brines of Permian or Triassic age (Hanor 1987) should give a strong imprint on the Sr-isotope composition of hydrothermal minerals. More specifically, considering the Sr-isotopic composition of Permo–Triassic sea-water with $^{87}Sr/^{86}Sr = 0.707$–0.708 (Martin & Macdougal 1995), such a contribution should drastically lower the $^{87}Sr/^{86}Sr$ ratio of hydrothermal phases. The composition of the low-temperature hydrothermal products, instead, shows an opposite trend with Sr-isotope signatures between those of Lower Palaeozoic host rocks and more radiogenic sources, probably crystalline rocks.

Conclusions

The concept of a 'crustal scale hydrothermal palaeofield' seems to be applicable to the post-orogenic mineralization in Western and Southern Europe. However, local age differences and background effects resulting from the pre-Variscan geological puzzle, should be taken into account when applied to each single ore district. In addition, more reliable isotopic ages are needed to trace hypothetical crustal-scale movements of post-orogenic hydrothermal fluids. In this sense, we expect that even isotopic dating (where possible) of the different Sardinian mineralization styles will yield ages that probably coincide with the temporal peaks of post-orogenic hydrothermal activity during the Permian and Mesozoic observed for Europe as a whole (Fig. 4a, b), which resulted in the formation of many important economic mineral deposits.

This work has been partly financed with the funds allowed by Napoli University (Italy) to Maria Boni in the years 1999 and 2000. We would like to thank the IGEA Company, Campo Pisano (Ca), for permitting the underground visits to their proportion, and especially the geologist R. Sarritzu for guidance and help. Thanks are also due to A. Iannace (Napoli) and T. Bechstädt (Heidelberg) for constant discussion and advice during the interpretation of the data, as well to two unknown referees, whose comments helped to improve the quality of the paper.

References

ALDERTON, D.H.M. & HARMON, R.S. 1991. Fluid inclusion and stable isotope evidence for the origin of mineralizing fluids in south-west England. Mineralogical Magazine, 55, 605–611.

ARRIBAS, A. JR & TOSDAL, R.M. 1994. Isotopic composition of Pb in Ore Deposits of the Betic Cordillera, Spain: Origin and Relationship to other European Deposits. *Economic Geology*, **89**, 1074–1093.

ARTHAUD, F. & MATTE, P. 1975. Les décrochements tardi-hercyniens du sud-ouest de l'Europe, Géometrie et essai de réconstruction des conditions de la déformation. *Tectonophysics*, **25**, 139–171.

ARTHAUD, F. & MATTE, P. 1977. Late-Palaeozoic strike-slip faulting in Southern Europe and Northern Africa: results of a right lateral shear zone between the Appalachians and the Urals. *Geological Society of America Bulletin*, **88**, 1305–1320.

ATZORI, P. & TRAVERSA, G. 1986. Post-granitic Permo-Triassic dyke magmatism in eastern Sardinia (Sarrabus p.p., Barbagia, Mandrolisai, Goceano, Baronie and Gallura). *Periodico di Mineralogia*, **55**, 203–231.

BAKOS, F. & VALERA, R. 1972. Le mineralizzazioni fluoritiche di S. Lucia (Sardegna sudoccidentale). Atti Giornate Studio Fluoriti Italiane. *Bollettino Associazione Mineraria Subalpina*, **3**, 299–323.

BECCALUVA, L., LEONE, F., MACCIONI, L. & MACCIOTTA, G. 1981. Petrology and tectonic setting of the Palaeozoic basic rocks from Iglesiente-Sulcis (Sardinia, Italy). *Neues Jahrbuch für Geologie und Mineralogie Abhandlungen*, **140**, 184–201.

BECHSTÄDT, T. & BONI, M. (EDS) 1994. *Sedimentological, stratigraphical and ore deposits field guide of the autochthonous Cambro-Ordovician of southwestern Sardinia*. Memorie Descrittive della Carta Geologica d'Italia, Servizio Geologico d'Italia, **48**.

BEHR, H.J. & GERLER, J. 1987. Inclusions of sedimentary brines in post-Variscan mineralisations in the Federal Republic of Germany - a study by neutron activation analysis. *Chemical Geology*, **61**, 65–77.

BEHR, H.J., GERLER, J., HEIN, U.F. & REUTEL, C.J. 1993. Tectonic Brines und Basement Brines in den mitteleuropäischen Varisziden: Herkunft, metallogenetische Bedeutung und geologische Aktivität. *Göttinger Arbeiten für Geologie und Paläontologi*, **58**, 3–28.

BOIRON, M.C., CATHELINEAU, M., BANKS, D.A., YARDLEY, B.W.D., NORONHA, F. & MILLER, M.F. 1996. P-T-X conditions of late Hercynian fluid penetration and the origin of granite-hosted gold quartz veins in northwestern Iberia: A multidisciplinary study of fluid inclusions and their chemistry. *Geochimica et Cosmochimica Acta*, **60**, 43–57.

BONHOMME, M.G., BÜHMANN, D. & BESNUS, Y. 1983. Reliability of K-Ar dating of clays and silicifications associated with vein mineralizations in Western Europe. *Geologische Rundschau*, **72**, 105–117.

BONHOMME, M.G., BAUBRON, J.C. & JEBRAK, M. 1987. Minéralogie, géochimie, terres rares et âge K-Ar des argiles associées aux minéralisations filoniens. *Chemical Geology*, **65**, 321–339.

BONI, M. 1985. Les gisements de type Mississippi Valley du Sud-Ouest de la Sardaigne (Italie); une synthèse. *ChroniqueRecherches Minières BRGM*, **479**, 7–34.

BONI, M. 1986. The Permo-Triassic vein and paleokarst ores in South-West Sardinia: contribution of fluid inclusion studies to their genesis and paleoenvironment. *Mineralium Deposita*, **21**, 53–62.

BONI, M. & AMSTUTZ, G.C. 1982. The Permo-Triassic paleokarst ores of south-west Sardinia (Iglesiente-Sulcis): an attempt at a reconstruction of paleokarst conditions. *In:* AMSTUTZ, G.C., EL GORESY, A., FRENZEL, G., KLUTH, C, MOH, G., WAUSCHKUHN, A. & ZIMMERMANN, R.A. (eds) *Ore Genesis, the State of the Art*. Springer Verlag, Berlin, 73–82.

BONI, M. & IANNACE, A. 1990. *Actes Colloque International 'Mobilité et concentration des métaux de base dans les couvertures sédimentaires'. Orléans, France, 28–30 mars 1988*. Documents du BRGM, **183**, 171–186.

BONI, M. & KÖPPEL, V. 1985. Ore-lead isotope pattern from the Iglesiente-Sulcis area (SW Sardinia) and the problem of remobilization of metals. *Mineralium Deposita*, **20**, 185–193.

BONI, M., IANNACE, A. & PIERRE, C. 1988. Stable isotope compositions of Lower Cambrian Pb-Zn-Ba deposits deposits and their host carbonates, south-western Sardinia, Italy. *Chemical Geology (Isotope Geoscience Section)*, **72**, 267–282.

BONI, M., RANKIN, A. & SALVADORI, M. 1990. Fluid inclusion evidence for the development of Zn-Pb-Cu-F skarn mineralization in SW Sardinia, Italy. *Mineralogical Magazine*, **54**, 279–287.

BONI, M., IANNACE, A., KÖPPEL, V., HANSMANN, W. & FRÜH-GREEN, G. 1992. Late- to post-Hercynian hydrothermal activity and mineralization in SW Sardinia. *Economic Geology*, **87**, 2113–2137.

BONI, M., IANNACE, A. & BALASSONE, G. 1996. Base metal ores in the Lower Palaeozoic of south-western Sardinia. *Economic Geology 75th Anniversary Volume*. Society for Economic Geology, Special Publications, **4**, 18–28.

BONI, M., BALASSONE, G. & VILLA, I.M. 1999. Age and evolution of granitoids from South West Sardinia: genetic links with hydrothermal ore bodies. *In:* STANLEY, C.J. *ET AL.* (eds) *Mineral Deposits: Processes to Processing*. Balkema, Rotterdam, 1255–1258.

BONI, M., PARENTE, G., BECHSTÄDT, T., DE VIVO, B. & IANNACE, A. 2000a. Hydrothermal dolomites in SW Sardinia (Italy): evidence for a widespread late-Variscan fluid flow event. *Sedimentary Geology*, **131**, 181–200.

BONI, M., IANNACE, A., BECHSTÄDT, T. & GASPARRINI, M. 2000b. Hydrothermal dolomites in SW Sardinia (Italy) and Cantabria (NW Spain): evidence for late- to post-Variscan fluid flow events. *Journal of Geochemical Exploration*, **69-70**, 225–228.

BONIN, B. 1988. From orogenic to anorogenic environments: Evidence from associated magmatic episodes. *Schweizerische Mineralogische und Petrographische Mitteilungen*, **68**, 301–312.

BOTTINGA, Y. & JAVOY, M. 1973. Comments on the oxygen isotope geothermometry. *Earth and Planetary Sciences Letter*, **20**, 250–265.

BOUCHOT, V., MILESI, J.P. & LEDRU, P. 2000. Crustal-scale hydrothermal palaeofield and related Variscan Au, Sb, W orogenic deposits at 310-305 Ma (French massif Central, Variscan belt). *SGA News*, **2000** (10 Dec), 1–8.

BOULADON, J. & DE GRACIANSKY, P.C. 1985. Les minéralisations dites de couverture (plomb, zinc, cuivre, uranium, barytine, fluorine) du Trias au

Pliocène, en France. *Chroniques du Recherches et Minières*, **480**, 17–33.

BRAY, C.J. & SPOONER, E.T.C. 1983. Sheeted vein Sn-W mineralization and greisenization associated with economic kaolinization, Goonbarrow China Clay Pit, St. Austell, Cornwall, England: Geologic relationships and geochronology. *Economic Geology*, **78**, 1064–1089.

BROCKAMP, O. & ZUTHER, M. 1985. K/Ar-Datierungen zur Alterseinstufung lagerstättenbildender Prozesse. *Naturwissenschaften*, **72**, 141–142.

CANALS, A. & CARDELLACH, E. 1993. Strontium and sulphur isotope geochemistry of low-temperature barite-fluorite veins of the Catalonian Coastal Ranges (NE Spain): A fluid mixing model and age constraints. *Chemical Geology*, **104**, 269–280.

CANALS, A., CARDELLACH, E., MORITZ, R. & SOLER, A. 1999. The influence of enclosing rock type on barite deposits, eastern Pyrenees, Spain: fluid inclusion and isotope (Sr, O, C) data. *Mineralium Deposita*, **34**, 199–210.

CARL, C. & DILL, H. 1985. Age of secondary uranium mineralizations in the basement rocks of northeastern Bavaria, FRG. *Chemical Geology*, **52**, 295–316.

CARMIGNANI, L., OGGIANO, G. & PERTUSATI, P.C. 1994. Geological outlines of the Hercynian basement of Sardinia. *In: Petrology, geology and ore deposits of the Paleozoic basement of Sardinia, Guidebook to the B3 Field excursion*. 16th General Meeting of the International Mineralogical Association, Pisa, 9–20.

CAROSI, R., GANDIN, A., GATTIGLIO, M. & MUSUMECI, G. 1992. Geologia della Catena Ercinica in Sardegna: Zona Esterna. *In: 'Gruppo Informale di Geologia Strutturale': Geologia della Catena Ercinica in Sardegna; Guida all'escursione sul Basamento Paleozoico della Sardegna, May 1992*. 43–60.

CASSINIS, G., CORTESOGNO, L., GAGGERO, L., PITTAU, P., RONCHI, A. & SARRIA, E. 1999. *Late Palaeozoic continental basins of Sardinia*. Field Trip Guidebook International Field Conference on 'The continental Permian of the Southern Alps and Sardinia (Italy). Regional Reports and general correlation. 15-25 September 1999, Brescia.

CHAREF, A. & SHEPPARD, S.M.F. 1988. The Malines Cambrian carbonate-shale-hosted Pb-Zn deposit, France: Thermometric and isotopic (H, O) evidence for pulsating hydrothermal mineralization. *Mineralium Deposita*, **23**, 86–95.

CHAURIS, L. 1978. Plate tectonics and ore deposits in Western Europe: The example of the Armorican Massif (France). *In: Proceedings of the 5th IAGOD symposium, Snowbird*, 401–413.

CHEN, Y., CLARK, A.H., FARRAR, E., WASTENEYS, A.H.P., HODGSON, M.J. & BROMLEY, A.V. 1993. Diachronous and independent histories of plutonism and mineralization in the Cornubian batholith, southwest England. *Journal of the Geological Society, London*, **153**, 1183–1191.

CHESLEY, J.T., HALLIDAY, A.N. & SCRIVENER, R.C. 1991. Sm-Nd direct dating of fluorite mineralization. *Science*, **244**, 949–951.

CLAYTON, R.N., O'NEIL, J.R. & MAYEDA, T.K. 1972.

Oxygen isotope exchange between quartz and water. *Journal of Geophysical Research*, **77**, 3057–3067.

CORTECCI, G., FONTES, J.C., MAIORANI, A., PERNA, G., PINTUS, E. & TURI, B. 1989. O, S, and Sr isotope and fluid inclusion studies of barite deposits from the Iglesiente-Sulcis mining districts, South-West Sardinia, Italy. *Mineralium Deposita*, **24**, 34–42.

COZZUPOLI, D., GERBASI, G., NICOLETTI, M. & PETRUCCIANI, C. 1984. Età K/Ar delle ignimbriti permiane di Galtellì (Orosei, Sardegna Orientale). *Rendiconti Società Italiana di Mineralogia e Petrologia*, **39**, 471–476.

CROWLEY, Q.G., FLOYD, P.A., WINCHESTER, J.A., FRANKE, W. & HOLLAND, J.G. 2000. Early Paleozoic rift-related magmatism in Variscan Europe: fragmentation of the Armorican Terrane Assemblage. *Terra Nova*, **12**, 171–180.

DARIMONT, A. 1983. Inclusions fluides dans les calcites associées à la minéralisations Pb-Zn de Poppelsberg (Est de la Belgique). *Mineralium Deposita*, **18**, 379–386.

DE VIVO, B., MAIORANI, A., PERNA, G. & TURI, B. 1987. Fluid inclusion and stable isotope studies of calcite, quartz and barite from karstic caves in the Masua mine, South-Western Sardinia, Italy. *Chemie der Erde*, **46**, 259–273.

DEL MORO, A., DI SIMPLICIO, P., GHEZZO, C., GUASPARRI, G., RITA, F. & SABATINI, G. 1975. Radiometric data and intrusive sequence in the Sardinian Batolith. *Neues Jahrbuch für Mineralogie Abhandlungen*, **126**, 28–44.

DIETZ, R.S. & HOLDEN, J.D. 1970. Reconstruction of Pangea: Breakup and dispersion of continents, Permian to present. *Journal of Geophysical Research*, **75**, 4939–4956.

DZIEDZIC, K. 1986. The Paleozoic rifting and volcanism in western Poland. *Zeitschrift geologischen Wissenschaften*, **14**, 445–457.

EDEL, J.B., MONTIGNY, R. & THIUZAT, R. 1981. Late Palaeozoic rotations of Corsica and Sardinia. New evidence from Paleomagnetic and K/Ar studies. *Tectonophysics*, **79**, 210–223.

FONTBOTÉ, L. & GORZAWSKI, H. 1987. *Petrographical and geochemical indicators for the exploration of hidden ore deposits in sedimentary rocks*. Final Report CEE, Project no. MSM-101D, 138–151.

FÖRSTER, B. & HAACK, U. 1995. U/Pb - Datierungen von Pechblenden und die hydrothermale Entwicklung der U-Lagerstätte Aue-Niederschlema (Erzgebirge). *Zeitschrift der geologischen Wissenschaften*, **23**(5/6), 581–588.

GAST, R.E. 1988. Rifting im Rotliegenden Niedersachsens. *Die Geowissenschaften*, **6**(4), 115–122.

GLEESON, S.A., WILKINSON, J.J., SHAW, H.F. & HERRINGTON, R.J. 2000. Post-magmatic hydrothermal circulation and the origin of base metal mineralization, Cornwall, UK. *Journal of the Geological Society London*, **157**, 589–600.

GROVES, D.I., GOLDFARB, R.J., GRBRE-MARIAM, M., HAGEMANN, S.G. & ROBERT, F. 1998. Orogenic gold deposits: a proposed classification in the context of the crustal distribution and relationship to other gold deposit types. *Ore Geology Review*, **13**, 7–27.

GUASPARRI, R., RICCOBONO, F. & SABATINI, G. 1984. Considerazioni sul magmatismo intrusivo ercinico e le connesse mineralizzazioni in Sardegna. *Rendiconti della Società Italiana di Mineralogia e Petrologia,* **32**, 17–52.

HAACK, U. & LAUTERJUNG, J. 1993. Rb/Sr Dating of hydrothermal overprint in Bad Grund by mixing lines. *In:* MÖLLER, P. & LÜDERS, V. (eds) *Formation of hydrothermal vein deposits-A case study on the Pb-Zn, barite and fluorite deposits of the Harz Mountains.* Monograph Series on Mineral Deposits, **30**, 103–114.

HALLIDAY, A.N. 1980. The timing of early and main stage ore mineralization in southwest Cornwall. *Economic Geology,* **75**, 752–759.

HALLIDAY, A.N. & MITCHELL, J.G. 1984. K-Ar ages of clay-size concentrates from the mineralisation of the Pedroches Batholith, Spain, and evidence for Mesozoic hydrothermal activity associated with the break up of Pangea. *Earth and Planetary Sciences Letters,* **68**, 229–239.

HANOR, J.S. 1987. *Origin and migration of subsurface sedimentary brines.* SEPM Short Course, **21**.

HEIJLEN, W., MUCHEZ, PH., BANKS, D. & NIELSEN, P. 2000. Origin and geochemical evolution of synsedimentary, syn- and post-tectonic high-salinity fluids at the Variscan thrust front in Belgium. *Journal of Geochemical Exploration,* **69-70**, 149–152.

HEIJLEN, W., MUCHEZ, PH. & BANKS, D.A. 2001. Origin and evolution of high-salinity, Zn-Pb mineralising fluids in the Variscides of Belgium. *Mineralium Deposita,* **36**, 165–176.

IHLEN, P.M., INESON, P.R. & MITCHELL, J.G. 1979. K/Ar dating of clay mineral alteration associated with ore deposition in the northern part of the Oslo region. *In:* NEUMANN, E.-R. & RAMBERG, I.B. (eds) *Petrology and geochemistry of continental rifts.* Elsevier, Amsterdam, 255–264.

INESON, P.R. & MITCHELL, J.G. 1972. Isotopic age determinations on clay minerals from lavas and tuffs of the Derbyshire orefield. *Geological Magazine,* **109**, 501–512.

INESON, P.R., MITCHELL, J.G. & VOKES, F.M. 1979. Further K/Ar determinations on clay mineral alteration associated with fluorite deposition in southern Norway. *In:* NEUMANN, E.R. & RAMBERG, I.B. (eds) *Petrology and geochemistry of continental rifts.* Elsevier, Amsterdam, 265–275.

JENSEN, M.L. & DESSAU, G. 1966. Ore deposits of South-Western Sardinia and their sulphur isotopes. *Economic Geology,* **61**, 917–932.

JOSEPH, D., BELLON, H., DERRE, C. & TOURAY, J.C. 1973. Fluorite veins dated in the 200 million years range at la Petite Verrière and Chavaniac, France. *Economic Geology,* **68**, 707–708.

LANCELOT, J.R., DE SAINT ANDRE, B. & DE LA BOISSE, H. 1984. Systematique U–Pb et évolution du gisement dúranium de Lodève (France). *Mineralium Deposita,* **19**, 44–53.

LAND, L.S. 1983. The application of stable isotopes to studies of the origin of dolomite and to problems of diagenesis in clastic sediments. *In:* ARTHUR, M.A. (ed.) *Stable isotopes in sedimentary geology.* Society of Economic Paleontologists and Mineralo-

gists, Short Courses, **10**, 4.1–4.22.

LEROY, J. & HOLLIGER, P. 1984. Mineralogical, chemical and isotopic (U-Pb-method) studies of Hercynian uraniferous mineralizations (Margnac and Fanay mines, Limousin, France). *Chemical Geology,* **45**, 121–134.

LIPPOLT, H.J., MERTZ, D.F. & ZIEHR, H. 1985. The late Permian Rb-Sr age of a K-Feldspar from the Wölsendorf mineralization (Oberpfalz, FR Germany). *Neues Jahrbuch für Mineralogie Mitteilungen,* 49–57.

LOMBARDI, G., COZZUPOLI, D. & NICOLETTI, M. 1974. Notizie geo-petrografiche e dati sulla cronologia K/Ar del vulcanesimo tardo-paleozoico sardo. *Periodico di Mineralogia,* **43**, 221–312.

LUDWIG, K.R., VOLLMER, R., TURI, B. & SIMMONS, K.R. 1989. Isotopic constraints on the genesis of base-metal ores in southern and central Sardinia. *European Journal of Mineralogy,* **1**, 657–666.

MARTIN, E.E. & MACDOUGAL, J.D. 1995. Sr and Nd isotopes at the Permian/Triassic boundary: A record of climate change. *Chemical Geology,* **125**, 73–99.

MASI, U., TURI, B. & VALERA, R. 1975. Composizione isotopica del quarzo e della calcite di ganga del Giacimento argentitifero' del Sarrabus (Sardegna sud-orientale) e sue implicazioni genetiche. *Rendiconti della Società Italiana di Mineralogia e Petrologia,* **31**, 467–485.

MATTHEWS, A. & KATZ, A. 1977. Oxygen isotope fractionation factors during the dolomitization of calcium carbonate. *Geochimica et Cosmochimica Acta,* **41**, 1431–1438.

MEYER, M., BROCKAMP, O., CLAUER, N., RENK, A. & ZUTHER, M. 2000. Further evidence of a Jurassic mineralizing event in central Europe: K/Ar dating in hydrothermal alteration and fluid inclusion systematics in wall rocks of the Käfersteige fluorite vein deposit in the northern Black Forest, Germany. *Mineralium Deposita,* **35**, 754–761.

MITCHELL, J.G. & HALLIDAY, A.N. 1976. Extent of Triassic/Jurassic hydrothermal ore deposits on the North Atlantic margins. *Transactions of the Institution of Mining & Metallurgy,* **B85**, 159–161.

MUCHEZ, PH., SLOBODNIK, M., VIAENE, W. & KEPPENS, E. 1994. Mississippi Valley-type Pb-Zn mineralization in eastern Belgium: indications for gravity-driven flow. *Geology,* **22**, 1011–1014.

MULLIS, J. & STALDER, H.A. 1987. Salt-poor and salt-rich fluid inclusions in quartz from two boreholes in Northern Switzerland. *Chemical Geology,* **61**, 263–272.

MÖLLER, P., WEISE, S.M., *ET AL.* 1997. Paleofluids and recent fluids in the upper continental crust: results from the German Continental Deep Drilling Program (KTB). *Journal of Geophysical Research,* **102**, 18233–18254.

NATALE, P. 1969. Il giacimento di Silius nel Gerrei. *Bollettino Associazione Mineraria Subalpina,* **6**(2), 1–35.

NORTHROP, D.A. & CLAYTON, R.N. 1966. Oxygen isotope fractionation in systems containing dolomite. *Journal of Geology,* **74**, 174–196.

O'NEIL, J.R., CLAYTON, R.N. & MAYEDA, T.K. 1969. Oxygen isotope fractionation in divalent metal carbonates. *Journal of Chemical Physics,* **51**, 5547–5558.

PAGEL, M. 1990. Le Permien et la métallogénie de l'uranium. *Chroniques Recherches Minières BRGM,* **499**, 57–68.

RAU, A. 1990. Evolution of the Tuscan domain between the Upper Carboniferous and the Mid-Triassic: a new hypothesis. *Bollettino della Società Geologica Italiana,* **109**, 231–238.

RESPAUT, J.P., CATHELINEAU, M. & LANCELOT, J.R. 1991. Multistage evolution of the Pierres-Plantées uranium ore deposit (Margeride, France): Evidence from mineralogy and U-Pb systematics. *European Journal of Mineralogy,* **5**, 85–103.

SALVADORI, I. & ZUFFARDI, P. 1973. Guida per l'escursione a Montevecchio e all'Arcuentu. Itinerari geologici, mineralogici e giacimentologici in Sardegna. *Ente Minerario Sardo,* **1**, 29–44.

SCHMITT, J.M., BAUBRON, J.C. & BONHOMME, M.G. 1984. Pétrographie et datations K-Ar des transformations minérales affectant le gite uranifère de Bertholène (Aveyron-France). *Mineralium Deposita,* **19**, 123–131.

SCHNEIDER, J. 2000. *Indirekte Rb-Sr Chronometrie postorogener Hydrothermalsysteme und assoziierter Gangmineralisationen im Rhenohercynikum.* PhD thesis, Universität Giessen, Germany.

SCHNEIDER, J. & HAACK, U. 2000. A different kind of Pb-Pb age. *In: Proceedings 78. Jahrestagung Deutsche Mineralogische Gesellschaft Meeting, September,* 2000, 188.

SCHNEIDER, J., HAACK, U., HEIN, U.F. & GERMANN, A. 1999. Direct Rb-Sr dating of sandstone-hosted sphalerites from stratabound Pb-Zn deposits in the northern Eifel, NW Rhenish massif, Germany. *In:* STANLEY, C.J. *ETAL.* (eds) *Mineral deposits: processes to processing.* Balkema, Rotterdam, 1287–1290.

SCHNEIDER, J., HAACK, U. & PHILIPPE, S. 2001. Precise Rb-Sr dating of binary isotopic mixing during hydrothermal wall rock alteration. *In:* PIESTZYNSKI, A. (ed.) *Mineral deposits at the beginning of the 21st century.* Proc. 6th biennal SGA-SEG meeting, Kraków, 205–208.

SCRIVENER, R.C., DARBYSHIRE, D.P.F. & SHEPHERD, T.J. 1994. Timing and significance of crosscourse mineralization in SW England. *Journal of the Geological Society, London,* **151**, 587–590.

SECCHI, F.A. & D'ANTONIO, M. 1996. Inferences of Sr, Nd and O isotopic tracers on the origin and evolution of a gabbronorite-granodiorite sequence from southern Hercynian chain of Sardinia. A case study from the Arburese igneous complex and its comparison with the earlier sequences of Sarrabus area. *Periodico di Mineralogia,* **65**, 257–273.

SECCHI, F.A., BROTZU, P. & CALLEGARI, E. 1991. The Arburese complex (SW Sardinia, Italy). An example of dominant igneous fractionation leading to peraluminous cordierite-bearing leucogranites or residual melts. *Chemical Geology,* **92**, 213–249.

SUBIAS, I., FANLO, I., YUSTE, A. & FERNÁNDEZ-NIETO, C. 1999. The Yenefrito Pb-Zn mine (Spanish

Central Pyrenees): an example of superimposed metallogenetic events. *Mineralium Deposita,* **34**, 220–223.

SWAINBANK, I.G., SHEPPERD, T.J., CABOI, R. & MASSOLI-NOVELLI, R. 1982. Lead isotopic composition of some galena ores from Sardinia. *Periodico di Mineralogia,* **51**, 275–286.

TORNOS, F., DELGADO, A., CASQUET, C. & GALINDO, G. 2000. 300 million years of episodic hydrothermal activity: stable isotope evidence from hydrothermal rocks of the Eastern Iberian Central System. *Mineralium Deposita,* **35**, 551–569.

VACCARO, C., ATZORI, P., DEL MORO, A., ODDONE, M., TRAVERSA, G. & VILLA, I.M. 1991. Geochronology and Sr-isotope geochemistry of late Hercynian dykes from Sardinia. *Schweizerische Mineralogische und Petrographische Mitteilungen,* **71**, 227–235.

VAI, G.B. 1991. Palaeozoic strike-slip pulses and palaeogeography in the circum-Mediterranean Tethyan realm. *Palaeogeography, Palaeoclimatology, Palaeoecology,* **87**, 223–252.

VALERA, R. 1967. *Contributo alla conoscenza dell'evoluzione tettonica della Sardegna.* Associazione Mineraria Sarda, **72**.

VALERA, R. 1974. Appunti sulla morfologia, termometria e composizione delle inclusioni fluide di fluoriti sarde. *Rendiconti della Società Italiana di Mineralogia e Petrologia,* **30**, 459–480.

WERNICKE, R.S. & LIPPOLT, H.J. 1997a. Evidence of Mesozoic multiple hydrothermal activity in the basement of Nonnenmattweiher (southern Schwarzwald), Germany. *Mineralium Deposita,* **32**, 197–200.

WERNICKE, R.S. & LIPPOLT, H.J. 1997b. (U+Th)-He evidence of Jurassic continuous hydrothermal activity in the Schwarzwald basement, Germany. *Chemical Geology,* **138**, 273–285.

WILKINSON, J.J., JENKIN, G.R.T., FALLICK, A.E. & FOSTER, R.P. 1995. Oxygen and hydrogen isotopic evolution of Variscan crustal fluids, south Cornwall, U.K. *Chemical Geology,* **123**, 239–254.

ZEEH, S. & BECHSTÄDT, T. 1994. Carbonate-hosted Pb-Zn mineralization at Bleiberg-Kreuth (Austria): compilation of data and new aspects. *In:* FONTBOTÉ, L. & BONI, M. (eds) *Sediment-Hosted Zn-Pb Ores.* Society of Geology Applied to Mineral Deposits, Special Publications, **10**, 271–296.

ZHENG, J.S., MERMET, J.F., TOUTIN-MORIN, N., HANES, J., GONDOLO, A., MORIN, R. & FÈRAUD, G. 1991. Datation ^{40}Ar-^{39}Ar du magmatisme et de filons minéralisès permiens en Provence orientale (France). *Geodinamica Acta,* **5**, 203–215.

ZIEGLER, P.A. 1988. Evolution of the Arctic-North Atlantic and the Western Tethys. *AAPG Memoir,* **43**.

ZIEGLER, P.A. 1993. Late Palaeozoic-early Mesozoic plate reorganization: Evolution and demise of the Variscan fold belt. *In:* VON RAUMER, J.F. & NEUBAUER, F. (eds) *Pre-Mesozoic geology of the Alps.* Springer, Heidelberg, 203–216.

The timing of W–Sn-rare metals mineral deposit formation in the Western Variscan chain in their orogenic setting: the case of the Limousin area (Massif Central, France)

M. CUNEY[1], P. ALEXANDROV[2], C. LE CARLIER DE VESLUD[2],
A. CHEILLETZ[2], L RAIMBAULT[3], G. RUFFET[4,5] & S. SCAILLET[6]

[1]UMR 7566 G2R-CREGU-Université Henri Poincaré, BP 239, 54506, Vandoeuvre-lès-Nancy, France (e-mail: michel.cuney@g2r.uhp-nancy.fr)

[2]CRPG-CNRS & ENSG, Rue du Doyen Marcel Roubault BP 40, 54500 Vandoeuvre-lès-Nancy, France

[3]Centre d'Informatique Géologique, Ecole des Mines, 35, rue St Honoré, 77305 Fontainebleau Cedex, France

[4]Géosciences Azur, UMR 6526 CNRS-UNSA-UPMC-IRD, Parc Valrose, 06108 Nice Cedex 2, France

[5]Géosciences Rennes, Université de Rennes, 1, Av. Général Leclerc, 35042 Rennes Cedex, France

[6]LSCE, CEA-CNRS, Avenue de la Terrasse, 91198, Gif sur Yvette, Cedex, France

Abstract: New $^{40}Ar/^{39}Ar$ dating performed on Rare Metal Granites and W $\pm$ Sn deposits in the northern Limousin has provided evidence of two metallogenic episodes. An Early Namurian episode (c. 325 Ma) was contemporaneous with the emplacement of the large peraluminous leucogranite bodies, which are associated with small W $\pm$ Sn deposits, but also with some larger deposits, at Puy-les-Vignes (323.4 ± 0.9 Ma) and Moulin-Barret (323.7 ± 0.8 Ma) formed at a shallower level above cryptic granite plutons. These new data indicate that the metallogenic potential of the Namurian leucogranites might have been underestimated. Most other W $\pm$ Sn deposits in the northern Limousin area are attributed to a Mid-Westphalian episode (c. 310 Ma), and are contemporaneous with the emplacement of all the Rare Metal Granites. Both episodes were related to leucogranite emplacement and associated fluid circulations, but in two different geodynamic contexts. The Early Namurian episode may be related to syncollisional extension of the Variscan belt, whereas the Mid-Westphalian one occurs during generalized extension and rapid exhumation of the belt associated with the granulite-facies metamorphism of the lower lithosphere probably related to the delamination of the lower lithosphere. Thus, W $\pm$ Sn and rare metals (Ta, Nb, Be, Li) deposits are clearly temporally and probably genetically related to leucogranitic magmatism.

The Limousin area is situated in the northwestern part of the French Massif Central which belongs to the inner orogenic domain of the mid-European Variscan belt. This area was subjected to protracted magmatic and hydrothermal activity from c. 330 Ma to 270 Ma associated with the formation of rare metals (Ta, Nb, Be, Li), W–Sn, Au and major U deposits. Most of these deposits are spatially associated with late-orogenic (325–305 Ma, Namurian–Westphalian) peraluminous granitoids and have been linked to a 'metalliferous peak' at c. 300 Ma (e.g. Marignac & Cuney 1999; Bouchot *et al.* 2000). However, only a few radiometric ages directly dating the mineralization

events exist, and the chronology of metalliferous events is only well constrained for U deposits (Leroy & Holliger 1984; Cuney *et al.* 1990; Scaillet *et al.* 1996a).

Under the auspices of the GéoFrance3-D national project '3-D Mapping and Metallogeny of the French Massif Central', a multi-disciplinary synthesis of ore deposits occurring in this area has been performed. The aim of the present paper is to present new $^{40}Ar/^{39}Ar$ thermochronological data on the micas associated with rare metals (Ta, Nb, Be, Li) and W–Sn deposits of the northern Limousin and to relate them to the magmatic and thermo-tectonic evolution of the mid-European Variscan belt.

From: BLUNDELL, D.J., NEUBAUER, F. & VON QUADT, A. (eds) 2002. *The Timing and Location of Major Ore Deposits in an Evolving Orogen*. Geological Society, London, Special Publications, **204**, 213–228. 0305-8719/02/$15.00 © The Geological Society of London 2002.

Geological setting

Regional geology

The Variscan belt is interpreted as a collision zone between the Gondwana and Baltica continental plates. The northern Limousin is situated in the northwestern part of the French Massif Central (Fig. 1), which belongs to the inner orogenic domain of the mid-European Variscan belt. Poly-phase Variscan deformation in the Limousin area may be separated into four main tectonic stages (Table 1; Ledru *et al.* 1989; Quenardel *et al.* 1991).

In the Limousin area (Fig. 1), shallow dipping thrust separated four main lithostructural metamorphic units or thrust nappes (Ledru *et al.* 1989) which are, from top to bottom: (i) the Upper Allochthon, containing low-grade Palaeozoic units; (ii) the Upper Gneiss Unit (UGU), characterized by metabasalts and basic metagreywackes with numerous high pressure relicts; (iii) the Lower Gneiss Unit (LGU), mainly consisting of acidic metagreywackes and orthogneisses; (iv) the Parautochthon, made up of micaschist, metagreywackes and rare orthogneisses.

Two major episodes of granitoid magmatism occurred in the Limousin (Table 1): (i) A Meso-Variscan (Late Devonian–Early Carboniferous) medium- to high-K calc-alkaline gabbro–dioritic to granitic magmatism (Peiffer 1986), which post-dates major Variscan thrusting, at the western margin of the Limousin area (360–349 Ma, U–Pb on zircon, Bertrand *et al.* 2001), synchronous with a peraluminous biotite ± cordierite granodioritic to granitic magmatism (Guéret type), emplaced at about 355 Ma (Berthier *et al.* 1979) all over the northern Limousin, excepted for the Auriat granite emplaced at 324 ± 1 Ma, and (ii) a Neo- to Late Variscan (Namurian–Westphalian) widespread peraluminous muscovite + biotite granite and leucogranite (Limousin type) magmatism, emplaced at 325 ± 10 Ma (Holliger *et al.* 1986; Williamson *et al.* 1996; Alexandrov 2000) associated with very small bodies of high-P peraluminous lepidolite ± muscovite leucogranites, rich in Ta, Nb, Sn, W and U (called Rare Metal Granites, hereafter referred to as RMG), but emplaced slightly later at 310 ± 3 Ma (Cheilletz *et al.* 1992; Alexandrov 2000; Raimbault 1999).

In the Limousin area almost no mineralization is associated with Eo- and Meso-Variscan geotectonic stages (Table 1). In contrast, most deposits (rare metals, W–Sn, and Au) are spatially related to Neo- to Late Variscan (325–305 Ma) peraluminous granites (Fig. 1) and have been linked to a 'metalliferous peak' at *c.* 300 Ma (Marignac & Cuney 1999; Bouchot *et al.* 2000).

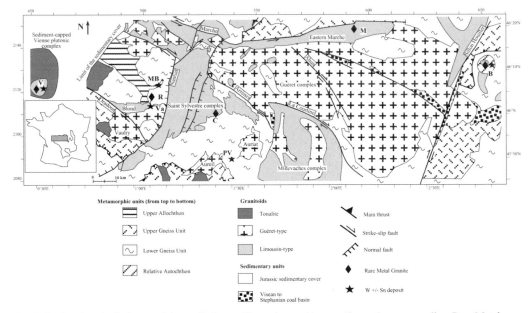

Fig. 1. Regional geological map of the studied area. The main granitic massifs are shown, as well as Rare Metal Granite and W ± Sn deposit occurrences. On the left and top margins, extended Lambert II coordinate system, a rectangular metric system, based on conformal conic projection and centred on the Paris meridian, is indicated. Key for Rare Metal Granite or W ± Sn deposits: B, Beauvoir; C, Chèdeville; M, Montebras; MB, Moulin-Barret; PV, Puy-les-Vignes; R, Richemont; V, Vienne; Va, Vaulry.

Table 1. *Schematic chart of tectonic and magmatic events occurred during Variscan orogeny in the Limousin area*

Main stages	Tectonics	Main magmatism
Eo-Variscan (Silurian, 430–400 Ma)	Closure of oceanic domains, subduction and formation of high pressure metamorphism	
Meso-Variscan (Devonian to Early Carboniferous, 400–340 Ma)	Continental collision and thrusting leading to crustal thickening	Episode 1 (360–345 Ma, Duthou 1978, Bertrand *et al.* 2001): medium- to high-K calc-alkaline association (so-called 'Tonalitic line', Peiffer 1986; peraluminous biotite ± cordierite bearing intermediate granitoids (Guéret type);
Neo-Variscan (Visean to Namurian. 340–310 Ma)	Transition from compressional to extensional tectonics, characterised by strike-slip and normal faults reflecting post-collisional tectonic readjustments	Episode 2: highly fractionated suite consisting of: peraluminous muscovite + biotite granites and leucogranites (Limousin type, 325 ± 10 Ma, Holliger *et al.* 1986; Williamson *et al.* 1996, Alexandrov 2000);
Late Variscan (Late Carboniferous to Early Permian, 310–280 Ma, Faure 1995)	General extension that corresponds to return to equilibrium of the thickened crust by normal faulting and erosion, together with the development of coal basins (Burg *et al.* 1994)	high-P, peraluminous, lepidolite ± muscovite leucogranites, rich in Sn, Ta, W and U (Rare Metal Granites, 310 ± 3 Ma Cheilletz *et al.* 1992, Alexandrov 2000, Raimbault 1999).

The Ta, Nb, Sn, Li, Be deposits are homogeneously disseminated within the Rare Metal Granites, which occur either as pegmatitic veins, or small granitic stocks or sheets, or rhyolitic dykes. When evaluated (e.g. Beauvoir Granite, Cuney *et al.* 1992), their emplacement depth is close to or less than 3–4 km. The rare metal mineralization has been interpreted as essentially of magmatic to late magmatic origin (Cuney *et al.* 1992; Raimbault *et al.* 1995). In contrast, W ± Sn deposits are located either in metamorphic rocks as stockworks or breccia pipes (considered as peribatholithic) or inside the granites as veins associated with greisen (intrabatholithic). The temporal and genetic relationships between W ± Sn deposits and these granites is still unclear (Marignac & Cuney 1999), owing to the lack of a direct radiometric date for the mineralization. Fluid inclusion studies in W ± Sn stockworks suggest that high temperature aquo-carbonic fluids, called 'pseudo-metamorphic' (fluids of unknown origin modified by equilibration with metamorphic rocks, Nesbitt & Muehlenbachs 1995; Fourcade *et al.* 2000) were diluted by aqueous fluids and were coeval with granite emplacement at pressures around 1 kbar or less (Marignac & Cuney 1999; Vallance *et al.* 2001). However, the Puy-les-Vignes deposit, the only breccia-type W ± Sn deposit of the French Massif Central, formed at 2.5 ± 0.5 kbar (Alikouss 1993).

Geology of the ore deposits

The samples analysed in the present study were taken from two already mined W ± Sn deposits of the Limousin: Puy-les-Vignes and Vaulry, one W + Sn deposit recognized by drillings: Moulin-Barret and two RMG occurrences: Richemont and Montebras. The results obtained on two other major RMG occurrences, Beauvoir and Chédeville, and on the Vienne peraluminous leucogranite and associated W ± Sn deposit were also included in the present paper because the data were only published in a local journal and are of major interest for the present study.

The Puy-les-Vignes W deposit. The deposit is a breccia pipe crosscutting the LGU, 20 km south of the Saint Sylvestre peraluminous leucogranite (324 ± 4 Ma, Holliger ± 1986), 5 km NE of the Auriat Guéret-type granite (324 ± 1 Ma, Gebauer *et al.* 1981) and 3 km SW of the Aureil Guéret-type granite (346 ± 14 Ma, Duthou 1978) (Fig. 1), but no granite body is known in the immediate vicinity of the deposit. The pipe has an oval shape (long axis 340 m, short axis 80 m, Fig. 2), and is formed by fragments of the host-gneisses cemented by quartz. The entire breccia is cut by wolframite bearing euhedral quartz veins with

A

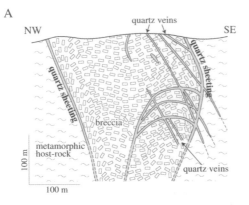

NW   SE

quartz veins

quartz sheeting

quartz sheeting

breccia

metamorphic host-rock

quartz veins

100 m

100 m

Fig. 2. Cross-section of Puy-les-Vignes W deposit (after Weppe 1951).

minor scheelite, pyrite, arsenopyrite and Bi, Pb and Zn minerals (Weppe 1951; Alikouss 1993). Wolframite belongs to the first of the four paragenetic stages and is associated with quartz, arsenopyrite and tourmaline. K-mica may be deposited either synchronous with (Alikouss 1993) or after wolframite (Weppe 1951). The micas chosen for this study were early K-micas deposited slightly before or synchronous with wolframite, as observed in Figure 3, where the extremity of a muscovite crystal is enclosed by

wolframite. According to fluid inclusion data (Alikouss 1993), the W deposit was formed at 400–450 °C and a depth probably less than 9 ± 1.8 km, as the author suggests that the relatively high pressure registered for this deposit may result from transient overpressures which led to the breccia pipe formation.

The Vaulry W deposit. The deposit is situated in the Blond Limousin-type granite at its eastern contact with the Vaulry Guéret-type granite (Fig. 1). The wolframite ore is hosted in centimetre-thick quartz veins enveloped by centimetre-scale greisens. One biotite and two muscovite grains were dated. The biotite comes from the walls of one larger quartz–wolframite bearing vein. The two muscovite grains come from the greisens. The micas being developed systematically at the wall of the wolframite bearing veins, the two minerals are considered to be synchronous (Alexandrov 2000). No intersection of the greisen by the quartz–wolframite bearing vein was observed. The depth of formation of the W deposit was estimated at 5.5 km from fluid inclusion data (Vallance *et al.* 2001).

The Moulin-Barret W ± Sn deposit. The mineralization is mainly hosted in amphibolites of the Lower Gneiss Unit and Late Devonian Guéret-type granite, 15 km north of the Blond peraluminous leucogranite (Fig. 1). The deposit, recognized in three boreholes, consists of millimetre- to

Fig. 3. Transmitted light microscopic photographs. Moulin-Barret (left) and Puy-les-Vignes (right). Q, quartz; W, wolframite; S, scheelite; M, muscovite; H, host-rock.

centimetre-thick quartz veinlets with scheelite, arsenopyrite, cassiterite, pyrite and chalcopyrite. In granite, cassiterite predominates over scheelite in unzoned veinlets, whereas in amphibolite, scheelite strongly predominates in zoned veinlets. In the latter case, scheelite and associated muscovite in the walls of the veinlets (Fig. 3a) were both interpreted as resulting from chemical reaction between hot hydrothermal fluids circulating in fractures, and Ca–Al-rich amphibolite wall rocks (Raimbault 1999). Occasionally, scheelite was found in the central part of the veinlets. Scheelite REE geochemistry shows that all three modes of scheelite occurrence originate from a single fluid. Higher Ta contents in lowermost scheelite crystals point to the derivation of the fluid from a hidden highly fractionated leucogranite cupola (Fig. 4). The Limousin-type peraluminous leucogranites are enriched in Ta (a few ppm to 400 ppm in the Beauvoir granite, Raimbault et al. 1995) whereas the Guéret-type peraluminous granites have generally less than 1 ppm Ta (M. Cuney, unpublished

data). The dated muscovite grain was obtained from the walls of a veinlet from the most mineralized borehole MBS1 (Fig. 5). No data are available on the depth of formation of this deposit or about the granite possibly related to it.

The Richemont RMG. The Richemont rhyolitic dyke (5 km long × 5 m width) crops out to the NNE of the Blond peraluminous leucogranite (Fig. 1) (Raimbault 1998a). The Blond leucogranite was dated at 319 ± 7 Ma by U–Pb on zircon (Alexandrov et al. 2000). The rhyolite belongs to the high phosphorous peraluminous RMG group with typically strong enrichment in Ta, Nb, Li, Be, F, P and extremely low REE and Th contents (Raimbault & Burnol 1998). Idiomorphic quartz and muscovite phenocrysts are dispersed in a devitrified homogeneous matrix. The very small thickness and the rhyolitic texture of the Richemont dyke imply a very fast cooling of the intrusion.

The Montebras RMG. The granite was emplaced in the Eastern Marche leucogranitic belt (Fig. 1). It was injected as a 150 m thick slab at the roof of a larger two mica leucogranite body. The Montebras granite is an albite–lepidolite granite presenting similar mineralogical and geochemical characteristics as the Beauvoir RMG (Aubert 1969). Lepidolite concentrates were extracted from an homogeneous granite sample from the 'Les Roches' open pit. The micas exhibit a strong zonation with a lepidolite core and a Li-muscovite rim. The Li-muscovite rim was interpreted as having grown at a late magmatic stage in response to an influx of meteoric fluid (Belkasmi et al. 1991). No indication is known for the emplacement level of the Montebras granite.

The Beauvoir RMG. The Beauvoir cassiterite–topaz–lepidolite leucogranite was investigated in detail thanks to a 900 m deep continuously cored drilling (Cuney et al. 1992; Raimbault et al. 1995). The upper part of the granite presents the strongest enrichment in rare metals among RMG granites of the Variscan belt. Fluid inclusion data indicates an emplacement depth of 3 km (Cuney et al. 1992). The lepidolite displays solid solutions between Li-muscovite, zinnwaldite and polylithionite.

The Chédeville pegmatites. These form a lepidolite–albite dyke swarm enriched in rare metals, located in the southern part of the Saint Sylvestre granite, in its uppermost structural level, close to enclosing metamorphic rocks (Raimbault 1998b). Lepidolite displays a limited solid solution between Li-muscovite and polylithionite. There is no evidence of their direct derivation from the Saint Sylvestre granite. Their geochemistry is comparable with that of other RMGs of the northern part of the French Massif central.

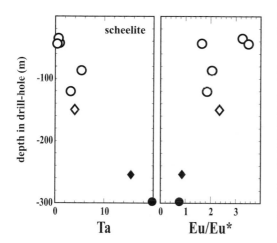

Fig. 4. Geochemical evolution of scheelite against depth in drill holes MBS1 (circles) and MBS2 (diamonds). As both holes have a similar inclination of *c.* 60 degrees, the true depth is approximately threequarters of the depth along the cores. Open symbols: scheelite from veinlets in amphibolite rocks, closed symbols: scheelite from veinlets in Ca-rich granites. Left, Ta content (in ppm) increases with depth from low values typical for scheelite far from granitic sources, to high values encountered in scheelites associated with evolved granites (Raimbault 1984). Right, the chondrite-normalized Eu anomaly is a measure of interaction between host rocks and hydrothermal fluids, which is at its maximum in shallow samples and is still present in the deepest samples in comparison with a usually strongly negative Eu anomaly in evolved granites. Together, both data sets point out to a deep-seated, evolved-granite source.

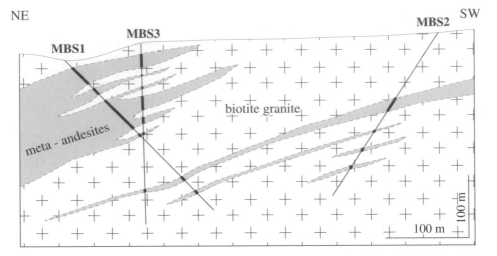

Fig. 5. Interpretative cross-section of the Moulin-Barret area, with position of the boreholes. The bold lines indicate the mineralized zones (after Raimbault 1998, 1999).

The Vienne leucogranite and W ± Sn deposit. In the Vienne area (Fig. 1) a Late Devonian medium and high-K calc-alkaline plutonic complex was discovered below a Mesozoic sedimentary cover by a borehole survey (Virlogeux *et al.* 1999; Cuney *et al.* 2001). Geophysical data indicated that a large peraluminous leucogranitic pluton intrudes the calc-alkaline complex. However, a highly fractionated peraluminous leucogranite was intersected in one of the boreholes. The leucogranite is enriched in rare metals but less than the Beauvoir RMG and is crosscut by wolframite ± cassiterite-bearing quartz and greisen veins. Muscovite occurs either intimately associated with the wolframite in the greisens or at the wall of the quartz–wolframite bearing veins and thus is considered as synchronous with wolframite deposition. Fluid inclusion data (Cathelineau *et al.* 1999*a*) indicate a rather shallow level of emplacement of 4 km for the Vienne leucogranites.

Analytical procedures

$^{40}Ar/^{39}Ar$ dating

Most analyses were performed on single muscovite and biotite grains. After crushing and sieving, muscovite and biotite alteration-free grains have been selected under the binocular microscope. Gases were extracted from single-grains by laser step-heating and analysed on a VG 3600 mass spectrometer. The isotopic ratios were corrected for background noise, blank value, Ca and K interference, instrumental mass discrimination and

atmospheric argon contamination according to the procedure described in Ruffet *et al.* (1997).

The Montebras RMG was analysed some years earlier using a different analytical procedure available at that time. After irradiation, a population of lepidolite grains was step heated using a Lindberg resistance furnace coupled to a purification line and analysed using a modified A.E. MS10 mass spectrometer at Queen's University, Canada (Hanes *et al.* 1992), according to the procedure described in Scaillet *et al.* (1996*a*).

A plateau age is defined when at least three consecutive steps, representing more than 70% of the extracted gas, give ages that overlap within the 1σ error margin. All errors on plateau ages are quoted at the 1σ confidence level, and include the error on the irradiation curve.

Oxygen isotopes

Oxygen isotope analysis was performed on a quartz hosting wolframite at Puy-les-Vignes by the classical fluorination method (Clayton & Mayeda 1963) using BrF_5. CO_2 was measured on a VG602D gas-source mass spectrometer. Results are expressed as $\delta^{18}O$ (‰) relative to the V-SMOW. Reproducibility for quartz standard is better than 0.2‰ and $\delta^{18}O = 9.60$‰ for NBS28.

Results

Table 2 summarizes all the dating performed in this study and other dating available in the literature, but mostly published in local journals. Two set of ages were obtained: a Namurian one

Table 2. *Summary of all dating available from the literature or from this study concerning magmatism and W ± Sn deposits in NW French Massif Central, sorted by period and then by area*

Period	Area	Location	Facies dated (mineral)	U–Pb (Ma)	$^{40}Ar/^{39}Ar$ Plateau age ± 1σ (Ma)	% of ^{39}Ar in plateau	Isochron age (Ma)	Integrated age (Ma)	Reference
Early Namurian events									
	Auriat	Auriat	Guéret-type granite (zircon and monazite)	324 ± 1					Gebauer et al. (1981)
		Puy-les-Vignes	Deposit (muscovite)		323.4 ± 0.9	57	324.4 ± 2.1	323.8 ± 1.8	This work
			Duplicate		323.3 ± 0.6	86	322.8 ± 3.0	323.5 ± 1.6	This work
		Saint Sylvestre	Limousin-type granite (zircon and monazite)	324 ± 4					Holliger et al. (1986)
	St Sylvestre Blond	Blond	Limousin-type granite (zircon)	319 ± 7					Alexandrov et al. (2000)
Mid-Westphalian events									
		Mouin-Barret	Deposit (muscovite)		323.7 ± 0.8	100	324.0 ± 2.0	323.7 ± 1.1	This work
		Vaulry	Deposit (biotite)		311.1 ± 0.7	61	312.5 ± 1.9	306.7 ± 0.5	This work
		Vaulry	Greisen (muscovite)		309.0 ± 1.4	72	310.3 ± 3.1	311.9 ± 1.7	This work
			Duplicate		312.0 ± 1.1	78	311.5 ± 2.0	312.1 ± 0.8	This work
	Blond	Richemont	Rare Metal dyke (muscovite)		313.4 ± 1.4	97	313.5 ± 7.0	313.5 ± 0.3	This work
	Saint Sylvestre	Chédeville	Rare Metal granite (lepidolite)		310 ± 1.5				Cheilletz et al. (1992)
	Echassières	Beauvoir	Rare Metal granite (lepidolite)		308 ± 2				Cheilletz et al. (1992)
	Eastern Marche	Montebras	Rare Metal granite (lepidolite)		309.9 ± 0.7	82	–	310.1 ± 1.3	This work
	Vienne	Vienne	Rare Metal granite (muscovite)		309 ± 1				Alexandrov (2000)
	Vienne	Vienne	Greisen (muscovite)		309 ± 1				Alexandrov (2000)

All the ages are interpreted, either as emplacement age for intrusions or as mineralization age for deposits. For the $^{40}Ar/^{39}Ar$ dating performed in this study, additional technical data are given in the right hand side of the table.

with the Moulin-Barret and Puy-les-Vignes W ± Sn deposits and a mid-Westphalian one with rare metals (Richemont, Montebras, Vienne, Beauvoir and Chèdeville) and W ± Sn (Vaulry and Vienne) deposits.

The Namurian (c. 325 Ma) ages

Puy-les-Vignes. Two muscovite grains yielded ages (Fig. 6a) of 323.4 ± 0.9 Ma (on 57% of the ^{39}Ar released; with an isochron age of 324.4 ± 2.1 Ma) and 323.3 ± 0.6 Ma (on 86% of the ^{39}Ar released; with an isochron age of 322.8 ± 3.0 Ma). This last analysis was interrupted by a technical problem resulting in the loss of 2.75% of the total Ar released, but the concordance of the two partial plateau ages indicates

that the final calculated age (323.3 ± 0.6 Ma) remains reliable.

Moulin-Barret. The muscovite (Fig. 6b) gave a plateau age of 323.7 ± 0.8 Ma (on 100% of released ^{39}Ar). The isochron age, which is very similar (324.0 ± 2.0 Ma), attests the quality of the age determination. In addition, a δ^{18}O of +11.6 ± 0.2‰ was measured on a quartz grain from a mineralized vein. The calculated δ^{18}O of the fluid in equilibrium with quartz, at temperatures of 400–450 °C (Alikouss 1993), and according to the fractionation curves of Matsuhisa *et al.* (1979), is +11.5‰. Thus, the signature of the fluid excludes a direct derivation of the fluid from meteoric waters, which should have a lighter composition (+10 to −40‰, Taylor 1968).

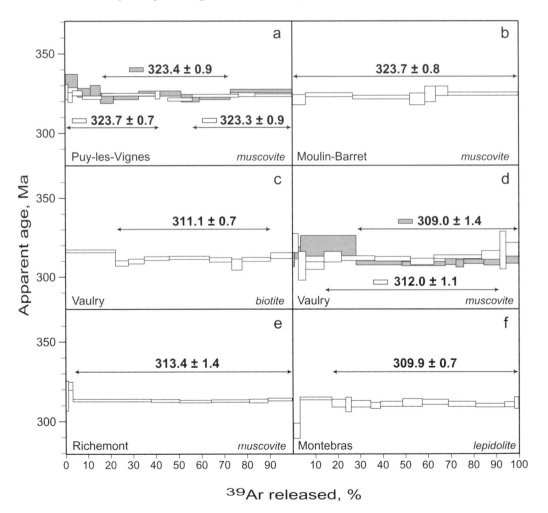

Fig. 6. ^{40}Ar/^{39}Ar spectra of several RMG or W deposits: (**a**) The Puy-les-Vignes deposit, (**b**) the Moulin-Barret deposit, (**c**) to (**d**) the Vaulry deposit, (**e**) the Richemont rhyolite, (**f**) the Montebras RMG. In each case, plateau age is indicated.

The Mid-Westphalian (c. 310 Ma) ages

Vaulry. $^{40}Ar/^{39}Ar$ dating of the micas from the Vaulry deposit provided flat age spectra (Fig. 6c–d). The plateau age is 311.1 ± 0.7 Ma for the biotite from the wall of the vein (on 71% of the released ^{39}Ar; isochron age of 312.5 ± 1.9 Ma). The two greisen-hosted muscovites yielded plateau ages of 309.0 ± 1.4 Ma (on 72% of the released ^{39}Ar, isochron age 310.3 ± 3.1 Ma) and 312.0 ± 1.1 Ma (on 78% of the released ^{39}Ar; isochron age of 311.5 ± 2.0 Ma) for the two muscovite grains from the greisen. These ages are similar to the $^{40}Ar/^{39}Ar$ plateau ages (313.2 ± 0.7 to 309.4 ± 1.4 Ma) determined on micas from Blond granite enclosing the Vaulry deposit (Alexandrov 2000).

Richemont. The $^{40}Ar/^{39}Ar$ spectrum is very flat (Fig. 6e) and defines a well-constrained plateau age (on 97% of the released ^{39}Ar) at 313.4 ± 1.4 Ma. This age corresponds closely to the isochron age (313.5 ± 0.7 Ma) and to the integrated age (313.5 ± 0.3 Ma), further supporting the reliability of the age determination.

Montebras. The $^{40}Ar/^{39}Ar$ spectrum (Fig. 6f) from the L36 sample defines a plateau age at 309.9 ± 0.7 Ma corresponding to 82% of ^{39}Ar liberated, very close to the integrated age (310.1 ± 1.3 Ma).

Former $^{40}Ar/^{39}Ar$ ages obtained on three other RMG and W ± Sn occurrences (the Beauvoir granite, the Chédeville pegmatite and the Vienne area) are reported in Table 2. Lepidolite crystals separated from the three Beauvoir units gave plateau ages of 308 ± 2 Ma. The ages obtained on the three units are very close despite the variation of composition of the micas from one unit to the other (Monier et al. 1987), which may partly result from variable subsolidus alteration. Lepidolite crystals from the Chédeville pegmatite yield a $^{40}Ar/^{39}Ar$ plateau age of 309 ± 0.9 Ma on two steps corresponding to 97% of ^{39}Ar liberated (Cheilletz et al. 1992). In the Vienne area (Alexandrov 2000), a muscovite separated from the leucogranite yielded an $^{40}Ar/^{39}Ar$ age of 309.5 ± 0.3 Ma (on 50% of the released ^{39}Ar) and a muscovite from spatially associated wolframite-bearing greisens gave an $^{40}Ar/^{39}Ar$ age of 309 ± 1 Ma (on 100% of the released ^{39}Ar).

Discussion

$^{40}Ar/^{39}Ar$ ages and regional cooling patterns

Textural observations discussed above suggest that the micas selected for $^{40}Ar/^{39}Ar$ age determination were synchronous with ore formation. Some of them may have crystallized slightly earlier than the ore minerals, because of their location at the

wall of the veins. However, the temperatures of the ore forming processes for rare metals and W ± Sn deposits (always higher than 400 °C, Marignac & Cuney 1999) were above the most commonly accepted Ar diffusion closure temperature of mica of about 300–350 °C (McDougall & Harrison 1999). Thus the $^{40}Ar/^{39}Ar$ age of the micas should have been reset at the time of ore deposition. Now, to discuss the meaning of the $^{40}Ar/^{39}Ar$ ages in terms of crystallization age or of cooling age it is necessary to know the temperature evolution of enclosing formations with respect to the Ar diffusion closure temperature of micas. Regional cooling patterns in the Limousin area, determined by Scaillet et al. (1996a) and Alexandrov (2000) mainly from muscovite $^{40}Ar/^{39}Ar$ dating, highlight two main features.

(1) All metamorphic formations within a few kilometres from the Saint Sylvestre leucogranitic complex gave $^{40}Ar/^{39}Ar$ ages of c. 340 Ma. A 340 Ma age is also observed over more than 100 km in the Limousin area and is interpreted as a regional cooling episode resulting from an early exhumation during syncollisional extension (Alexandrov 2000).

(2) In the Saint Sylvestre complex itself, emplaced at 10.5 ± 1.5 km, $^{40}Ar/^{39}Ar$ ages range from 315 Ma at the roof of the complex, to 300 Ma towards the base of the complex. Thermal modelling has shown that such age variation resulted from a fast exhumation episode (with exhumation rate of c. 1.5 mm a^{-1}, Scaillet et al. 1996b), which started at about 305 Ma during the generalized late Variscan extension stage, centred around the leucogranites (Faure & Pons 1991; Faure 1995).

Thus, the metamorphic formations located around the Saint Sylvestre complex represent formerly higher structural levels, laterally down-warped during the fast exhumation episode. For instance, the formations west of the Saint Sylvestre complex, hosting the Moulin-Barret deposit, are separated from the complex by the normal brittle Nantiat fault (Fig. 1) which has a minimal vertical offset of about 3 km (Le Carlier de Veslud et al. 2000). Numerical modelling, assuming a single major intrusion and conductive heat transfer, shows that the small leucogranite bodies and associated mineralized systems emplaced at such structural levels will equilibrate with enclosing rocks very rapidly. For example, for the Blond leucogranite (15 km × 5 km large and 1 km thick in average according to gravimetric data inversion), one of the largest leucogranites emplaced around the Saint Sylvestre complex, the time duration for its thermal equilibration with surrounding rocks is only about 0.3 Ma (Le Carlier de Veslud et al. 2000). This duration is well within the error of $^{40}Ar/^{39}Ar$ age determinations. For hectometric-sized RMG intrusions, this time will be only a few thousand years (Cathles & Erendi 1997). In addition, thermo-barometric data when available, confirm that most RMG and W–Sn deposits were emplaced at shallow depth (see above). Only the Puy-les-Vignes deposit was formed at greater depth, but $^{40}Ar/^{39}Ar$ dating of muscovite in enclosing metamorphic rocks, a few kilometres from the deposit, shows that the 300–350 isotherm was already crossed at 340 Ma. Moreover, former fluid inclusion and isotopic studies (given above) have shown intensive early interaction of the mineralized systems with meteoric fluids. The additional advective heat transfer induced by such fluid circulation must have led to a significant reduction of the estimated cooling time, up to one order of magnitude according to Cathles & Erendi (1997). Therefore, it can be considered that the micas give $^{40}Ar/^{39}Ar$ ages close to the emplacement ages for the RMG and thus of associated rare metal mineralization and of ore formation for the W–Sn deposits.

The origin of W

Several lines of evidence support a direct granitic origin for the W ± Sn deposits in the Limousin.

(i) *Temporal relationships.* The two age periods for W deposition obtained in the present paper coincide with the two periods of leucogranitic peraluminous magmatism in northern Limousin (Table 2): first the Namurian period 325–315 Ma with the main generation of leucogranites such as Saint Sylvestre and Blond; secondly the mid-Carboniferous period 310 ± 3 Ma with the generation of the small RMG bodies.

(ii) *Geochemical relationships.* In the Limousin area, the average W abundance in meta-igneous and metasedimentary rocks (M. Cuney, unpublished data), and Late Devonian Guéret-type granitoids (Cuney et al. 1999) is similar to the upper crustal Clarke value (1–2 ppm). In contrast, the two generations of Limousin-type peraluminous granites present high W contents: 5–20 ppm for Saint Sylvestre or Blond, and up to 100 ppm for the RMG granites (Cuney et al. 1992; Raimbault 1998a). Therefore, the Limousin-type peraluminous granites may represent a major source of W for the ore deposits. Thus, the Puy-les-Vignes W breccia pipe, enclosed in metamorphic rocks, is more probably related to a hidden peraluminous leucogranite similar to Saint Sylvestre rather than to the nearby Auriat Guéret-type granite. A more direct evidence of such a relation between Namurian Limousin type peraluminous leucogranites and W mineralization is provided by the increasing Ta concentration with depth in scheelite (Fig. 4) in

the Moulin-Barret deposit (see discussion above in the section on the geology of the ore deposits). However, in the case of an extreme fractionation of the magma and early oxidation by meteoric fluids, the extraction of W from the melt may be strongly reduced, as was demonstrated in the case of the Beauvoir RMG (Fig. 1, Cuney et al. 1992).

(iii) *Spatial relationships.* Marignac & Cuney (1999) highlighted the role of pseudo-metamorphic fluids on the formation of the W ± Sn ore deposits in Limousin. However, in the regional shear zones, where early (probably Namurian–Westphalian) pseudo-metamorphic fluid circulation occurred in association with regional As anomalies (Cathelineau et al. 1999b), no stream-sediment W anomaly nor W ± Sn deposit is observed. In contrast, well-correlated stream-sediment W anomalies and W deposits occur at the contacts between the large peraluminous leucogranitic laccoliths or RMG bodies and their enclosing rocks, especially close to their apical and most fractionated parts (Le Carlier de Veslud et al. 2000). The thermal anomalies and the fractured aureoles induced by the emplacement and the cooling of such granites are known to cause significant intrusion-centred hydrothermal circulation (Furlong et al. 1991). Circulation is generally focused on apical parts of the intrusions, where heat and mass transfers between granites and surrounding rocks are enhanced. Thus, W mineralization might be related to the fluid circulation induced by the emplacement of Limousin type peraluminous granites.

The W ± Sn deposits in their geodynamic context

During the time span within which W ± Sn deposition occurred (325–310 Ma) major changes in the tectonic regime in the French Massif Central occurred. Therefore, it is of major interest to place W ± Sn deposit genesis in the magmatic–tectonic framework of the French Massif Central and the related evolution of fluid circulation patterns in the crust.

The Early Namurian (c. 325 Ma) metallogenic episode.. The Late Visean–Westphalian (335–310 Ma) corresponds to a 'syn-convergence' extension stage in the Western part of the Variscan belt (Burg et al. 1994; Faure 1995). During this period, the large Namurian Limousin-type leucogranites, emplaced at deep structural level (> 9 km, Scaillet et al. 1996a; Le Carlier de Veslud et al. 2000). The ductile deformation regime prevailing at such depths, led to a low permeability of the rocks (Furlong et al. 1991), which precluded large hydrothermal fluid transfer and meteoric fluid infiltration. However, the emplacement of the granites, at such depths, may have induced temporary magmatic and metamorphic fluid circulation, restricted to the envelope of these granites (Furlong *et al.* 1991). This is for instance the case of the Saint Sylvestre leucogranite laccolith (Le Carlier de Veslud *et al.* 2000), where minor W ± Sn showings were formed in the vicinity of the contact between the most fractionated parts of the granite and enclosing metamorphic rocks (Fig. 7a). In contrast, the Puy-les-Vignes, and the Moulin-Barret W deposits, formed at the same age, were located at a higher structural level (6–9 km depth) and presumably above some apices of similar leucogranitic intrusions. At such locations, hydraulic brecciation of the enclosing metamorphic rocks was already possible during the Namurian, and intensive but spatially limited hydrothermal fluid circulation was generated, as attested by the strong and localized As and W stream-sediment anomalies observed around the Puy-les-Vignes deposit (Le Carlier de Veslud *et al.* 2000).

The Mid-Westphalian (c. 310 Ma) metallogenic episode (Table 2).. At *c.* 310–290 Ma, an important thermal event caused granulite-facies metamorphism of the entire lower crust (Pin & Vielzcuf 1983; Costa & Rey 1995). The breakdown of micas produced F–Li–Sn–W–U-rich fluids, which may have contributed to the genesis of the Rare Metal Granites (Cuney *et al.* 1990; Marignac & Cuney 1999). The thermal anomaly may have resulted from basic magma underplating as suggested by the nature of the xenoliths brought to the surface by Cenozoic volcanoes (Downes *et al.* 1991; Williamson *et al.* 1996). However, basic magma underplating seems to have occurred over a long period (360–305 Ma, Downes *et al.* 1991; Williamson *et al.* 1996), whereas the lower crust granulite-facies metamorphism event occurred within a discrete period at the end of the Variscan orogeny and required an exceptionally large thermal anomaly that affected the entire lower crust. This large thermal anomaly is therefore more likely to have been caused by a lithospheric delamination, as suggested by Pin & Vielzeuf (1983) and Marignac & Cuney (1999). The time-limited effect of this thermal event is especially well illustrated by the remarkable homogeneity of the rare metal and W ± Sn deposit ages at the scale of the French Massif Central, most of them having occurred in the 310 ± 3 Ma period for the northern French Massif Central (Table 2) and 306 ± 3 Ma for the southern French Massif Central (Bouchot *et al.* 2000; Monié *et al.* 2000; Lerouge *et al.* 1999).

In the upper crust, this period also corresponds to the final collapse of the western part of the Variscan orogen and the onset of the generalized extension stage (Faure 1995) with enhanced uplift rates (1.5 mm a^{-1}, Fig. 7c) associated with tectonic denudation dated at about 305 Ma by Scaillet *et al.* (1996b). The uplift of the crustal blocks led to an increased vertical permeability and extensive fluid percolation in the Limousin upper crust, especially along reactivated Early Namurian structures (fractured aureoles of leucogranitic laccoliths and regional shear zones, Boiron *et al.* 2000). The increased permeability permitted the mixing of metamorphic fluids with an increasing amount of meteoric fluids (Fourcade *et al.* 2000; Fig. 7b). In this context, the emplacement of highly fractionated peraluminous leucogranites at a depth of 3–5 km, in brittle tectonic conditions, acted as hot points, leading to fracturing of the surrounding rocks and magma degassing at the apex of the intrusions, followed by an input of pseudo-metamorphic or meteoric fluids (Cathelineau *et al.* 1999b; Fourcade *et al.* 2000). For example, such a pattern of fluid mixing was revealed by combined studies (Alexandrov 2000; Le Carlier de Veslud *et al.* 2000; Vallance *et al.* 2001) in the Blond massif area where several RMG bodies are known.

Comparison of W ± Sn, Au and U depositional conditions in Limousin

Other ore deposits, such as gold and uranium, display strong spatial relationships with W ± Sn deposits (Fig. 8; Le Carlier de Veslud *et al.* 2000). Therefore, a comparison of the timing of deposition between all these deposits may provide additional information about their genesis.

Ages, together with their geological and geochemical evidence provided in this paper, indicate that W ± Sn and rare metals deposits are genetically and temporally related to leucogranitic magmatism. It is also shown for the first time that favourable conditions for W ± Sn mineralization existed at least during two events in Limousin, in relation to the generation of two main peraluminous magmas of Limousin- and Rare Metal-type during the Late Carboniferous thermo-tectonic evolution of the Variscan belt (Figs 7a, b and 8). In contrast, the metallogenic processes leading to the formation of the Au and U deposits cover several tens of millions of years and involve several stages (Fig. 8; Scaillet *et al.* 1996a, b; Bouchot *et al.* 2000; Cuney *et al.* 1990; Le Carlier de Veslud *et al.* 2000).

Despite the lack of direct dating, Bouchot *et al.* (2000) proposed a 310–305 Ma interval for the

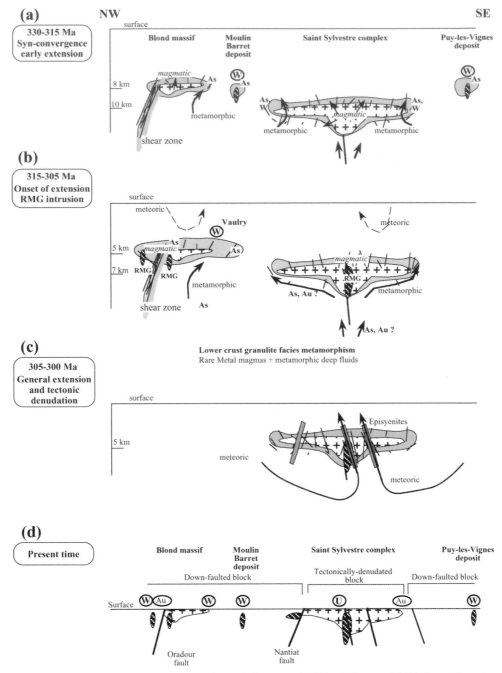

Fig. 7. Schematic patterns of fluid circulation (after Cathelineau *et al.* 1999*a*, Boiron *et al.* 2000, Fourcade *et al.* 2000) in the northern Limousin area, and implications for mineral deposition. (**a**) The 330–315 Ma period, with limited fluid circulation in shear zones and temporarily in the vicinity on leucogranites. (**b**) The 315–305 Ma period, with both lower crust granulite-facies metamorphism at depth and onset on general extension. This stage corresponds to the emplacement of RMG, which focused mixing of fluid of different origins toward their apices. (**c**) The 305–300 Ma period corresponds to the tectonic collapse of the belt, which is characterized in Limousin by tectonic denudation of the Saint Sylvestre complex by downwarping of its metamorphic or granitic cover. (**d**) At the present time, to point out the horst configuration of the Saint Sylvestre complex, surrounded by down-faulted blocks. The location of all known mineral deposits is indicated: U (economic), W ± Sn (sub-economic) and Au (minor).

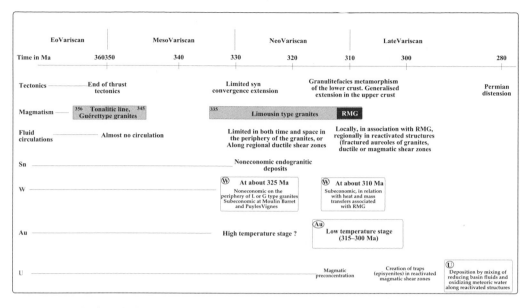

Fig. 8. Schematic chart, depicting the tectono-magmatic evolution of the northern Limousin area and the formation of related ore deposits.

main episode of gold deposition. Therefore, Au mineralization seems to be related to the same thermal event as the second W ± Sn episode and rare metal mineralization, corresponding to the granulite-facies metamorphism of the lower crust. All the metals may ultimately have been derived from the granulitized lower crust. However, according to Bouchot et al. (2000), Au was probably transported from the lower crust by fluids, not by granitic melts as for U, W, or rare metals. The spatial and temporal relationships between Au and W ± Sn deposits in the 310–300 Ma period are best explained by the fact that permeable crustal zones (regional shear zones, crustal lineaments used as feeding zones for the Limousin-type plutonic complex and, to a lesser extent, fractured aureoles around these granites) have been used by magmas and by fluid generated at different levels in the crust in response to the same thermal anomaly and extensional event (Fig. 7b, c).

Of the metallogenic processes leading to the formation of the uranium ore deposits, two major episodes were synchronous with the two W ± Sn episodes documented in the present paper. The first episode corresponded to the formation of the uraninite preconcentrations in relation to specialized fine-grained granite intrusion synchronous with the major Limousin-type laccolith emplacement at about 325 Ma (Cuney et al. 1990) and the Puy-les-Vignes and Moulin-Barret W ± Sn deposit formation, but which occurred at a lower structural level. Towards the end of the second

W ± Sn episode, fluid percolation, induced by the fast uplift regime, produced hydrothermal quartz leaching of the granite ('episyenite') at 305 ± 1 Ma (Scaillet et al. 1996b) at about the same structural level as the W ± Sn and rare metal deposits of this period (El Jarray et al. 1994). The episyenites represent the major trap for uranium ore deposition that occurred during a later extension episode, at 280–270 Ma (Leroy & Holliger 1984).

Conclusion

This paper provides new $^{40}Ar/^{39}Ar$ dating concerning the timing of W ± Sn and rare metal mineralization in the northern Limousin. Two metallogenic episodes occurred in this area: an Early Namurian episode and a Mid-Westphalian one. If the second episode corresponds with previous available results in the northern (Cheilletz et al. 1992) and southern (Bouchot et al. 2000) part of the French Massif Central, the first episode is demonstrated for the first time. Such timing of mineralization is directly related to the tectono-magmatic evolution of the northern Limousin area during Neo- to Late Variscan stages.

More specifically, the two W ± Sn and rare metal metallogenic episodes found in the northern Limousin are both related to peraluminous leucogranite emplacement, but in two different geodynamic contexts.

(i) The Early Namurian episode (*c.* 325 Ma) is contemporaneous with the deep emplacement of Limousin-type peraluminous leucogranite laccoliths during syncollisional extension (Faure & Pons 1991), associated with non-economic deposits at the same structural level as the granites (9–12 km), but also with larger deposits, at Puy-les-Vignes and Moulin-Barret, formed at shallower level (6–9 km), above cryptic apices of similar leucogranites. These early deposits were preserved in the blocks downfaulted during the late Variscan generalized extension stage. This unexpected result indicates that the W ± Sn metallogenic potential of the Namurian peraluminous leucogranites might have been more important than initially proposed.

(ii) The Mid-Westphalian episode (*c.* 310 Ma) corresponds to the generalized extension and is coeval with the thermal event corresponding to the devolatilization of the lower crust that induced the emplacement of more fractionated peraluminous leucogranites as the RMG at shallow structural levels (3–5 km). At the scale of the French Massif Central, a major part of the W ± Sn deposits results from this episode.

In the northern Limousin, no W ± Sn deposits shallower than 3–5 km are known. Comparatively, in the Erzgebirge province, granitic magma emplacement occurred at depths of only 1–2 km (Durissova 1988). Such shallow depths have permitted the development of more intensive fracturing of enclosing rocks and fluid circulation, and thus the formation of larger stockworks (Cuney *et al.* 1992). Consequently, in the Limousin, larger W ± Sn deposits might have existed during the Neo- to Late Variscan stage, at upper structural levels, but were eroded; secondly some of them may be preserved in down-faulted blocks and may represent interesting targets for W ± Sn exploration.

This work was supported by the GéoFrance 3D programme '3-D Mapping and Metallogeny of the French Massif Central' (contribution no. 118). It is also the CRPG contribution no. 1558. The authors thank the referees (R. Clayton and an anonymous referee) and the editor (D. Blundell) for very constructive and helpful comments on the manuscript.

References

ALEXANDROV, P. 2000. *Géochronologie U/Pb at* [40]*Ar/* [39]*Ar de deux segments Calédoniens et Hercyniens de la chaine Varisque: Haut Limousin et Pyrénées Orientales.* PhD thesis, INPL Nancy.

ALEXANDROV, P., CHEILLETZ, A., DELOULE, E. & CUNEY, M. 2000. 319 ± 7 Ma age for the Blond granite (northwest Limousin, French Massif Central) obtained by U/Pb ion-probe dating of zircons.

Comptes Rendus, Académie des Sciences, **330**, 1–7.

ALIKOUSS, S. 1993. *Contribution à l'étude des fluids crustaux: approche expérimentale et analytique.* PhD thesis, INPL, Nancy, France.

AUBERT, G. 1969. *Les coupoles granitique de Montebras et d'Echassières (Massif central français) et la genèse de leur minéralisations en étain, lithium, tungstène et béryllium.* Mémoire BRGM, **46**.

BELKASMI, M., CUNEY, M., RAIMBAULT, L. & POLLARD, P. 1991. Chemistry of the micas from the Yashan rare metal granite (SE China). A comparison with Variscan exemples. *In:* PAGEL, M. & LEROY, J. (eds) *Proc. 25 years SGA meeting: Source, Transport and deposition of metals.* Balkema, Rotterdam, **25**, 729–732.

BERTHIER, F., DUTHOU, J.-L. & ROQUES, M. 1979. Datation Rb-Sr du granite de Guéret: âge fini-Dévonien de la mise en place de l'un de ses faciès types. *Bulletin du Bureau de Recherche Géologique et Minière*, **1**, 59–72.

BERTRAND, J.M., LETERRIER, J., CUNEY, M., BROUAND, M., STUSSI, J.M., DELAPIERRE, E. & VIRLOGEUX, D. 2001. Géochronologie U-Pb sur zircons de granitoïdes du Confolentais, du massif de Charoux-Civray (seuil de Poitou) et de Vendée. *Géologie de la France*, **1-2**, 167–189.

BOIRON, M.C., CATHELINEAU, M., BANKS, D., VALLANCE, J., FOURCADE, S. & MARIGNAC, C. 2000. Behaviour of fluid in Variscan crust during Late carboniferous uplifting and fluid mixing, and their consequences on metal transfer and deposition. *In: A Geode-Geofrance 3D workshop on orogenic gold deposits in Europe with emphasis on the Variscides.* Documents BRGM, **297**, 58–59.

BOUCHOT, V., MILÉSI, J.P. & LEDRU, P. 2000. Crustal-scale hydrothermal palaeofield and related Variscan Au, Sb W orogenic deposits at 310-305 Ma (French Massif Central, Variscan Belt). *Society for Geology Applied to Mineral Deposits News*, **10**, 1–12.

BURG, J.P., VAN DEN DRIESSCHE, J. & BRUN, J.P. 1994. Syn- to post-thickening extension: mode and consequences. *Comptes Rendus, Académie des Sciences, Série II*, **319**, 1019–1032.

CATHELINEAU, M., CUNEY, M., BOIRON, M.C., COULIBALY, Y. & AYT OUGOUGDAL, M. 1999a. Paléopercolations et paléointeractions fluides-roches dans les plutonites de Charroux-Civray. *In: Etude du massif de Charroux-Civray.* Actes des Journées Scientifiques CNRS/ANDRA, Poitiers, 13 et 14 Octobre 1997, 151–170.

CATHELINEAU, M. & MARIGNAC, C. 1999b. Sources of fluids and regimes of fluids during the Late Carboniferous uplift of the Variscan crust and consequences on metal transfer and deposition. *In: Colloque GéoFrance 3D: results and perspectives.* Documents BRGM, **293**, 46–48.

CATHLES, L.M. & ERENDI, A.H. 1997. How long can a hydrothermal system be sustained by a single intrusive event? *Economic Geology*, **92**, 766–771.

CHEILLETZ, A., ARCHIBALD, D., CUNEY, M. & CHAROY, B. 1992. Ages [40]Ar/[39]Ar du leucogranite à topaze-lépidolite de Beauvoir et des pegmatites sodolithiques de Chédeville (Nord Massif Central, France). Signification pétrologique et géodynamique. *Comp-*

tes Rendus, Académie des Sciences, Série II, **315**, 329–336.

CLAYTON, R. & MAYEDA, T. 1963. The use of bromine pentafluoride in the extraction of oxygen from oxides and silicates for isotopic analysis. *Geochimica et Cosmochimica Acta*, **27**, 43–52.

COSTA, S. & REY, P. 1995. Lower crustal rejuvenation and growth during post-thickening collapse: Insight from a crustal cross section through a Variscan metamorphic core complex. *Geology*, **23**, 905–908.

CUNEY, M., FRIEDRICH, M., BLUMENFELD, P., BOURGUIGNON, A., BOIRON, M.C., VIGNERESSE, J.L. & POTY, B. 1990. Metallogenesis in the French part of the Variscan orogen, Part I: U preconcentration in pre-Variscan and Variscan formations - a comparison with Sn, W and Au. *Tectonophysics*, **177**, 39–57.

CUNEY, M., MARIGNAC, C. & WEISBROD, A. 1992. The Beauvoir topaz-lepidolite albite granite (Massif Central, France): the disseminated magmatic Sn-Li-Ta-Nb-Be mineralisation. *Economic Geology*, **87**, 1766–1794.

CUNEY, M., BROUAND, M., STUSSI, J.M. & VIRLOGEUX, D. 2001. Le complexe plutonique de Charroux-Civray (Vienne): témoin du magmatisme infra-carbonifère dans le segment occidental de la chaîne varisque européenne. *Géologie de la France*, **1-2**, 143–166.

DOWNES, H., KEMPTON, P.D., BRIOT, D., HARMON, R.S. & LEYRELOUP, A. 1991. Pb and O isotope systematics in granulite facies xenoliths, French Massif Central: Implications for crustal processes. *Earth and Planetary Science Letters*, **102**, 342–357.

DURISSOVA, J. 1988. Diversity of fluids in the formation of ore assemblages in the Bohemian Massif (Czechoslovakia). *Bulletin de Minéralogie*, **111**, 177–492.

DUTHOU, J. 1978. Les granitoïdes du Haut Limousin (Massif central français) chronologie Rb/Sr de leur mise en place; le thermo-métamorphisme carbonifère. *Bulletin de la Société Géologique de France*, **20**, 229–235.

EL JARRAY, A., BOIRON, M.C. & CATHELINEAU, M. 1994. Percolation microfissurale des vapeurs aqueuses dans le granite de Pény (Massif de Saint Sylvestre, Massif Central): relations avec la dissolution du quartz. *Comptes Rendus, Académie des Sciences, Série II*, **318**, 1095–1102.

FAURE, M. & PONS, J. 1991. Crustal thinning recorded by the shape of Namurian-Westphalian leucogranite in the Variscan belt of the northern Massif Central, France. *Geology*, **19**, 730–733.

FAURE, M. 1995. Late orogenic carboniferous extensions in the Variscan French Massif Central. *Tectonics*, **14**, 132–153.

FOURCADE, S. & BOIRON, M.C. 2000. Fluides and Late Carboniferous Variscan gold mineralizations in the French Massif Central; the bearing of stable isotopes. *In: A Geode-Geofrance 3D workshop on orogenic gold deposits in Europe with emphasis on the Variscides*. Documents BRGM, **297**, 58–59.

FURLONG, K.P., HANSON, R.B. & BOWERS, J.R. 1991. Modelling thermal regimes. *In:* KERRICK, D.M. (ed.) *Contact metamorphism*. Reviews in Mineral-ogy, **26**, 437–505.

GEBAUER, H., BERNARD-GRIFFITHS, J. & GNÜNEN-FELDER, M. 1981. U/Pb zircon and monazite dating of mafic-ultramafic complex and its country rocks. Example: Sauviat-sur-Vige, French Massif Central. *Contributions to Mineralogy and Petrology*, **76**, 292–300.

HANES, J.A., ARCHIBALD, D.A., HODGSON, C.J. & ROBERT, F. 1992. Dating of Archean quartz vein deposits in the Abitibi greenstone belt, Canada: [40]Ar/[39]Ar evidence for a 70–100-m.y.-time gap between plutonism-metamorphism and mineralization. *Economic Geology*, **87**, 1849–1861.

HOLLIGER, P., CUNEY, M., FRIEDRICH, M. & TURPIN, L. 1986. Age carbonifère de l'unité de Brâme du complexe granitique peralumineux de St Sylvestre (N.O. Massif Central) défini par les données isotopiques U-Pb sur zircon et monazite. *Comptes Rendus, Académie des Sciences*, **303**, 1309–1314.

LE CARLIER DE VESLUD, C. & CUNEY, M. 2000. Relationships between granitoids and mineral deposits: 3-D modelling of the Variscan Limousin province (NW French Massif Central). *Transactions of the Royal Society of Edinburgh, Earth Sciences*, **91**, 283–301.

LEDRU, P., LARDEAUX, J.M., SANTALLIER, D., AUTRAN, A., QUÉNARDEL, J.M., FLOC'H, J.P., LEROUGE, G., MAILLET, N., MARCHAND, J. & PLOQUIN, A. 1989. Ograve; sont passées les nappes dans le Massif central français? *Bulletin de la Société Géologique de France*, **8**, 605–618.

LEROUGE, C., FOUILLAC, A.M., ROIG, J.Y. & BOUCHOT, V. 1999. Stable isotope contraints on the formation temperature and origins of the Late Variscan As-W mineralization at La Chataigneraie, Massif Central, France. *EUG 10. Journal of Conference Abstracts*, **4**, 483.

LEROY, J. & HOLLIGER, P. 1984. Mineralogical, chemical and isotopic (U-Pb method) studies of the Hercynian uraniferous mineralizations (Fanay and Margnac mines, Limousin, France). *Chemical Geology*, **45**, 121–134.

MARIGNAC, C. & CUNEY, M. 1999. Ore deposits of the French Massif Central: insight into the metallogenesis of the Variscan collision belt. *Mineralium Deposita*, **34**, 472–504.

MATSUHISA, Y., GOLDSMITH, J.R. & CLAYTON, R.N. 1979. Oxygen isotope fractionation in the system quartz-albite-anortite-water. *Geochimica & Cosmochimica Acta*, **43**, 1131–1140.

McDOUGALL, M. & HARRISON, T.M. 1999. *Geochronology and thermochronology by the [39]Ar/[40]Ar method*. Oxford University Press, New York.

MONIÉ, P., RESPAUT, J.P., BRICHAUD, S., BOUCHOT, V., FAURE, M. & ROIG, J.Y. 2000. [40]Ar/[39]Ar and U-Pb geochronology applied to Au-W-Sb metallogenesis in the Cévennes and Châtaigneraie districts (Southern Massif central, France). *In: A Geode-Geofrance 3D workshop on orogenic gold deposits in Europe with emphasis on the Variscides*. Documents BRGM, **297**, 77–79.

MONIER, G., CHAROY, B., CUNEY, M., OHNENSTETTER, D. & ROBERT, J.L. 1987. Evolution spatiale et temporelle de la composition des micas du granite albitique à topaze-lépidolite de Beauvoir. *Géologie Profonde de la France*, **1**, 179–188.

NESBITT, B. & MUEHLENBACHS, K. 1995. Geochemical studies of the origins and effects of synorogenic crustal fluids in the southern Omineca Belt of British Columbia, Canada. *Geological Society of America Bulletin*, **107**, 1033–1050.

PEIFFER, M.T. 1986. La signification de la ligne tonalitique du Limousin. Son implication dans la structuration varisque du Massif Central français. *Comptes Rendus, Académie des Sciences, Série II*, **303**, 305–310.

PIN, C. & VIELZEUF, D. 1983. Granulites and related rocks in Variscan median Europe: A dualistic interpretation. *Tectonophysics*, **93**, 47–74.

QUENARDEL, J.M., SANTALLIER, D., BURG, J.P., BRIL, H., CATHELINEAU, M. & MARIGNAC, C. 1991. Le Massif Central. *Sciences Géologiques, Bulletin*, **44**, 105–206.

RAIMBAULT, L. 1984. *Géologie, pétrographie et géochimie des granites et minéralisations associées de la région de Meymac (Haute Corrèze, France)*. PhD thesis, Ecole des Mines, Paris.

RAIMBAULT, L. 1998a. *Minéralisations Sn-W et granites à métaux rares en Nord-Limousin. Zonalité géochimique du prospect de Moulin-Barret et du massif granitique de Blond*. GéoFrance 3D Report, BRGM, **LHM/RD/98/56**.

RAIMBAULT, L. 1998b. Composition of complex lepidolite-type pegmatites and of constituent columbite-tantalite, Chèdeville, Massif Central, France. *The Canadian Mineralogist*, **36**, 563–583.

RAIMBAULT, L. 1999. Tin-tungsten vein mineralisation at Moulin-Barret, France. *In:* STANLEY, C.J. *ET AL.* (eds) *Mineral deposits: processes to processing*. Balkema, Rotterdam, 417–420.

RAIMBAULT, L. & BURNOL, L. 1998. The Richemont rhyolite dyke, Massif Central, France: a subvolcanic equivalent of Rare Metal granites. *The Canadian Mineralogist*, **36**, 265–282.

RAIMBAULT, L., CUNEY, M., AZENCOTT, C., DUTHOU, J.L. & JORON, J.L. 1995. Geochemical evidence for a multistage magmatic genesis of Ta-Sn-Li mineralization in the granite at Beauvoir, French Massif

Central. *Economic Geology*, **90**, 548–576.

RUFFET, G., GRUAU, G., BALLÈVRE, M., FÉRAUD, G. & PHILIPOT, P. 1997. Rb/Sr and ^{40}Ar/^{39}Ar laser probe dating of high-pressure phengites from Sesia zone (Western Alps): underscoring of excess argon and new age constraints on the high pressure metamorphism. *Chemical Geology*, **141**, 1–18.

SCAILLET, S., CHEILLETZ, A., CUNEY, M., FARRAR, E. & ARCHIBALD, A.D. 1996a. Cooling pattern and mineralisation history of the Saint Sylvestre and Western Marche leucogranite plutons, French Massif Central: I.^{40}Ar/^{39}Ar isotopic constraints. *Geochimica et Cosmochimica Acta*, **60**, 4653–4671.

SCAILLET, S., CUNEY, M., LE CARLIER DE VESLUD, C., CHEILLETZ, A. & ROYER, J.J. 1996b. Cooling pattern and mineralisation history of the Saint Sylvestre and Western Marche leucogranite plutons, French Massif Central: II. Thermal modelling and implications for the mechanisms of U-mineralization. *Geochimica et Cosmochimica Acta*, **60**, 4673–4688.

TAYLOR, H.P. 1968. The oxygen isotope geochemistry of igneous rocks. *Contributions to Mineralogy and Petrology*, **19**, 1–71.

VALLANCE, J., CATHELINEAU, M., MARIGNAC, C., BOIRON, M.C., FOURCADE, S., MARTINEAU, F. & FABRE, C. 2001. Microfracturing and fluid mixing in granites: W-(Sn) ore deposit at Vaulry (NW French Massif Central). *Tectonophysics*, **336**, 43–62.

VIRLOGEUX, D., ROUX, J. & GUILLEMOT, D. 1999. Apport de la géophysique à la connaissance du massif de Charroux-Civray et du socle poitevin. *In: Etudes du Massif de Charroux-Civray, Journées scientifiques CNRS/ANDRA, Poitiers, 13 et 14 octobre 1997*. EDP sciences, Les Ulis, 33–62.

WEPPE, M. 1951. *Contribution à l'étude des gîtes de tungstène français: Puy-les-Vignes (Haute Vienne), La Châtaigneraie (Cantal)*. Société d'Impressions typographiques, Nancy.

WILLIAMSON, B.J., SHAW, A., DOWNES, H. & THIRLWALL, M.F. 1996. Geochemical constraints on the genesis of Hercynian two-mica leucogranites from the Massif central, France. *Chemical Geology*, **127**, 25–42.

Discrimination criteria for assigning ore deposits located in the Dinaridic Palaeozoic–Triassic formations to Variscan or Alpidic metallogeny

IVAN JURKOVIĆ[1] & LADISLAV A. PALINKAŠ[2]

[1]Faculty of Mining, Geology and Oil Engineering, Univ. of Zagreb, Pierottijeva 6, Croatia
[2]Faculty of Natural Sciences, Univ. of Zagreb, Horvatovac bb, Croatia
(e-mail: lpalinka@public.srce.hr)

Abstract: Among the ore deposits located in the Palaeozoic–Triassic formations of the Dinarides, siderite ($\pm$ ankerite) and barite mineralization hosted by Carboniferous and Lower Permian sediments as well as polymetallic deposits in the Mid-Bosnian Schist Mts (MB), caused a long standing dispute concerning their metallogenic affiliation.

Geochemical data obtained for ore minerals from the deposits of ambiguous affinity delivered certain patterns, which might be useful as discrimination criteria. The mean $\delta^{34}S$ value of sulphate sulphur in barites from Palaeozoic hydrothermal deposits (+10.79‰) is close to that of Upper Permian gypsum–anhydrite deposits (+11.02‰). Triassic hydrothermal deposits bear barites with a mean $\delta^{34}S$ value of +21.7‰, which corresponds to that of Triassic gypsum–anhydrite deposits ($\delta^{34}S$ = +20.45‰). The mean content of $SrSO_4$ (3.62%) of barites in Palaeozoic rocks differs significantly from that obtained for Triassic barites (2.36% $SrSO_4$). Tectonic settings of the Palaeozoic ore deposits are antithetic to those of the Alpidic cycle.

Paragenetic characteristics and the chemical composition of ore-forming fluids can be used for the same purpose. Parageneses of ore deposits hosted by Palaeozoic rocks are essentially distinct from those hosted by Triassic rocks. Fluid inclusion microthermometry reveals differences between ore fluids related to the Variscan and Alpidic metallogenic cycles, respectively: fluid inclusions in minerals from 'Variscan' deposits are characterized by great salinity variations (5–35 NaCl wt% equiv), abundant daughter minerals, and homogenization temperatures between 150–450 °C. The microthermometric data point to $NaCl-H_2O$, $CaCl_2$ ($\pm MgCl_2$)–H_2O and $NaCl-KCl-H_2O$ fluid systems. In contrast, Triassic ore fluids have low salinities (3–10 NaCl wt.% equiv.) and homogenization temperatures (60–180 °C). Their chemical composition is dominated by $NaCl-CaCl_2-H_2O$.

The aim of this paper is to summarize the growing number of geochemical data on the ore deposits in the Dinarides, in order to distinguish the time of their genesis between two metallogenic periods, Variscan and Alpidic, respectively.

There is a good agreement in genetic interpretations for those Variscan genetic types, which have corroborated stratiform signatures and are hosted in Palaeozoic formations. These are first the rift related Late Silurian (S_3), Early Devonian (D_1) and early Mid-Devonian ($^2D_2$) iron deposits of Mt Medvednica (Šinkovec *et al.* 1988; Fig. 1) and the area of Ključ (Una-Sana, US), Dusina (MB, Jurković 1957) and Tajmište (western Macedonia, Cissarz 1956); secondly, polymetallic quartz breccias (Okoška Gora, Slovenia) and bedded veins (Remšnik, Slovenia), both of them with Fe, Cu, Pb and Zn sulphides, Drovenik *et al.* (1980).

There is also the reconciled opinion for post-collisional stratiform and stratabound deposits in the Mid-Permian Groeden formation, deposited in intramontane basins. These are the U-peneconcordant deposit Žirovski Vrh (pitchblende–coffinite, Sava folds, SF, Fig. 1: Omaljev 1982; Palinkaš 1986), Cu-red bed type Škofje (chalcopyrite, bornite, chalcocite and pyrite, Sava Folds, SF), and Ustiprača (SEB, Kulenović 1987); then bedded, low-manganese, SEDEX, hematite deposits at Hrastno (Sava Folds, SF), Rude (Samoborska gora, Mt, SG) and Bukovica (Petrova Gora, P; Cop *et al.* 1998). The widespread stratiform, lower Upper Permian, gypsum–anhydrite deposits display a conspicuous distinction from Scythian stratiform gypsum–anhydrite deposits, as indicated by the difference in $\delta^{34}S$ value of sulphate

From: BLUNDELL, D.J., NEUBAUER, F. & VON QUADT, A. (eds) 2002. *The Timing and Location of Major Ore Deposits in an Evolving Orogen.* Geological Society, London, Special Publications, **204**, 229–245. 0305-8719/02/$15.00 © The Geological Society of London 2002.

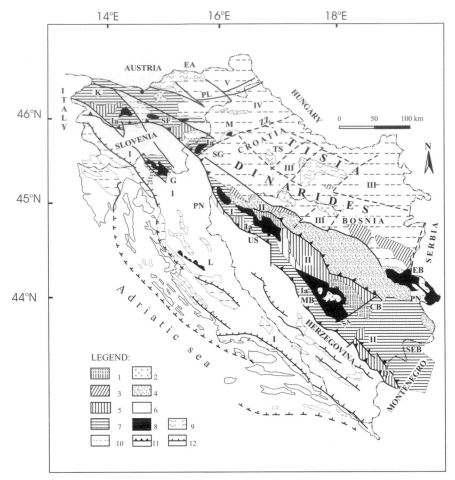

Fig. 1. Geological sketch-map of the central and northwestern Dinarides and the southwestern parts of Tisia showing the location of Palaeozoic complexes (black), (Mioč 1984; Pamić & Jurković 1997) with index-map (Neubauer & von Raumer, 1993). 1, Tertiary; 2, Palaeogene metamorphic rocks and granitoids; 3, Upper Cretaceous–Palaeogene flysch (active continental margin); 4, Dinaridic Ophiolite zone, mostly mélange; 5, Jurassic to Upper Cretaceous sequences (passive continental margin); 6, Adriatic–Dinaridic carbonate platform; 7, allochthonous Triassic sequences (CB, Central Bosnia); 8, allochthonous Palaeozoic sequences and rhyolites (x); 9, Palaeozoic metamorphic rocks and granitoids (Eastern Alps, EA and Tisia, TS); K, Karawanken; SF, Sava Folds; G, Gorski Kotar; SG, Samoborska Mt; M, Medvednica Mt; L, Lika; P, Petrova Mt; T, Trgovska Mt; US, Una-Sana; MB, Middle Bosnian Schist Mts; EB, East Bosnia; SEB, SE Bosnia; 10,faults; 11, interterrane thrust; 12, intra-terrane thrust-klipe or window. **I,** External Dinarides; **Ia,** Sava nappe; **II,** Internal Dinarides; **III,** Pannonian Basin underlain by Tisian basement; **IV,** Zagorje–Midtransdanubian zone; **A,** Austroalpine domain (Eastern Alps); **DN,** Durmitor nappe; **PN,** Pannonian nappe; **MBSM,** Mid-Bosnian Schist Mts. Large transform faults: **PL,** Periadriatic; **ZZ,** Zagreb–Zemplen, **SA,** Sarajevo

sulphur (Jurković & Šiftar 1995). On the other hand, there is no dilemma about the affiliation of all mineral deposits hosted by Lower and Middle Triassic sedimentary and igneous rocks, which are genetically closely related to the Triassic rifting magmatism.

The ambiguity in affiliation arises, however, within three large groups of deposits hosted by Carboniferous and Permian rock formations. The first group comprises stratabound deposits of

siderite $\pm$ ankerite $\pm$ Cu, Pb, Zn sulphides, located in Middle and Upper Carboniferous (C_2 and C_3) sediments (Jurković 1961; Jurić 1971; Drovenik *et al.* 1980; Palinkaš 1988). The second group of deposits in dispute includes veins and irregular barite bodies located in Carboniferous and Lower Permian strata (Jurković 1956, 1957, 1958, 1987). The third group of disputable mineral deposits is located within the Mid-Bosnian Schist Mts, spatially connected with extrusion of Palaeozoic rhyolites (Jurković 1956).

Controversial interpretation of their origin exists. One opinion leads the origin of the deposits (hosted by Palaeozoic formations) to the Variscan orogeny (Cissarz 1956; Jurković 1956, 1957, 1961, 1995; Janković 1967, 1982a; Ramović 1976, 1979; Kubat et al. 1980; Drovenik et al. 1980; Jurković & Palinkaš 1996; Jurković & Pamić 1999). However, based on more recent investigations, Janković (1982b; 1987; 1990) and Kubat (2001) believe that these deposits belong to the Alpidic metallogenic cycle.

The intention of this paper is to define firmer discrimination criteria for affiliation of the deposits to the Variscan or Alpidic metallogenic cycle. In this respect the following characterics were applied: (1) tectonic setting; (2) isotopic composition of sulphate sulphur; (3) $SrSO_4$ content of barite; (4) paragenesis of ore deposit; (5) fluid inclusion data.

Basic geological characteristics of Palaeozoic and Triassic formations and ore deposits

The Dinarides, a mountain chain about 700 km long striking NW–SE, are characterized by a regularly zoned pattern of distinct Mesozoic–Palaeogene tectonostratigraphic units: the Adriatic–Dinaridic carbonate platform; the Dinaridic Ophiolite zone; and the Upper Cretaceous–Palaeogene flysch formations (Pamić & Jurković 1997; Pamić et al. 1998; Fig. 1a). This consistent pattern is disturbed by allochthonous Palaeozoic–Triassic nappes, which are thrust onto the Alpine units of the Internal Dinarides and the northeastern margin of the External Dinarides. During Pliocene tectonism, the Dinaridic segment was thrust onto the South Tisia block, which is situated within the Pannonian Basin and comprises Palaeozoic crystalline rocks (Tari & Pamić 1998). The most important ore deposits related to Variscan and Alpidic cycles are presented in Figure 2.

Variscan formations

The geodynamic evolution of the Dinaridic Palaeozoic complexes is related to the Prototethys evolution between Gondwana and Laurussia (Jurković & Pamić 2001; Pamić & Jurković 2002). Rifting processes started in Silurian time on a pre-Variscan basement. The rifting is indicated by the occurrence of metabasalts interlayered within the Cambrian–Ordovician–Silurian formations of the Drina–Ivanjica Palaeozoic terrane, the NE Dinarides and the Mid-Bosnian

Schist Mts (MB). Numerous small occurrences of iron-quartzites and bedded magnetite–hematite and manganese oxide deposits in Western Macedonia and in the Drina–Ivanjica are related to this magmatic activity (Djoković 1985; Popović 1984). Opening of the Prototethys started probably in Late Silurian/Early Devonian time. The occurrence of Early Palaeozoic platform carbonates in the southwestern parts of the allochthonous Variscan complexes of the Internal Dinarides implies that their southern terranes were related to the North Gondwana margin. In the shelf realms of this passive continental margin, the following types of ore deposits were formed: (1) iron-quartzites in the Medvednica Mts (M, Fig. 1; Šinkovec et al. 1988), (2) thinly stratified beds of magnetite–chamosite and pyrite–chamosite beds in the Ključ area, Central Bosnia, and (3) stratified beds of hematite ores in the Mid-Bosnian Schist Mts (Jurković 1995).

An active margin developed in the northern part of the Prototethys, along South Laurussia. The clastic formations of the South Tisia block are interlayered with tholeiitic basalts (orthoamphibolites) of back-arc basin affinity. This feature and the occurrence of Alpine-type ultramafics and I-type granites suggest a subduction-related geotectonic setting. Subduction processes were taking place at the end of the Devonian and at the beginning of the Carboniferous (343 Ma). The main Variscan deformation and metamorphism occurred in the Westphalian (Jurković & Palinkaš 1996; Jurković & Pamić 2001).

This geological process leads to metamorphic sequences of Barrovian-type (greenschists and amphibolite facies), migmatites and syn-kinematic S-type granites. Lenses and irregular pegmatite bodies cross-cut gneisses, migmatites and S-type granites of the Papuk Mts. Small-sized I-type granite bodies contain juvenile pegmatites, and locally skarns and syn-kinematic and post-kinematic quartz secrete veins and lenses in the schists. Numerous small deposits of graphitite, meta-antracite and D_2 graphite occur in the Psunj Mts (Slavonian Mts, TS, Fig. 1; Jurković 1995). Small ortho-amphibolites of back-arc basin affinity and Alpine-type ultramafics in South Tisian Barrovian-type sequences provide evidence for the Prototethyan oceanic crust generation. The main compressional event took place by the end of Namurian and Westphalian time (310–320 Ma), followed by the closure of the Prototethys, uplift and exhumation of the Variscan orogenic belt and termination of the Variscan cycle. The final phase is characterized by subsidence of the carbonate platform and accumulation of syn-orogenic flysch (Jurković & Pamić 2001; Pamić & Jurković 2002).

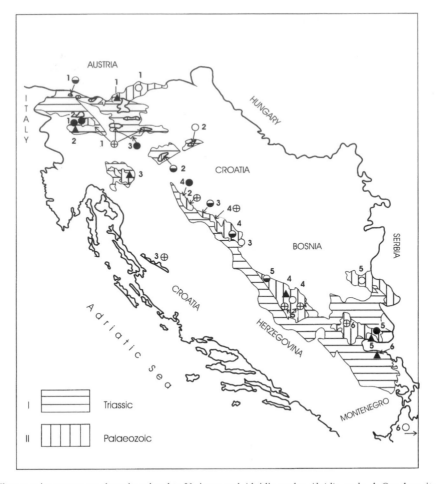

Fig. 2. The most important ore deposits related to Variscan and Alpidic cycles. *Alpidic cycle.* ▲ Ore deposits related to Early and Mid-Triassic rifting: 1, Mežice; 2, Idrija; 3, Gorski Kotar; 4, Radovan Mts, 5, Kovač Mts; 6, Šuplja Stijena; 7, Vareš; 8, Čevljanovići. *Variscan cycles.* ● Intramontane trough deposits: 1, Žirovski vrh; 2, Škofje; 3, Hrastno; 4, Bukovica; 5, Goražde–SEB. ⊕ Post-collision deposits:.1, Sava–Fold; 2, Petrova G. Mts; 3, Lika; 4, Una–Sana; 5, Kreševo –MBS Mts; 6, Prača. △ Subduction collision deposits: 1, Jesenice; 2, Samoborska G. Mts; 3, Trgovska G. Mts; 4, Ljubija; 5, Sinjakovo MBS Mts; ○ Rifting phase deposits: 1, Remšnik; 2, Medvednica; 3, Ključ; 4, Dusina MBS Mts; 5, Drina–Ivanjica; 6, Tajmište (western Macedonia).

Post-Variscan formations

The Variscan basement is disconformably overlain by younger post-Variscan formations. The post-Variscan formations of the Dinarides and the adjacent South Tisia display a prominent facies differentiation. Tectonic activity, which took place after the main Variscan deformation, gave rise to regional, differentiated regression and to transcurrent faulting in the exhumed Variscan orogeny.

Middle Carboniferous (C2) hosts numerous stratabound siderite ± ankerite deposits and Upper Carboniferous (C₃) rocks (shales, silts, sandstones and carbonates) as beds, lenses, bedded veins or massive bodies (Jurković 1961; Jurić 1971; Palinkaš 1988; Jurković & Palinkaš 1996). The most important ones are found in the Ljubija ore district in the Una–Sana Palaeozoic terrane (C₂, US, Fig. 1). Siderite deposits with chalcopyrite occur in Sinjakovo (C₂, MB) and in the Trgovska Gora Mt. (C₂, Gradski Potok deposit, T, Figure 1), silver-bearing galena was mined in Zrinj deposit (C₂, T, Fig. 1), and medium-sized siderite deposits occur at Jesenice and Vitanje (C₃, Slovenia, Jurković 1995).

Beside siderite deposits, the largest barite deposits of the Dinarides are found in post-Variscan formations. Devonian carbonate rocks and pre-Devonian schists of the Mid-Bosnian Schist Mts host the most important ores. The gold-silver-bearing, mercurian tetrahedrite (schwazite) is the main ore mineral and contains $20-50 \mathrm{g\,t}^{-1}$ Au, $0.08-0.21$ wt% Ag and $1.5-16.0$ wt% Hg (Jurković 1960). Barite deposits in the Devonian carbonate rocks contain $BaSO_4$ as the major constituent along with abundant tetrahedrite. Some of the deposits in the pre-Devonian schists locally contain siderite in significant amounts. In such deposits tetrahedrite is the main ore mineral, with $25-65$ wt% in the siderite crude ore and with $10-15$ wt% in the barite crude ore (Mačkara and Mračaj copper deposits near Gornji Vakuf, Central Bosnia; Jurković 1956, 1960).

Other barite deposits, mostly monomineralic, occur in SE Bosnia, in the area of Prača-Foča (SEB, Fig. 1) and in the Una–Sana Palaeozoic terrane (US, Fig. 1). Palaeozoic sediments in the adjacent Trgovska gora Mts (T, Fig. 1) contain small, almost monomineralic barite deposits located only in Carboniferous strata. Large, monomineralic barite veins were found in the eastern part of the Petrova Gora Mts (P, Fig. 1; Jurković 1958). Concordant and discordant lead, zinc, mercury, antimony and iron ore veins with different proportions of quartz, calcite and barite as gangue minerals, occur along the 100 km long Sava Folds Zone (SF, Slovenia, Fig. 1). These veins were generated prior to sedimentation of Groeden clastics and are genetically related to the presumed Lower Permian rhyolitic magmatism (Drovenik et al. 1980; Mlakar 1993). In the Lika area (L, Fig. 1), Middle Carboniferous shales contain primary, syn-sedimentary barite–pyrite deposits. Upper Carboniferous limestones are cross-cut by lenses, irregular bodies and veins of remobilized barite (coarse-grained) and dolomite (Palinkaš 1988).

Of particular interest are small and middle-sized epigenetic polymetallic ore deposits in the area of Busovača–Fojnica–Kiseljak–Kreševo (MB) mostly with pre-Devonian host rocks. The ore deposits are closely associated in space and time with calc-alkaline rhyolitic magmatites. Extrusions and hypabyssal intrusions formed during the uplift of the Mid-Bosnian Schist Mts from Early Carboniferous through Early Permian. Katzer (1925), Jurković (1956), Jeremić (1963) and Hrvatović (1996) confirmed the presence of rhyolitic pebbles in the Middle–Upper Permian basal conglomerates in the Mid-Bosnian Schist Mts.

Mid–Late Permian ore-forming events

There is a group of post-Variscan deposits whose genesis time is not in dispute, but is worth mentioning. Peneconcordant, epigenetic, sedimentary, red-bed type deposits of uranium (Žirovski Vrh, Slovenia) and copper (Škofje, Slovenia), were formed within braided river environment in Mid-Permian time (Drovenik et al. 1980; Omaljev 1982; Palinkaš 1986). Stratiform, low-manganese deposits near Hrastno (Slovenia) and the Bukovica hematite deposit in the Petrova gora Mt (Croatia) were formed at the same time (Čop et al. 1998). The lower part of the Upper Permian sediments, below the Bellerophon limestone formation, comprises numerous, stratiform gypsum–anhydrite deposits in Croatia and Bosnia–Herzegovina (Jurković & Šiftar 1995).

Alpidic formations

Alpine rifting magmatism took place on the Variscan basement of Pangaea. The magmatic activity produced intrusive rocks, mostly diorite and albite–syenite, with subordinate gabbro, albite-syenite and granite–granodiorite. More abundant volcanics are andesites, dacites and basalts, accompanied by volcanic breccias and tuffs. They are interlayered with marine Upper Permian and Scythian, primarily clastic sediments and Anisian to Norian, predominantly carbonate, alternating with shales, cherts and subordinate sandstones. The most intense volcanic activity occurred during the Ladinian (Pamić 1984).

The Triassic ore deposits are located in hypabyssal intrusive and the volcano-sedimentary formations as well as in adjacent rocks composed of Palaeozoic, Scythian and Anisian sediments. The ore deposits predominantly contain iron, manganese, barium, mercury, lead and zinc ores, with minor content of copper and fluorine minerals (Janković, 1955; Jurković & Pamić 1999; Palinkaš et al. 1993). The most important genetic and paragenetic types of the deposits, which are hosted by Triassic rocks, are presented in Figure 3.

Discrimination criteria

Differences in geotectonic settings between ore deposits hosted by Triassic and Palaeozoic rocks

Triassic host rocks.. During Late Permian and Early Triassic times, an extensive intra-continental rifting phase took place and finished in the development of multiple graben–horst structures and parallel-sided rift valleys. The Scythian stage in the Dinarides is characterized by shallow-water sedimentation in horst–graben structures. The Anisian stage is characterized by subsidence and widening of the narrow and elongated horst–graben structures and locally by uplift of separate blocks. Subsidence continued into Ladinian time, followed by deposition locally of carbonates and shales. Strong igneous activity occurred along rift faults, often associated with the formation of volcano-sedimentary complexes.

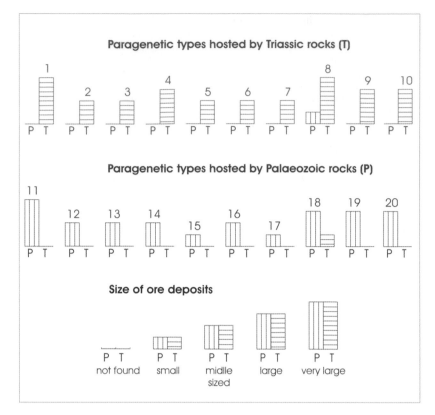

Fig. 3. Paragenetic types hosted by Triassic and Palaeozoic rocks.(**a**) Triassic rocks: 1, Ladinian–Anisian stratiform siderite + hematite ± barite deposits, type Vareš (CB); 2, gabbro–diorite skarns with magnetite + hematite + pyrite ± pyrrhotite (Radovan Mts, Tovarnica—MBS Mts and Čajniče—SEB); 3, stratiform hematite ± manganese oxides (Konjic, Prozor, Bugojno—MBS Mts); 4, stratiform manganese oxides, type Čevljanovići (CB), Bužim (US); 5, stratiform barite + pyrite + sphalerite deposits in cherts, shales, tuffites, type Rammelsberg (Rupice—MBS Mts); 6, intraformational polymictic breccia with barite + galena + sphalerite (Veovača, Srednje, Maine—CB); 7, stratabound mineralized dolomite horizons with quartz + barite + galena + sphalerite + pyrite (Borovica, Maine, Srednje—CB); 8, hydrothermal cinnabar deposits (veinlets, nests impregnations) (Podljubelj—K, Idrija SF), Tršće—G, Draževići— CB, Kreševo —MBS Mts); 9, volcanic–hydrothermal veins and disseminated galena–sphalerite–pyrite deposits within quartz–keratophyre (Šuplja Stijena—Montenegro); 10, volcanic ± volcanic sedimentary hydrothermal massive sphalerite + galena + pyrite deposits (Brskovo—Montenegro) located conformably in layered tuff–pyroclastics– volcanic sequence of keratophyre. (**b**) Palaeozoic rocks: 11, stratabound (C$_2$) siderite ± ankerite (Ljubija—US); 12, stratabound (C$_2$) siderite + chalcopyrite ± galena ± sphalerite (Gradski Potok—T, Sinjakovo—MBS Mts); 13, Stratabound (C$_2$) siderite + silver -bearing galena ± sphalerite ± chalcopyrite (Zrin, Čatrnja—T); 14, gold-bearing pyrite–siderite veins in metarhyolites and Pz-schists; 15, veins and nests of quartz + siderite + sphalerite + galena + magnetite + molybdenite + cassiterite + stannite (Fojnica—MBS Mts); 16, quartz veins in ottrelite schist with antimonite + sphalerite + wolframite + silver-bearing lead–antimony sulphosalts (Čemercica —MBS Mts); 17, quartz veins and nests with realgar + orpiment + fluorite (Hrmza—MBS Mts); 18, barite veins and irregular bodies ± quartz ± siderite ± accessory copper–iron–lead–zinc sulphides (Petrova Gora Mts—P, Blagaj—US, Jezero, Sabeljine pećine, Smiljeva kosa, Ljetovik—MBS Mts, Prača -Foča—SEB, Kovač Mts—Montenegro); 19, siderite + barite veins in pre-Devonian schists ± calcite + gold–silver-bearing mercurian tetrahedrite (Gornji Vakuf MBS Mts); 20, barite metasomatic irregular bodies and veins in Devonian dolomites with calcite ± fluorite ± siderite ± pyrite + gold–silver-bearing mercurian tetrahedrite (Kreševo —MBS Mts).

Associated groups of mineral deposits include:

1   Mineral deposits related to plutonic–hypa-
    byssal intrusions

Skarns within the contact zone of intrusion and surrounding rocks (Fig. 3, No. 2)
Hydrothermal replacement deposits situated some distance from intrusion (Fig.
3, No. 3)

Hydrothermal vein deposits (Fig. 3, No. 9)
Volcano-sedimentary deposits (SEDEX)
  Manganese oxides (Fig. 3, No. 4)
  Pyrite–barite–siderite–hematite, Vareš type, (Fig. 3, No. 1)
  Massive Fe, Pb, Zn sulphides (Fig. 3, No. 10, Janković 1955, 1987)
  Cinnabar (Fig. 3, No. 8)
  Polymetallic barite (Fig. 3, Nos 5, 6, 7).The majority of deposits belonging to the Alpidic cycle mark parageneses rich in sulphosalts of Cu, Pb, Zn, Sb and As (Jurković & Pamić 1999; Jurković & Palinkaš 1996).

These deposits are very rarely found in the adjacent Upper Palaeozoic rocks that occur almost within the Lower and Middle Triassic rocks (tectonic situation).

Palaeozoic host rocks. The most prominent feature of mineralization within the Palaeozoic Mid-Bosnian Schist Mts is a lateral zoning of mineral parageneses controlled by thermal gradients around huge masses of calc-alkaline rhyolites. There is a broad span of temperature zones, from pneumatolytic–katathermal mineralization with Sn, W, Mo, Li, F, B accessory minerals, mesothermal with silver-bearing antimonite, Pb–Sb sulphosalts and gold-bearing pyrite, and epithermal deposits with As-sulphides, melnicovite pyrite and hexahedral fluorite. It indicates a decreasing temperature profile away from the intrusive rhyolitic body. Unusual genetic and paragenetic types of ore deposit, unique in the Dinarides, hosted by Devonian dolomites, pre-Devonian schists and by Upper Palaeozoic rhyolites, are barite–siderite vein and replacement deposits with Au–Ag–Hg tetrahedrite as the main mineral. This Variscan mineralization can be distinguished easily from the neighbouring Alpidic mineralization, which occurs within the Triassic gabbro–diorite intrusion as skarn-related and distinctive hydrothermal iron deposits.

The siderite ± ankerite deposits are characterized by evidence of an affiliation during the Variscan orogeny based on the following observations: (a) stratabound position within Carboniferous sediments, (b) conformable beds, lenses, lensoid veins, (c) relics of primary sedimentary structures in the Ljubija deposits (UnaSana Palaeozoic, US, Fig. 1; Jurković 1961; Jurić 1971), (d) simple parageneses, (e) association with rhyolite sills cross-cut by Triassic intrusions of diabase composition (Sinjakovo deposit, MB, Fig. 1; Katzer 1925; Jurković 1995).

Numerous Pb, Zn, Cu, Fe and Sb veins with different proportions of quartz, calcite, and barite as gangue minerals (Sava Folds region, SF, Fig. 1) are situated in the b_2 unit (Westphalian A-zone) of Mid-Carboniferous age, below the Asturian unconformity (Mlakar 1993). Barite and barite–siderite deposits in the Petrova Gora Mt spatially belong to Lower Permian sandstones and to polymictic conglomerates (Jurković 1958).

Isotope composition

Isotopic compositions of sulphate sulphur of stratiform gypsum–anhydrite deposits in Upper Permian and in Middle Triassic sediments are compared with the values of sulphate sulphur of hydrothermal barites from deposits hosted by Palaeozoic and Triassic rocks. Figure 4a presents $\delta^{34}S$ values of 65 Upper Permian gypsum–anhydrite samples taken from over 20 localities in Croatia, Bosnia and Herzegovina. The arithmetic mean value is $+11.02‰$. Figure 4b shows the results of 117 barite samples from barite deposits from Palaeozoic sediments and metarhyolites. Samples were taken from deposits located in the Sava Folds terrane (SF, Slovenia,), the Samoborska Gora Mt. (SG), the Petrova Gora Mt. (P) and Lika (L) all in Croatia; in the Una-Sana Palaeozoic terrane (US), in SE Bosnia (SEB), and in the Mid-Bosnian Schist Mts. (MB, Fig. 1). The arithmetic mean value is $+10.79‰$ (Šiftar 1981, 1984, 1988; Kubat *et al.* 1980). Figure 4c presents the results of 31 samples of Scythian and Middle–Upper Scythian gypsum–anhydrite deposits from Dalmatia and Bosnia. Their mean $\delta^{34}S$ value is $+20.45‰$ (Jurković & Šiftar 1995). Figure 4d shows $\delta^{34}S$ values of 56 barite samples of deposits located within Triassic rocks. Their mean value is $+21.69‰$ (Šiftar 1982, 1986).

The following persuasive conclusions can be drawn from the sulphur isotope data.

(1) Barites hosted by Triassic strata tend to increase to heavier sulphur isotope values than barites in Palaeozoic rocks. This is illustrated by three examples. (i) In the Petrova Gora Mt (P), numerous barite vein deposits, hosted by Lower Palaeozoic rocks give a mean $\delta^{34}S$ value of $+8.7‰$ (Šiftar 1984). An epigenetic barite vein, however, cutting the stratiform hematite deposits, located in the Middle Permian Groedden formation at Bukovica (the same region), has a value of $+19.5‰$. (ii) Mono-mineralic barite veins in the Vareš region, Central Bosnia (CB), within a small Palaeozoic window, have $\delta^{34}S$ values ranging between $+12$ and $14‰$, while surrounding polymetallic barite deposits, hosted by Triassic rocks, yield $\delta^{34}S$ values from $+20$ to $+24‰$. (iii) Barite deposits hosted by Palaeozoic rocks (from Silurian through Lower Permian) in SE Bosnia (SEB) have $\delta^{34}S$ values around $+12‰$, while barite deposits,

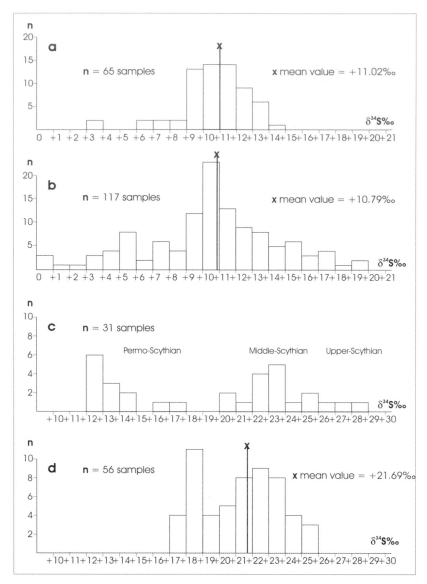

Fig. 4. $\delta^{34}S$ values of: (**a**) Upper Permian evaporites; (**b**) barites hosted by Palaeozoic rocks; (**c**) Triassic evaporites and (**d**) barites hosted by Triassic rocks.

hosted by Triassic rocks, in the very vicinity of the former ones (the Kovač Mt) have markedly heavier sulphur isotopes with $\delta^{34}S$ values of +22‰.

(2) Sulphur from buried Permian evaporates, Permian seawater, or partially evolved Permian evaporitic brines, may include a contribution with the isotopic composition of the post-Variscan barite deposits. A possible alternative would be a mixture of sedimentary sulphur in the barites, with a strong mode at +11‰, derived from the Palaeozoic pile. This possibility, however, is unlikely,

since it should have a regional characteristic, persistently constant in the post-Variscan barite deposits, stretching along the Dinaridic belt for more than 700 km. The conformable sulphur isotope composition of the barite and evaporate deposits is interpreted as a result of ore-forming and evaporitic processes occurring at the same time, or with a slight delay in their operation.

Figure 4c shows a bimodal distribution of isotope values for Permo–Scythian and Middle–Upper Scythian evaporates. The data set increasing from +12 to +28‰ in a geologically short time is the well-known dramatic change of the seawater S-isotope values at the Permian–Triassic transition. Clearly, an irregular frequency distribution has been achieved by gathering biased data that do not cover the whole time-span from Permo–Scythian to Mid–Late Scythian. Figure 4d presents two data sets with bimodal distribution, which correspond to the Lower (Gorski Kotar, G) and Middle Triassic barite deposits (Central Bosnian). The first group (δ^{34}S +19.0‰) consists of 11 samples of the Lowermost Triassic stratabound pyrite–barite deposit in Gorski Kotar, Croatia (Sabkha type, Palinkaš 1988; Palinkaš et al. 1993) and Middle Triassic stratiform Pb–Zn–Fe sulphide deposits in Central Bosnia dominate the second group. Their mean δ^{34}S value of +21.15‰ (39 samples) is close to those evaporates of Mid–Late Scythian age (Jurković & Pamić 1999).

Comparing the results (δ^{34}S values) from Figures 4a–d for the barite deposits in dispute, one important conclusion leads to the interpretation that the two distinctive groups of deposits with sulphur isotopes constrained to Permian and Triassic seawater is represented by time-equivalent evaporates.

SrSO₄ content in barites

The content of $SrSO_4$ in barites may be another discrimination criterion obtained by optical emission spectroscopy and wet chemistry and PIXE (Jurković 1987; Šiftar 1988; Slovenec et al.1997). The Palaeozoic rocks in the Mid-Bosnian Schist Mts (MB) contain the largest deposits of barite. Less important (in quantity) are barite deposits within Triassic rocks in the Central Bosnia (Ramović 1976; Jurković & Pamić 1999). The mean value of 120 samples from the former barite district is 5.20% $SrSO_4$. Some of the barite deposits, south and east of Kreševo, contain increasing $SrSO_4$ content (celestobarite): Gusta Šuma 8.1 8.6 mole%, Dubrave–Dugi Dol 8.7 mole%, Martinovac 9.6 mole%, Bijele Jame 12.8 mole% and Kolovoje 17.6 mole%. In contrast, the barite deposits (29 samples) from Triassic rocks in Central Bosnia show a mean value of 2.7 mole% $SrSO_4$ (Slovenec et al. 1997; Jurković & Pamić 1999). The varying strontium content reflects their different evolution probably in two tectonic settings. The former are related to the Permian rhyolites and Palaeozoic host rocks, while the latter may be part of the Triassic volcano-sedimentary sequences.

Parageneses of ore deposits

There is a great difference between the mineral parageneses of the Palaeozoic and Triassic depos-its (Fig. 3). The main paragenetic differences between Variscan and Alpidic cycles can be summarized:

1 manganese, as well as larger lead-zinc, magnetite, hematite deposits are missing within Palaeozoic rocks

2 Triassic rocks are devoid of copper, antimony, silver-bearing galena, Pb–Sb sulphosalts, gold–silver bearing pyrite and gold–silver–mercury bearing tetrahedrite deposits; the last mentioned is exclusively in the Palaeozoic host rocks (Jurković 1957)

3 lack of Sn, W, Mo, Li, B and F-bearing minerals in Triassic host rocks

4 barite deposits within Palaeozoic rocks have a simple paragenesis—barite (90–99 wt% $BaSO_4$), quartz and siderite are scarce, accessories are Cu, Pb, Zn sulphide (exception is Au–Ag–Hg tetrahedrite in the Mid-Bosnian Schist Mts)

5 in contrast, barite within Triassic rocks occurs only as subordinate gangue mineral (10–15 wt% $BaSO_4$, rarely more than 30–50 wt%) associated with dolomite, calcite, quartz, and pyrite; ore minerals are galena and sphalerite with numerous accessory sulphides and sulphosalts

6 the Triassic rocks contain stratiform and stratabound deposit types, such as beds, bedded lenses, intercalated breccias and ore-bearing dolomite horizons; the Palaeozoic rocks are characterized by veins and irregular replacement bodies.

Physico-chemistry of ore forming fluids

Ore-forming fluids circulating during late Variscan and early Alpidic metallogenic periods differed due to the highly diversified genetic condition in different geotectonic environments. While the former were primarily sub-terrestrial, the latter were mostly products of the advanced rifting as volcano-sedimentary deposits. The contributions of connate, meteoric, magmatic and metamorphic waters resulted in a wide spectrum of ore fluids. In spite of the complexity of fluid chemical compositions, microthermometric data revealed two distinctive groups of solution systems of regional extent, as already indicated by the δ^{34}S values of barite. Influences of Permian and Triassic seawaters and their evaporitic products on the formation of ore fluids are conceivable. A systematic fluid inclusion (FI) analysis of ore and gangue minerals from the major Dinaridic deposits of both metallogenic cycles provides a convincing criterion for their discrimination.

Variscan fluid systems. The most prominent late Variscan and post-Variscan mineralization in the Internal Dinarides consists of numerous siderite (± ankerite) deposits in Palaeozoic sedimentary host rocks and mercurian tetrahedrite-bearing siderite–barite deposits related to metarhyolites in the Mid-Bosnian Schist Mts (MB). Among the best-studied deposits of the former are those in the Petrova Gora Mt (P), the Trgovska Gora Mt (T), and the Ljubija ore district (US), and Kreševo area (MB) (Palinkaš & Jurković 1994; Palinkaš 1988).

Fluid inclusions in quartz from the Trgovska Gora have the following characteristics (Fig. 5a): Liquid (L) plus Vapour (V) type predominates with minor metastable FIs (L); degree of filling is uniform, and density of fluids is close to 1 g cm^{-3}. Temperature of first melting, T_{fm}, is between -18 and $-33\,°C$, indicating a relatively pure NaCl– H_2O fluid system with observed meta-stable eutectic temperature of halite. Temperature of last melting of ice, T_{mice}, is between 0 and $-9\,°C$ with a maximum frequency at -5 to $-8\,°C$. Homogenization temperature, T_h, is between $+\,90$ and $+250\,°C$, with two maxima at $+140$ and $+240\,°C$. Salinity is between 0 to 12 wt% NaCl equivalent, with maxima between 5 and 11 wt% NaCl equivalent. Salinity is higher in the ubiquitous FIs with halite daughter minerals (carbonates are present as well). The data are comparable to other Variscan fluid systems involved in the formation of the barite–siderite deposits in Central Europe (Behr & Gerler 1987; Behr *et al.* 1987; Klemm 1994). Fluid inclusions in quartz from the Petrova Gora Mt. (P, Fig. 5a), give similar results. The presence of ($L_w + L_{CO2} + V$) is recorded as well. Salinity ranging between 2–17 and 19–22 wt% NaCl equivalent suggests involvement of saline brines. Daughter minerals are the same as in the Trgovska Gora Mt.

Quartz crystals occurring with texturally different siderite ores (dark massive, zebra, light sparry in veins), from the Ljubija ore district, within the neighbouring Una–Sana Palaeozoic terrane, show a broad range of salinity between 2 and 23.4 wt% NaCl equivalent, and even higher due to the presence of halite daughter minerals. This indicates mixing of different fluid types, probably formational brines and meteoric water (Fig. 5b).

The deposits in the Samoborska Gora Mt (SG) and Bistra (M, Fig. 1), being of the same type of siderite–barite mineralization, located in the extension of the Internal Dinarides into the Zagorje– Mid-Transdanubian zone, show similar characteristics.

Leachates, determined by ion chromatography on minerals from all the deposits mentioned above, expressed as Cl/Br and Na/Br ratios, suggest a contribution of halite-saturated waters derived from Permian evaporitic lagoons or pools, rather than water percolating through buried Permian evaporates. A ternary diagram (Fig. 6), of Ca, Na, K × 10 in mol%, shows an increasing trend of Ca content from the Variscan to the Triassic ore-forming fluids.

The ore deposits of Mid-Bosnian Schist Mts (MB), Central Bosnia, are represented by the Kreševo and Čemernica mineral deposits (Fig. 5c). FI studies were performed on quartz, barite and fluorite crystals. There are two distinctive fluid systems. (1) A metamorphic, retrogressive one, generated during exhumation of the Dinarides in Miocene time, overprinting the FI system of primary ore related fluids. The primary FIs in the Kreševo ore deposit show the following characteristics (Palinkaš & Jurković 1994): the presence of FIs (L), (L + V), (L + V + S_1 + S_2 + S_3). The solid phases S_1 & S_2 are carbonates and S_3 is an opaque mineral. The degree of filling (L/(L + V)) is 0.92 and the T_h is190–310 °C with data grouped around +250 °C. All inclusions homogenize into the liquid phase. T_{fm} shows three maxima, at -51, -30 and $-21\,°C$. The eutectic temperatures, together with a pale brown colouration of the frozen content of FIs, obtained after metastable cooling, suggest a NaCl–$CaCl_2$– (±$MgCl_2$)– H_2O fluid system. The last melting phase was hydrohalite and T_m is between -22 and $+4\,°C$. Salinity is between 24.2 and 26.3 wt% NaCl equivalent. The presence of chloride daughter minerals, which failed to dissolve on heating due to sample decrepitation, confirms a high salinity of the fluids.

The fluids in quartz from the Čemernica deposit are aqueous and aqueous-carbonic. Their $P–T–X$ characteristics reveal a complex geological history of the area. The primary ore-forming fluids are expressed as aqueous (L + V) FIs, with a uniform

Fig. 5. Fluid inclusion microthermometry reveals differences between ore fluids related to the Variscan and Alpidic metallogenic cycles. (**a, b**) Variscan siderite–barite deposits Petrova Gora Mt., Trgovska Gora Mt and Ljubija, characterized by great salinity variations, abundant daughter minerals, and homogenization temperature between 150 and 450 °C. (**c**) Characteristics of fluid inclusions in minerals from the Mid-Bosnian Schist Mts, Kreševo and Čemernica. Aqueous-carbonic retrogression, operating during exumation of the Dinarides in Miocene time, overprints the Variscan ore forming fluid system. (**d**) Characteristics of typical Triassic SEDEX deposits, Idrija and Vareš. Ore fluids have low salinity and low homogenization temperature (60–180 °C). Their chemical composition is dominated primarily by NaCl–$CaCl_2$–H_2O.

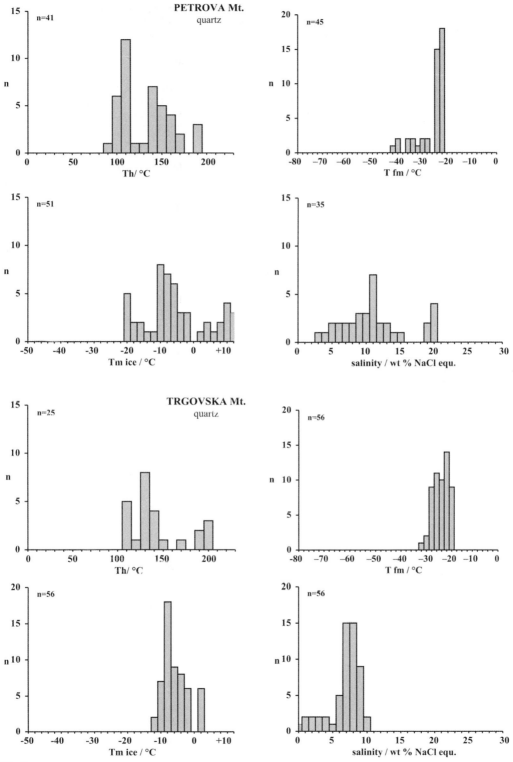

Fig. 5.

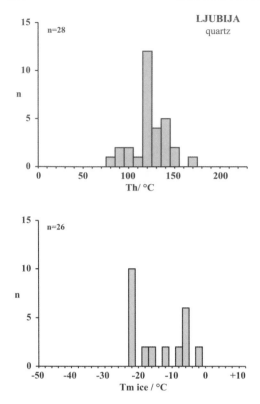

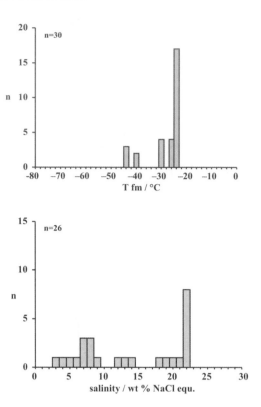

Fig. 5. (continued)

degree of fill and T_{fm} between -24 and $-32\,°C$ (maximum at $-25\,°C$). A perfectly clear content of FIs (after freezing) and eutectic temperature (T_e) point to the presence of fluids in the NaCl–KCl–H_2O system. The ion chromatography data in the neighbouring deposit of Raštelica confirmed the observation (Fig. 6). T_{mice} between -16.6 and $-11\,°C$ is equivalent to between 14.0 and 20.2 wt% NaCl. T_h into the liquid phase is between 150 and 230 °C (max. 190 °C). The secondary carbonic FIs, NaCl–H_2O–CO_2 (+CH_4, N_2), were formed during exhumation of the Dinarides by retrogressive metamorphic fluids in Miocene time, as determined on the minerals from the Alpine veins (Strmić *et al.* 2000).

Alpidic fluid system. The metallogenic evolution in the Dinarides during the Triassic reflects advanced Tethyan rifting and favoured the formation of SEDEX type deposits. The Idrija mercury mine in Slovenia as well as the Vareš siderite–hematite–barite–pyrite mine in Central Bosnia represent two well-known SEDEX-type deposits. The Idrija deposit is a stratiform, monomineralic mer-

cury deposit in Middle Triassic volcano-sedimentary formations with well-developed feeder zones within Upper Palaeozoic and lower Triassic sediments. FIs in quartz and cinnabar crystals from Idrija have the following characteristics (Fig. 5d): (L + V + S_{anisot}), (L + V), (L) and solid inclusion ($S_{cinnabar}$) types predominate. T_{fm} is around $-52\,°C$. T_{mhyd} spans between -20.5 and $-25.7\,°C$ and T_{mice} is between -1.5 and $-8.9\,°C$, while the salinity is varying between 2.1 and 5.8 wt% NaCl equivalent. The NaCl/$CaCl_2$ ratio is between 0.57 and1.58. T_{htot} into the liquid phase shows an interval between 160 and 218 °C.

The chemical composition of ore-forming fluids in Idrija is rather simple. They are low or a moderately saline $CaCl_2$–NaCl–H_2O fluids, which may be a result of seawater circulation between sediments and basalts. Fluids with the same characteristics were found within hydrothermal veins cross-cutting metabasalts and in the Triassic volcano-sedimentary rifting formation of the Hruškovec, Zagorje–Mid-Transdanubian zone. The stratiform pyrite–barite–siderite–hematite deposit of Vareš, exposed in a volcano-sedimen-

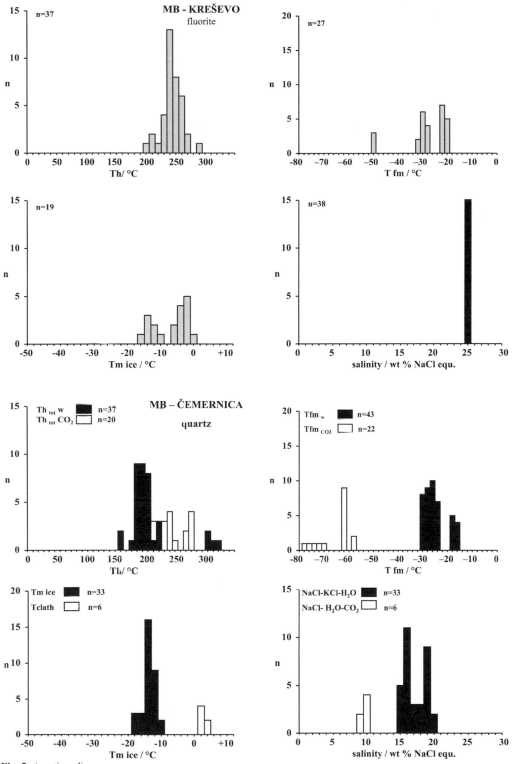

Fig. 5. (continued)

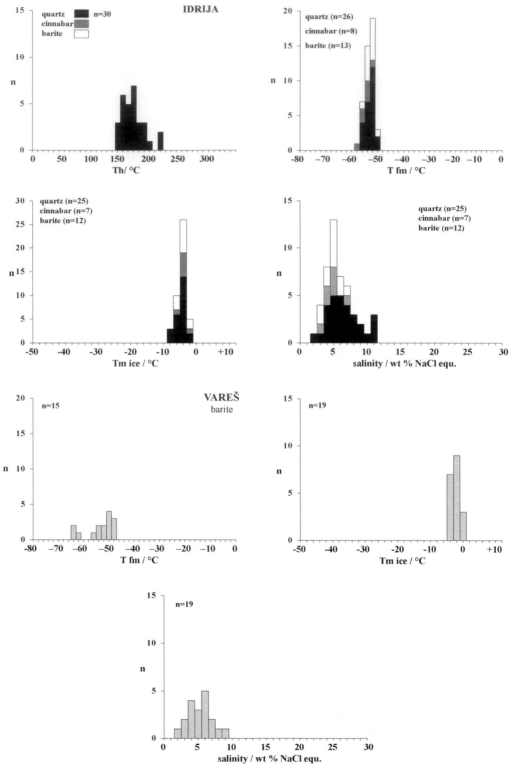

Fig. 5. (continued)

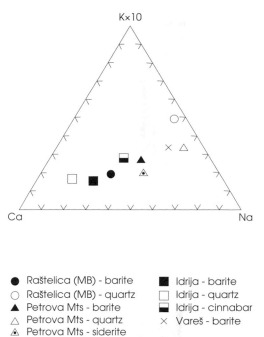

K×10

Ca
Na

● Raštelica (MB) - barite   ■ Idrija - barite
○ Raštelica (MB) - quartz   □ Idrija - quartz
▲ Petrova Mts - barite   ◨ Idrija - cinnabar
△ Petrova Mts - quartz   × Vareš - barite
⧄ Petrova Mts - siderite

Fig. 6. Ternary diagram Ca, Na, K × 10, in mole%, shows an increasing trend of Ca content from the Variscan to the Triassic ore forming fluids.

deposits, and their proper affiliation to the main metallogenic events in the Dinarides. This paper has used geochemical and paragenetic differences to classify ore deposits formed in Palaeozoic or Triassic times. These characteristics have been applied as discrimination criteria: tectonic settings of the deposits, sulphur isotope δ^{34}S values in hydrothermal barite deposits, content of SrSO$_4$ in barites, composition of ore parageneses, physico-chemistry of ore-forming fluids in the fluid inclusion systems of ore and gangue minerals. The discrimination method may need further discussion. Probably more detailed studies of specific ore deposits will lead to improved interpretation of the ore formation processes.

The authors wish to express their gratitude to the reviewers. Many thanks go to Jens Schneider for careful examination of the manuscript, valuable suggestions, corrections and improvement of the language, which made the text more transparent and readable.

tary sequence of Ladinian age in the Internal Dinarides of Central Bosnia, represents another important type of the Triassic deposits.

FIs in barite can be classified into: (i) primary (L + V) with a degree of fill between 0.75 and 0.9. T_{fm} between −59.0 and −52.0 °C indicates a CaCl$_2$−NaCl−H$_2$O system. T_{mhyd} is between −28.5 and −25.0 °C, T_{mice} −2.5 and −1.2 °C and salinity between 2.1 to 4.2 wt% NaCl equivalent and (ii) rear (L+V) FIs with a filling degree of 0.5. UV microscopy confirmed the presence of hydrocarbons within the FI. T_h measured in sphalerite and barite is between +60 and +160 °C. Ion chromatography on fluid leachates confirmed a fluid chemistry close to the seawater, with respect to the Cl/Br ratio.

Conclusions

The increased quantity of geochemical data for ore deposits in the Dinarides enables new insight and interpretation into the geotectonic processes (in space and time) as well as for the ore formation processes. The Dinaridic geotectonic units are placed close together, often in allochthonous positions, and were intensively modified by thrust and nappe tectonics. This leads to problems in determining the time of formation of ore

References

BEHR, H.J. & GERLER, J. 1987. Inclusions of sedimentary brines in post-Variscan mineralization in the Federal Republic Germany - A study by neutron activation analysis. *Chemical Geology*, **61**, 65−77.

BEHR, H.J., HORN, E.E., FRENTZEL-BEYME, K. & REUTEL, CH. 1987. Fluid inclusion characteristics of the Variscan and post-Variscan mineralising fluids in the Federal Republic Germany. *Chemical Geology*, **61**, 273−285.

CISSARZ, A. 1956 Lagerstättenbildung in Jugoslawien in ihren Beziehungen zu Vulkanismus and Geotektonik. *Vijesti Zavoda za geolosko-geofizicka istrazivanja R Srbije*, **6**, 3−152.

ČOP, M., JURKOVIĆ, I. & SLOVENEC, D. 1998. Sedimentary low-manganese hematite deposits of the Bukovica area in the northwestern Mt. Petrova Gora, Central Croatia. *Rudarsko-geolosko-naftni zbornik, Zagreb*, **10**, 1−24.

DJOKOVIĆ, I. 1985. [*The use of structural analysis in determining the fabric of Palaeozoic formations in the Drina-Ivanjica region.*] PhD Thesis, Geoloski anali Balkanskog poluostrva, **49**, 11−160. [in Serbian].

DROVENIK, M., PLENIČAR, M. & DROVENIK, F. 1980. [Origin of ore deposits in SR Slovenia]. *Geologija, Ljubljana*, **23**, 1−157. [In Slovenian].

HRVATOVIĆ, H. 1996. [*Analysis of structures and facies of the northeastern part of the Mid-Bosnian Schist Mts.*] PhD thesis, University of Tuzla. [In Bosnian].

JANKOVIĆ, S. 1955. [*Geology and metallogeny of the lead-zinc deposit Suplja Stijena Montenegro*]. Zbornik radova Geolosko rudarskog fakulteta u Beogradu. Special Publications, **1**. [In Serbian].

JANKOVIĆ, S. 1967. [*Metallogenic epochs and ore-bearing regions in Yugoslavia*]. Rudarsko-geoloski zbornik, Belgrade, [in Serbian].

JANKOVIĆ, S. 1982a. Yugoslavia. *In:* DUNNING, F.W. (eds) *Mineral deposits of Europe. Vol.II. Southeast*

Europe. Mineralogical Society and IMM, London, 143–202.

JANKOVIĆ, S. 1982b. Major Metallogenic Units and Epochs in Yugoslavia. *Earth Evolution Sciences, Berlin,* **1**, 41–47.

JANKOVIĆ, S. 1987. Genetic types of major Triassic deposits of the Dinarides, Yugoslavia. *In:* JANKOVIĆ, S. (ed.) *Mineral deposits of the Tethyan Euroasian metallogenic belt between the Alps and the Pamirs.* Faculty of Mining and Geology, Belgrade, 11–33.

JANKOVIĆ, S. 1990. *[Ore deposits of Serbia].* Republic Foundation for Geologic Investigation and Faculty of Mining and Geology, Belgrade [in Serbian].

JEREMIĆ, M. 1963. [Metallogeny of barite deposits in Bosnia]. *Arhiv za tehnologiju,* **1**(1/2), 1–63. [in Serbian].

JURIĆ, M. 1971. *[Geology of the Una-Sana Palaeozoic in the northwestern Bosnia].* Special publications, Geoloski glasnik, Sarajevo, **11**.

JURKOVIĆ, I. 1956. *[Mineral parageneses in Mid-Bosnian Ore Mountains with special respect to tetrahedrite].* PhD Thesis, University of Zagreb, [in Croatian].

JURKOVIĆ, I. 1957. The basic characteristics of the metallogenic region of the Mid-Bosnian Ore Mountains. *In: Proceedings of the 2nd Congress of Yugoslav geologists.* 504–519.

JURKOVIĆ, I. 1958. [Metallogeny of the Petrova Gora Mt. in the southwestern Croatia]. *Geoloski vjesnik, Zagreb,* **11**, 143–213. [in Croatian].

JURKOVIĆ, I. 1960. Quecksilberfahlerz vom Mackaragang bei Gornji Vakuf in Bosnien (Jugoslawien). *Neues Jahrbuch der Mineralogie, Festband Ramdohr,* **94**, 539–558.

JURKOVIĆ, I. 1961. [Mineralogical investigation of iron-ore deposit Ljubija near Prijedor in Bosnia]. *Geoloski vjesnik, Zagreb,* **14**, 161–220. [in Croatian].

JURKOVIĆ, I. 1987. Barite deposits on Mount Medjuvrsje south of and south-east of the town of Kreševo, Bosnia. *Geoloski vjesnik, Zagreb,* **40**, 313–336.

JURKOVIĆ, I. 1995. [Metallogeny of Dinaridic Palaeozoic in Slovenia, Croatia, Bosnia and Herzegovina, Montenegro and Western Macedonia]. *In: Proceedings of the 1st Croatian Geological Congress 1995.* **1**, 275–280. [In Croatian].

JURKOVIĆ, I. & PALINKAŠ, A.L. 1996. Late Variscan, Middle Upper Permian, Post-Variscan and Middle Triassic Rifting Related Ore Deposits in the Northwestern and Central Dinarides. *UNESCO Project No 356.* University of Mining and Geology St. Ivan Rilski, Sofia, **1**, 19–27.

JURKOVIĆ, I. & PAMIĆ, J. 1999. Triassic rifting-related magmatism and metallogeny of the Dinarides. *Acta geologica, Croatian academy of sciences and arts, Zagreb,* **26/1**, 1–26.

JURKOVIĆ, I. & PAMIĆ, J. 2001. Geodynamics and metallogeny of Variscan complexes of the Dinarides and South Tisia as related to plate tectonics. *Nafta, Zagreb,* **59**(9), 267–294.

JURKOVIĆ, I. & ŠIFTAR, D. 1995. Sulphate sulphur isotope composition of evaporates in the western Dinarides. *In:* PAŠAVA, J., KRUBEK, B. & ŽAK, A.A. (eds) *Mineral deposits from their origin to their environmental impact.* Balkema, Rotterdam, 581–

584.

KATZER, F. 1925. *Geologie von Bosnien und Herzegovina.* Museum of Bosnia and Herzegovina, Sarajevo.

KLEMM, W. 1994. Chemical Evolution of Hydrothermal Solution During Variscan and Post-Variscan Mineralization in the Erzgebirge, Germany. *In:* SELTMANN, R., KÄMPF, H. & MÖLLER, P. (eds) *Metallogeny of Collisional Orogens.* Czech Geological Survey, Prague, 1–158.

KUBAT, I. 2001. *Metallogeny.* Uiversity of Tuzla, [in Bosnian], **214**.

KUBAT, I., RAMOVIĆ, E., TOMIČEVIĆ, D., PEZDIĆ, J. & DOLENEC, T. 1980. Isotopic composition of carbon, oxygen and lead in some ore deposits and occurrences in the Bosnia and Herzegovina (in Serbo-Croatian). *Geoloski glasnik, Sarajevo,* **24-25**, 61–84. [in Serbo-Croatian].

KULENOVIĆ, E. 1987. [Metallogeny in Palaeozoic of South-east Bosnia]. *Geoloski glasnik, Sarajevo,* **31/32**, 84–95. [in Croatian].

MIOČ, P. 1984. *[Geology of the area between Southern and Western Alps in Slovenia].* PhD thesis, University of Zagreb, [in Slovenian].

MLAKAR, I. 1993. [The Litija ore field]. *Geologija, Ljubljana,* **3b**, 249–328. [in Slovenian].

NEUBAUER, F. & VON RAUMER, J.F. 1993. The Alpine basement-linkage between Variscides and East-Mediterranean mountain belt. *In:* VON RAUMER, J.F. & NEUBAUER, F. (eds) *Pre-Mesozoic geology in the Alps.* Springer Verlag, Berlin, 641–644.

OMALJEV, V. 1982. *[Metallogenic features of the Žirovski Vrh uranium deposit].* Special publications, Geoloski institut, Belgrade, **170**. [In Serbian].

PALINKAŠ, A.L. 1986. Geochemical Facies Analysis of Groeden Sediments and Ore Forming Processes in Žirovski Vrh Uranium Mine, Slovenia, Yugoslavia. *Acta Geologica, Zagreb,* **16/2**, 43–82.

PALINKAŠ, A.L. 1988. *Geochemical characteristics of Palaeozoic metallogenic region: Samoborska gora Mt., Gorski Kotar, Lika, Kordun and Banija.* PhD thesis, University of Zagreb. [In Croatian].

PALINKAŠ, A.L., PEZDIĆ, J. & ŠINKOVEC, B. 1993. Lokve Barite Deposit, Croatia: an Example of Early Diagenetic Sedimentary Ore Deposits. *Geologia Croatica, Zagreb,* **46**, 84–95.

PALINKAŠ, A.L. & JURKOVCIĆ, I. 1994. Lanthanide geochemistry and fluid inclusion peculiarities of the fluorite from the barite deposits south of Kreševo, Bosnia. *Geologia Croatica, Zagreb,* **47/1**, 103–116.

PAMIĆ, J. 1984. Triassic magmatism of the Dinarides in Yugoslavia. *Tectonophysics,* **226**, 503–518.

PAMIĆ, J. & JURKOVIĆ, I. 1997. Bosnia and Herzegovina. *In:* MOORES, E.M. & FAIRBRIDGE, R.W. (eds) *Encyclopedia of European and Asian Regional Geology.* Chapman & Hall, London, 86–93.

PAMIĆ, J. & JURKOVIĆ, I. 2002. Palaeozoic tectonostratigraphic units of the northwest and central Dinarides and the adjoining South Tisia. *International Journal of Earth Sciences,* **91**, 538–554.

PAMIĆ, J., GUŠIĆ, I. & JELASKA, V. 1998. Geodynamic evolution of the Central Dinarides. *Tectonophysics,* **297**, 251–268.

POPOVIĆ, R. 1984. *[Metallogeny of metamorphic Drina district (western Serbia)].* Geoinstitut, Belgrade,

Special publications, **9**. [In Serbian].

RAMOVIĆ, M. 1976. [Barite]. *In:* MILOJEVIC, R. (ed.) *[Mineral resources of Bosnia and Herzegovina].* Geoinzinjering, Sarajevo [in Croatian], **1**, 358–379.

RAMOVIĆ, M. 1979. [Base metal ore deposits.] *In:* MILOJEVIC, R. (ed.) *Mineral resources of Bosnia and Herzegovina*. **2**, 7–251. [In Bosnian].

ŠIFTAR, D. 1981. [On the chemism of barite and on the same conditions of the barite deposit formation in Gorski Kotar and Lika]. *Geoloski vjesnik, Zagreb*, **34**, 95–107. [in Croatian].

ŠIFTAR, D. 1982. [Sulphur isotope composition and the age of the evaporates with examples from the Dinaride territory in southern Croatia]. *Nafta*, **33**(4), 177–183. [in Croatian].

ŠIFTAR, D. 1984. On the chemism of barite from Petrova Gora Mt. and its comparison with the chemism of barite from other deposits in Croatia. *Geoloski vjesnik, Zagreb*, **37**, 197–204.

ŠIFTAR, D. 1986. {Age of evaporates in the Sinj area and in the upper course of the Una river]. *Geoloski vjesnik, Zagreb*, **39**, 55–60. [in Croatian].

ŠIFTAR, D. 1988. The chemical characteristics of barite from some Bosnian deposits. *Rudarsko-metalurski zbornik, Ljubljana*, **35**, 75–89.

ŠINKOVEC, B., PALINKAŠ, A.L. & DURN, G. 1988. [Ore occurrences in the Medvednica Mt]. *Geoloski vjesnik, Zagreb*, **41**, 395–405. [in Croatian].

SLOVENEC, D., ŠIFTAR, D., JAKŠIĆ, M. & JURKOVIĆ, I. 1997. Strontium Dependence of the Lattice Constants of Barites from the Kreševo Area in Central Bosnia (Bosnia and Herzegovina). *Geologia Croatica*, **50/1**, 27–32.

STRMIĆ, S., PALINKAŠ, A.L., JURKOVIĆ, I. & HRVATOVIĆ, H. 2000. Quartz from the Middle Bosnia. *Proceedings of the 2nd Congress of Croatian Geologists, Cavtat, Institute of Geology, Zagreb*, 415–419.

TARI, V. & PAMIĆ, J. 1998. Geodynamic evolution of the northern Dinarides and southern part of the Pannonian Basin. *Tectonophysics*, **297**, 269–281.

Example of a structurally controlled copper deposit from the Hercynian western High Atlas (Morocco): the High Seksaoua mining district

A. CHAUVET[1], L. BARBANSON[1], A. GAOUZI[2],
L. BADRA[2], J. C. TOURAY[1] & S. OUKAROU[3]

[1]ISTO, UMR 6113, Université d'Orléans, Bâtiment Géosciences, B.P. 6759, 45067 Orléans
Cédex 2, France (e-mail: Alain.Chauvet@univ-orleans.fr)

[2]Faculté des Sciences, Université Moulay Ismaïl, B.P. 4010 Beni - M'Hamed, Meknès,
Morocco

[3]SNAREMA, 300 rue Mostapha L. Maâni, Casablanca 01, Morocco

Abstract: The mineralized district of the High Seksaoua (Western High Atlas, Morocco) is characterized by a lithological succession marked by an alternation of schists and limestones attributed to the Cambrian and affected by at least five deformational events. The copper mineralization herein analysed is systematically localized close to a dolomite/black schist level in which a top-to-the-NNW décollement-type tectonics (D_d) has been identified. We demonstrate that the economic mineralization is a syn-tectonic stockwork formed in response of this top-to-the-NNW shearing event (D_d). This tectonic event can be reasonably correlated with the late Hercynian tectonics responsible, in the same area, for the Tichka granite emplacement at *c.* 291 ± 5 Ma, also under the control of a NW-SE shortening direction. Indeed, the Tichka granite represents a good candidate to explain the origin of the mineralized fluid. Such an hypothesis is confirmed by the Permo-Triassic age (*c.* 270 Ma) given by $^{40}Ar/^{39}Ar$ dating realized on white micas related to the stockwork formation (this study). This important result questions the syn-genetic interpretation accepted until now for this mineralization and allows us to propose a new model of formation for this kind of deposit that could contribute to exploration programmes.

The more popular model for explaining polymetallic or metallic mineralization which generally occurs interbedded between two sedimentary levels was that of the massive sulphide deposit subdivided into volcanic-associated massive sulphide deposits (e.g. Franklin *et al.* 1981; Lydon 1984) and sedimentary exhalative massive sulphide deposits (e.g. Russell *et al.* 1981; Large 1983; Lydon 1996). These kinds of deposits were frequently linked with huge and important economic concentrations within the world (i.e. the Iberian Pyrite Belt, e.g. Leistel *et al.* 1998; Carvalho *et al.* 1999; the Kidd Creek deposit, Ontario, Canada, e.g. Barrie *et al.* 1999; Jackson & Fyon 1991, the Windy Craggy Besshi-type deposit, British Columbia, Canada, Peter & Scott 1999, or the Urals deposits, Prokin & Buslaev 1999). Moreover, every occurrence of stockwork or mineralized veins within the vicinity of such deposits has been systematically interpreted in terms of a feeder zone (e.g. Lydon 1984, 1988; Gibson & Kerr 1993; Nehlig *et al.* 1998). Because this kind of mineralization formed relatively early during the geological history of a selected area, it was inevitably affected by subsequent tectonics and metamorphism, which cannot be excluded from the determination of the typology (geometry, remobilization) of the deposit. This notion has not been considered in numerous works, even though several studies highlight the significant role of deformation within the formation process of polymetallic deposits (e.g. Brill 1989; Aerden 1994; Nicol *et al.* 1997). The importance of significant remobilization due to such mechanisms appears to be frequently underestimated, mainly because economic geologists focus their analysis on stratiform and huge mineralization that predated the deformation.

We present, in this work, a case-study of copper mineralization from which the model of formation cannot be understood without the intervention of a late-orogenic tectonic event. This example represents for us a typical case in which the syn-genetic model cannot explain the concentration of copper.

From: BLUNDELL, D.J., NEUBAUER, F. & VON QUADT, A. (eds) 2002. *The Timing and Location of Major Ore Deposits in an Evolving Orogen*. Geological Society, London, Special Publications, **204**, 247–271.
0305-8719/02/$15.00 © The Geological Society of London 2002.

It also represents an instructive study in which the presence of stockwork cannot be systematically linked with the occurrence of feeder zones. A syntectonic stockwork formation associated with hydrothermal and remobilization processes seems here to be responsible for the economic copper concentration. This concept appears to be a very crucial and significant goal of mineral deposit studies within their orogenic context and its recognition is essential for the identification of this kind of orogenesis-related deposit. Effects on exploration are also very important because ore body distribution could then depend on the tectonic style of the deformation related to the mineralizing event.

Geological setting

General geology

The High Seksaoua district is located within the western High Atlas, part of the Hercynian domains of Morocco (Fig. 1a, b). This area is reputed for its structural complexity and multiple occurrences of barite and copper mineralization (Fig. 1c) even if very few studies precisely concerned with these deposits are available. Barite occurs in two distinct forms; stratiform (the Ousaga area) and near vertical veins (the Ifri domain and surroundings) whereas copper is supposed to occur as stratiform layers and related stockworks (Rchid 1996; Rchid et al. 1996). Copper has been exploited since the Almohades times (twelfth century) and re-worked by the SNAREMA Company during the 1960s. It is now intensively studied by the SNAREMA – ANGLO-AMERICAN consortium (CMS, Mining Company of Seksaoua) in order to estimate reserves and to undertake an intense and industrial exploitation. The vertical barite veins, mined by the SNAREMA Company, presently represent the main activity of the area. Stratiform barite veins of the Ousaga area have been abandoned as exploration targets.

The Ifri copper mineralization is the largest and more economic deposit of the area (Fig. 2a). Four ancient galleries are still open and allow the analysis of good outcrops. Other galleries are available along the Bourichira copper rich layer that corresponds to the same level as the Ifri deposit. Additional copper occurrences visible in the study area are represented by the M'Tili, Tansmahkt and Amerdoul indices. All these systems are indicated within the general structural map of the High Seksaoua district (Fig. 1c).

Lithological succession

The detailed lithological succession of the study area shown on the map in Figure 2(a) is sketched in Figure 3. From the base to the top, we distinguish:

1 a lower calc-schist unit that is the lowest one observed in the study area;
2 a discontinuous level of black schist systematically overlain by a grey dolomite layer 2 m thick;
3 an upper calc-schist sequence with intercalations of andesite sills and, towards the top, limestone levels;
4 a thick pelitic-sandstone sequence;
5 a sericite-schist unit that forms the top of the sequence.

All these units are cut by late NE–SW dolerite dykes not represented in Figure 3 and supposed to be late Hercynian in age (Ouazzani 2000). The Tichka granite, also assumed to be late Hercynian in age and emplaced during a tectonic shortening event of NW–SE trend (Gasquet 1991; Lagarde & Roddaz 1983), constitutes the southeastern limit of our study area (Fig. 1c). This granite gives ages of 291 ± 5 Ma by Rb–Sr and 283 ± 24 Ma by Sm–Nd (Gasquet 1991). The age of 291 Ma is retained in the literature because of lowest uncertainty.

Copper mineralization occurs everywhere in the vicinity of the dolomite/black schist levels. We will describe in more detail the forms of occurrence in the section describing the ore bodies. However, we can note that this systematic location within or in the vicinity of a specific lithological unit such as the grey dolomite has formed the basis for previous interpretations which conclude that the copper mineralization of the Ifri domain is syn-genetic (Hmeurras 1995; Rchid 1996; Rchid et al. 1996).

Controversy exists concerning the age of the lithological sequence presented here. Based on facies analogy, these units have generally been attributed to the Lower or Middle Cambrian (Cornée et al. 1987; Schaer 1964; Termier & Termier 1966). With the lack of fossil occurrences, it remains difficult to confirm this assumption. A recent geochemical study has refuted this age and interprets these units as Neoproterozoic deformed during the Panafrican orogeny (c. 600–500 Ma) (Ouazzani et al. 1998, 2001). Based only on lithological and geochemical comparisons, this interpretation remains questionable. We do not present arguments in this study to resolve this problem of age. However, we will see that deformations described here are more easily compar-

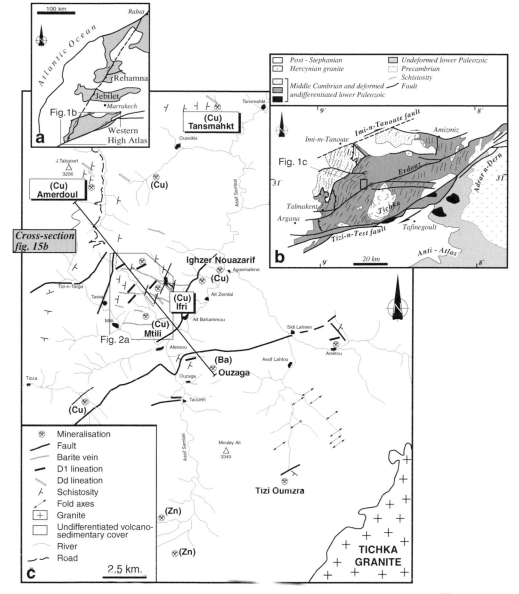

Fig. 1. (**a**) Location of the western High Atlas within the Hercynian domains of Morocco. (**b**) Simplified structural map of the western High Atlas showing the localization of Palaeozoic terrains between the Imi-n-Tanoute and Tizi-n-Test wrench faults (after Cornée *et al.* 1987). (**c**) Simplified geological and structural map of the High Seksaoua area and surroundings. The locations of copper occurrences concerned in this study are indicated as well as the position of the late Hercynian Tichka granite, southeastward of the study area.

able with the ones that affect the area during the Hercynian cycle, in accordance with the works of Cornée *et al.* (1987), Cornée (1989) and Tayebi (1989). This point will be discussed in more detail in the conclusions of the paper but a Cambrian age for the lithological units appears to be the more acceptable interpretation at this state of knowledge.

Structural evolution of the High Seksaoua and surrounding areas

The units described above are exposed within a domain limited by two major extensional faults: the Sembal fault to the south and the Tassa one to the north (Fig. 2a). Although these faults presently exhibit a normal motion, they could have had a

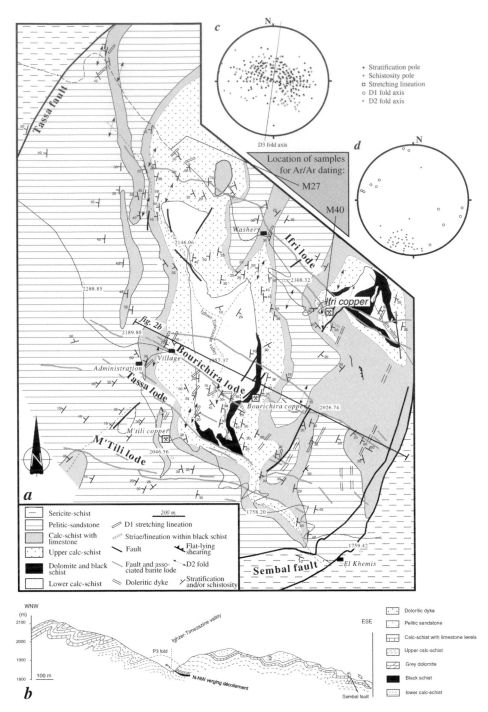

Fig. 2. (**a**) Detailed geological and structural map of the central High Seksaoua showing the geological environment of the Ifri and Bourichira deposits. The locations of the two samples selected for $^{40}Ar/^{39}Ar$ geochronology are indicated. (**b**) East–west cross-section through the study area showing the geometry of the foliation and the importance of folding. (**c**) Distribution of the schistosity and stratification within a lower hemisphere Schmidt diagram. The effects of a north–south-oriented large-scale fold (D$_3$) are detectable. (**d**) Distribution of the D$_1$-related stretching lineation and fold axes within a lower hemisphere Schmidt diagram.

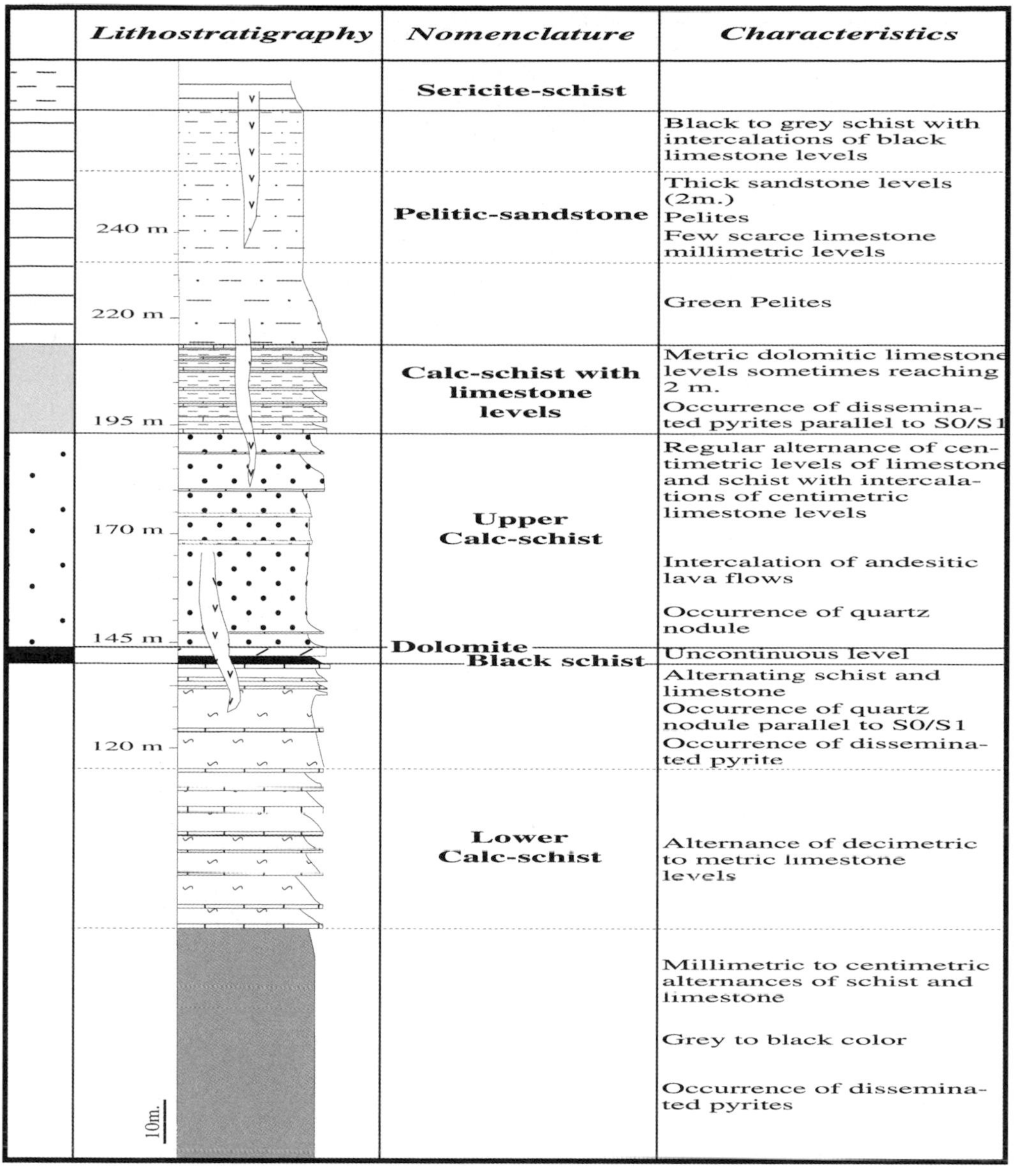

Lithostratigraphy	Nomenclature	Characteristics
	Sericite-schist	
	Pelitic-sandstone	Black to grey schist with intercalations of black limestone levels
240 m		Thick sandstone levels (2m.) Pelites Few scarce limestone millimetric levels
220 m		Green Pelites
195 m	Calc-schist with limestone levels	Metric dolomitic limestone levels sometimes reaching 2 m. Occurrence of disseminated pyrites parallel to S0/S1
170 m	Upper Calc-schist	Regular alternance of centimetric levels of limestone and schist with intercalations of centimetric limestone levels
		Intercalation of andesitic lava flows
		Occurrence of quartz nodule
145 m	Dolomite — Black schist	Uncontinuous level
120 m	Lower Calc-schist	Alternating schist and limestone Occurrence of quartz nodule parallel to S0/S1 Occurrence of disseminated pyrite
		Alternance of decimetric to metric limestone levels
10m.		Millimetric to centimetric alternances of schist and limestone

Grey to black color

Occurrence of disseminated pyrites |

Fig. 3. Lithological succession of the High Seksaoua area. Thicknesses of the different units are indicated even though the real base of the entire column is not known.

more complex history of successive movements during the polyphase tectonics that seems to have affected the area. The location of one of these faults with respect to the position of the Ifri mineralized domain and barite veins trending N120°E is indicated in the panoramic view of Figure 4.

We have distinguished five types of tectonic structures that we will describe in the following section before attempting to define a relative

chronology among them. These events have been essentially recognized within the domain limited by the two major faults (see above) and represented on the detailed structural map of Figure 2(a). The effects of such tectonic events within other mineralized areas such as the Amerdoul and Tansmahkt domains will be discussed in the appropriate part of the paper. A first remark may focus on the fact that the area is not intensively deformed because stratification can be easily

Fig. 4. Large-scale view of the domain concerned in our study. The main barite lodes and large-scale faults are indicated. The Ifri mine is marked.

recognized in several places (Fig. 2a, c) even though a significant number of tectonic events exist (at least five).

Structural analysis

Major ductile tectonics (D_1). A first ductile tectonic event was responsible for the formation of the regional cleavage or schistosity (S_1) that can obliterate the S_0 stratification. Associated with this event, a stretching lineation developed with a preferred orientation close to NE–SW (Fig. 2a, d) in areas not affected by subsequent folds. This lineation does not present a strongly penetrative character and does not reveal, at first inspection, an intense shearing rate. However, it appears well expressed within carbonaceous levels in which intense isoclinal-shape folds could be found (Fig. 5a). Numerous shear criteria are associated with this D_1-related stretching lineation assigned to a transport direction. The more common are shear bands, which are everywhere indicators of a top-to-the-SW verging motion (Fig. 5b) whereas asymmetric porphyroclasts were also encountered with a consistent vergence (Fig. 5c). Metamorphic conditions are difficult to define because of the poor mineralogical expression. The rocks commonly exhibit quartz,

carbonates and white mica and only few places show the occurrence of biotite and scarce garnets (i.e. Tansmahkt area). Indeed, metamorphic conditions close to the upper greenschist facies are supposed to prevail during D_1 deformation.

East-verging knee-folds (D_2). The second type of structures is knee-folds with very regularly oriented north–south to N30°E axes (Fig. 2d). They are systematically overturned towards the east (Fig. 5d) and are very characteristic structures of the study area. These structures are particularly well developed within the lower and upper calc-schist units even though some could be seen within other lithologies in the area. Because a first assumption may attribute these folds to the D_1 event presented above, we have carefully analysed relationships between knee-folds and top-to-the-SW shear criteria as expressed in Figure 6. Both shear criteria and stretching lineations are affected by the knee-folds and found within each flank of the folds. These relations demonstrate that folding occurred later than the D_1 ductile shearing. Indeed, a sequence of deformation that initiates with ductile shearing towards the SW followed by knee-fold formation looks to be the more realistic with the available structural data.

Fig. 5. Macro-structures illustrating the successive tectonic events that affect the study area. (**a**) Isoclinal-shaped folds within carbonaceous levels revealing the intensive effects of the D_1-related tectonic event. (**b**) Shear band indicators of SW-verging movements developed within the upper calc-schist unit of the High Seksaoua district. (**c**) Asymmetrical porphyroclasts indicative of top-to-the-SW movement. (**d**) D_2-related knee-fold developed within lower calc-schist units and systematically overturned towards the east. (**e**) Right micro-folds related to the development of large-scale north–south-trending folds due to east–west contraction event (D_3). D_1-related asymmetrical porphyroclasts are still recognizable. (**f**) Large-scale view of the Sembal fault showing the 2 m thick quartz filling. D_2-related knee-folds are present in the pelitic-sandstone rocks that formed the footwall of the fault. (**g**) Normal fault bends along the Sembal fault. Note the significant quartz filling that underlies the Sembal fault. (**h**) Extensional feature along barite vein trending N100°E. Note that barite fills open jog structures, thus demonstrating the contemporaneity between barite and normal motion. Ba, barite.

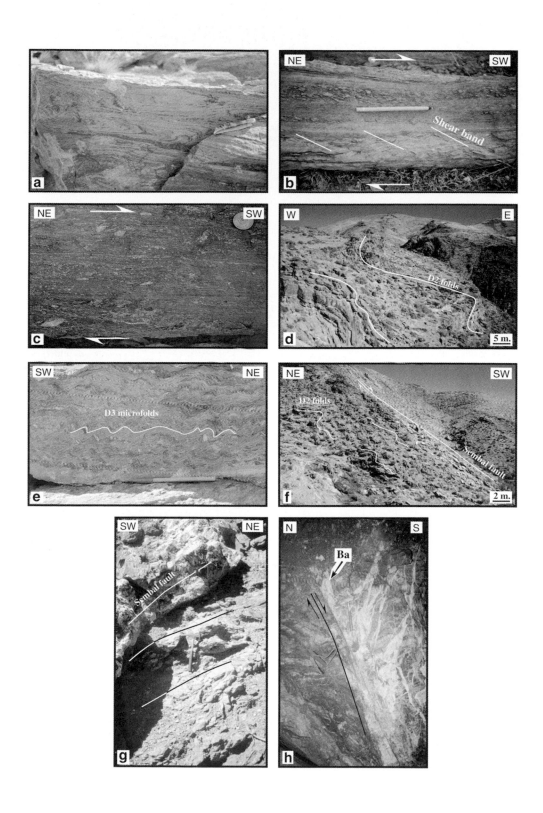

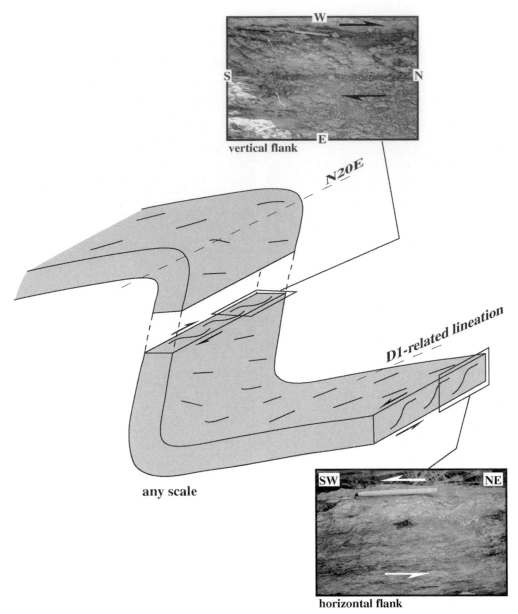

Fig. 6. Relationships between D_2 east-verging knee-folds and D_1 top-to-the-SW shear criteria demonstrating that the fold developed after the ductile shear bands.

Large-scale north–south contraction event (D₃). This tectonic event remains the least significant and characteristic of the study area. It is detectable with difficulty in the field and its existence is essentially deduced from the observation of the stereonet diagram of Figure 2(c) in which the distribution of S_1 cleavage poles and stratification reveals the effects of a large-scale, north–south-trending, D_3 right fold. If the fold is

not visible in the field because of its large scale, it is visible on the general cross section through the study area (Fig. 2b). Micro-folds with near-vertical axial planes and north–south-trending axes are found in several locations (Fig. 5e). They affect and fold the D_1 related shear structures (Fig. 5e). These structures, which cannot be mistaken for the D_2-related knee-folds, are interpreted as small-scale expressions of the large fold.

Late extension (D₄). Numerous features reveal the existence of a late brittle extension within the High Seksaoua district. First, the regional-scale Tassa and Sembal faults are related to this event (Fig. 5f). Evidence of normal movement is shown, for the Tassa fault, by indirect criteria such as the displacement between lithological units cropping out on each side of the fault. Conversely, the Sembal fault exhibits direct evidence of normal movements demonstrated by striations on fault mirrors and fault bends (Fig. 5g). A huge quantity of quartz emplaced along this fault underlies this structure in side view (Figs 4 and 5f, g). The multiple barite veins that occur in the area located between the two normal faults described above are also associated with normal faults. These faults trend east–west to NW–SE and contain the barite mineralization (Fig. 2a). A close view across one of these structures demonstrates the relationships between barite filling and extensional structures (Fig. 5h). This is in favour of a coeval formation of normal fault movements and barite veins. These barite occurrences need to be distinguished from those forming the Ousaga stratiform deposit cropping out more to the south (Fig. 1c) and not concerned in this study.

The relations between the two normal features herein described (NE–SW large-scale faults and N110°E-trending barite-bearing ones) have also been analysed in order to establish a relative chronology. Because the barite veins cross-cut the quartz injection of the Sembal fault and are displaced by NE-SW faults (Fig. 2a), we conclude that both are contemporaneous. Their age of formation will be discussed later in this paper.

Numerous NNE–SSW- to NE SW-trending faults are also encountered, especially within the surroundings of the Ifri mines (Fig. 2a). It remains difficult to estimate the time of creation of these faults. We can just observe that they displace some of the near-vertical barite veins in a left-lateral motion (Fig. 2a). Because of their specific locations close to the Ifri mineralization and their increasing concentration in this area, we can question whether these faults play a significant role during the process of copper concentration.

It should be stressed that all the structures described here have certainly been reactivated during Alpine tectonics from which the effects are supposed to be relatively mild within the study area. Numerous successive events are then assumed to affect the area. This may explain the difficulty in understanding and identifying different fault motions because several faults, certainly formed earlier in the structural evolution, are constantly re-activated during successive tectonic events.

Décollement-type tectonics (D_d)

The dolomite/black schist level appears as a discontinuous level that particularly crops out in zones in which copper mineralization was described (Fig. 2a). Indeed, the best outcrops of these two levels are found within the galleries of the Ifri and Bourichira mines and within the core of the Ifri and Ighzer-Timezazine valleys (see location in Fig. 2a). Copper occurrences are systematically localized in the vicinity of these domains. A positive correlation is well defined between copper occurrences and the existence of the dolomite/black schist levels. This is also the case for the copper index of Amerdoul and Tansmahkt that will be briefly described at the end of this paper.

The microtectonic analysis of the black schist demonstrates that this level is affected by an additional tectonic event that gives the general impression of a mylonitic shear zone. One might suppose that this could be a strong expression of the regional deformation within low competence schist level. However, the existence of unambiguous shear criteria inconsistent with the numerous events described above questions such an assumption. For a better understanding of this point, a schematic block diagram that summarizes outcrops found within the Ifri and Bourichira galleries is presented and discussed (Fig. 7).

The black schist generally exhibits on its foliation planes a north-south-oriented striation that could not be confused with the regional and N050°E-trending D₁-related stretching lineation. Such a north–south lineation is systematically associated with the development of brittle–ductile top-to-the-south criteria of movement typically represented by low-angle normal faults (Figs 7 and 8a). The observation in a few places of criteria indicative of a contrasting NW vergence (Fig. 8b) suggests that this level could be affected by polyphase tectonics. Within several places, criteria indicative of a ductile shearing towards the NW have been suspected, but the corresponding stretching lineation could not be found, probably because of a late overprint by the north–south-trending extensional features. In order to clarify this point, three thin sections have been systematically prepared, oriented north–south, N050°E and N150°E, within the black schist and dolomite units. Based on observations at more than 15 locations, it was found that ductile shear criteria are best expressed within sections oriented N150°E. Indeed, the ductile shearing that affects the black schist level is attributed to a NW-verging tectonic event (Fig. 7).

Within the dolomite level, the expression of this ductile shearing is not visible certainly because of

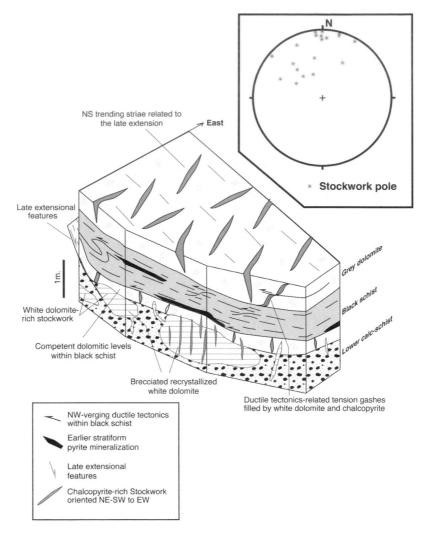

NS trending striae related to the late extension

East

Stockwork pole

Late extensional features

1m.

Grey dolomite

Black schist

White dolomite-rich stockwork

Competent dolomitic levels within black schist

Lower calc-schist

Brecciated recrystallized white dolomite

Ductile tectonics-related tension gashes filled by white dolomite and chalcopyrite

NW-verging ductile tectonics within black schist

Earlier stratiform pyrite mineralization

Late extensional features

Chalcopyrite-rich Stockwork oriented NE-SW to EW

Fig. 7. Block diagram showing the distribution of the different metallogenic features summarized after observations made within the four galleries of the Ifri mine and in the Bourichira mine. We note that stockwork veins are particularly well developed within competent grey and white dolomite levels. The stereonet diagram (Schmidt diagram, lower hemisphere) shows the stockwork orientation within the Ifri mine.

Fig. 8. Photographs of the black schist/dolomite level. (**a**) Outcrop of the Ifri gallery no. 1 showing a late top-to-the-south normal fault associated with northsouth striations along schistosity planes. (**b**) Top-to the-NW shear criteria within the black schist level (Ifri gallery of the ancients). (**c**) Correlation between ductile deformation within the black schist and brittle behaviour of the more competent dolomite level (Ifri gallery no. 2). (**d**) Aspect of stockwork within white dolomite levels (Ifri gallery no. 1). (**e**) Thin-section of sample M40 showing a parallelism between top-to-the-NW shear band and a stockwork vein filled by quartz and calcite. Note the increase of shear band distribution approaching the vein. (**f**) Example of relationships between mineralization and tectonics. The geometry of chalcopyrite-filled cavities is the result of pull-apart formation along a NW verging shear plane. (**g**) Top-to-the-NNW shearing deformation and coeval formation of opened jog filled by chalcopyrite. (**h**) Example of mineralized tension gashes formed in response of NW verging shearing within the pillar of Ifri gallery no. 1. Chalcopyrite grains constitute the centre of the structure. The cartoon illustrates the process of formation of the mineralized tension gashes during ductile shearing. Symbols: Py II, pyrite (II); Cp, chalcopyrite; Qtz, quartz.

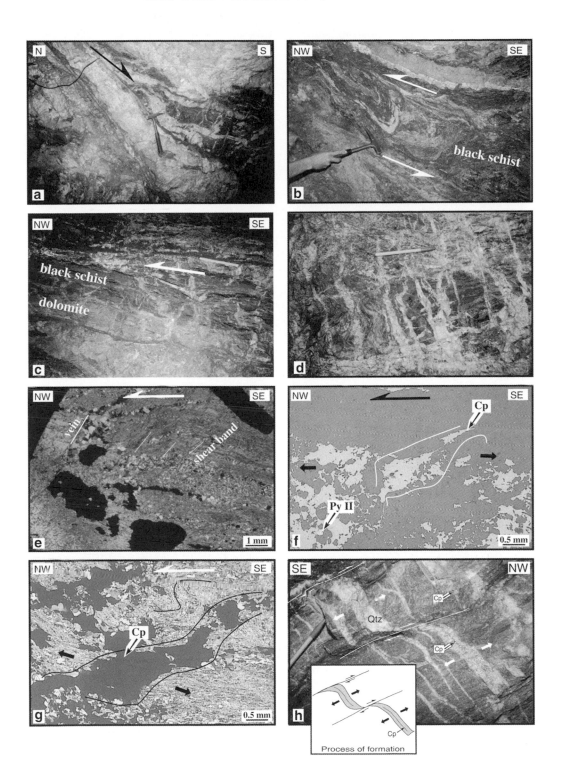

the competence of this level. Moreover, the identification of tension gashes commonly oriented N050°E to N090°E (Fig. 7) has been reported within outcrops of the gallery no. 1 of Ifri (Fig. 8c, d). These tension gashes have been interpreted as the brittle expression of the ductile shearing within the black schist. In fact, their formation is the result of the ductile tectonic accommodation within the more competent dolomite levels (Figs 7 and 8c). We will see in the appropriate section of this paper that these tension gashes contain the main sulphides and thus could be assimilated within a stockwork representative of the main ore bodies (Fig. 8h).

Accurate observation of the relations between stockwork veins and ductile shearing towards the north demonstrates the contemporaneity between the two structures. The parallelism between the two sets of structure (Fig. 8e), the increasing occurrences of shear bands towards the stockwork vein (Fig. 8e) and the specific geometry shown by sulphide rich traps (Fig. 8f, g) are arguments in favour of a coeval development of ductile shear bands within schist, and mineralized veins within dolomite. Similar evidence has been found within the pillar of the Ifri gallery no. 1 in which the specific geometry of quartz–carbonate–sulphide-rich stockwork is directly related to the control by NW-verging shear planes during its formation (Fig. 8h).

Because the ductile shearing expressed within the black schist is restricted to this level and does not affect the over- and underlying calc-schist units, it has been defined as 'décollement'-type tectonics. We will see in the final sections of this paper the strong significance of such a tectonic style for the regional development of copper mineralization.

The copper mineralization

The copper mineralization systematically occurs at the interface between dolomite and black schists. Because this lithological level was the preferential site for the development of the décollement-type tectonics presented in this study, a

new attention needs to be addressed to this mineralization that looks, from an initial inspection, to consist mainly of stockwork-type ore bodies. A detailed geometrical and mineralogical description of each copper mineralization that formed the High Seksaoua district is now presented. Because it is representative, the Ifri mine study is more detailed than the Amerdoul and Tansmahkt areas. However, we will see that most of the conclusions deduced from the study of the Ifri mine could easily be extended to the other two mines and that all could be interpreted to have occurred during a single and similar process of formation.

The Ifri deposit

Geometry and main characteristics. Two kinds of sulphur-bearing features are observable within the four Ifri galleries. One consists of massive and stratoid pyrite levels (Fig. 9a) and the other is related to chalcopyrite-rich stockwork (Fig. 8d, h). The massive pyrite levels are essentially developed at the top and the bottom of the black schist, although several disseminated lodes can be found within the upper calc-schist unit. Conversely, the stockwork broadly cross-cuts the grey dolomite, the black schist and the lower calc-schist units. In fact, it is better developed within some white dolomite lenses from which the development is interpreted to represent the first stage of the hydrothermalism. The best expression of the stockwork within this white dolomite is explained by the strong competence of this latter lithology. For a better understanding of the mineralizing process that affects the study area, features need to be described chronologically. Three main hydrothermal and/or mineralized events are recognized. The first one is represented by the occurrence of first and earlier pyrite for which the origin (epigenetic or syngenetic) needs to be discussed. The two others exhibit typical epigenetic character and are related to the emplacement of white dolomite by replacement of the grey one and by the stockwork formation.

Fig. 9. Photographs of metallogenic features. (**a**) Massive pyrite mineralization parallel to the S_0/S_1 cleavage. (**b**) Disseminated pyrite (I) showing post-kinematic overgrowths of pyrite (II). (**c**) General aspect of the massive pyrite (I). Note the weakly developed schistosity and the occurrence of chalcopyrite-fill late fractures.(**d**) Relict of melnikovite (collomorph) texture indicative of pyrite (0). (**e**) Automorphic grain of pyrite (II) with phyllite inclusions cross-cut by growth band. (**f**) Pyrite (II) within micro-fissure showing the difference between pyrite within (free of phyllite inclusions) and out (included the schistosity) of the fissures. Note the chalcopyrite filling fissures of the pyrite. (**g**) Association between pyrite (II), gersdorffite, sphalerite and chalcopyrite within the mineralized stockwork. (**h**), euhedral crystal of gersdorffite (black zones) and ullmannite (white zones). Symbols: Py 0, earlier pyrite (0); Py I, pyrite (I); Py II, pyrite (II); S0, stratification, S1, schistosity; Cp, chalcopyrite; gb, growth band; Ph, phyllite; Qtz, quartz; Sph, sphalerite; Ge, gersdorffite; Ul, ullmannite. All the views are back-scattered electron images except (a). Samples in photographs (b), (d), (e) and (f) are etched with nitric acid.

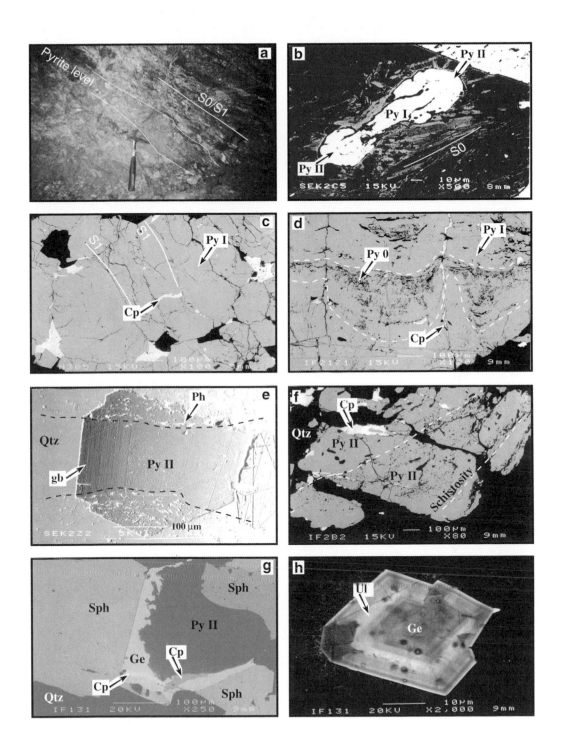

First and earlier pyrites occur under two distinct forms. One is disseminated within the black schists levels and massive pyrite levels interlayered within the stratification represented by the other. These levels are 40 cm thick on average and are distributed at the top and base of the black schist level (Fig. 9a).

White dolomite lenses occur within the four galleries of the Ifri mine. In fact, this rock unit appears as large stratoid lenses emplaced at the interface between black schist and lower calc-schist units (Fig. 7). The internal texture consists of a cockade breccia with black schist, grey dolomite and primary massive pyrite. Along the border of the large-scale lens, the white dolomite occurs as a stockwork formed by an intense network of white dolomite veins cutting the surrounding black schist, calc-schist and grey dolomite lithologies. The formation of the white dolomite, certainly by replacement of the grey one, is attributed to the initial fluid that marks the beginning of the hydrothermalism.

The last hydrothermal pulse appears to form a *mineralized stockwork* that represents, in our opinion, the main mineralized structures of the study area. This stockwork is developed from the upper part of the lower calc-schist unit upward to the grey dolomite (Fig. 7). Veins can be observed in the upper part of the lithostratigraphic succession but not in sufficient number to represent potential ore bodies. This stockwork is particularly well developed within competent levels such as dolomite (Fig. 8d, h). Its density appears to be more intense within these specific lithologies. Veins are oriented east–west to NE–SW (Fig. 7) and are generally a few centimetres thick. Because they contain the essential chalcopyrite grains (see the next section), they are assumed to represent the ore bodies.

Texture and mineralogy. Massive sulphide levels (Fig. 9a) mainly contain pyrite, herein called pyrite (I). Where it is disseminated, pyrite (I) occurs as elongated sub-automorphic minerals that are parallel to the stratification and in place surrounded by pressure shadows (Fig. 9b). Within massive levels, pyrite (I) is arranged in large-scale domains composed by mosaic grains joined at 120° and rarely affected by a very weak schistosity (Fig. 9c). These zones can exhibit in a few locations the development of micro-breccias and the presence of collomorphic texture (melnikovite texture, Fig. 9d). Except within the melnikovite zones, etching with nitric acid does not reveal the existence of any kind of internal texture, either within disseminated or massive pyrite grains. All these facts, characteristic of deformed pyrite grains affected by metamorphism (Craig & Vokes

1993; Lianxing & McClay 1992; McClay & Ellis 1983), demonstrate that these levels have been overprinted by the regional deformational events. The collomorphic features could then be interpreted as relics of the initial pyrite, prior to the deformation. It should be stressed that very few occurrences of chalcopyrite are found associated with this primary pyrite. Only a few grains, intercalated between growth bands of pyrite (I), could be interpreted as primary chalcopyrite.

The *stockwork veins* contain abundant chalcopyrite grains that are mainly associated with quartz, pyrite (II), ankerite, siderite, galena, sphalerite, gersdorffite, ullmannite, brannerite, pyrrhotite, native Bi, bismuthinite and emplectite (Fig. 10). Accessory grains of stannite, scheelite, wolframite, cassiterite, electrum, native silver and hessite are also found. The established relative chronology shows first the formation of pyrite (II) followed by the formation of chalcopyrite and other minerals (Fig. 10).

Like pyrite (I), pyrite (II) exhibits two different habits: disseminated within the matrix and within the stockwork veins. However, the following characteristics allow pyrite (II) to be distinguished from pyrite (I). Pyrite (II) grains are systematically automorphic and exhibit typical growth bands revealed by etching with nitric acid (Fig. 9e). Pyrite (II) grains contain numerous inclusions of chalcopyrite (Fig. 9f), sphalerite, pyrrhotite and, more rarely, inclusions related to the lithological nature of the surroundings (Fig. 9e). They are generally associated with quartz and carbonate. Both disseminated and stockwork-related pyrite (II) grains have typical growth bands that include and cut the host-rock cleavage (Fig. 9e, f), thus attesting for their post-deformation timing.

Chalcopyrite is the main copper-rich mineral that explains the relatively high economic potential of the Ifri deposit. It occurs first as inclusions within pyrite (II) grains (Fig. 9f) but the best occurrences are chalcopyrite grains filling fissures cross-cutting pyrite (I) and pyrite (II) (Fig. 9c, d, f). Chalcopyrite also contains pyrite inclusions generally aligned along micro-fissures; a fact that demonstrates that pyrite crystallization was continuous during the entire mineralizing process. Sphalerite seems to be associated with the earliest chalcopyrite whereas brannerite is coeval with the last one (Fig. 10). The Ni-rich minerals are commonly automorphic (Fig. 9g) and occur within the late fissure network, like the Bi-rich ones. We determine that ullmannite and gersdorffite predate the formation of galena and sphalerite (Fig. 9g, h). The Sn–W-rich minerals also occur as inclusions within quartz, pyrite (II) and chalcopyrite. Ag is expressed under the form of hessite that appears in contacts between pyrite (II) and chalcopyrite

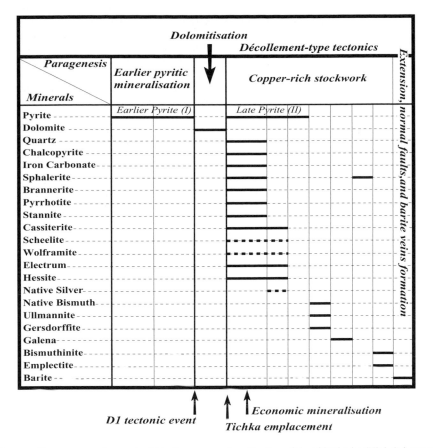

Fig. 10. Paragenetic sequences observed within the copper mineralization of the Ifri District (High Seksaoua).

and also within micro-fissures in pyrite (II). Native silver is also present. Gold has been observed as very rare electrum grains (1–20 μm. in size) included within chalcopyrite and micro-fissures. Very scarce barite occurs at the end of the mineralogical history and classical minerals characteristic of supergene alteration are found, such as digenite, covellite, malachite, azurite, tenorite and native copper.

Carbonates have been analysed in order to highlight the chemical variations between those of the stockwork and the primary ones that form the sedimentary grey dolomites. Results are shown within an Fe + Mn v. Mg diagram (Fig. 11). Carbonates related to the stockwork or included within sulphur minerals show the highest Fe-values even though some of them occur in the field of the primary ones (Fig. 11). Within a few of these carbonates, the occurrence of alternating bands of different colours can be recognized. These bands, detected by MEB analysis, have

been correlated with variable Fe-contents, demonstrating a heterogeneous distribution of Fe within carbonates. No photograph of this phenomenon is available. This fact is consistent with the existence of a Fe-rich metasomatism of primary and secondary grey and white dolomite, probably associated with stockwork formation and consequently with the concentration of the mineralization.

Geochronology. Preliminary $^{40}Ar/^{39}Ar$ dating has been attempted in order to assess the timing of the deformation and of the hydrothermalism responsible for stockwork formation.

Analytical method and sample characteristics. Two samples located on Figure 2(a) have been selected. The first (sample M27), comes from the upper calc-schist unit and corresponds to white mica-rich calc-schists affected by the ductile D_1 deformation. The second (sample M40). was collected within the Ifri gallery no. I and corresponds to Figure 8(e). This sample exhibits the

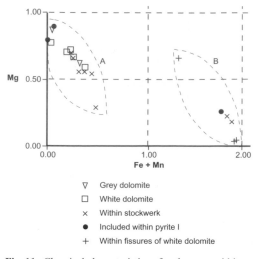

∇   Grey dolomite

□   White dolomite

×   Within stockwerk

●   Included within pyrite I

+   Within fissures of white dolomite

Fig. 11. Chemical characteristics of carbonates within a Mg v. Fe + Mn diagram. Compositions are determined by EDS analyses on a SEM.

effects of the NNW-verging décollement-type tectonics as illustrated by the multiple shear bands underlain by tiny white micas (Fig. 8e). The dating of micas within such a sample is particu-

larly exciting because of the possibility to precisely date micas developed within the shear bands and, consequently, to estimate the age of the deformation responsible for the formation of these structures. Analyses have been realized *in situ* on polished thin-sections by laser probe fusion dating techniques according to the procedure described by Monié *et al.* (1994). The conventional step-heating dating method has not been applied because of the impossibility of separating grain populations. Ages obtained for each laser analysis are shown in Table 1. Results are also presented in the form of an isotope correlation diagram (Fig. 12, $^{36}Ar/^{40}Ar$ v. $^{39}Ar/^{40}Ar$ diagram, Roddick *et al.* 1980; Turner *et al.* 1971). Within this type of diagram, points representative of the isotopic composition of each analysed mineral are supposed to be aligned along a straight line correlation from which the intercept with the horizontal axis (lower intercept) allows the determination of an average age by the use of the $^{39}Ar/^{40}Ar$ ratio. Conversely, the upper intercept is representative of the initial atmospheric $^{40}Ar/^{36}Ar$ ratio, supposed to be close to 295.5 for analyses of value.

Results. A first inspection of the two diagrams reveals that points are not well aligned along an isochron, thus indicating analytical problems. The

Table 1. *Results of individual laser analyses made on white micas in samples M27 and M40 (High Seksaoua district, High Atlas, Morocco)*

Analysis no.	40Ar/ 39Ar	36Ar/40Ar × 1000	39Ar/40Ar	37Ar/39Ar	38Ar/39Ar × 100	% Atm.	Age ± 1 s (Ma)
Sample M27		J = 0.016844					
1	9.612	0.061	0.1021	1.025	0.08	1.81	270.7 ± 1.8
2	9.508	0.111	0.1017	0.000	0.01	3.28	268.0 ± 2.1
3	9.667	0.096	0.1004	0.157	0.17	2.84	272.2 ± 3.9
4	9.566	0.101	0.1013	0.052	0.11	3.01	269.5 ± 3.4
5	9.704	0.087	0.1003	0.023	0.00	2.59	273.1 ± 2.2
6	9.506	0.195	0.0991	1.485	0.55	5.78	268.0 ± 4.1
7	9.752	0.158	0.0977	4.058	1.16	4.67	274.4 ± 3.3
8	9.682	0.128	0.0993	0.000	0.18	3.79	272.6 ± 1.2
9	9.732	0.057	0.1009	0.000	0.01	1.71	273.9 ± 1.0
10	9.781	0.067	0.1000	0.134	0.06	1.99	275.1 ± 1.7
11	9.920	0.044	0.0994	0.000	0.00	1.33	278.8 ± 1.3
12	9.831	0.048	0.1002	0.000	0.57	1.43	276.5 ± 1.3
Sample M40			Total age = 272.4 ± 2.5				
1	10.716	0.178	0.0883	18.521	24.84	5.29	299.4 ± 3.8
2	9.568	0.126	0.1005	3.923	2.77	3.74	269.6 ± 2.3
3	10.067	0.676	0.0794	24.369	34.04	19.99	282.6 ± 4.1
4	10.147	0.289	0.0901	18.615	18.72	8.56	284.7 ± 3.5
5	10.912	0.992	0.0647	41.718	85.54	29.32	304.4 ± 6.3
6	9.534	0.146	0.1003	2.710	1.79	4.34	268.7 ± 1.8
7	9.637	0.119	0.1000	2.206	1.93	3.53	271.4 ± 4.2
8	9.319	0.216	0.1004	4.957	4.58	6.39	263.1 ± 4.0
9	9.789	0.182	0.0965	6.931	7.68	5.39	275.4 ± 2.4
10	9.859	0.295	0.0925	13.110	11.76	8.73	277.2 ± 2.4
11	9.705	0.090	0.1002	3.189	2.05	2.66	273.2 ± 1.3
		Total age = 274.1 ± 2.5					

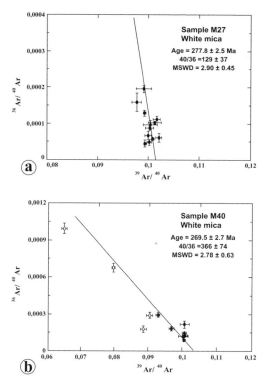

Fig. 12. Correlation diagrams for the white micas of samples M27 (**a**) and M40 (**b**).

M27 calc-schist sample gives ages between 268 and 279 Ma (Table 1) with an average value around 272.4 ± 2.5 Ma. The lower intercept age deduced from the diagram (Fig. 12a) is c. 277.8 ± 2.5 Ma. This age can be accepted for muscovite of sample M27 but the fact that the upper intercept value gives an initial ratio close to 129 ± 37 Ma implies that this date should be considered with great caution.

For the sample representative of the hydrothermal zone (M40), results are of better value because of the correct upper intercept value close to 366 ± 74 Ma. The calculated ages range between 263 and 304 Ma (Table 1). The oldest ages, correlated with the highest $^{37}Ar/^{39}Ar$ and $^{38}Ar/^{39}Ar$ ratios, could result from excess Ar due to some possible contamination by Ca and Cl. These elements could be captured during the laser ablation from neighbouring carbonate grains. Analyses from which such an excess Ar has not been identified (lower $^{37}Ar/^{39}Ar$ and $^{38}Ar/^{39}Ar$ ratios) allow to define an average age around 271.7 ± 2.5 Ma. An age calculated from the isotope correlation diagram (269.5 ± 2.7 Ma, Fig. 12b) is consistent with this result.

Discussion. Rigorously, $^{40}Ar/^{39}Ar$ dating may be interpreted as reflecting cooling ages through a temperature of c. 400 °C that represents the temperature necessary for intracrystalline retention of argon within muscovite grains (Wagner *et al.* 1977; Robbins 1972). Added to the fact that such closing temperatures were constantly discussed (Harrison *et al.* 1985), the results may only be considered as minimum ages.

It is stressed that the age given by sample M40 could reflect a minimum age close to the age of the hydrothermal event responsible for the formation of the copper mineralization. This is argued because the analysed tiny white micas are in a very specific position with respect to the deformation and that their crystallization appears coeval with stockwork formation as discussed above (Fig. 8e). Because white micas involved in this dating are small crystals only represented within shear bands parallel to the stockwork (Fig. 8e), they cannot be confused with older micas developed during regional metamorphism and re-oriented within shearing structures. Applying the restriction cited above, it is suggested that the result is a minimum age representative of the cooling of the white micas from higher temperature coeval with the initiation of a hydrothermal system.

The age given by sample M27 remains difficult to interpret because of the many uncertainties given by the badly defined diagram. It appears impossible to correlate this age to the regional metamorphism that affects the study area because it is very close in value to the age of initiation of the hydrothermal system given by the other sample. Taking into account the strong uncertainties due to unreliable analytical results (problem of upper isotope correlation age), we decided to not relate this age to a geological event. However, the good age concordance in the two samples remains surprising and one can question whether the late hydrothermal system responsible for copper formation could effect a rejuvenation of the Ar system within the micas of the study area.

Other copper deposits of the High Seksaoua

During our work, we also analysed two other copper occurrences of the High Seksaoua domain, located in Figure 1(c) (Amerdoul and Tansmahkt areas). Even though few data have been collected, they are totally consistent with those of the Ifri domain and allow us to propose a systematic model that seems crucial to the understanding of the copper mineralization of the High Seksaoua domain. Only geometric and structural characteristics are described in this work for the two other areas studied. More details concerning the precise mineralogical evolution of these two mineral deposits are given in another paper presently in preparation. There exist other mineral occurrences (essentially Zn and Ba mineral deposits) within the study area (Fig. 1c). These prospects were not integrated within this work because of the lack of data even though preliminary observations suggest that Zn occurrences are also associated with a stockwork, similar to the Ifri copper mineralization.

Amerdoul. The Amerdoul prospect is located to the north of the Ifri domain and occurs within a lithological unit that we can correlate with the calc-schists (lower and upper units) of the Ifri domain but composed of volcanic rocks (mainly andesite). Geometrically, the mineral deposit consists of a near-vertical quartz lode strictly parallel to an andesite dyke (Fig. 13a, d). Within neighbouring rocks, specific geometric relationships give evidence of thin-skin tectonic features (Fig. 13a, b). The first crop out north of the road exhibits flat-lying thrust planes that create individual host-rock lenses in which the schistosity is nearly vertical and sigmoidal (Fig. 13a, c). This outcrop, typical of duplex-style thrust tectonics, is the key to understanding the Amerdoul mineralization. We suggest that the vertical mineralized lode of Amerdoul represents, on a larger scale, the equivalent of the vertical schistosity interlayered between two flat-lying thrust planes. This defines a typical ramp and flat geometry that is strongly consistent with the existence of décollement-type tectonics related to the formation of the mineralization. Micro-scale ramp and flat geometry is also observed NE of the Amerdoul main vein (Fig. 13e). The ramps are usually underlain by elongated holes, clear evidence of ancient crystallization coeval with the formation of schistosity and now dissolved away (Fig. 13a, e).

Further towards the WNW, other near-vertical dolomite levels are encountered. Four levels have been determined towards the WNW, whilst only two of them occur on the other side of the Amerdoul lode, towards the ESE (Fig. 13b). Two alternative solutions can be proposed in order to explain this repetition. First, isoclinal folds intensively affected the dolomite level. Secondly, the dolomite level was affected by a décollement-type thrusting event that implies its multiple occurrences, as illustrated by the interpretative cross section of Figure 13(b). We adopt the second alternative. It must be highlighted that, more to the south, the succession of dolomite levels was achieved by the occurrence of a low-angle dipping thrust plane (Fig. 13f) along which a N120°E-trending stretching lineation has been observed and measured. Overturned folds towards the NNW could also be observed (Fig. 13b).

Tansmahkt. Very few data are available for the copper mineralization of the Tansmahkt area that occurs NE of our area of investigation (Fig. 1c). The lithological succession is similar to the one of the Ifri domain. Apparently, a first ductile tectonic event was also responsible for the formation of the regional cleavage. This cleavage is steep-dipping, and mainly trends N20°E, and the very weak lineation is sub-horizontal. Kinematic indicators are scarce but everywhere consistent with a sinistral sense of shearing. These criteria can be correlated to the D_1-related criteria of the Ifri domain if we consider that they have been affected and made vertical by the large-scale, north–south-trending D_3 folds defined previously. Such folds are better developed at Tansmahkt and give a general vertical aspect to the structures.

Mineralization occurs within a zone marked by dolomites and cleavage dipping at low-angle, resulting certainly from the occurrence of thrust tectonics. The surrounding rocks appear to have been affected by an intense silicification and a N110°E-trending stretching lineation is observed on the flat-lying schistosity planes. No kinematic indicator is detectable but the general attitude (cleavage bends) suggests a reverse motion towards the NNW. The intensity of the alteration does not allow fresh samples to be collected for further work on the mineralogy.

Like the Amerdoul mineralization, ore bodies systematically appear in a zone characterized by the occurrence of flat-lying thrusts towards the NNW. We recognize here a feature that also characterizes the Ifri copper mineralization. This determining point will be discussed in the conclusions that follow.

Discussion

Main results

The main results provided by our structural, mineralogical, metallogenic and geochronological study of the High Seksaoua district are as follows.

The High Seksaoua domain underwent a complex and polyphase deformational history in which three major tectonic events have been identified (Fig. 14). The first is related to a SE-verging ductile shearing and was responsible for the formation of the regional foliation. During a second event, apparently east verging knee-folds were developed whilst a third event implies the occurrence of NNW décollement-type tectonics restricted to the grey dolomite/black schist levels. A general north–south-trending extensional event coeval with barite formation and an east–west-trending shortening event have also been identified.

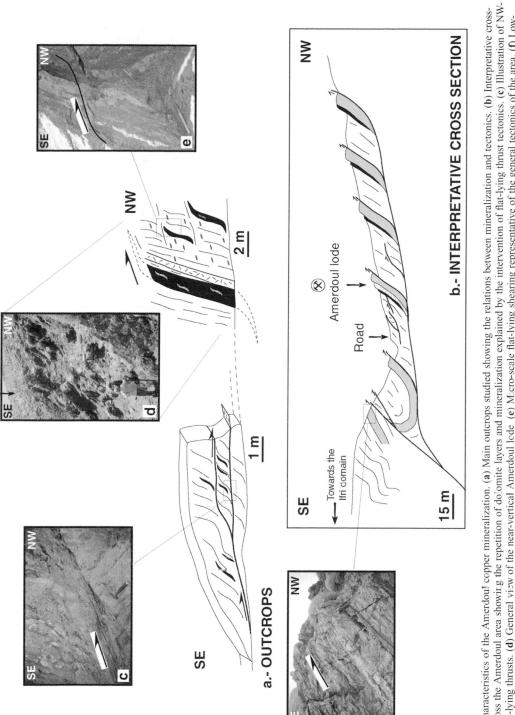

Fig. 13. Characteristics of the Amerdoul copper mineralization. (**a**) Main outcrops studied showing the relations between mineralization and tectonics. (**b**) Interpretative cross-section across the Amerdoul area showing the repetition of dolomite layers and mineralization explained by the intervention of flat-lying thrust tectonics. (**c**) Illustration of NW-verging flat-lying thrusts. (**d**) General view of the near-vertical Amerdoul lode. (**e**) Micro-scale flat-lying shearing representative of the general tectonics of the area. (**f**) Low-angle mylonite marking the SE limit of the Amerdoul area.

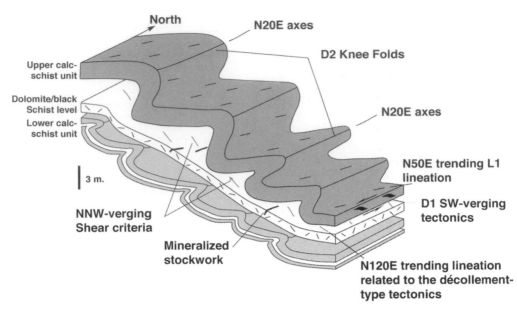

a. Block diagram summarizing the relationships between structural features

b. Illustration of the figure 13a within the Ifri Valley

Fig. 14. (**a**) Schematic block diagram summarizing geometrical relationships between the D_1, D_2 and D_d tectonic events. (**b**) View of the Ifri Valley (see location in Fig. 2a) showing the difference in tectonic styles between calc-schist and dolomite levels.

The economic copper mineralization is essentially contained within a complex stockwork in which veins are mainly oriented NE–SW to east–west. This stockwork development is restricted to the dolomite/black schist lithologies even though a few veins could be encountered within the lower part of the Upper calc-schist unit and the upper parts of the lower calc-schist one. Veins were essentially developed within the competent white and grey dolomite levels.

The hydrothermal event responsible for copper concentration began with the formation of white dolomite lenses in replacement of the sedimentary grey dolomite. Ore bodies formed during the development of the décollement-type ductile event by first the crystallization of quartz and dolomite followed by sulphur-rich minerals (Fig. 10).

Mineralogically, pyrite is a ubiquitous phase present from the earlier mineralization stages (parallel to the stratification) until the final ones (stockwork). Chalcopyrite, that represents the economic mineralization, essentially occurs within microfissures in the pyrite grains. The main associated mineral phases are sphalerite, galena, Ni- and Bi-rich minerals and carbonates.

Preliminary ^{40}Ar/^{39}Ar dating does not allow precise dating of the mineralization even though an age of *c.* 270 Ma could represent the age of the hydrothermal system responsible for the formation of the copper-rich stockwork.

Structural evolution

The structural evolution of the High Seksaoua district remains complex even though rocks look, from an initial inspection, to be unaffected by a strong ductile deformation. Five distinct tectonic events are recognized without taking into account the multiple brittle faults created and constantly re-activated during the polyphase evolution. Although the extension (D_4) responsible for emplacement of barite lodes has been attributed on the basis of general criteria to the Permo-Trias (Rchid 1996) or the Trias (Jaillard 1993), the place and the significance of the so-called D_2 folding and D_d décollement-type events remain more difficult to integrate. Within the literature, a complex tectonic story has been proposed to explain the structuration of the domain situated directly to the west of the High Seksaoua (Cornée 1989). The main Hercynian tectonics have been described as an east–west-trending shortening event whereas the late Hercynian evolution appears as a succession of extensional and compressional events. On a larger scale, the entire eastern High Atlas has been interpreted as a large-scale dextral pull-apart structure controlled by the Imin-Tanoute and Tizi-n-Test faults to the north and south respectively (Cornée *et al.* 1987) (Fig. 15a). Within such a setting, a N100°E-trending direction of shortening induces the development of N60°E dextral shearing along NE–SW-trending schistosity (Cornée *et al.* 1987). Moreover, the Tichka emplacement has been interpreted by Lagarde & Roddaz (1983) as resulting from a combination of tangential and strike-slip tectonics. Taking into account these constraints and our own results summarized in Figure 14(a), we propose that the entire ductile deformation observed within our study area developed during a long-lived Hercynian tectonic event controlled by a NW–SE direction of shortening. Such an interpretation, consistent with the previous ones available in the literature (Cornée *et al.* 1987; Lagarde & Roddaz 1983), allows us to integrate within a single event all the structures observed in the field. As previously suggested, the model implies the combination of NE–SW dextral strike-slip tectonics (D_1

event) and NNW verging tangential tectonics (D_d event) (Fig. 15a). We assume that this unexpected style of deformation results from the peculiar geometry defined by the large-scale transpression zone (Fig. 15a). Field relations demonstrate that the strike-slip tectonics took place prior to the tangential event. Indeed, this last tangential event could have developed at the end of the long-lived tectonic event and correlate with the late Hercynian evolution. The D_3 large-scale folds could also have resulted from the east–west to NW–SE shortening direction. Only the D_2 folding event remains difficult to integrate. These folds presumably developed as antithetic folds in response to décollement-type thrust tectonics. This assumption is illustrated in Figure 14(a), showing the relations between D_2 folds and décollement-type tectonics within the Ifri area. We note that there exists a contrast in behaviour between calc-schist and dolomite units with respect to the D_2 folds (Fig. 14b). A model in which folds were created in response of the décollement-type tectonics could explain this fact. Such a model is supported by the general interpretative cross-section through the High Seksaoua in which D_2 knee-folds seem to be particularly well developed on top of the zones where the main flat-lying thrusts are recognized (Fig. 15b).

This particular model of deformation adopted here for the High Seksaoua may explain the relative complexity yielded by the Hercynian domains of the Eastern High Atlas. This complexity appears as a paradox compared with the weak intensity of deformation shown in these domains. This model remains hypothetical but appears to us to be the only one that can resolve the tectonic evolution of such an external area without implying multiple deformation stages with unrealistic stress direction rotation.

Model of formation of the copper mineralization

Concerning the process of formation of the economic copper mineralization, whatever the adopted structural scenario, we demonstrate here its relationship with the stockwork rather than its occurrence within stratified sulphide levels essentially composed by the first pyrite. As the stockwork occurs systematically close to the dolomite-schist level even in the case of the neighbouring mineralization of Amerdoul and Tansmahkt areas, its formation can be reasonably correlated with the décollement-type deformation (D_d) that only affects this specific lithological level. Indeed, the formation of the stockwork is related to the occurrence of this NNW verging tectonic event

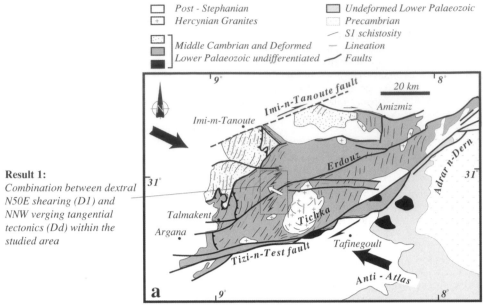

Modified after Cornée et al. 1987

Result 2: *The copper mineralization formation was mainly controlled by NNW verging décollement-type tectonics*

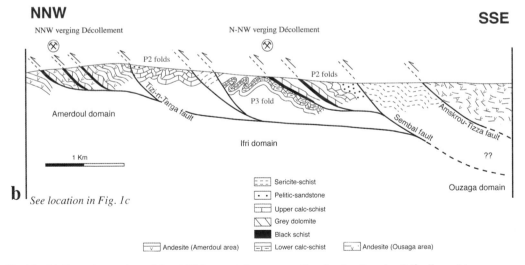

Fig. 15. (**a**) General tectonic model and (**b**) interpretative cross-section showing the role of décollement-type tectonics in the formation of High Seksaoua copper mineralization.

within all three domains concerned in this study (Fig. 15b). We suggest that vein formation is the response, within competent dolomite levels, of the ductile deformation that affected the black schist.

This model is strongly supported by evidence in Figure 14(a, b) that illustrates the contrast in behaviour between the two lithologies with respect to the deformation. Moreover, the fact that the stockwork and the top-to-the-NW ductile deformation were contemporaneous is demonstrated by the perfect parallelism between vein and shear bands (Fig. 8e) and also by the specific geometry exhibited by chalcopyrite-filled open spaces (Fig. 8f, g). With this interpretation, we re-address the classical massive sulphide deposit model that has been proposed for the copper mineralization of the High Seksaoua even though a pre-concentration within stratiform sulphide levels cannot be totally excluded.

Age and origin of the mineralization

Our study suggests that copper mineralization formed around c. 270 Ma associated with a tectonic regime controlled by shortening in an east–west to NW–SE direction. It could be reasonably suggested that these two events occur during a continuous tectono-hydrothermal period achieved at the end of the Hercynian cycle in Morocco, coeval with the emplacement of the Late Hercynian Tichka granite to the SE of the study area (Fig. 1c). This assumption may imply that this granite, also supposed to be emplaced under the control of a NW–SE shortening (Lagarde & Roddaz 1983), represents a suitable heat source to explain the origin of the mineralized fluid. As the age of the granite is of 291 ± 5 Ma (Gasquet 1991), this indicates a time interval of 20 Ma between granite emplacement and mineralization formation. Little documentation is available on this concept of the time lapse between granite emplacement and formation of mineralization. Although some studies argue that this delay may reach c. 1 Ma (Cathles et al. 1997), a time lapse of between 10 and 20 Ma has been frequently suggested within mining areas where attempts have been made to establish the relations between granite and mineralizing event. Examples can be found within the Cévennes area of France (e.g. Charonnat 2000), the Rio Itapicuru greenstone belt of Brasil (e.g. Alves da Silva et al. 1998), the Meguma terrane of eastern Canada (e.g. Kontak et al. 1990) or the Mother Lode of California (e.g. Bohlke & Kistler 1986). We thus consider that, in spite of the relative poor constraint given by the ages obtained in this study, the tectono-hydrothermal event responsible for the formation of the copper mineralization within the High Seksaoua can be reasonably correlated with the late-orogenic emplacement of the Tichka granite. This is consistent with the fact that this granite is assumed to be responsible for numerous polymetallic and gold metallogenic occurrences (e.g. Bastoul 1992; Jouhari 1989).

This later interpretation re-addresses the problem of the origin of the copper. Two solutions are available: either copper was provided from the remobilization of a sediment-hosted base metal deposit (represented by the earlier pyrite-rich layer in our case) or copper was driven by magmatic fluids and thus directly links with an additional hydrothermal process. It remains difficult to choose between the two alternatives. If our first assumption is that the granite could provide the heat source for the fluids that re-worked copper disseminated within a syn-sedimentary polymetallic pre-concentration, the total lack of copper and copper-rich phases within such a pyrite-rich earlier layer, excludes this hypothesis. We thus suggest that copper enrichment was partly related to fluids with a magmatic origin as is evident from the occurrence of stannite, cassiterite, scheelite and wolframite within the mineralized stockwork.

Conclusions

We have demonstrated in this study that copper mineralization of the High Seksaoua domain formed during a ductile tectonic event that is, on a regional scale, the last ductile event that affected the area. The formation of the mineral deposits has been attributed to the development of a décollement-type ductile tectonic event (Fig. 15b) that is poorly described in the literature. Particularly within the Amerdoul area, duplex-like structures have been identified. This case represents an example of structurally controlled copper concentration, demonstrating that the classical model of massive sulphide deposit is not the only one that can explain economic metallic concentration. As recently described within the Iberian Pyrite Belt (Onézime et al. 1999; Onézime 2001) and for other polymetallic districts of the world (e.g. Aerden 1991; Subba Rao & Naqvi 1997), the circulation of mineralized fluid during a subsequent tectonic event that affects previously formed massive sulphide deposits needs to be carefully reconsidered. Although this process did not create significant deposits in the case of the Iberian Pyrite Belt, we have demonstrated in this study that economic ore bodies could be created in response of such a process. Indeed, the authors intend to highlight the crucial importance of the recognition of relationships between tectonics and mineralization in order to identify this kind of syntectonic mineral deposits and to be able to estimate their volumes and specific geometries.

This work has been undertaken with the help of the Franco-Moroccan programmes 'Action Intégrée no. 1014/95 and 222F/STU/00' that provided funds for field and laboratory analyses. The SNAREMA mining company and specially all the members of the Lazrak family, A. Mahboud and B. Driss are thanked for permission to conduct this study, logistical, constant help and fruitful discussions. $^{40}Ar/^{39}Ar$ ages have been obtained within the laboratory of Geochronology of the University of Montpellier (France) by P. Monié. The CNRS Research Federation FR09 provided funds for these dating analyses. The manuscript has benefited from constructive criticism by R. Moritz and C. Gauert.

References

AERDEN, D.G.A.M. 1991. Foliation-boudinage control on the formation of the Rosebery Pb-Zn orebody, Tasmania. *Journal of Structural Geology*, **13**, 759–775.

AERDEN, D.G.A.M. 1994. Microstructural timing of the Rosebery massive sulphides, Tasmania: evidence for a metamorphic origin through mobilization of disseminated base metals. *Journal of Metamorphic Geology*, **12**, 505–522.

ALVES DA SILVA, F.C., CHAUVET, A. & FAURE, M. 1998. General features of the gold deposits in the Rio Itapicuru Greenstone Belt (RIBG, NE Brazil), discussion of the origin, timing and tectonic model. *Revista Brasileira de Geociências*, **28**, 377–390.

BARRIE, C.T., HANNINGTON, M.D. & BLEEKER, W. 1999. The giant Kidd Creek volcanic-associated massive sulfide deposit, Abitibi Subprovince, Canada. *In:* BARRIE, C.T. & HANNINGTON, M.D. (eds) *Volcanic-associated massive sulfide deposits: processes and examples in modern and ancient settings.* Reviews in Economic Geology, **8**, 247–259.

BASTOUL, A. 1992. *Origine et évolution des fluides hydro-carbo-azotés dans les formations métamorphiques: relation avec les minéralisations associées (U, Au, graphite).* PhD thesis, University of Nancy.

BOHLKE, J.K. & KISTLER, R.W. 1986. Rb-Sr, K-Ar, and stable isotope evidence for the ages and sources of fluid components of gold-bearing quartz-veins in the northern Sierra Nevada foothills metamorphic belt, California. *Economic Geology*, **81**, 296–322.

BRILL, B.A. 1989. Deformation and recrystallization microstructures in deformed ores from the CSA mine, Cobar, N.S.W., Australia. *Journal of Structural Geology*, **11**, 591–601.

CARVALHO, D., BARRIGA, F.J.A.S. & MUNHA, J. 1999. Bimodal-siliciclastic systems - the case of the Iberian Pyrite Belt. *In:* BARRIE, C.T. & HANNINGTON, M.D. (eds) *Volcanic-associated massive sulfide deposits: processes and examples in modern and ancient settings.* Reviews in Economic Geology, **8**, 375–408.

CATHLES, L.M., ERENDI, A.H.J., THAYER, J.B. & BARRIE, T. 1997. How long can a hydrothermal system be sustained by a single intrusion event? *Economic Geology*, **92**, 766–771.

CHARONNAT, X. 2000. *Les minéralisations aurifères tardi-hercyniennes des Cévennes (Massif central français). Cadre structural, gîtologie et modélisation 3D.* PhD thesis, University of Orléans.

CORNÉE, J.J. 1989. *Le Haut-Atlas occidental paléozoïque: un reflet de l'histoire hercynienne du Maroc occidental.* PhD thesis, University of Aix - Marseille III.

CORNÉE, J.J., FERRANDINI, J., MULLER, J. & SIMON, B. 1987. Le Haut-Atlas occidental paléozoïque: un graben cambrien moyen entre deux décrochements dextres N60°E hercyniens (Maroc). *Comptes Rendus de l'Académie des Sciences Paris*, **305**(II), 499–503.

CRAIG, J.R. & VOKES, F.M. 1993. The metamorphism of pyrite and pyritic ores: an overview. *Mineralogical Magazine*, **57**, 3–18.

FRANKLIN, J.M., LYDON, J.W. & SANGSTER, D.F. 1981. Volcanic-Associated Massive Sulfide Deposits. *Economic Geology, 75th Anniversary Volume*, 485–627.

GASQUET, D. 1991. *Genèse d'un pluton composite hercynien - Le massif du Tichka (Haut-Atlas occidental, Maroc).* PhD thesis, University of Nancy.

GIBSON, H.L. & KERR, D.J. 1993. Giant volcanic-associated massive sulfide deposits: with emphasis on Archean deposits. *In: Giant Ore Deposits.* Society of Economic Geologists, Special PaperS, **2**, 319–348.

HARRISON, T.M., DUNCAN, I. & McDOUGALL, I. 1985. Diffusion of ^{40}Ar in biotite: Temperature, pressure and compositional effects. *Geochimica et Cosmochimica Acta*, **49**, 2461–2468.

HMEURRAS, M. 1995. *Etude gîtologique des minéralisations sulfurées du Haut-Seksaoua.* Ministère de l'Energie et des Mines, Ministère de l'Energie et des Mines, Rabat.

JACKSON, S.L. & FYON, J.A. 1991. The western Abitibi subprovince in Ontario. *In: The western Abitibi subprovince in Ontario.* Ontario Geological Survey Special Volumes, **4**, 405–484.

JAILLARD, L. 1993. Distribution des filons de barytine du Haut-Atlas occidental. *Géologie et Energie*, **54**, 111–120.

JOUHARI, A. 1989. *Au-Mo-Cu mineralization in eastern Tichka massif, western High-Atlas, Morocco, and its geological setting.* PhD thesis, University of Grenoble.

KONTAK, D.J., SMITH, P.K., REYNOLDS, P. & TAYLOR, K. 1990. Geological and $^{40}Ar/^{39}Ar$ geochronological constraints on the timing of quartz vein formation in Meguma group lode-gold deposits, Nova Scotia. *Atlantic Geology*, **26**, 201–227.

LAGARDE, J.L. & RODDAZ, B. 1983. Le massif plutonique du Tichka (Haut Atlas Occidental, Maroc): un diapir syntectonique. *Bulletin de la Société Géologique de France*, **25**, 389–395.

LARGE, R.R. 1983. Sediment-hosted massive sulphide lead-zinc deposits: an empirical model. *In:* SANGSTER, D.F. (ed.) *Short Course in Sediment-Hosted Stratiform Lead-Zinc Deposits.* Mineralogical Association of Canada, Victoria, 1–29.

LEISTEL, J.M., MARCOUX, E., THIÉBLEMONT, D., QUESADA, C., SÁNCHEZ, A., ALMODÓVAR, G.R., PASCUAL, E. & SÁEZ, R. 1998. The volcanic-hosted massive sulphide deposits of the Iberian Pyrite Belt. Review and preface to the thematic issue. *Mineralium Deposita*, **33**, 2–30.

LIANXING, G. & McCLAY, K.R. 1992. Pyrite deformation in the stratiform lead-zinc deposits of the Canadian Cordillera. *Mineralium Deposita*, **27**, 169–181.

LYDON, J.W. 1984. Volcanogenic massive sulphide deposits, part I: a descriptive model. *Geosciences Canada*, **11**, 195–202.

LYDON, J.W. 1988. Volcanogenic massive sulphide deposits, part II: genetic models. *Geosciences Canada*, **15**, 43–65.

LYDON, J.W. 1996. Sedimentary exhalative sulphides (SEDEX). *In:* ECKSTRAND, O.R., SINCLAIR, W.D. & THORPE, R.I (eds) *Geology of Canadian mineral deposit types.* Geology of Canada, Geological Survey of Canada, **8**, 130–152.

MCCLAY, K.R. & ELLIS, P.G. 1983. Deformation and recristallization of pyrite. *Mineralogical Magazine*, **47**, 527–538.

MONIÉ, P., SOLIVA, J., BRUNEL, M. & MALUSKI, H. 1994. Les cisaillements mylonitiques du granite de Millas (Pyrénées, France). Age Crétacé ^{40}Ar/^{39}Ar et interprétation tectonique. *Bulletin de la Société Géologique de France*, **165**, 559–571.

NEHLIG, P., CASSARD, D. & MARCOUX, E. 1998. Geometry and genesis of feeder zones of massive sulphide deposits: constraints from the Rio Tinto ore deposit (Spain). *Mineralium Deposita*, **33**, 137–149.

NICOL, N., LEGENDRE, O. & CHARVET, J. 1997. Les minéralisations Zn-Pb de la série paléozoïque de Pierrefitte (Hautes-Pyrénées) dans la succession des événements tectoniques hercyniens. *Comptes Rendus de l'Académie des Sciences Paris*, **324**(II), 453–460.

ONÉZIME, J. 2001. *Environnement structural et géodynamique des minéralisations de la Ceinture Pyriteuse Sud Ibérique: leur place dans l'évolution hercynienne.* PhD thesis, University of Orléans.

ONÉZIME, J., CHAUVET, A., CHARVET, J. & FAURE, M. 1999. Syn- to late-deformation stockwork in the Iberian Pyrite belt, south Portuguese Zone, Spain - Structural constraints. *In:* STANLEY, C.J. *ET AL.* (ed.) *Mineral Deposits: Processes to Processing.* Balkema, Rotterdam, 1337–1340.

OUAZZANI, H. 2000. *Le paléovolcanisme des secteurs de Guedmioua et du Haut-Seksaoua (massif ancien du Haut-Atlas occidental, Maroc): témoin d'un contexte convergent.* PhD thesis, University of Meknès.

OUAZZANI, H., BADRA, L., POUCLET, A. & PROST, A.E. 1998. Mise en évidence d'un volcanisme d'arc néoprotérozoïque dans le Haut-Atlas occidental (Maroc). *Comptes Rendus de l'Académie des Sciences Paris, Sciences de la terre et des planètes*, **327**, 449–456.

OUAZZANI, H., POUCLET, A., BADRA, L. & PROST, A. 2001. Le volcanisme d'arc du massif ancien de l'ouest du Haut-Atlas occidental (Maroc), un témoin de la convergence de la branche occidentale de l'océan panafricain. *Bulletin de la Société Géologique de France*, **172**, 587–602.

PETER, J.M. & SCOTT, S.D. 1999. Windy Craggy, northwestern British Columbia: the world's largest Besshi-type deposit. *In:* BARRIE, C.T. & HANNINGTON, M.D. (eds) *Volcanic-associated massive sulfide deposits: processes and examples in modern and ancient settings.* Reviews in Economic Geology, **8**, 261–295.

PROKIN, V.A. & BUSLAEV, F.P. 1999. Massive copper-zinc sulphide deposits in the Urals. *Ore Geology Reviews*, **14**, 1–69.

RCHID, S. 1996. *Etudes géologiques et métallogéniques de la minéralisation de Barytine et de cuivre du Haut-Seksaoua (Haut-Atlas occidental).* Third cycle thesis, Université Mohammed V, Rabat.

RCHID, S., AZZA, A., HMEURRAS, M., BELHAJ, O., FADLI, M., FDIL, M., MECHICHE, A.M. & MAKKOUDI, D. 1996. Les minéralisations cuprifères et barytiques du Haut-Seksaoua. *Mines, Géologie et Energie, Rabat*, **55**, 163–174.

ROBBINS, C.S. 1972. *Radiogenic argon diffusion in muscovite under hydrothermal conditions.* Masters thesis, Brown University, Providence, R.I.

RODDICK, J.C., CLIFF, R.A. & REX, D.C. 1980. The evolution of excess argon in Alpine biotites. A ^{40}Ar/^{39}Ar analysis. *Earth and Planetary Science Letters*, **48**, 185–208.

RUSSELL, M.J., SOLOMON, M. & WALSHE, J.L. 1981. The genesis of sediment-hosted, exhalative zinc and lead deposits. *Mineralium Deposita*, **16**, 113–127.

SCHAER, J.P. 1964. Volcanisme cambrien dans le massif ancien du Haut-Atlas occidental. *Comptes Rendus de l'Académie des Sciences Paris*, **258**, 2114–2117.

SUBBA RAO, D.V & NAQVI, S.M. 1997. Geological setting, mineralogy, geochemistry and genesis of the Middle Archaean Kalyadi copper deposit, western Dharwar craton, southern India. *Mineralium Deposita*, **32**, 230–242.

TAYEBI, M. 1989. *Etude géologique d'un tronçon du couloir de cisaillement hercynien de l'Ouest marocain dans le Haut-Atlas occidental paléozoïque (Maroc).* PhD thesis, University of Aix-Marseille III.

TERMIER, H. & TERMIER, G. 1966. Le Cambrien inférieur au voisinage du massif granito-dioritique du tichka (Haut-Atlas marocain). *Comptes Rendus de l'Académie des Sciences Paris*, **262**, 843–845.

TURNER, G., HUNEKE, J.C., PODOSEK, F.A. & WASSERBURG, G.J. 1971. ^{40}Ar/^{39}Ar ages and cosmic ray exposure ages of Apollo 14 samples. *Earth and Planetary Science Letters*, **12**, 19–35.

WAGNER, G.A., REIMER, G.M. & JÄGER, E. 1977. Cooling ages derived from apatite fission, mica Rb-Sr and K-Ar dating: the uplift and cooling history of the central Alps. *Padovia University Institute of Geology and Mineralogy Memoirs*, **30**, 1–27.

The Baikalide–Altaid, Transbaikal–Mongolian and North Pacific orogenic collages: similarity and diversity of structural patterns and metallogenic zoning

ALEXANDER YAKUBCHUK

Centre for Russian and Central Asian Mineral Studies, Department of Mineralogy, Natural History Museum, Cromwell Road, London SW7 5BD, UK (e-mail: a.yakubchuk@nhm.ac.uk)

Abstract: The Baikalide–Altaid, Transbaikal–Mongolian and North Pacific orogenic collages consist of several oroclinally bent magmatic arcs separated by accretionary complexes and ophiolitic sutures located between the major cratons. The tectonic and metallogenic patterns of these collages are principally similar as they were formed as a result of rotation of the surrounding cratons and strike-slip translation along the former convergent margins.

The Altaid and North Pacific collages have principally the same distribution of metallogenic belts. In particular, the mid–late Palaeozoic belts of porphyry and epithermal deposits in the Altaids occupy the same position as the Mesozoic–Cenozoic metallogenic belts of the North Pacific collage. The Ural platinum belt occupies a similar position to the belt of platinum-bearing intrusions in Alaska. Major mineralizing events producing world-class intrusion-related Au, Cu–(Mo)-porphyry and VMS deposits in the Altaid and North Pacific collages coincided with plate reorganization and oroclinal bending of magmatic arcs. Formation of major porphyry, epithermal and Alaska-type PGM deposits took place simultaneously with oroclinal bending.

The tectonic setting of the orogenic gold deposits in the Tien Shan and Verkhoyansk–Kolyma provinces, hosting world-class hard rock gold deposits, is also similar, especially the distribution of their gold endowments. Major orogenic gold deposits occur within the sutured backarc basins. The craton-facing passive margin rock sequences, initially formed within backarc basins and now entrapped within such oroclines, represent favourable locations for emplacement of orogenic gold deposits.

The land-locked Altaid orogenic collage occurs between cratons separated or bounded by the fragments of the Neoproterozoic fold belts, e.g. the Baikalides and their equivalents, and Phanerozoic fold belts of the Tethysides, whereas the North Pacific orogenic collage and its splay into the Transbaikal–Mongolian orogenic collage are fragments of the Circum-Pacific belt formed between the subducting oceanic plates of the Pacific Ocean and the continents of Eurasia and North America (Fig. 1). Both regions host numerous economically important copper, gold, lead, zinc, PGM, nickel and other deposits, many of world class (White *et al.* 2001). I had a unique opportunity to work across these regions of Eurasia since 1984, participating in the regional geological surveys, international academic projects and exploration works. This experience helped me to recognize some similarity and diversity between them, on the one hand, and the links between geodynamics and metallogeny, on the other. A comprehensive study of tectonics and metallogeny of the Russian Far East and Alaska was recently completed by Nokleberg *et al.* (1993, 1998) and Nokleberg & Diggles (2001), which represents an ideal combination for the purposes of this article. The key works by Şengör & Natal'in (1996 a, b) analysed the tectonic evolution of all Asia, including both the Russian Far East and Central Eurasia. These latter works revealed the complex tectonics of Asia that consists of multiple, oroclinally bent magmatic arcs and accretionary complexes (Şengör & Natal'in 1996a, b). However, metallogenic aspects of Central Eurasia were only recently analysed by Yakubchuk *et al.* (2001). All these studies employ plate tectonics to explain the evolution and metallogeny of these orogens, but they sometimes recognize different tectonic units, use different terminolgy and disagree with each other in details. This paper intends to compare and to revise the existing interpretations of the geodynamic evolution of the two regions in relation to the setting of selected types of mineral deposits in the orogenic collages of Central Asia and northern segments of the Circum-Pacific belt, aiming to reveal their metallo-tectonic features.

From: BLUNDELL, D.J., NEUBAUER, F. & VON QUADT, A. (eds) 2002. *The Timing and Location of Major Ore Deposits in an Evolving Orogen*. Geological Society, London, Special Publications, **204**, 273–297.
0305-8719/02/$15.00 © The Geological Society of London 2002.

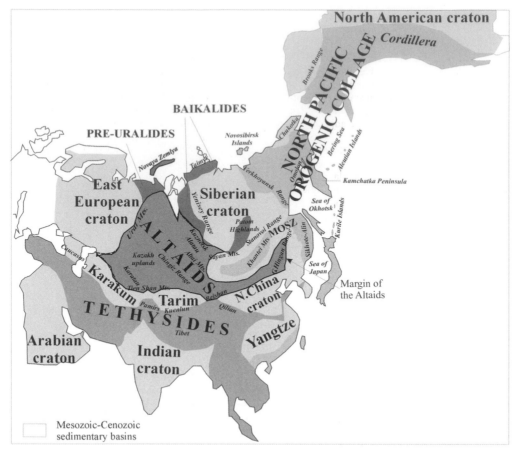

Fig. 1. First-order tectonic units of Eurasia and western North America (modified after Yakubchuk *et al*. 2001).

North Pacific and Transbaikal–Mongolian orogenic collages

Tectonics

This article considers the North Pacific orogenic collage as a system of orogenic belts extending from the Sikhote–Alin Mountains and Kurile Islands in the Russian Far East via mountainous ranges of northeast Russia and Alaska to the North American Cordillera in western Canada and USA. A detached fragment of the North Pacific orogenic collage constitutes the late Palaeozoic–early Mesozoic Transbaikal–Mongolian orogenic collage of central Mongolia (Yakubchuk & Edwards 1999; Yakubchuk *et al*. 2001). These orogenic collages were formed mostly in the Mesozoic–Cenozoic due to subduction of the oceanic plates of the Pacific Ocean (Scotese & McKerrow 1990; Nokleberg *et al*. 1993; Sengör & Natal'in 1996*a, b*). However, these collages also host fragments of now dismembered Palaeozoic

orogenic belts, which formed in response to subduction of oceanic plates of the Palaeo-Pacific Ocean of Nokleberg *et al*. (1993) and Sengör & Natal'in (1996a, b) or Panthalassic Ocean of Scotese & McKerrow (1990).

The North Pacific orogenic collage is usually separately described as part of the North American and Eurasian continents. During the last few years, an international project, which synthesized the data on geology and mineral deposits on both sides of the northern Pacific Ocean, was completed (Nokleberg *et al*. 1993; Nokleberg & Diggles 2001). This was a significant breakthrough into understanding the tectonic evolution in this region and showed how tectonics links to mineralization.

The North Pacific orogenic collage occupies a unique position at the triple junction of the North American, Eurasian and Pacific plates, accompanied by several microplates. The active spreading axes of the Atlantic–Arctic and Pacific Oceans, as well as the extinct spreading axis of the Amerasian basin, also occur nearby. A simplified tectonic scheme of the North Pacific collage shows Siberian and North American cratons, including their deformed passive margins, and several adjacent magmatic arcs (Fig. 2). Each arc has generated a subduction-accretionary complex. Ophiolitic sutures separate these magmatic arcs from each other and from major adjacent cratons. Nokleberg *et al.* (1993) recognized major tectonic units, i.e. terranes and sutures, in the North Pacific orogenic collage.

The North Pacific orogenic collage consists of several discordant oroclines (Sengör & Natal'in 1996*a, b*). These are Kolyma orocline, Alaskan orocline, Sakhalin orocline, Okhotsk orocline, Koryak orocline and several oroclines in the Bering Sea area (Fig. 2). The pattern of these

oroclines reflects a varying direction of strike-slip translation along the North Pacific margins during the accretionary growth of the North American and Eurasian continents. The principal structural markers within this collage are magmatic arcs. However, in several cases the tracing of their continuity is difficult, not only due to the obscuring Cenozoic sedimentary basins, especially in the central Kolyma area or in the Bering and Okhotsk Seas, but also due to tectonic 'shuffling' and repetition of the same accretionary complex along the giant strike-slip faults (Natal'in & Borukaev 1991; Khanchuk 2000).

Analysis of various tectonic schemes shows that in the present structure one can identify three fossil and still active late Palaeozoic to Cenozoic arc–backarc systems on the basis of synchroneity of magmatic arc, rifting and spreading events.

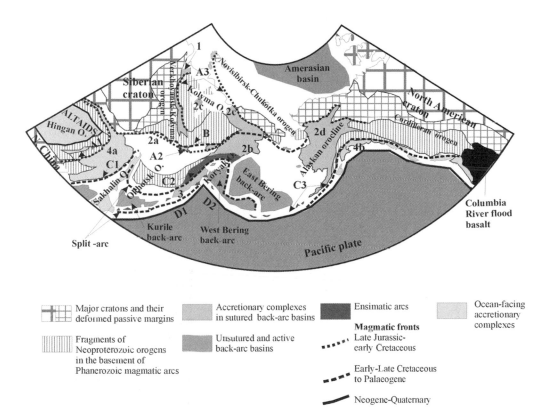

Fig. 2. Tectonics of the North Pacific orogenic collage (recompiled using data by Newberry *et al.* 1995; Nokleberg *et al.* 1997; Prokopiev 1998; Oksman *et al.* 2001). Magmatic arcs: A, Hingan and Omolon arcs (A1, Greater Hingan segment; A2, Uda–Murgal segment; A3, Alazeya–Oloy segment); B, Okhotsk–Chukotka arc; C, Sikhote–Alin–Aleutian arc (C1, Sikhote–Alin segment; C2, Kamchatka–Koryak segment; C3, Aleutian segment); D, Kurile–Komandor arc (D1, Kurile–East Kamchatka segment; D2, Komandor segment). Accretionary complexes in sutured backarc basins and ocean-facing subduction zones: 1, Chersky zone (former Oimyakon basin); 2, Okhotsk–Alaska backarc basin (2a, Okhotsk segment; 2b, Koryak segment; 2c, South Anyui segment; 2d, Yukon–Kuskokwim segment); 3, Kamchatka intra-arc basin; 4, ocean-facing accretionary complex (4a, Sikhote–Alin segment; 4b, Chugach segment).

These are the Verkhoyansk–Chukotka, Okhotsk–Alaska, and Kurile–Komandor systems (Fig. 3). In addition, there is an extinct Transbaikalian magmatic arc with an attached subduction-accretionary complex entrapped in the core of the Transbaikal–Mongolian orogenic collage. Although the former backarc basin of the Verkhoyansk–Chukotka system is totally sutured, backarc basins in the other systems were sutured only in part, and major inactive portions in the eastern Bering Sea remain unsutured. In addition, the backarc basins in the western Bering Sea, in the Kurile basin, and in the Sea of Japan are still actively spreading.

The Verkhoyansk–Chukotka arc–backarc system as described in this article, consists of the Verkhoyansk–Kolyma and Novosibirsk–Chukotka collisional orogens (Zonenshain *et al.* 1990; Bogdanov & Tilman 1992; Sengör & Natal'in 1996*a, b*). Nokleberg *et al.* (1993) and Sokolov *et al.* (1997) use the term super-terrane for these orogens. They formed as a result of several deformational events during the late Jurassic–Early Cretaceous collision of magmatic arcs with the adjacent Siberian and North American cratons against the background of their relative clockwise rotation with respect to each other and subduction of the Pacific plates (Parfenov 1991, 1995; Sengör

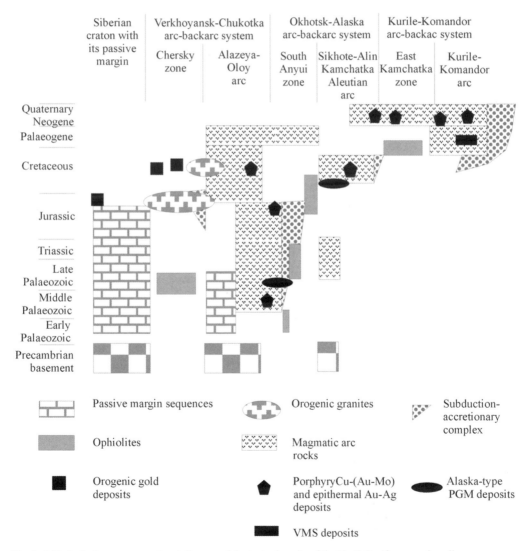

Fig. 3. Lithological sequences and metallogeny of the tectonic units of the North Pacific orogenic collage.

& Natal'in 1996*b*; Oksman *et al.* 2001). These multiple deformations were also responsible for oroclinal bending in the North Pacific orogenic collage.

In its external portion, the *Verkhoyansk–Kolyma* orogen, deformed into the Kolyma orocline, consists of craton-facing Palaeozoic to Mid-Jurassic passive margin shelf limestone and turbidite rocks (Fig. 3), which are now incorporated into imbricated allochthonous sheets thrust onto the rest of the Siberian craton (Prokopiev 1998). In the core of the Kolyma orocline there is an Alazeya–Oloy arc (Fig. 2, unit A3) consisting of Palaeozoic to Mesozoic rocks of magmatic and subduction-accretionary affinity. The arc remains attached to the Siberian craton via its Omolon segment (Fig. 2). The oldest magmatic rocks of the arc are Devonian to Carboniferous felsic and intermediate volcanics and intrusives found in the Alazeya and Omolon segments, and also in the Uda-Murgal segment (Fig. 2, unit A2) (Nokleberg *et al.* 1993; Sengör & Natal'in 1996b).

In their present structure, magmatic arc and passive margin units of the Siberian craton are separated by the Chersky zone (Fig. 2, unit 1), which bears the Palaeozoic ophiolites (Oksman *et al.* 2001). Late Jurassic collisional granites were emplaced within the suture zone, but formed a wider belt.

This orogen hosts several provinces of orogenic gold deposits constituting the Yana–Kolyma metallogenic belt (Fridovsky 2000), hosting medium to large and even world-class gold deposits (Nokleberg *et al.* 1993). The belt consists of several metallogenic provinces (Fig. 4), which form two sub-parallel strips in the external and internal portions of the orogen. Fridovsky (2000) classified the Verkhoyansk, Allakh–Yun and Ulakhan–Sis–Solur provinces as early collisional ore fields formed in the Late Jurassic to Neocomian, whereas the South Verkhoyansk, Kular and Adycha–Nera–Central Kolyma provinces are late collisional ore fields formed in the Early Cretaceous. The orientation of the provinces generally follows the strike of the orogen, but individual metallogenic clusters, especially those of orogenic gold deposits in the Verkhoyansk–Kolyma orogen, usually display an oblique orientation with respect to the general strike of the orogen.

The *Novosibirsk–Chukotka* orogen, forming the western part of the Chukotalaskides of Sengör & Natal'in (1996a, b), extends for approximately 2000 km from the Novosibirsk Islands in the Arctic Ocean towards northern Alaska. In the south, the Novosibirsk–Chukotka orogen is bound by the Alazeya–Oloy magmatic arc rocks (Fig. 2, unit A3), which were thrust northward onto Triassic passive margin rocks. In between there is a South Anyui suture (Fig. 2, unit 2c) hosting late Palaeozoic and Mid–Late Jurassic ophiolites (Sengör & Natal'in 1996*b*; Oksman *et al.* 2001).

The Verkhoyansk–Kolyma and Novosibirsk–Chukotka orogens were oroclinally bent and amalgamated in the early Late Cretaceous in response to the opening of the Amerasian basin (Nokleberg & Diggles 2001). This deformation also entrapped accretionary complexes of the Okhotsk–Alaska arc–backarc system in the core of the Kolyma orocline. Since the Mid-Cretaceous, the Okhotsk–Chukotka magmatic arc (Fig. 2, unit B) started to form above the amalgamated fragments of the two orogens. This magmatic arc system extended from Transbaikalia and the southern margin of the Siberian craton towards Alaska (Sengör & Natal'in 1996*a, b*).

The magmatic events related to the evolution of this latter arc are considered to be responsible for emplacement of some orogenic gold deposits, such as Nezhdaninskoe and Natalka, in its rear part in the Central Kolyma province of the Verkhoyansk–Kolyma orogen and a majority of orogenic gold deposits in the Novosibirsk–Chukotka orogen (Nokleberg *et al.* 1993). The arc developed on top of the older Devonian arc, which hosts epithermal gold deposits, e.g. Kubaka. The Cretaceous arc hosts small to medium epithermal Au deposits, e.g. Pokrovskoe, as well as medium to large Cu-porphyry deposits and occurrences such as Peschanka (Fig. 4). They host hypogene mineralization and lack supergene enrichment blankets (Nokleberg *et al.* 1993). In Alaska, the late Palaeozoic arc-related ultramafic rocks host platinoid mineralization (Nokleberg *et al.* 1993). Carlin-type gold mineralization was recognized in the southeastern portion of the Siberian craton in the rear of this arc (Yakubchuk 2000), where the Allakh–Yun province of the Sette–Daban orogen is considered to be prospective for this type of mineralization (Eirish 1998).

Transbaikal–Mongolian orogenic collage Here (Fig. 5), the accretionary complexes occur in the Khantei zone, and then they can be traced via the 500 km long Mongol–Okhotsk suture into the North Pacific orogenic collage (Yakubchuk & Edwards 1999). Sengör *et al.* (1993) and Sengör & Natal'in (1996a, b) recognized this link, but they considered the orogens of central Mongolia to be a part of the Altaid orogenic collage. The late Palaeozoic to early Mesozoic differentiated magmatic rocks form a Transbaikalian arc (Fig. 2, unit A1; Fig. 4, unit G), which is bent into the Central Mongolian and Hingan oroclines and extends from the Greater Hingan in NE China via southern and central Mongolia to Russian Transbaikalia and the Stanovoy Range on the southern

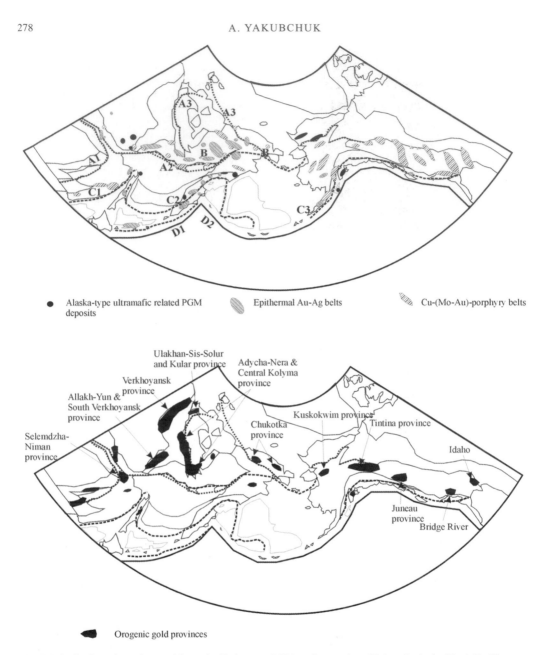

Fig. 4. Distribution of porphyry, epithermal, Alaska-type PGM, and orogenic gold deposits in the North Pacific orogenic collage. See Figure 2 for arc symbols.

rim of the Siberian craton. This magmatic arc represents a continuation of the synchronous magmatic arcs of the Verkhoyansk–Chukotka arc–backarc system. These structures are now almost completely tectonically isolated from the rest of the North Pacific orogenic collage.

In central Mongolia, the Transbaikalian arc rests on Neoproterozoic to early Palaeozoic rocks, consisting of turbidite sequences, ophiolites, and late Proterozoic to mid-Palaeozoic magmatic arcs. The tectonic affinity of the Neoproterozoic ophiolites and early to middle Palaeozoic terrigenous rocks in central Mongolia is not quite clear. In particular, previous studies (Sengör *et al.* 1993; Sengör & Natal'in 1996*a, b*) considered them as part of an accretionary complex of the Tuva–

Fig. 5. Tectonics of the Baikalide–Altaid orogenic collage and adjacent cratons. Semi-arrows show direction of strike-slip displacement of major blocks. Magmatic arcs: A, Kipchak arc (A1, Kokchetav–North Tien Shan segment; A2, Bozshakol–Chingiz segment; A3, Salair–Kuznetsk Alatau segment); B, Tuva–Mongol arc (B1, Tuva segment; B2, Bureya segment); C, Kazakh–Mongol arc (C1, mid-Palaeozoic segment; C2, late Palaeozoic segment); D, Mugodzhar–Rudny Altai arc (D1, Tagil segment; D2, Magnitogorsk segment; D3, Rudny Altai segment); E, Valerianov–Beltau–Kurama arc; F, Qilian arc; G, Transbaikalian arc; H, Hingan arc. Backarc rifts and backarc basin sutures: 1, Baikonur–Karatau backarc rift; 2, Khanty–Mansi backarc suture (2a, South Tien Shan segment; 2b, East Urals segment; 2c, Irtysh–Zaissan segment); 3, Sakmara backarc suture. MOSZ, Mongol–Okhotsk suture zone.

Mongol arc (Fig. 5, units B1–B2), which was framed by synchronous accretionary complexes on both sides.

All pre-Carboniferous differentiated magmatic rocks within the Transbaikal–Mongolian collage can be attributed more or less definitely to the Tuva–Mongol or Kazakh–Mongol arcs whose magmatic rocks were produced due to subduction in the Altaids.

The Transbaikalian magmatic arc consists of late Palaeozoic to early Mesozoic rocks, occurring on top of the older arc rocks and accretionary complexes to the west. It is rimmed by a late Palaeozoic to Triassic accretionary complex to the east, in the core portion of the Central Mongolian orocline. This allows subduction to be restored from the east, and if the link of this unit with the North Pacific collage is accepted, the accretionary complex might have been produced by subduction of the Palaeo-Pacific Ocean prior to the oroclinal bending of the arc in the Jurassic that caused collision of its present eastern part in the Hingan segment with the Stanovoy unit, producing the Mongol–Okhotsk suture zone and isolating the Transbaikal–Mongolian orogenic collage from the other Circum-Pacific orogens.

The Transbaikalian arc hosts large Triassic Cu-porphyry (Erdenet), Mo-porphyry (Zhireken), and large epithermal Au (Baley) deposits (Kirkham & Dunne 2000; Sotnikov & Berzina 2000; Yakubchuk et al. 2001). The Mesozoic (Cretaceous?) magmatism in the Stanovoy Range might be responsible for emplacement of large Carlin-type deposits of the Kuranakh group (Yakubchuk 2000).

The Okhotsk–Alaska arc–backarc system extends from the Korean Peninsula in the south to Alaska in the NE. Various units, such as the Tuva–Mongol arc of the Altaids, the Siberian craton, the Verkhoyansk–Chukotka arc–backarc system, and Transbaikalian arc and Khantei accretionary complex of central Mongolia, occur behind the magmatic arcs that form a discontinuous belt (Fig. 3) that starts in the Sikhote–Alin mountains in the south of the Russian Far East (Fig. 2, unit C1), then follows two oroclinal bends of Sakhalin and Okhotsk, and can be traced further using magnetic data on the Okhotsk shelf towards western Kamchatka (Fig. 2, unit C2). From Kamchatka, it continues northward to the Koryak segment. It constitutes several oroclines in the Bering Sea, finally extends into the Aleutian arc (Fig. 2, unit C3) and links the magmatic arcs of the North American Cordillera. This long magmatic arc rests in its various parts on Precambrian slivers and Palaeozoic accretionary complexes or consists of immature ensimatic magmatic arc fragments. The accretionary complexes of late Palaeozoic to Cretaceous age mark part of this arc. The oldest differentiated magmatic rocks of late Palaeozoic age are found in western Kamchatka above the Precambrian metamorphic rocks (Nokleberg et al. 1993), constituting the only fragment of the sialic-type basement in this arc. Sengör & Natal'in (1996a, b) suggest that it is an expelled fragment of the Palaeozoic orogens in northeast China, which was incorporated into this generally ensimatic arc, formed almost in the middle of the Pacific Ocean, and separated the floor of its northern part from the rest of its plates (Nokleberg &

Diggles 2001). This northern part was then subducted under Eurasia and North America and the arc docked to these continents as suggested in the reconstructions by Nokleberg et al. (1998), Parfenov (1995) and Nokleberg & Diggles (2001).

In the Greater Hingan (Fig. 2, unit A1), there are Jurassic to Early Cretaceous magmatic arc rocks. The origin of this magmatism is debatable due to its location more than 500 km from the nearest palaeotrench. Sengör & Natal'in (1996a) suggested that strike-slip duplication of Precambrian slivers in front of this arc might significantly increase the width of its Pacific-facing accretionary complex in the late Mesozoic. The growth of a wide accretionary wedge could then have stopped subduction-related magmatism in the Greater Hingan and have been responsible for the eastward migration of the magmatic front in the Cretaceous to form the bulk of magmatic arc rocks in Cretaceous to Palaeogene times.

The rocks that can be identified with the back-arc basin settings of the Okhotsk–Alaska system occur either in isolated sedimentary basins in the rear flank of the arc in eastern Mongolia and NE China or were incorporated into subduction-accretionary complexes accumulated in the front of the arcs of the Verkhoyansk–Chukotka system. Oceanic crustal rocks of this backarc basin are Late Cambrian to Cretaceous ophiolites within accretionary complexes in the Russian Far East and NE, which represent a mixture of oceanic crust generated in the backarc basin and the main ocean (Sokolov et al. 1997). The East Bering backarc basin with oceanic crust remains unsubducted. Its Mesozoic basaltic floor is now buried under deep sea and terrigenous sediments.

Metallogenically, this arc–backarc system hosts several minor Late Cretaceous to Palaeogene Cu-porphyry deposits and occurrences (Fig. 4) in the Kamchatka Peninsula and Sikhote–Alin (Nokleberg et al. 1993; Gusev et al. 2000). They host low-grade hypogene mineralization and lack supergene enrichment blankets. In Sikhote–Alin, there are medium-size economically viable Palaeogene–Neogene Au–Ag epithermal deposits such as Mnogovershinnoe (Khanchuk & Ivanov 1999). The latter province is also a host to important tin deposits (Khanchuk 2000). In central Alaska, the arc rocks host a Tintina gold province with major granitoid-related gold deposits (Goldfarb et al. 2001). Various workers emphasize difficulties in classifying these deposits as Au-porphyry or granitoid-related orogenic gold deposits (Lang et al. 2000). The Late Cretaceous immature arc rocks in northern Kamchatka host Alaska-type ultramafic intrusives, which are a bedrock source of significant platinum-producing placers in northern Kamchatka.

The Kurile–Komandor arc–backarc system extends from the Japanese Islands along the Kurile Islands to eastern Kamchatka and the Komandor Islands as a presently active system of magmatic arcs (Fig. 2, units D1–D2). Its Neogene–Quaternary volcanics rest on the Mesozoic arc or accretionary complex rocks accumulated due to subduction of the Pacific oceanic crust. In the rear of this arc, there is a system of still actively spreading backarc basins in the Kurile basin and in the western Bering Sea. However, in eastern Kamchatka, there is a suture with Cretaceous ophiolites (Fig. 2, unit 3), which separates the magmatic arcs of the Kurile–Komandor and Okhotsk–Alaska systems, thus representing an intra-arc suture.

The arc rocks host minor VMS deposits known in the Kurile and Komandor islands. Medium-size epithermal Au–Ag deposits of Neogene–Quaternary age are a potential source of these metals in Kamchatka. Some deposits of this type located in the Kurile Islands were gold producers in the past.

Distribution of mineralization in the North Pacific and Transbaikal–Mongolian collages

It was shown in numerous previous works that in the active Circum-Pacific orogens there is a regional-scale correlation between tectonic setting and principal types of mineral deposits (e.g. Nokleberg et al. 1993; Khanchuk 2000). In particular, the Cu–(Au) porphyry and epithermal Au deposits formed within the magmatic arcs with sedimentary hosted Cu and Pb–Zn deposits in backarc settings, whereas orogenic gold deposits form during or after collision in the orogens. A slab window theory was recently employed to explain clustering of giant Cu–(Mo–Au) porphyry, epithermal Au and Sn deposits in various orogens around the Circum-Pacific rim (Kirkham 1998; Khanchuk 2000). Similar factors may control the distribution of orogenic gold deposits.

Figure 4 shows the distribution of principal metallogenic belts of Cu–(Au) porphyry, epithermal Au, and Alaska-type ultramafic related PGM deposits, as well as orogenic Au deposits in the North Pacific collage. Analysis of the orogenic gold endowment using the data collected by Goldfarb et al. (2001) reveals that the most significant gold accumulations (>30 Moz Au) occur in the Yana–Kolyma, Verkhoyansk, Chukotka, Selemdzha–Niman, Kuskokwim, Tintina and Juneau provinces. Of these, the Yana–Kolyma province is the most anomalous region, which originally contained about 150 Moz of gold. Only placer operation in the Yana–Kolyma province

produced in excess of 125 Moz of gold during the past 55 years (Goldfarb et al. 2001). In addition, the Verkhoyansk–Chukotka system hosts several world-class and large orogenic gold deposits (Fig. 4), e.g. Nezhdaninskoe (16 Moz) in the South Verkhoyansk province, Natalka (>6 Moz) in the Central Kolyma province, Kyuchus (>10 Moz) in the Ulakhan–Sis–Solur province, Maiskoe (>8 Moz) in the Chukotka province, numerous medium-size deposits and multiple prospective targets (Nokleberg et al. 1993; Goldfarb et al. 2001). The adjacent Okhotsk–Alaska and Kurile–Komandor systems host much smaller orogenic Au deposits (Goldfarb et al. 2001). However, they host medium to large epithermal Au deposits in western Canada and USA and in Kamchatka in Russia, large Au-porphyry deposits (Fort Knox, Tintina province in Alaska; Lang et al. 2000), and also large Carlin-type Au deposits in the Great Basin (Ludington et al. 1993) and in the Kuranakh area located in the Aldan shield of the Siberian craton in South Yakutia.

The search for what might be a reasonable explanation for such an irregular distribution of orogenic gold deposits in the North Pacific collage reveals two principal factors, which identify unique features in the Verkhoyansk–Chukotka system. These factors include the occurrence of black-shale-bearing passive margin sedimentary rocks, on the one hand, and a time affinity to major collisional/suturing events in the respective backarc basin, on the other. Several gold deposits, e.g. Fort Knox and Pogo in the Tintina belt in Alaska reveal an affinity to granitoid intrusives (Smith et al. 1999; Lang et al. 2000), a factor that may come to be recognized as very common with further progress of geological studies. These passive margin sediments accumulated on the craton-facing side of the Verkhoyansk–Chukotka backarc basin, and emplacement of orogenic deposits and granitoids also took place in the rear part of its frontal magmatic arc in the Late Jurassic. The orogenic gold deposits of the Verkhoyansk–Chukotka system form several clusters in the Yana–Kolyma belt. This region of the former backarc basin coincides with the area where its frontal magmatic arc was attached to the rear craton. The Early Cretaceous Okhotsk–Chukotka Andean-type magmatic belt amalgamated the oroclinally bent late Palaeozoic to Jurassic magmatic arcs of the Verkhoyansk–Chukotka system (Oksman et al. 2001). Orogenic gold deposits in other parts of the North Pacific collage, e.g. Mongol–Okhotsk, Chukotka, and Alaska, were also emplaced in the rear parts of the other arcs in the Early to Mid-Cretaceous, but they occur within former accretionary complexes and not in the deformed passive margin sediments. Their gold endowment is smaller by an order of magnitude. I speculate that this difference in tectonic setting might control the amount of gold emplaced or regenerated within the system. In addition, it seems to be important that the largest gold endowment is concentrated in the backarc basin area where a magmatic arc is attached to its craton.

Although orogenic gold deposits form metallogenic belts which are parallel to the strike of deformed arc–backarc systems, the distribution of the metallogenic provinces is often oblique with respect to the extent of such structures. They are clearly superimposed onto already amalgamated tectonic units forming several major provinces within the belt. The distribution of these provinces seems to be controlled by local structural factors.

This suggests searching for similar localities in other fossil backarc basins of the North Pacific orogenic collage. It appears that, with the exception of the Verkhoyansk–Chukotka arc–backarc system, the other structures host in general much smaller orogenic gold deposits within other arc–backarc systems of the North Pacific orogenic collage. Some areas, such as the East Bering basin, may represent one of the most favourable targets after its suturing.

Baikalides and Altaids

The Altaid orogenic collage of Palaeozoic age (Sengör et al. 1993), also known as the Ural–Mongolian, Ural–Okhotsk, or Central Asian fold belt (Coleman 1989; Zonenshain et al. 1990; Mossakovsky et al. 1993), is framed by the Neoproterozoic orogens of the Baikalides (Fig. 5). Many workers consider the Baikalides and their analogues as part of the Ural–Mongolian fold belt, whereas others identify them as an independent orogenic system (Milanovsky 1996).

The Baikalides and the Altaids consist of Neoproterozoic–Palaeozoic rocks forming an orogenic collage lying between the East European and Siberian cratons in the west and NE respectively. The Karakum, Alai–Tarim and North China blocks in the south (the intermediate units of Sengör & Natal'in 1996a, b) separate the Altaids from the Tethysides (Sengör & Natal'in 1996a, b). Traditional interpretations suggest that Palaeozoic structures of central Mongolia also constitute a part of the Altaids and join the Circum-Pacific belt via the Mongol–Okhotsk suture zone (Milanovsky 1996; Sengör & Natal'in 1996a, b). However, in western and southern Mongolia and in NE China, there are numerous Precambrian slivers that form the basement of the Neoproterozoic–Palaeozoic magmatic arc, known as the Tuva–Mongol arc (Fig. 5, units B1–B2) (Sengör et al. 1993; Yakubchuk et al. 2001). This arc

everywhere separates the accretionary complexes of the Transbaikal–Mongolian and North Pacific orogenic collages, on the one hand, and the Altaid orogenic collage, on the other. In the east the Precambrian slivers in the basement of this arc have a T-shaped junction with the North China craton, thus providing a natural barrier between the Altaids and the Circum-Pacific belts. On this basis, I suggest using the Precambrian units in the basement of the Tuva–Mongol arc as a boundary between the Altaids and the Circum-Pacific belt and therefore to exclude the Palaeozoic structures of central Mongolia (Transbaikal–Mongolian orogenic collage) from the Altaids.

The absence of large Precambrian massifs in the east, if the narrow Precambrian slivers in Mongolia are excluded, allowed a reconstruction of the Neoproterozoic–Palaeozoic Palaeo-Asian Ocean as an embayment of the Palaeo-Pacific Ocean (Zonenshain et al. 1990; Mossakovsky et al. 1993; Sengör et al. 1993), in place of the Altaid orogenic collage. This assumption was based on the idea that the Alai–Tarim and North China cratons constitute a single block, whereas the significance of the Precambrian slivers in Mongolia was underestimated. However, as was shown above, these latter units can be considered as a boundary between the Altaid and Transbaikal–Mongolian orogenic collages, whereas the Alai–Tarim and North China cratons are clearly separated in their present structure by a Beishan orogen and, therefore, their identification as a single block is not correct. I suggest that this latter area and the Qinling orogen can be used as a link between the Altaids, to the north, and the Tethysides, to the south (Fig. 5).

Sengör et al. (1993) and Sengör & Natal'in (1996a, b) indicated that the stucture of the Altaids consists of several oroclines. They recognized that all early Palaeozoic magmatic arc and accretionary complex rocks in the Altaids are very similar across this vast region and suggested on this basis that they might be formed in front of a former single magmatic arc, which they named as the Kipchak arc (Fig. 5, units A1–A3). They interpreted almost all ophiolites as fragments of the main ocean-facing accretionary complex and largely ignored their possible position in the sutures of the backarc basins. They suggested that the Precambrian slivers found in the basement of this arc were rifted off the combined Eastern Europe–Siberia forming the Khanty–Mansi Ocean, which spread behind the Kipchak arc. According to Sengör et al. (1993) and Sengör & Natal'in (1996a, b), the Kipchak arc and its accretionary complex were later multiply repeated along giant strike-slips and oroclinally bent against the background by clockwise rotation of

Siberia relative to Eastern Europe. Palaeozoic magmatic arcs of the Urals were considered as an independent system formed during subduction of the oceanic crust of the Khanty–Mansi Ocean under Eastern Europe and Siberia, thus suggesting the presence of a second arc behind the Kipchak arc. In their work, these authors used magmatic arc fronts as structural markers, which helped them to identify the direction of the accretionary growth in the entire collage.

However, significant portions of the Neoproterozoic–early Palaeozoic basement of the Altaids are obscured under Mesozoic–Cenozoic sedimentary basins (Fig. 1) and middle to late Palaeozoic magmatic arcs (Fig. 5). This creates significant difficulties for correlation and understanding how different parts of the Altaids link to each other and how they strike under the sedimentary basins. The airborne magnetic data (National Geophysical Data Center 1996) employed in this study allow the orientation of the belts under the basins to be deciphered.

In addition, not all Neoproterozoic to early Palaeozoic volcanic and terrigenous rocks represent an accretionary complex, e.g. Baikonur–Karatau zone (Fig. 5, unit 1), which hosts alkaline basalts and chert to clastic sedimentary rocks, lacks ophiolites and might be better interpreted as a rift structure (Mossakovsky et al. 1993) behind the Kipchak arc. Its sedimentary rocks host vanadium–molybdenum deposits, which makes them metallogenically distinct from all other chert-terrigenous sequences in the area, where they might be interpreted as an accretionary complex. In addition, some ophiolites occur in sutures between the former magmatic arcs, reveal petrological signature of supra-subduction origin (Yakubchuk & Degtyarev 1991; Yakubchuk 1997; Degtyarev 1999) and are coeval with the differentiated magmatic rocks in the adjacent magmatic arcs, whereas others occur as slivers within long-lived, viz. longer than 200 Ma, and very wide accretionary wedges that might therefore face a former major ocean. The regional aeromagnetic data (National Geophysical Data Center 1996) allow magnetically distinct magmatic arcs and ophiolitic sutures to be traced, on the one hand, and non-magnetic accretionary complexes and passive margin sedimentary sequences, on the other. The orientation and relationships between various units recognized on this basis (Fig. 5) appears to differ from the maps suggested by Sengör & Natal'in (see Sengör & Natal'in 1996b, fig. 21.18) and therefore the details of the tectonic evolution of the Altaid orogenic collage can be also viewed differently.

However, this study recognizes the importance of Precambrian slivers and magmatic fronts in the same manner as suggested initially by Sengör et al. (1993). In the Baikalides and Altaids, there are several generations of arc magmatism from the Riphean ($<1650>680$ Ma) to the Mesozoic (Fig. 6). Recognition of the synchroneity of some ophiolites, largely neglected by Sengör et al. (1993) and Sengör & Natal'in (1996a, b), and magmatic arc rocks was the basis for a new tectonic interpretation of the Baikalides and the Altaids suggested recently by Yakubchuk et al. (2001).

Each fragment of the Baikalides consists of one magmatic arc separated by an ophiolitic suture from the craton-facing passive margin sequences, whereas the Altaid collage consists of two sub-parallel major island arcs of Vendian (<680 Ma) to early Palaeozoic age (Figs 5 and 6): Kipchak arc in Central Kazakhstan and Altai (Fig. 5, units A1–A3) and Mugodzhar–Rudny Altai in the Urals (Fig. 5, units D1–D2) and Altai (Fig. 5, unit D3), constituting the Kazakh orocline in the western half of the Altaids. The sutures, which trace former backarc basins, separate the Kipchak (Fig. 5, units 2a–2c) and Mugodzhar–Rudny Altai arcs (Fig. 5, unit 3) from the cratons and each other. The Tuva–Mongol arc in the east, an intermediate unit between the Altaids, on the one hand, and the Transbaikal–Mongolian and the North Pacific collages, on the other hand, is deformed into several en echelon oroclines. In the mid-Palaeozoic, the Kipchak and Tuva–Mongol arcs amalgamated into a single Kazakh–Mongol arc (Fig. 5, units C1–C2).

In their present structural configuration, there are difficulties and uncertainties in recognizing the affinity of accretionary complexes to the respective arcs. If the above-mentioned recognition that magmatic arc complexes of the Urals and Rudny Altai form a single structure is correct and that it is sub-parallel with the Kipchak arc, then it is difficult to interpret them as a result of strike-slip repetition of the same structure, because, on the one hand, they are not synchronous and, on the other hand, there would be a significant space problem if they constituted a single arc. One can constrain this problem using the migration of magmatic fronts with respect to magmatic arcs. Accretionary complexes might also develop within a system of several sub-parallel magmatic arcs, which may be considered as an alternative to a single, but complexly deformed arc. If there were several evolving arc–backarc systems in the western part of the Altaid collage, e.g. Ural–Altai, Kazakh–Khanty Mansi, each of them might generate its own accretionary complex, a magmatic arc and a backarc basin. This option does not require 'construction' of a single arc, whose length would exceed the length of the combined

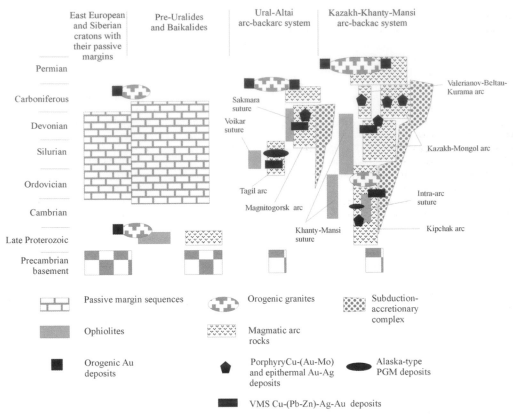

Fig. 6. Lithotectonic units and mineralization in the Baikalide–Altaid orogenic collage.

margin of united Eastern Europe–Siberia. In addition, each of these units reveals quite a distinct subduction-related metallogeny, which better fits into a system of more than one arc, and only late Palaeozoic orogenic gold deposits are superimposed onto various pre-orogenic tectonic units. The following describes an alternative tectonic interpretation of the Altaids employing the presence of several magmatic arcs in its western portion.

The Baikalides

The Baikalides and their analogues, such as the Pre-Uralides, were identified in the Patom Highlands, in the southwestern margin of the Siberian craton in the Eastern Sayan, in the Yenisey Range, in the Taimyr Peninsula, and in the Pechora Lowlands. In all these locations they bound either the Siberian or East European cratons. It is suggested that the Baikalides may constitute significant areas of the Arctic shelf (Shipilov & Tarasov 1998). Their analogues are also known as Neoproterozoic

crustal slivers within the Altaid orogenic collage in the Urals, Altai, and in the Kazakh uplands.

The Baikalides and Pre-Uralides host sutures with late Proterozoic ophiolites (Vernikovskiy et al. 1996, 1999). The sutures represent traces of former backarc basins between late Proterozoic magmatic arcs and craton-facing passive margin sedimentary sequences. Around the Siberian craton, the late Proterozoic passive margin sedimentary rocks in the Baikalides host giant orogenic Au–(PGM) (Fig. 6) and large Pb–Zn deposits.

The Neoproterozoic ophiolitic sutures can be more or less clearly traced on the flanks of the East European and Siberian cratons. In orogenic belts in Transbaikalia and Mongolia, it is possible that ophiolitic sutures of the Patom orocline in northern Transbaikalia and the Tuva–Mongol arc might be dextrally offset by 1000 km (Yakubchuk et al. 2001). This indicates that a province of medium-size mesothermal gold deposits in south Transbaikalia may represent an offset continuation of the Lena orogenic gold province.

Other fragments of the Riphean magmatic arcs can also be found as slivers in the basement of younger magmatic arcs in the Kuznetsk Alatau, West Sayan, in the Kazakh uplands, and in the Northern Tien Shan.

The Altaid orogenic collage

The Altaid orogenic collage consists of the Palaeozoic orogens of the Urals, Kazakh uplands, Tien Shan and Altai. In their present structural configuration, these systems occur as apparently independent orogens separated by vast oil-bearing Mesozoic–Cenozoic sedimentary basins in western Siberia and Central Asia. However, the airborne magnetic data (National Geophysical Data Center 1996) suggest that they represent exposed fragments of a single orogenic collage.

The Tuva–Mongol arc can be traced from southern Transbaikalia to western and southern Mongolia and NE China. It now constitutes the Central Mongolian and the Hingan oroclines. The arc separates Vendian–Palaeozoic accretionary complexes of the Altaids from the mid–late Palaeozoic accretionary complexes of the Transbaikal–Mongolian collage. Within the Tuva–Mongol arc, there are intra-arc sutures with Vendian–Early Cambrian ophiolites in Transbaikalia, northern and central Mongolia.

The early Palaeozoic magmatic rocks of the ensimatic portion of this arc in Tuva and Transbaikalia host Cu–Pb–Zn–Ag–Au VMS deposits (Kovalev *et al.* 1998) and Cu-porphyry occurrences (Sotnikov & Berzina 2000) in its western portion in Tuva and western Mongolia (Figs 6 and 7). There are granite-related orogenic gold deposits emplaced after suturing of the intra-arc basins in the south Transbaikalia province (Yakubchuk *et al.* 2001).

The Kazakh–Khanty–Mansi arc–backarc system. consists of the Vendian–early Palaeozoic Kipchak arc and its Palaeozoic accretionary complex in the core of the Kazakh orocline, and a suture of the Khanty–Mansi backarc basin whose fragments are exposed in the southern Tien Shan, east Urals and Irtysh–Zaissan segments. The Kipchak differentiated magmatic rocks occur above the heterogenous basement in several ensialic fragments in the Gorny Altai, Salair, and Kuznetsk Alatau in Russia (Fig. 5, unit A3), the ensimatic Bozshakol–Chingiz segment in Kazakhstan (Fig. 5, unit A2), and the ensialic Stepnyak–Betpakdala segment in Kazakhstan and Kyrgyzstan (Fig. 5, unit A1). Even in their present structure the fragments of this arc remain attached to the Siberian craton in the east. Its opposite 'end' stops in the Kyrgyz and Chinese Tien Shan. The Palaeozoic magmatic arc rocks in the Altyn and Kunlun mountains on the southern rim of the Alai–Tarim, Karakum and perhaps including the Qaidam block may also represent equivalents of this arc. In this case, the Alai–Tarim and Karakum blocks could have been part of the Kipchak arc since the Late Proterozoic, in contrast to their traditional interpretation as fragments of Gondwana docked to Laurasia only in the late Palaeozoic (Scotese & McKerrow 1990; Zonenshain *et al.* 1990; Sengör & Natal'in 1996a). This means that the Altaids may have been formerly related to the Palaeo-Tethys Ocean and not to the Palaeo-Pacific Ocean.

The backarc structures of the Kipchak arc include a 1000 km long Vendian–early Palaeozoic Baikonur–Karatau backarc rift (Fig. 5, unit 1) in west Central Kazakhstan (Mossakovsky *et al.* 1993).

Within the Kipchak arc, there are several 500–1000 km long sutures with Cambrian–Ordovician ophiolites. However, even in their present structural configuration they appear as a system of en echelon sutures, which may represent a system of former intra-arc basins. Their ensimatic segments host early Palaeozoic VMS deposits in Central Kazakhstan (Maikain) and Altai (Salair group) and also Cu-porphyry deposits in Northern Kazakhstan (Bozshakol) (Heinhorst *et al.* 2000). Suturing and related strike-slip deformation at the end of the Ordovician led to the emplacement of Late Ordovician granitoid plutons which host orogenic granite-related gold deposits in the North Kazakhstan province (Shatov *et al.* 1996; Heinhorst *et al.* 2000) and in the Kuznetsk Alatau province in Russia (Distanov & Obolensky 1993; Yakubchuk *et al.* 2001). Their emplacement took place synchronously with the above-mentioned similar orogenic deposits in the South Transbaikal province.

In the rear part of the Kipchak arc, there is a very long suture marked by Ordovician–Devonian ophiolites, which strike from the South Tien Shan (Fig. 5, unit 2a) to the East Urals (Fig. 5, unit 2b) and then ends at the Irtysh–Zaissan segment (Fig. 5, unit 2c) (Yakubchuk *et al.* 2001). This suture marks the Khanty–Mansi backarc basin that started to open in the Latest Proterozoic to the early Palaeozoic.

The Kazakh–Mongol arc–backarc system formed as a result of reorganization at the Ordovician–Silurian transition. The mid-Palaeozoic ophiolites occur in the suture of the Khanty–Mansi backarc basin and the Kazakh–Mongol magmatic arc extending from Central Kazakhstan to Mongolia and NE China (Fig. 5, units C1 and C2). Its very wide accretionary complex occurs in the core of the Kazakh orocline and then extends to South Mongolia (Fig. 5). The synchronicity of ophiolites

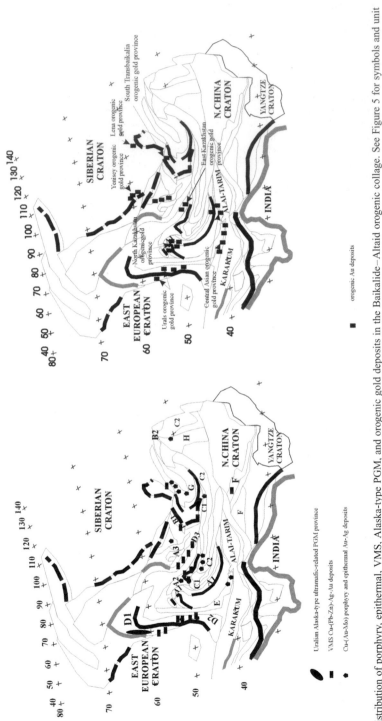

Fig. 7. Distribution of porphyry, epithermal, VMS, Alaska-type PGM, and orogenic gold deposits in the Baikalide–Altaid orogenic collage. See Figure 5 for symbols and unit numbers.

and magmatic arc rocks suggests an island arc setting for this arc.

The Silurian–Devonian Kazakh–Mongol magmatic arc (Fig. 5, unit C1) amalgamated fragments of the older Kipchak and Tuva–Mongol arcs. The mid-Palaeozoic arc rocks host several VMS and important porphyry Cu deposits in Central Kazakhstan (Nurkazghan), Mongolia (Oyu Tolgoi), and northeast China (Duobaoshan) (Fig. 7). The accretionary complex of this arc is not spectacular in terms of mineralization, but it is a good structural marker, which can be traced from the core of the Kazakh orocline via south Mongolia to the core of the Hingan orocline.

In the Early Carboniferous, there was a key transitional event in the tectonic and metallogenic evolution of Central Asia. In the Early–Mid-Carboniferous, the Kazakh orocline began to collide with the Mugodzhar–Rudny Altai arc to form a 3000 km long suture extending from the South Tien Shan to the East Urals and Irtysh–Zaissan zones (Puchkov 1993). Most porphyry deposits in Kazakhstan and Central Asia formed during this time (Heinhorst et al. 2000).

In the western part of the Kazakh–Mongol arc, a collision caused migration of the magmatic front and emplacement of numerous Cu–Mo-porphyry, skarn and epithermal deposits in Kazakhstan. In western Central Kazakhstan and Uzbekistan there is the Early–Mid-Carboniferous *Valerianov–Beltau–Kurama* arc. This arc magmatism produced large Fe-skarn deposits in northwest Kazakhstan and very large porphyry (Kalmakyr–Dalnee) and epithermal deposits in Uzbekistan (Kirkham & Dunne 2000). The giant sedimentary-Cu deposits of Dzhezkazghan formed between these two magmatic arcs. In the Late Carboniferous, Mo–W deposits controlled by pull-apart structures (Heinhorst et al. 2000) were emplaced in Central Kazakhstan.

In southern Mongolia and northeast China the Kazakh–Mongol arc produced large volumes of Early Carboniferous to Permian magmatic rocks (Fig. 5, unit C2) lacking significant mineralization. This magmatism buried the mineralized Silurian–Devonian arc, whose rocks are now found only in erosion windows. The frontal part of the arc is clearly marked by a late Palaeozoic accretionary complex striking from Central Kazakhstan to NE China.

If this is accepted, it implies a dextral (westward) displacement of the North China craton for 2000 km, its juxtaposition with the Tarim block, and oroclinal bending of central Mongolia, Hingan and the Yangtze craton with respect to North China. This event completely separated the Altaids from the Tethysides, and Mesozoic magmatic arcs and subduction-accretionary complexes began

to grow on the northern periphery of the Palaeo-Tethys Ocean.

Collision of the Kazakh–Mongol arc with the Mugodzhar–Rudny Altai arc produced a 3000 km long suture in place of the Khanty–Mansi backarc basin. It hosts the Tien Shan and East Kazakhstan orogenic gold provinces with world-class deposits. The provinces broadly coincide with this suture, but their pattern indicates that they are clearly superimposed onto the suture and adjacent structures. The East Kazakhstan province hosts the Bakyrchik deposit; the Tien Shan province, extending from Muruntau to Kumtor and then to the Chinese Tien Shan, is the world's second largest gold province after the Witwatersrand (White et al. 2001). The host rocks of these deposits are early Palaeozoic black shale-bearing passive margin sedimentary sequences accumulated either on or near the adjacent crustal blocks. However, formation of economic gold deposits in these belts took place in the Permian (Drew et al. 1996) or even in the Triassic (Wilde et al. 2001).

Many orogenic-type gold deposits in the Tien Shan are located within late Palaeozoic granitoid intrusions or within their contact metamorphic aureoles (Shayakubov et al. 1999). Where radiometric dates have been obtained, mineralization is found to be broadly coincident with magmatism (Boorder 2000; Cole 2000). A number of recent studies, particularly in the Tien Shan, have developed geochemical, isotopic and fluid structural models that implicate highly evolved syntectonic late Palaeozoic I-type granitoids as the source of fluids and metals for spatially associated orogenic-type gold deposits. In these examples, the gold–quartz vein systems represent only part of a larger magmatic–hydrothermal system that commonly includes earlier scheelite (±Au) skarn mineralization.

Ural–Altai arc–backarc system In this system, there are several generations of early to middle Palaeozoic island arcs extending from the South Urals to Rudny Altai (Fig. 5). It is obscured under Mesozoic–Cenozoic sediments of western Siberia, and it is now exposed in the two apparently separate locations in the Urals and Rudny Altai, whose affinity to a single system is based on magnetic data (Yakubchuk et al. 2001). In the Altai, it unites with the Kazakh–Mongol arc, thus making a system of two parallel, external and internal, arcs, which were oroclinally bent together. An accretionary complex is known on the eastern flank of the arc in the Urals and on the southwestern flank in Rudny Altai. It was interpreted as a subduction product of the Khanty–Mansi backarc basin (Yakubchuk et al. 2001).

The Urals portion of this arc–backarc system is better exposed. Its structure is traditionally simplified as a western slope hosting passive margin sediments of the East European craton and an eastern slope with magmatic arcs and intra-arc sutures (Puchkov 1993). It is commonly assumed that the Main Uralian fault represents a principal boundary between the western and eastern slope units. Some workers (Necheukhin 2001) recognized that the orientation of individual tectonic units of the eastern slope is oblique with respect to the Main Uralian fault. In particular, the Silurian Tagil arc is well known only in the Middle and North Urals and does not trace well to the South Urals. Its analogues constitute allochthonous fragments in the South Urals. The Devonian Magnitogorsk arc is exposed in the South Urals. Its southern continuation is obscured under Cenozoic sediments. Analysis of the present structural pattern suggests that the structure of the Urals is controlled by north–south-trending late Palaeozoic faults (Sazonov et al. 2001), which sinistrally offset some fragments of this arc.

In the South Urals, the Sakmara ophiolitic suture separates the Magnitogorsk arc segment and the East European craton. It can be traced to the south on the basis of magnetic data for 200–300 km. Its northern continuation is believed to coincide with the Main Uralian fault continuing towards the North Urals, representing a main trace of the sutured basin with the early–middle Palaeozoic oceanic crust (Puchkov 1993). Everywhere to the east this suture is bounded by the early to middle Palaeozoic magmatic arc rocks. This supports interpretation of these ophiolites as products of backarc spreading behind this arc. However, the Devonian–Early Carboniferous history of the closure of this basin is viewed as an east-dipping subduction of its oceanic crust under the Mugodzhar arc (Matte et al. 1993; Puchkov 1993) before collision with Eastern Europe and exhumation of the ultra-high pressure rocks in the Early Carboniferous (Matte et al. 1993).

Airborne magnetic patterns suggest that the Mugodzhar arc might have remained attached by its southwestern edge to the East European craton. This arc can be relatively well traced along the Urals orogen but, from the Polar Urals, this arc can be further traced southeastward under West Siberian Mesozoic–Cenozoic sediments towards Rudny Altai.

In the Urals, the main collision between the Mugodzhar–Rudny Altai arc and the East European craton took place in the Middle Carboniferous, with total suturing of the Sakmara backarc basin by the Early Permian (Puchkov 1993; Mossakovsky et al. 1993). This collision was associated with sinistral strike-slip translation, which provoked emplacement of a number of granite-related orogenic gold deposits in the Urals (Sazonov et al. 2001) and offset some tectonic units, possibly for up to 300 km.

The Urals and Rudny Altai host famous VMS deposits in the Middle Devonian differentiated volcanic rocks (Figs 6 and 7) in the Magnitogorsk arc of the Urals (Gusev et al. 2000) and in the same age rocks in Rudny Altai (Popov 1995, 1997; Yang 1994). There are also middle Palaeozoic Mo–(Cu)-porphyry sub-economic deposits in this arc in the Urals (Gusev et al. 2000; Kirkham & Dunne 2000) and in the Altai-Sayan region (Sotnikov & Berzina 2000). The Silurian Tagil arc segment hosts VMS deposits and numerous ultramafic massifs of the so-called Ural platinum belt associated with the Alaska-type intrusions. These intrusions contain minor to medium-size hard rock PGE deposits, which are a source of PGE placers (Dodin et al. 2000).

Paikhoi–Novaya Zemlya, North Barents and North Caspian basins The structures of Paikhoi and Novaya Zemlya orogens consist of Palaeozoic passive margin equivalents of the Urals' western slope. These orogens occur at the northern closure of the Kazakh orocline. Their setting is disputed. Some believe that it is a 'degenerated' northern continuation of the Urals whose further offset continuation is suggested to be found in the southern zones of the Taimyr Peninsula (Milanovsky 1996) or in the Arctic shelf (Zonenshain et al. 1990). These structures, however, may represent a part of another (en echelon?) system of backarc basins which developed due to stretching of the continental crust of the East European craton and Pre-Uralides further behind the Voikar backarc basin.

Similar basins remained undeformed in the North Barents and in the North Caspian areas. They host petroliferous sedimentary rocks that accumulated on oceanic crust from at least middle Palaeozoic time until the Mesozoic, in the North Caspian basin and, until the Quaternary, in the North Barents Sea. The thickness of these sequences now exceeds 15 km (Shipilov & Tarasov 1998). In the North Caspian basin, there are Permian evaporites that represent a fragment of the evaporite belt extending from Central Europe to the South Urals. The famous sedimentary rock-hosted base metal deposits of East Germany and Poland occur on the flanks of this basin, and their equivalents are found in the Ural fore-deep as far as Timan. Similar mid–late Palaeozoic sedimentary copper and lead–zinc deposits were discovered on the Novaya Zemlya archipelago on the periphery of the North Barents basin (Evdokimov et al. 2000).

Distribution of mineralization in the Baikalides and Altaids

The distribution of orogenic and pre-orogenic metallogenic belts in the Baikalides and Altaids is irregular. Porphyry and epithermal mineralization within magmatic arcs constituting the Baikalides and the Altaids occurs mostly within mature arcs (Fig. 6) and mostly within their oroclinally bent fragments (Fig. 7). The VMS and Alaska-type PGM deposits occur within immature magmatic arcs of the Urals and Rudny Altai.

Orogenic gold deposits formed at the final stages of suturing of the backarc basins between the magmatic arcs and cratons. As the Baikalide–Altaid collage is a product of several collided arc–backarc systems there are several metallogenic belts of orogenic gold deposits of various ages ranging from the Late Proterozoic to late Palaeozoic, which were superimposed onto sutured structures. Their calculated total gold resource exceeds 300 Moz (Goldfarb et al. 2001; White et al. 2001). However, analysis of the distribution of their gold endowment shows that the most significant deposits are clustered within the Yenisey and Lena provinces of Late Proterozoic to Late Palaeozoic age, containing in excess of 70 Moz of gold each, and in the Tien Shan gold province of late Palaeozoic age, containing in excess of 250 Moz of gold, whereas early to middle Palaeozoic orogenic gold provinces of the Urals, northern Central Kazakhstan and Altai host medium-size deposits. Analysis of the factors controlling distribution of such deposits reveals the same result as in the North Pacific orogenic collage. The deposits with the largest gold endowment are superimposed onto deformed black-shale-bearing craton-facing passive margin sedimentary sequences. In contrast to the North Pacific orogenic collage, they more clearly associate with the granitoid intrusions in the Tien Shan.

Tectonic evolution of the Altaid, Transbaikal–Mongolian and the North Pacific collages

The above description of the Altaid, Transbaikal–Mongolian and North Pacific orogenic collages reveals many similar features in their structural patterns and distribution of metallogenic belts. It is especially intriguing that both areas host world-class orogenic Au provinces in similar setting and timing of emplacement relative to the evolution of their respective orogens. The following discussion will compare their tectonic and metallogenic evolution.

Evolution of the Altaid and Transbaikal–Mongolian orogenic collages

In the 1990s, several plate-tectonic interpretations were proposed to explain the evolution of the Altaid orogenic collage (Zonenshain et al. 1990; Puchkov 1993; Mossakovsky et al. 1993; Sengör et al. 1993; Sengör & Natal'in 1996a; Yakubchuk 1997). Zonenshain et al. (1990) and Mossakovsky et al. (1993) suggested that Precambrian blocks occurring in the internal part of the Altaids and also Karakum and Alai–Tarim were rifted off the Gondwana super-continent in the Latest Proterozoic; then, they drifted across the Panthalassic (Palaeo-Pacific) Ocean and docked to Siberia and Eastern Europe to produce the orogenic collage.

An alternative interpretation by Sengör et al. (1993) suggested that the Precambrian slivers occurring in the internal portion of the Altaids were rifted off combined Eastern Europe–Siberia during the Neoproterozoic to produce the basement of the Kipchak magmatic arc. A similar mechanism was recently suggested to explain the Tuva–Mongol arc (Yakubchuk et al. 2001) as a structure, which might have rifted off combined Siberia–Laurentia together with the North China craton to form the Palaeo-Pacific ocean, which initially evolved as a backarc basin. The model by Sengör et al. (1993) suggested that clockwise rotation of the Siberian craton with respect to the East European craton caused oroclinal bending and strike-slip duplication of all arcs and their accretionary complexes, which then collided with the framing cratons.

All models suggested after 1993 employed the palaeomagnetic data of Torsvik et al. (1992) for the major cratons, which showed that the East European and Siberian cratons were attached to each other by their present northern margins during the early Palaeozoic. Since the Late Ordovician, Siberia began clockwise rotation with respect to Eastern Europe due to spreading events between Siberia and Laurentia (Fig. 8). This rotation continued until the late Palaeozoic when the Altaids had been finally amalgamated in the form of the present orogenic collage.

Yakubchuk et al. (2001) suggested a new interpretation for the tectonic evolution of the Altaids showing the setting and timing of emplacement of mineral deposits of VMS, porphyry, epithermal, Alaska-type PGM and orogenic Au types. This latter interpretation is in general agreement with the model by Sengör et al. (1993), but it recognizes a greater number of arc–backarc basins within the Altaid collage, puts more emphasis on collision, suggests lesser amplitude of displacement during strike-slip duplication of the structures and differs in the understanding how the

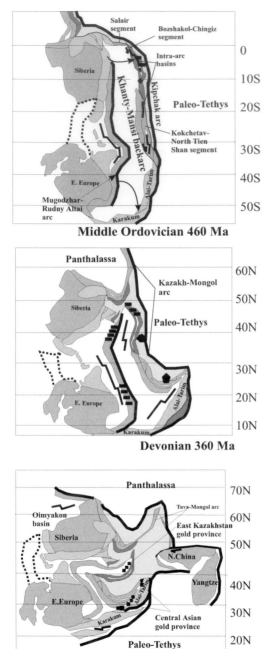

Middle Ordovician 460 Ma

Devonian 360 Ma

Permian 255 Ma

ALTAIDS

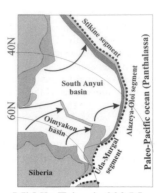

Middle Triassic 230 Ma

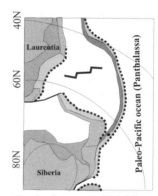

Early Jurassic 210 Ma

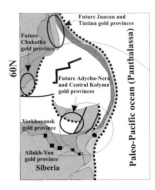

Late Jurassic 160 Ma

NORTH PACIFIC

Fig. 8. Geodynamic evolution of the Altaid and North Pacific orogenic collages. Both orogenic systems evolved through backarc spreading in several basins and following collision of magmatic arcs with each other and adjacent cratons. Epithermal, porphyry, VMS, and Alaska-type PGM deposits formed within magmatic arcs. The amalgamation of orogenic collages took place due to clockwise rotation of adjacent cratons and subsequent arc–arc and arc–continent collisions with associated emplacement of orogenic gold deposits. See Figures 4 and 7 for deposit symbols and province names. Compiled using Parfenov (1995) and Yakubchuk *et al.* (2001).

oroclines were deformed into their present pattern.

According to this latter model, several episodes of rotation of the major cratons caused spreading in the two sub-parallel, external and internal, Khanty–Mansi and Sakmara backarc basins behind the Kipchak and Mugodzhar–Rudny Altai arcs, respectively, and also in the smaller intra-arc basins within these arcs (Fig. 8a). The external Kazakh–Khanty Mansi arc–backarc system started to form in the latest Proterozoic, whereas the internal Urals–Rudny Altai arc–backarc system began to evolve during the Late Ordovician–Silurian. This intra-arc rifting and spreading was associated with emplacement of the early Palaeozoic VMS and porphyry deposits in the Kipchak arc (Fig. 8a). The suturing of its intra-arc basins at the end of the Ordovician generated emplacement of granite-related orogenic gold deposits in northern Kazakhstan, North Tien Shan and also in southern Transbaikalia and Kuznetsk Alatau. The same rotation simultaneously stimulated the beginning of formation of the Mugodzhar–Rudny Altai arc with its VMS and Alaska-type PGM deposits in intra-arc basins in the Urals. Spreading in these basins resulted in oceanic crust containing Cr–(Os–Ir) deposits now found in the ophiolites, which were emplaced into the Ural orogen after suturing.

In the mid-Palaeozoic, the continuing rotation was then responsible for oroclinal bending of the Kipchak and Mugodzhar–Rudny Altai arcs and suturing of their intra-arc basins (Fig. 8b). Simultaneously the Tuva–Mongol arc was oroclinally bent as well. This process led to amalgamation of the Kipchak and Tuva–Mongol arcs into a single Kazakh–Mongol arc in the mid-Palaeozoic, but spreading continued in the Khanty–Mansi backarc basin and in the Sakmara backarc basin.

Spreading events and subduction against the continuing clockwise rotation of Siberia and oroclinal bending of the new Kazakh–Mongol magmatic arcs in the Devonian coincided with emplacement of the porphyry deposits of Central Kazakhstan and Mongolia and a new episode of VMS mineralization in the Mugodzhar–Rudny Altai arc. The oroclinal bending of the Kazakh–Mongol arc caused its intrusion between Tarim–Karakum and Siberia towards the East European craton. This created temporary subduction on the present western flank of the Kazakh–Mongol arc in the Early Devonian, but since the Mid-Devonian the evaporite-bearing and molasse-filled rift-related backarc basins started to cover the previously amalgamated fragments. In the early Carboniferous, continued clockwise rotation of major cratons caused southeastward migration of the western part of the Kazakh–Mongol arc and a

new episode of bending in the Kazakh orocline, pushing it further towards the East European craton along its bounding strike-slip faults. This coincides with emplacement of main porphyry deposits in Central Kazakhstan and a new episode of temporary subduction to form the Valerianov–Beltau–Kurama arc, which produced its large porphyry, skarn and epithermal deposits.

In the Early to Mid-Carboniferous, the Kazakh orocline collided with the Mugodzhar–Rudny Altai arc to form a 3000 km long suture extending from the South Tien Shan to the East Urals and Irtysh–Zaissan zones (in place of the Khanty–Mansi backarc basin) and the Main Ural suture (in place of the Sakmara backarc basin) between the Mugodzhar arc and the East European craton. By the end of the late Palaeozoic, the continuing rotation of major cratons caused collision of the Kipchak and Mugodzhar–Rudny Altai arcs with each other and with the East European and Siberian cratons (Fig. 8c). These collisional events continued throughout the late Palaeozoic. At the final stage of amalgamation in the late Palaeozoic, this rotation formed an orogenic collage of the Altaids and produced transpressive strike-slip deformation in its external part in the Urals. This suturing was an important event in the structural preparation of the region, which later produced world-class orogenic gold provinces in the Tien Shan, eastern Kazakhstan and Urals.

The Tuva–Mongol arc in the eastern portion of the Altaids was bent into the Central Mongolian and Hingan oroclines mostly during the early Mesozoic, after generation of the Transbaikal arc. Since the Cretaceous, it was affected by Yanshanian arc magmatism of the Circum-Pacific belt, and then the Indian collision produced Eurasia.

Evolution of the North Pacific orogenic collage

Most publications about the early stages of evolution of the North Pacific orogenic collage suggest that the magmatic arcs were generated in the Pacific Ocean and then accreted to Laurentia and Eurasia (e.g. Sokolov et al. 1997). Similar to the Altaids, an alternative viewpoint (Sengör & Natal'in 1996b) suggests that first an arc was rifted off combined Siberia–Laurentia and then accreted back to the cratons.

In the Alaska–Kolyma portion, the shaping of the North Pacific orogenic collage started in the mid–late Palaeozoic with backarc rifting and spreading behind the Uda–Murgal, Alazeya–Oloy and Stikene magmatic arc segments in the Verkhoyansk–Chukotka arc–backarc system (Fig. 8d, e). This stimulated accumulation of the clinoforms on the craton-facing passive margins in the back-arc basins. The arcs host Cu–Mo porphyry, epithermal Au and Alaska-type PGM occurrences. Spreading in the Amerasian basin caused rifting of the Chukotka–Brooks Precambrian crustal block off Laurentia and its migration towards Siberia to form the Verkhoyansk–Kolyma and Novosibirsk–Chukotka collisional orogens by the Late Jurassic to Early Cretaceous (Fig. 8f) (Nokleberg et al. 1993; Nokleberg & Diggles 2001). This collision was responsible for generation of the orogenic gold deposits in the Yana–Kolyma metallogenic belt.

In the Transbaikal–Mongolian collage, the late Palaeozoic–early Mesozoic Transbaikalian arc was the first arc, which can be explained via subduction from the Palaeo-Pacific Ocean. It continued to develop until the Early Cretaceous, but only Triassic arc magmatism is responsible for economic Cu–(Mo)-porphyry and epithermal Au mineralization in Mongolia and Transbaikalia. The magmatic products of the Tuva–Mongol and Kazakh–Mongol arcs of the Altaids occur now in its basement. Prior to oroclinal deformation, this arc might have formed a single structure with the North China craton.

The Sikhote–Alin–West Kamchatka–Aleutian arc and its backarc basin started to develop in the late Palaeozoic–early Mesozoic. Some workers (Nokleberg et al. 1998; Nokleberg & Diggles 2001) suggested that this arc formed in the centre of the Pacific Ocean and then docked to Siberia and Laurentia. In the Mid-Cretaceous–Palaeogene, there were two magmatic arcs, the internal Okhotsk–Chukotka and external Sikhote–Alin–West Kamchatka–Aleutian arc, which developed above the amalgamated portions of the Verkhoyansk–Chukotka and Transbaikal–Mongolian orogens. The front of the Okhotsk–Chukotka arc retreated towards the continent with respect to the Jurassic fronts in the same manner as in the Chilean segment of the Andean belt. This subduction-related magmatism produced medium-size epithermal Au–Ag and Cu-porphyry deposits in the Okhotsk–Chukotka arc. The Sikhote–Alin–West Kamchatka–Aleutian arc produced VMS, skarn, minor porphyry, and Alaska-type PGM deposits. Medium to large orogenic gold provinces were superimposed onto collisional and accretionary orogens in the rear parts of the two arcs.

Evolution of the North Pacific orogenic collage since the Cretaceous until the present can be considered as a gradual oceanward rollback of subduction zones with subsequent opening of a series of backarc basins and accretion of ensimatic arcs generated in the central parts of the Pacific Ocean. The beginning of spreading in each new backarc basin coincided with suturing of the previous backarc basin in the rear part of the former (Nokleberg & Diggles 2001). In the Palaeogene, the strike-slip translation along the active margins of the Pacific Ocean caused oroclinal bending of the Sikhote–Alin–West Kamchatka–Aleutian arc and its collision with the rest of Asia and Alaska. This collision stopped magmatism in the Okhotsk–Chukotka arc and sutured the Okhotsk–Alaska backarc basin, but its East Bering portion has remained unsutured. This subduction produced the Au-porphyry deposits of Alaska. At the same time, the intra-arc spreading split the Sikhote–Alin–West Kamchatka–Aleutian arc or intra-oceanic subduction produced a separate Kurile–Komandor arc, which collided by the end of the Palaeogene. In the Neogene–Quaternary, a still active magmatic arc developed on top of the previously collided structures in Kurile and Kamchatka. This subduction-related magmatism produced numerous medium size epithermal deposits in the Kamchatka Peninsula and some in the Kurile Islands.

Discussion

The interpretation of tectonic patterns and models of geodynamic evolution of the Baikalide–Altaid and North Pacific orogenic collages described above shows both similarities and differences between the two collages. The common features include similar tectonic patterns and similar distribution of Cu-porphyry, Au–Ag epithermal, Alaska-type PGM, and orogenic Au metallogenic belts. Besides the age, the differences relate mostly to the sequence in opening of backarc basins. For instance, in the Altaids the internal magmatic arcs (e.g. Mugodzhar–Rudny Altai) began to develop after and behind the external arcs (e.g. Kipchak), whereas in the North Pacific collage there is a subsequent oceanward younging of magmatic arcs. The consequence of this difference in evolution, along with many other individual differences, seems to be not critical for metallogeny, but it is important to emphasize that in both collages the largest gold deposits formed during major orogenic (collisional) events and occur predominantly within the deformed former passive margin black shale-bearing sedimentary sequences. This affinity of orogenic gold deposits to black shale host rocks has been a reason for intense discussions. At present, one can assume that initial enrichment of sediments in gold is as important as the later influence of collisional tectonism and granitoid magmatism.

Analysis of the global setting of the two orogenic collages shows that each of them evolved in a similar position relative to the two adjacent major oceans (Fig. 9). In particular, the North

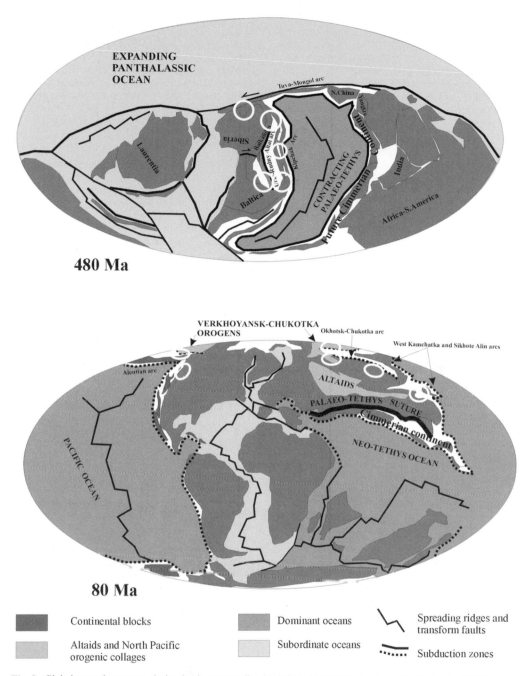

Fig. 9. Global tectonic patterns during backarc spreading in the Altaid (480 Ma ago) and North Pacific (80 Ma ago) orogenic collages (major cratons are shown after Scotese & McKerrow 1990). The arc–backarc systems in both collages initially evolved through rifting, which separated crustal fragments from adjacent cratons into arc basements to form several sub-parallel arc–backarc systems occurring on the flanks of the two dominant oceans surrounded by subduction zones on all margins, e.g. Palaeo-Tethys and Pacific Oceans respectively, whereas the margins of subordinate or expanding oceans remain largely inactive. White circles show the location of orogenic gold provinces with the most significant metal endowment. In the Altaid and North Pacific collages they are concentrated in a backarc setting, in areas where magmatic arcs remained attached to the cratons.

Pacific orogenic collage developed on the northern flank of the Pacific Ocean, which displays the most active spreading along its ridge, the most dominant spreading system on Earth today. This ridge extends from North America towards Australia and Antarctica and then continues in the Indian Ocean towards the Red Sea. Subduction zones bound oceanic plates of these two oceans. It is a global subduction zone, which can be subdivided into two major en-echelon segments. One of them extends from the Mediterranean region towards the Himalaya, Indonesia and SW Pacific region; the other starts near the Philippines and continues along the Asian island arcs towards the western coast of the Americas. The magmatic arcs developed near these zones are a host to all principal deposit types of Mesozoic–Cenozoic age. In contrast, the mostly passive margins of the Atlantic Ocean, except the Caribbean region, do not host Mesozoic–Cenozoic mineral deposit types considered in this review.

Analysis of the geodynamic setting of the Altaids against the global Palaeozoic plate tectonic pattern is difficult due to uncertainty with the existing global plate tectonic reconstructions for the Palaeozoic. These uncertainties mostly relate to the understanding of the relationships between the Palaeo-Tethys and Palaeo-Pacific Oceans (Scotese & McKerrow 1990). It is traditionally considered that the Altaids formed due to subduction of the oceanic crust of the Palaeo-Asian Ocean, which is believed to represent an embayment of the Palaeo-Pacific Ocean (Zonenshain et al. 1990; Mossakovsky et al. 1993; Sengör et al. 1993; Sengör & Natal'in 1996a). It is assumed that in Meosozoic–Cenozoic times, the latter developed into the present Pacific Ocean, and subduction of its oceanic plates produced the orogens of the Circum-Pacific rim. However, it was shown above that the structures of the Altaid orogenic collage cannot be directly traced into the orogens of the Circum-Pacific belt as the former and the latter are everywhere separated by Precambrian slivers in the basement of the Tuva–Mongol magmatic arc. The dominant polarity of this arc in Late Proterozoic–mid-Palaeozoic times was SW facing. This means that the Late Proterozoic–early Palaeozoic sequences occurring in the core of the Central Mongolian orocline must be considered as a portion of the Palaeo-Pacific Ocean, which automatically places the latter in a backarc position with respect to the Tuva–Mongol arc. Subduction-related magmatic arc complexes in the Eurasian portion of the North Pacific orogenic collage started to form in the Devonian and continued in the late Palaeozoic, whereas older rocks are mostly of passive margin or ocean-floor origin. This suggests that the Palaeo-

Pacific Ocean evolved as a large backarc basin or Atlantic-type Ocean during its initial phases of opening and began to generate its own subduction zones when it expanded significantly.

On the other hand, the subduction-accretionary complexes of the Altaids might be considered as equivalents of similar complexes in the Tethysides, and the entire Altaid collage may simply represent a tectonically isolated fragment of the Tethysides.

Therefore, global Palaeozoic plate tectonics can be considered in the following way. Based on the distribution of the early–mid-Palaeozoic magmatic arcs one can suggest that by analogy with the present Pacific–Indian Ocean, dominant Palaeozoic spreading ridges were located in the Palaeo-Tethys Ocean, whereas the Palaeo-Pacific Ocean could be analogous to the present Atlantic Ocean. This means that the North Pacific and Altaid orogenic collages occupied almost exactly the same positions relative to the major cratons during their respective phases of development of the Pacific and Palaeo-Tethys Oceans, and therefore their global settings were principally the same at the respective times of their evolution.

Conclusion

The Baikalide–Altaid and North Pacific orogenic collages each consist of several oroclinally bent magmatic arcs separated by accretionary complexes and ophiolitic sutures located between the major cratons. The tectonic patterns of the two collages are principally similar as they were formed as a result of rotation of the framing cratons and strike-slip translation along the former convergent margins. Sometimes this includes such details as similarity of the oroclinal structural patterns of the Alaska orocline in the North Pacific collage and West Sayan orocline in the Altaid collage.

As a consequence, the two collages have principally the same distribution of metallogenic belts. In particular, the mid–late Palaeozoic belts of porphyry and epithermal deposits in the Altaids occupy the same position as the Mesozoic–Cenozoic belts of the North Pacific collage. The Ural platinum belt occupies the same position as the belt of platinum-bearing intrusions in Alaska. Major mineralizing events producing world-class intrusion-related Au, Cu–(Mo)-porphyry and VMS deposits in the Altaid and North Pacific collages coincided with plate re-organization and oroclinal bending of magmatic arcs. Formation of porphyry, epithermal and Alaska-type PGM deposits took place simultaneously with the episodes of oroclinal bending.

The tectonic setting of the orogenic gold belts in the Tien Shan and Verkhoyansk–Kolyma provinces, hosting large and giant hard rock gold deposits, is also similar, especially the distribution of their gold endowments. Major orogenic gold deposits occur within the sutured backarc basins. The craton-facing passive margin rock sequences initially formed within backarc basins. As they were entrapped within such oroclines, they represent favourable locations for emplacement of orogenic gold deposits.

I thank Alaster Edwards, Andy Wilde, Jeffrey Hedenquist, Noel White, Reimar Seltmann, Rich Goldfarb, Rod Kirkham and many other colleagues for inspiring virtual and real discussions during many years. The tectonic and metallogenic understanding of these regions benefited a lot from discussions with Anatoly Nikishin, Evgeny Milanovsky, Vladimir Buryak, Brian Windley, Celal Sengör, Boris Natal'in, Valentin Burtman, Victor Puchkov, Tatiana Kheraskova, Leonid Parfenov and Alexander Khanchuk. I thank Celal Sengör and Walter Kurz for useful comments provided during revision of the manuscript.

References

BOGDANOV, N.A. & TILMAN, S.M. 1992. *Tektonika i geodinamika Severo-Vostoka Azii.An explanatory note to the tectonic map of northeast Asia, scale 1:500 000*. Institute of Lithosphere, Russian Academy of Sciences, Moscow [In Russian].

BOORDER, H. DE 2000. Major structural elements of the Urals and linkage to epigenetic deposits. GEODE Urals Workshop 2000, 14-15 April 2000. Natural History Museum, London. URL: http://www.gl.rhbnc.ac.uk/geode/wkshop.html

COLE, A. 2000. *Genesis of granitoid-hosted gold-tungsten mineralisation, Jilau, Tajikistan*. Doctoral thesis, Univ. London, UK.

COLEMAN, R.G. 1989. Continental growth of northwest China. *Tectonics*, **8**, 621–635.

DEGTYAREV, K.E. 1999. *Tektonicheskaya evolyutsiya rannepaleozoiskoi aktivnoi okrainy v Kazakhstane*. Nauka, Moscow [In Russian].

DISTANOV, E.G. & OBOLENSKY, A.A. 1993. Some problems of metallogeny of the Central Asian mobile belt according to their geodynamic evolution. *In:* DOBRETSOV, N.I. & BERZIN, N.A. (eds) *Fourth international symposium on geodynamic evolution of the Palaeoasian ocean, Abstracts*. IGCP Project 283 Reports, **4**, 189–191.

DODIN, D.A., CHERNYSHOV, N.M. & YATSKEVICH, B.A. 2000. *[Platinum-metal deposits of Russia]*. Nauka, St Petersburg [in Russian].

DREW, L.J., BERGER, B.R. & KURBANOV, N.K. 1996. Geology and structural evolution of the Muruntau gold deposit, Kyzylkum desert, Uzbekistan. *Ore Geology Reviews*, **11**, 175–196.

EIRISH, L.V. 1998. [Perspectives on discovery of the Carlin-type deposits in the Russian Far East]. *Tikhookeanskaya Geologiya*, **17**(4), 72–79. [in Russian].

EVDOKIMOV, A.N., KALENICH, A.P., KRYUKOV, V.D., LASTOCHKIN, A.V. & SEMENOV, Y.P. 2000. [No-

vaya Zemlya as a new prospective resource target on the Barents-Kara shelf]. *Razvedka i Okhrana Nedr*, **12**, 40–43. [in Russian].

FRIDOVSKY, V.Y. 2000. Collisional gold metallogeny of the Verkhoyansk-Kolyma region. *In:* MEZHELOVSKY, N.V, MOROZOV, A.F, GUSEV, G.S & POPOV, V.S (eds) *Gedynamics and metallogeny: theory and implications for applied geology*. GEOKART, Moscow, 389–397.

GOLDFARB, R.J., GROVES, D.I. & GARDOLL, S. 2001. Orogenic gold and geologic time: a global synthesis. *Ore Geology Reviews*, **18**, 1–75.

GUSEV, G.S., GUSHCHIN, A.V., ZAYKOV, V.V., MASLENNIKOV, V.V., MEZHELOVSKY, N.V., PEREVOZCHIKOV, B.V., SURIN, T.N., FILATOV, E.I. & SHIRAI, E.P. 2000. Geology and metallogeny of island arcs. *In:* MEZHELOVSKY, N.V., MOROZOV, A.F., GUSEV, G.S & POPOV, V.S (eds) *Geodynamics and metallogeny: theory and implications for applied geology*. GEOKART, Moscow, 213–295.

HEINHORST, J., LEHMANN, B., ERMOLOV, P., SERYKH, V. & ZHURUTIN, S. 2000. Palaeozoic crustal growth of Central Asia: evidence from magmatic-hydrothermal ore systems of Central Kazakhstan. *Tectonophysics*, **328**, 69–87.

KIRKHAM, R.V. 1998. *Tectonic and structural features of arc deposits*. British Columbia Geological Survey Open-File reports, **1998-8**.

KIRKHAM, R.V. & DUNNE, K.P.E. (COMPILERS) 2000. *(compilers) World distribution of porphyry, porphyry-associated skarn, and bulk-tonnage epithermal deposits and occurrences, Scale 1:35,000,000*. Geological Survey of Canada, Open File Report, **3792**.

KHANCHUK, A.I. 2000. [Paleogeodynamic analysis of the formation of ore deposits in the Russian Far East]. *In:* KHANCHUK, A.I./eds;*Rudnye mestorozhdeniya kontinentalnykh okrain*. Dalnauka, Vladivostok [in Russian], 5–34.

KHANCHUK, A.I. & IVANOV, V.V. 1999. Geodinamika vostoka Rossii v mezo-kainozoe. *In:* KHANCHUK, A.I./eds;*[Geodynamics and metallogeny]*. Dalnauka, Vladivostok [in Russian], 7–30.

KOVALEV, K.R., DISTANOV, E.G. & PERTSEVA, A.P. 1998. Variations in isotopic composition of sulphur during volcanic-sedimentary ore formation and metamorphism of the Ozernoe ore district, western Transbaikalia, *Geology of Ore Deposits*, **40**, 336–353. [in Russian].

LANG, J.R., BAKER, T., HART, C.J.R. & MORTENSEN, T.K. 2000. An exploration model for intrusion-related gold systems. *SEG Newsletter*, **40**, 6–15.

LUDINGTON, S., COX, D.R., SINGER, D.A., SHERLOCK, M.G., BERGER, B.R. & TINGLEY, J.V. 1993. Spatial and temporal analysis of precious-metal deposits for a mineral resource assessment of Nevada. *In:* KIRKHAM, R.V., SINCLAIR, W.D, THROPE, R.I. & DUKE, J.M (eds) *Spatial and temporal analysis of precious-metal deposits for a mineral resource assessment of Nevada*. Geological Association of Canada Special Papers, **40**, 31–40.

MATTE, P., MALUSKI, H., CABY, R., NICOLAS, A., KEPEZHINSKAS, P. & SOBOLEV, S. 1993. Geodynamic model and ^{39}Ar-^{40}Ar dating for generation and emplacement of the high pressure metamorphic rocks in SW Urals. *Comptes Rendus de l'Academie de Sciences Paris, Series II*, **317**, 1667–1674.

MILANOVSKY, E.E. 1996. *Geologiya Rossii i blizhnego zarubezhya*. Moscow University, [in Russian].

MOSSAKOVSKY, A.A., RUZHENTSEV, S.V., SAMYGIN, S.G. & KHERASKOVA, T.N. 1993. Central Asian fold belt: geodynamic evolution and history of formation. *Geotektonika*, **6**, 3–33. [in Russian].

NATAL'IN, B.A. & BORUKAEV, C.B. 1991. Mezozoiskie sutury na yuge Dalnego Vostoka. *Geotektonika*, **1**, 84–96. [in Russian].

National Geophysical Data Center 1996. *National Geophysical Data Center Magnetic anomaly data of the former Soviet Union*. CD-ROM, National Geophysical Data Center, Boulder, CO.

NECHEUKHIN, V.M. 2001. Akkretsionno-kollizionnaya tektonika Uralskogo orogena. *In: Tektonika neogeya: obshchie i regionalnye aspekty*. Geos, Moscow [in Russian], **2**, 71–74.

NEWBERRY, R.J., MCCOY, D.T. & BREW, D.A. 1995. Plutonic-hosted gold ores in Alaska: igneous vs. metamorphic origins. *Resource Geology Special Issue*, **18**, 57–100.

NOKLEBERG, W.J. & DIGGLES, M.F. (EDS) 2001. *Dynamic computer model for the metallogenesis and tectonics of the Circum-North Pacific*. Department of the Interior, US Geological Survery Open File Reports, **01-261**.

NOKLEBERG, W.J., BUNDTZEN, T.K. & GRYBECK, D.K. ET AL. 1993. Metallogenesis of mainland Alaska and the Russian Northeast: Mineral deposit maps, models, and tables, metallogenic belt maps and interpretation, and references cited. *US Geological Survey Open-File Report*, 93–339.

NOKLEBERG, W.J., PARFENOV, L.M. & MONGER, J.W.H. 1998. *Phanerozoic tectonic evolution of the Circum-North Pacific*. US Department of the Interior, US Geological Survey, Open-File Reports, **98-754**.

OKSMAN, V.S., BONDARENKO, G.E. & SOKOLOV, S.D. 2001. Kollizionnye poyasa Verkhoyano-Chukotskoi orogennoi oblasti (severo-vostok Azii). *In: Tektonika neogeya: obshchie i regionalnye aspekty*. Geos, Moscow [in Russian], **2**, 86–89.

PARFENOV, L.M. 1991. Tectonics of the Verkhoyansk-Kolyma Mesozoides in the context of plate tectonics. *Tectonophysics*, **139**, 319–342.

PARFENOV, L.M. 1995. Terreiny i istoriya formirovaniya mezozoiskikh orogennykh poyasov Vostochnoi Yakutii. *Tikhookeanskaya Geologiya*, **14**(6), 32–43. [in Russian].

POPOV, V.V. 1995. [Geological localization of large polymetallic deposits in Rudny Altai]. *Geology of Ore Deposits*, **37**, 371–389. [in Russian].

POPOV, V.V. 1997. Regional factors in origin of large concentrations of polymetallic ore in the Urals. *Geology of Ore Deposits*, **39**, 465–476. [in Russian].

PROKOPIEV, A.V. 1998. The Verkhoyansk-Chersky collisional orogen. *Tikhookeanskaya Geologiya*, **8**, 3–10. [in Russian].

PUCHKOV, V.N. 1993. Palaeo-oceanic structures of the Urals. *Geotektonika*, **3**, 18–33. [in Russian].

SAZONOV, V.N., VAN HERK, A.H. & DE BOORDER, H. 2001. Spatial and temporal distribution of gold deposits in the Urals. *Economic geology*, **96**, 685–703.

SCOTESE, C.R. & MCKERROW, W.S. 1990. Revised world maps and introduction. *In:* MCKERROW, W.S. & SCOTESE, C.R. (eds) *Palaeozoic Palaeogeography and Biogeography*. Geological Society, London, Memoirs, **12**, 1–21.

SENGÖR, A.M.C., NATAL'IN, B.A. & BURTMAN, V.S. 1993. Evolution of the Altaid tectonic collage and Palaeozoic crustal growth in Eurasia. *Nature*, **364**, 299–307.

SENGÖR, A.M.C. & NATAL'IN, B.A. 1996a. Turkic-type orogeny and its role in the making of the continental crust. *Annual Review in Earth and Planetary Sciences*, **24**, 263–337.

SENGÖR, A.M.C. & NATAL'IN, B.A. 1996b. Palaeotectonics of Asia: fragments of a synthesis. *In:* AN YIN, & HARRISON, T.M. (eds) *The tectonic evolution of Asia*. Cambridge University Press, Cambridge, 486–640.

SHATOV, V., SELTMANN, R., KREMENETSKY, A., LEHMANN, B., POPOV, V. & ERMOLOV, P. (EDS) 1996. *Granite-related ore deposits of Central Kazakhstan and adjacent areas*. Glagol Publishing House, St Petersburg.

SHAYAKUBOV, T., ISLAMOV, F., KREMENETSKY, A. & SELTMANN, R. (EDS) 1999. *Au, Ag, and Cu deposits of Uzbekistan*. Excursion guidebook to the International Field Conference of IGCP-373. International Field Conference of IGCP-373, Excursion B6 of the Joint SGA-IAGOD Symposium. London/Tashkent: 27/28 August-4 September 1999, **August-4**.

SHIPILOV, E.V. & TARASOV, G.A. 1998. *[Regional geology of the oil-bearing sedimentary basins of the Russian West-Arctic shelf]*. Apatity, [in Russian].

SMITH, M., THOMPSON, J.F.H., BRESSLER, J., LAYER, P., MORTENSEN, J.K., ABE, I. & TAKAOKA, H. 1999. Geology of the Liese Zone, Pogo Property, East-Central Alaska. *SEG Newsletter*, **38**, 12–21.

SOKOLOV, S.D., DIDENKO, A.N., GRIGORIEV, V.N., ALEKSYUTIN, M.V., BONDARENKO, G.E. & KRYLOV, K.A. 1997. Palaeotektonicheskie rekonstruktsii severo-vostoka Rossii: problemy i neopredelennosti. *Geotektonika*, **6**, 72–90. [in Russian].

SOTNIKOV, V.I. & BERZINA, A.P. 2000. Porphyry Cu-(Mo) ore-magmatic systems of Siberia and Mongolia. *In:* KREMENETSKY, A.A., LEHMANN, B. & SELTMANN, R. (eds) *Ore-bearing granites of Russia and adjacent countries*. IMGRE, Moscow, 263–279.

TORSVIK, T.H., SMETHURST, M.A. & VAN DER VOO, R. 1992. Baltica. A synopsis of Vendian-Permian paleomagnetic implications. *Earth Science Reviews*, **33**, 133–152.

VERNIKOVSKIY, V.A., VERNIKOVSKAIA, A.E., CHERNYKH, A.I. & MELGUNOV, M.S. 1996. [Petrology and geochemistry of the Riphean ophiolites in northern Taimyr]. *Geologiya i Geofizika*, **37**(1), 113–129. [in Russian].

VERNIKOVSKIY, V.A., VERNIKOVSKAIA, A.E., SALNIKOVA, E.B. & KOTOV, A.B. 1999. [New U-Pb data on age of the paleo-island arc complex of the Predivinsky Terrane, Yenisey Range]. *Geologiya i geofizika*, **40**(2), 255–259. [in Russian].

WHITE, N.C., HEDENQUIST, J.W. & KIRKHAM, R.V. 2001. Asia: the waking giant. *Mining Journal supplement*, **March 2001**, 1–12.

WILDE, A.R., LAYER, P., MERNAGH, T. & FOSTER, J. 2001. The giant Muruntau gold deposit: geologic, geochronologic, and fluid inclusion constraints of ore genesis. *Economic Geology*, **96**, 633–644.

YAKUBCHUK, A. 1997. Kazakhstan. *In:* MOORES, E.M. & FAIRBRIDGE, R.W. (eds) *Encyclopedia of European and Asian Regional Geology.* Chapman and Hall, New York, 450–465.

YAKUBCHUK, A. 2000. Bodaibo mesothermal goldfields in Siberia and Carlin-type gold mineralization in South China: a similarity of regional structural pattern. *In:* CLUER, J.K, PRICE, J.G, STRUHSACKER, E.M., HARDYMAN, R.F & MORRIS, C.L. (eds) *Geology and ore deposits 2000: the Great Basin and beyond.* Geological Society of Nevada Symposium Proceedings, May 15-18, 2000, 539–547.

YAKUBCHUK, A.S. & EDWARDS, A. 1999. Auriferous Palaeozoic accretionary terranes within the Mongol-Okhotsk suture zone, Russian Far East. *In: Proceedings Pacrim'99, 10-13 October 1999, Bali, Indonesia.* The Australasian Institute of Mining and Metallurgy Publication Series, **4/99**, 347–358.

YAKUBCHUK, A.S. & DEGTYAREV, K.E. 1991. O kharaktere sochleneniya Chingizskogo i Boshchekulskogo napravleniy v kaledonidakh severo-vostoka Tsentralnogo Kazakhstana. *Doklady AN SSSR*, **317**, 957–962. [in Russian].

YAKUBCHUK, A.S., SELTMANN, R., SHATOV, V.V. & COLE, A. 2001. The Altaid orogenic collage: tectonic evolution and metallogeny. *SEG Newsletter*, **46**, 7–14.

YANG, K. 1994. Volcanogenic massive sulfide deposits in China. *International Geology Review*, **36**, 293–300.

ZONENSHAIN, L.P., KUZMIN, M.I. & NATAPOV, L.M. 1990. *Geology of the USSR: a plate tectonic synthesis.* American Geophysical, Washington, DC.

Tectonics, geodynamics and gold mineralization of the eastern margin of the North Asia craton

VALERIY YU. FRIDOVSKY[1] & ANDREI V. PROKOPIEV[2]

[1]*Yakut State University, Belinsky St., 58, Yakutsk, 677000, Russia (e-mail: fridovsky@sitc.ru)*
[2]*Diamond and Precious Metal Geology Institute, Siberian Department, Russian Academy of Sciences, Lenin Av., 39, Yakutsk, 677891, Russia (e-mail: prokopiev@diamond.ysn.ru)*

Abstract: The North Asia craton is a crustal block including the Siberian platform and marginal fold-and-thrust belts. On the eastern margin of the North Asia craton there is the Verkhoyansk fold-and-thrust belt making up the western part of the Verkhoyansk–Chersky collisional orogenic belt extending for 2000 km from the Laptev Sea in the north to the Sea of Okhotsk in the south. A system of frontal thrusts separates the belt from the platform structures. The frontal part of the belt is mainly made of Carboniferous and Permian terrigenous rocks of palaeodeltas and submarine fans which grade eastward into Triassic and Jurassic sediments of the continental slope. The front of the belt is characterized by thrusting and strike-slip faulting with large horizontal displacements. The largest anticlinoria at the front of the belt have a duplex structure. Formation of the major gold deposits and fold and fault structures, as well as igneous activity in the region, are related to the collision of the North Asia craton with the Kolyma–Omolon superterrane and the Okhotsk terrane in the Late Jurassic–Neocomian. The collision occurred in two stages: the early Neocomian frontal collision and the late Neocomian oblique collision.

The principal gold potential of the western part of the Verkhoyansk–Chersky orogen (Republic of Sakha, Yakutia, Russian Federation) was created during the late Mesozoic collision of the North Asia craton (Siberian platform and its Verkhoyansk passive margin in the east) with the Kolyma–Omolon microcontinent (large composite terrane, or superterrane) and the Okhotsk microcontinent (terrane with Archaean basement). The early frontal collision took place in the Late Jurassic and Neocomian and the late oblique collision occurred in the Early Cretaceous.

Two groups (early and late collisional) of gold–quartz deposits are recognized, localized in five metallogenic zones. The early collisional group is related to the Verkhoyansk, Alakh–Yun, and Kular metallogenic zones; the late collisional group is localized in the Adycha–Nera and South Verkhoyansk metallogenic zones. These metallogenic zones are within the Yana–Indigirka metallogenic belt (Fig. 1). Some of the deposits described in this paper were mentioned earlier in overviews by Nokleberg *et al.* (1996, 1997).

The formation of the early group was related to the tectono-metamorphic transformation of gold-bearing sedimentary rocks on the shelf of a passive continental margin and to the transport of ore components by solutions expelled from source rocks up to the ore-localizing barriers in shear zones of the thrust structures (Fridovsky & Prokopiev 1998). Deposits of the late group are considered to be derived from ore-magmatic systems related to the main granite batholith belt (Parfenov & Kuz'min 2001) and localized in the thrust and strike-slip zones. The deposits considered in this paper mainly belong to the type of 'Au deposits in shear zones and quartz veins' (after Nokleberg *et al.* 1996). The characteristics of the gold deposits are presented in Table 1.

Verkhoyansk–Chersky orogen

Main geological features of the orogen

Collisional orogens (orogenic belts) are mountain-fold structures resulting from collision of large continental blocks of the Earth's crust, such as continents and microcontinents or superterranes (Hatcher 1990; Twiss & Moores 1992). They are distinct from accretionary orogenic belts formed during accretion of comparatively small blocks of the Earth's crust of different origin to the continental or superterrane margin along the subduction zone.

The Verkhoyansk–Chersky orogen comprises all tectonic structures east of the Siberian platform including the Zyryanka Basin (Figs 2 and 3). It also covers the western part of the Verkhoyansk–

From: BLUNDELL, D.J., NEUBAUER, F. & VON QUADT, A. (eds) 2002. *The Timing and Location of Major Ore Deposits in an Evolving Orogen*. Geological Society, London, Special Publications, **204**, 299–317.
0305-8719/02/$15.00 © The Geological Society of London 2002.

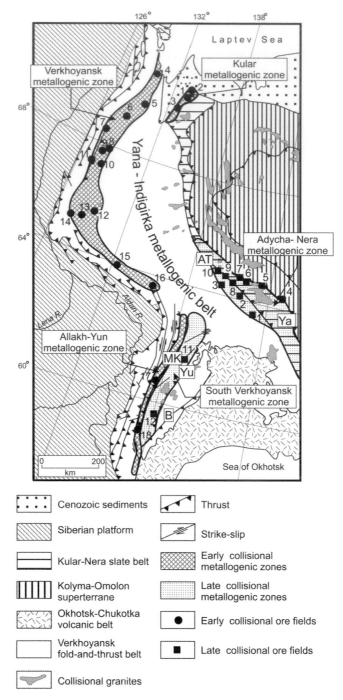

Fig. 1. Metallogenic zones of the Yana–Indigirka metallogenic belt.Faults (letters in boxes): Ya, Yana–Indigirka; AT, Adycha–Taryn; B, Billyakchan; MK, Minor,Kiderikin; Yu, Yudoma. Early collisional ore fields (numerals in figure): 1, Kyllakh; 2, Burguat; 3, Dzhuotuk; 4, Dyandi,Okhonosoi; 5, Meichan; 6, Dzhardzhan; 7, Balagannakh; 8, Sereginsky; 9, Aialyr; 10, Sudyandalakh; 11, Imtachan; 12, Kygyltas; 13, Sagandzha; 14, Kitin; 15, Barain; 16, Kharangak; 17, Bular–Onocholokh; 18, Yursky–Brindakit. Late collisional ore fields: 1, Yakutsky; 2, Saninsky; 3, Badran; 4, Kurun–Agalyk; 5, Sokh; 6, Tuora–Tas; 7, Ol'chan; 8, Talalakh; 9, Bazovsky; 10, Zhdaninsky; 11, Nezhdaninskoe; 12, Zaderzhinsky.

Table 1. *Significant gold deposits and occurrences of the Yana–Indigirka metallogenic belt*

Groups of gold deposits	Main fault kinematics	Orebody morphology	Mineral deposits and occurrences	Deposit type	Metallogenic zone	Ore field	Major metals
Early collisional	Thrust faults	Saddle vein- at crests and sheet veins at limbs of the folds	Yurskoe*	Au in shear zone and quartz vein	Allakh–Yun	Yursky–Brindakit	Au
			Nekur		Allakh–Yun	Yursky–Brindakit	Au
			Duet		Allakh–Yun	Yursky–Brindakit	Au
		Sheet, lenticular, and ladder veins	Fin	Au in shear zone and quartz vein	Allakh–Yun	Yursky–Brindakit	Au
			Burguat		Kular	Burguat	Au
			Emelyanovskoe		Kular	Burguat	Au
			Kyllakh		Kular	Kyllakh	Au
			Estakadnoe		Kular	Kyllakh	Au
		Sheet and lenticular veins, stratiform stockworks mineralized crush zones	Srednee	Au in shear zone and quartz vein	Kular	Kyllakh	Au
			Kieng–Yuryakh		Kular	Dzhuotuk	Au
			Verkhnee		Kular	Dzhuotuk	Au
			Onocholokh		Allakh–Yun	Bular–Onocholokh	Au
			Dyandi		Verkhoyansk	Dyandi–Okhonosoi	Au
			Balbuk		Verkhoyansk	Barain	
			Anna–Emeskhin		Verkhoyansk	Meichan	
Late collisional	Thrust faults	Mineralized crush zones, concordant and cross-cutting veins, stockwoks, lenticular bodies	Mastakh	Au in shear zone and quartz vein	Adycha–Nera	Yakutsky	Au
	Strike-slip faults	Concordant lenticular bodies, cross-cutting veins and veinlets at limbs and crests of folds with steeply dipping hinges	**Yakutskoe**		Adycha–Nera	Yakutsky	Au, Ag
			Zhdannoe	Au quartz vein	Adycha–Nera	Zhdaninsky	Au
			Tuora–Tas		Adycha–Nera	Tuora–Tas	Au
			Sokhatinoe		Adycha–Nera	Tuora–Tas	
			Venera		Adycha–Nera	Tuora–Tas	
			Sokh–Bar		Adycha–Nera	Tuora–Tas	
			Kellyam		Adycha–Nera	Sokh	
		Mineralized crush zones, concordant and cross-cutting veins, stockwoks, lenticular bodies	Bazovskoe	Au in shear zone and quartz vein	Adycha–Nera	Bazovsky	Au
			Badran		Adycha–Nera	Badran	Au
			Nezhdaninka		South Verkhoyansk	Nezhdaninskoe	Au, Ag
			Zaderzhinskoe		South Verkhoyansk	Zaderzhinsky	Au

* Deposits in production

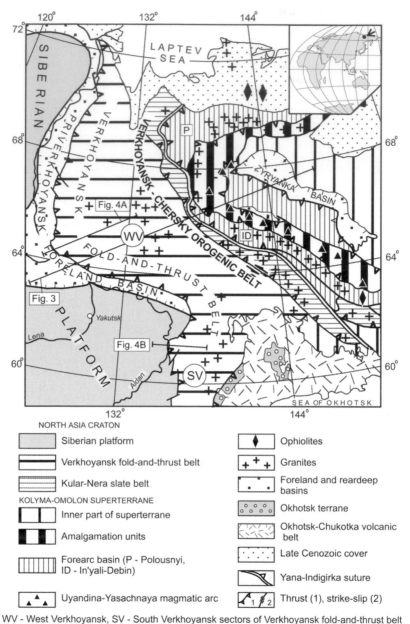

Fig. 2. Tectonic sketch-map of the Verkhoyansk–Chersky collisional orogen (orogenic belt).

Kolyma Mesozoides and resulted from collision between the North Asia craton and the Kolyma–Omolon superterrane at the end of the Late Jurassic–Neocomian, and from the closing of the smaller Oimyakon ocean basin. The formation of the Kolyma–Omolon superterrane (microcontinent) is linked to amalgamation of several terranes at the end of the Mid-Jurassic, part of which had been earlier separated from the North Asia craton in the course of Late Palaeozoic rifting (Parfenov 1991, 1995; Prokopiev 1998). The Verkhoyansk–Chersky orogen bears all the characteristic features found in collisional belts. The structure of the Verkhoyansk–Chersky collisional orogen has bean subdivided into outer, inner and rear zones (Prokopiev 1998; Fig. 3).

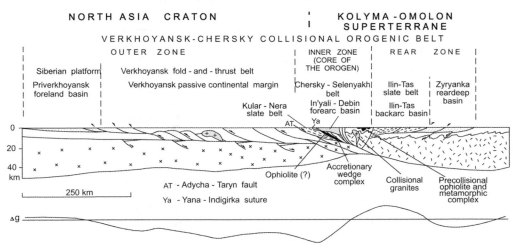

Fig. 3. Geological section through the central part of Verkhoyansk–Chersky collisional orogen. Section line shown in Figure 2.

Outer zone of the orogen

The zone is represented by the Priverkhoyansk foreland basin and the Verkhoyansk fold-and-thrust belt. The Priverkhoyansk foreland basin, consisting of Late Jurassic–Cretaceous sedimentary sequences, stretches for 1100 km at the front of the fold-and-thrust belt. Continental units of Cretaceous age conformably overlie shallow-water marine deposits in the north and continental Upper Jurassic units in the south. Their maximum thickness (3–4 km) is at the front of the fold-and-thrust belt. The Verkhoyansk fold-and-thrust belt has a typical miogeoclinal structure and is subdivided into the West Verkhoyansk and South Verkhoyansk sectors. The sediments of the fold-and-thrust belt were deposited on the Verkhoyansk passive continental margin of the North Asia craton.

Close to the platform, the West Verkhoyansk sector is made up of predominantly Carboniferous and Permian rocks, which to the east are replaced by Triassic and Jurassic units. The sediments make up a thick wedge (up to 15 km) of fragmental coastal-marine, deltaic and shelf rocks of the Verkhoyansk terrigenous complex. To the west, within the Siberian platform, they are replaced by synchronous coastal-marine and alluvial accumulations, and to the east grade into turbidites and deep-sea black shale deposits.

In the frontal zone of the West Verkhoyansk sector fault-propagation folds, back thrusts and intercutenous wedges are mainly developed. The detachment is confined to clay horizons at the base of the Triassic, shifting eastward along the fault onto Permian clay horizons and, apparently,

Middle–Upper Devonian gypsum (Parfenov et al. 1995). The width of the frontal zone reaches 100 km in the central part of the West Verkhoyansk sector, decreasing both northward and southward. The structure of the central zone is characterized by a blind duplex in a carbonate Late Precambrian–Mid-Palaeozoic complex, and by imbricate fans and back thrusts of the Late Palaeozoic–Mesozoic Verkhoyansk terrigenous complex (Fig. 4a). The detachment there is displaced onto the basement of a sedimentary sequence, is traced by the roof of the crystalline basement, and is the floor thrust of the duplex. The roof thrust of the duplex is a floor thrust for deformation in the Verkhoyansk complex. Duplex development in the carbonate complex determined the formation of the largest anticlinoria of the fold-and-thrust belt. The culminations of the anticlinoria correspond to the most uplifted parts of the duplex structures (Prokopiev 1998). Transverse belts of Cretaceous granite massifs intrude the fold structures of the sector.

The South Verkhoyansk sector along the boundary with the platform is made up of Upper Precambrian, Lower and Middle Palaeozoic terrigenous-carbonate shelf sediments, which are replaced eastward by coeval slope units. In the central part of the sector, Middle–Upper Devonian rift-related rocks are exposed. The Lower Carboniferous–Lower Permian terrigenous deposits on the east of the sector are turbidites typical of deep-sea slope fans. The Upper Permian–Mesozoic sediments, consisting mainly of sandstone units, are of a deltaic nature. In the South Verkhoyansk sector, three tectonic zones of

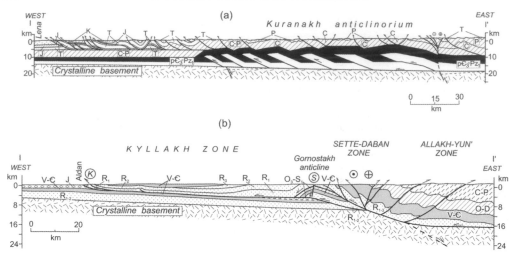

Fig. 4. Geological sections of the central part of the West Verkhoyansk sector (**a**) and of the central South Verkhoyansk sector (**b**), Verkhoyansk fold-and-thrust belt. K, Kyllakh thrust; S, Svetlyi fault (Prokopiev 1998). For locations see Figure 2.

north–south strike are distinguished, differing from each other in the style of deformation (Fig. 4b). Along the boundary with the platform is the Kyllakh zone, which has a fold-and-thrust structure. The structure of the zone is determined by the Kyllakh thrust with an amplitude of horizontal displacement of 90 km, the largest thrust in the fold-and-thrust belt (Parfenov *et al.*1998). The rear part of the zone is a ramp anticline formed above a large truncated duplex. The Sette–Daban zone, occupying an axial position within the sector, has a 'flower' or a 'palm tree' structure in its transverse section. The formation of a fan-shaped structure and an approximately north–south metamorphic belt is linked to transpressional sinistral displacements along the zone axis. The strike-slip deformations are superimposed on the early thrust structures exposed in the Kyllakh zone. The Allakh–Yun zone is characterized by tight folds with a cleavage of the axial plane of eastern vergence, which are associated with steep thrusts (Prokopiev 1989; Parfenov *et al.*1995). The inner and the rear zones of the Verkhoyansk–Chersky orogen include rocks of the Kular–Nera slate belt, southwestern and northeastern flanks of the Kolyma–Omolon superterrane.

The inner zone (core) of the orogen

This zone is represented by the Chersky–Selennyakh belt encompassing the Kolyma–Omolon superterrane and the Kular–Nera slate belt (Figs 2 and 3). The Kular–Nera slate belt is made up of intricately and multiply deformed deep-sea black

shale units of Permian, Triassic and Lower Jurassic age, which are distal accumulations of the continental slope of the Verkhoyansk passive margin (North Asia craton). The units of the slate belt are separated from the outer shelf deposits of the passive margin by the NW-striking Adycha–Taryn fault zone with indicators of substantial thrust and subsequent sinistral strike-slip displacements. It is presumed that the black shale units are overthrust for 150 km in a westerly direction (Norton *et al.*1994).

The peripheral assemblages of the Kolyma–Omolon superterrane are represented predominantly by shallow water, dominantly carbonate deposits of Ordovician, Silurian and Devonian age, with, at the base, cherty-clay deep-sea deposits from the Carboniferous, Permian, Triassic and Lower Jurassic. All these strata are intensely deformed by thrusts and strike-slip faults. The Ordovician–Upper Devonian carbonate assemblages of the orogen core are akin to the coeval units of the Sette–Daban zone (southern part of the orogen outer zone) (Bulgakova 1997). In the Early–Mid-Palaeozoic they were located at the margin of the North Asia craton, became separated from it as part of the terranes during Mid-Palaeozoic rifting and were subsequently moved for a distance of 1500–2000 km with respect to the craton during the Carboniferous, Permian and Triassic. This resulted in the formation of the Oimyakon minor ocean basin (Parfenov 1995). Now the tectonic blocks consisting of Lower–Middle Palaeozoic, dominantly carbonate rocks are separated from the Upper Palaeozoic–Mesozoic deposits of the superterrane periphery by systems of gentle thrusts and strike-slip faults.

Within the orogen core, scattered ophiolitic fragments are known. These, of presumably Early Palaeozoic age, make up tectonic nappes and have been described as the Chersky Range ophiolite belt by Oxman et al.(1995). The obduction of these ophiolites and subsequent metamorphism occurred at the end of the Mid-Jurassic during amalgamation of terranes and formation of the Kolyma–Omolon superterrane (Parfenov 1995). During the collision, these ophiolites were deformed by late thrusts and strike-slip faults. The Upper Jurassic calc-alkaline volcanites of the Uyandina–Yasachnaya arc, which unconformably overlie Palaeozoic and Early Mesozoic sediments, formed above the subduction zone in which the oceanic crust of the Oimyakon basin moves under the Kolyma–Omolon superterrane during its approach to the North Asia craton. The Polousnyi and In'yali–Debin synclinoria are made up of thick, intricately deformed flysch units of Mid-Late Jurassic age and have been interpreted as assemblages of forearc basins of the Uyandina–Yasachnaya arc (Parfenov 1991). The assemblages of the orogen core are intruded by collisional granites of Late Jurassic–Cretaceous age.

Units of the Kular–Nera slate belt are assigned to the accretionary wedge of the Uyandina–Yasachnaya volcanic arc (Fig. 3; Parfenov 1991, 1995). Deep-sea Lower Carboniferous–Lower Jurassic cherty-clay deposits of the superterrane slope and foot occurring as tectonic sheets and wedges along the periphery of the carbonate unit outcrop in the orogen core. The sediments of the Polousnyi and In'yali–Debin forearc basins were moved during collision in a westerly direction and overthrust on the assemblages of the Kular–Nera slate belt along the Yana–Indigirka suture. They are less deformed than the underlying, more ancient deposits of the accretionary wedge. The Yana–Indigirka suture separates the accretionary wedge and forearc basins, and includes the Chai–Yureinskiy, Charky–Indigirka and Yana faults, recognized earlier (Gusev 1979), which successively replace each other towards the northwest. The Debin outcrops of ophiolites in the SE of the In'yali–Debin synclinorium are, in all likelihood, relics of the oceanic crust of the Oimyakon Basin (Oxman et al.1995). Their presence in the middle reaches of the Khroma and Berelyokh Rivers can also be presumed, where they are overlapped by Cenozoic rocks of the Primorsk lowland. Here, they are indicated by intense positive gravity and magnetic anomalies (Spektor & Dudko 1983; Fig. 2). Details of the structure of the accretionary wedge complex remain debatable because it has as yet been poorly studied. The absence of sub-

stantial ophiolite outcrops may be evidence that the oceanic crust in the Oimyakon Basin was nearly completely absorbed in the subduction zone. However, the formation of the orogen in the environment of development of an A-subduction zone cannot be excluded (Fujita & Newberry 1982).

An intense gradient in the gravity field corresponds to the accretionary wedge and the orogen core (Fig. 3). The gravitational model (Norton et al.1994), computed on the basis of a profile transverse to the strike of the Verkhoyansk–Chersky orogen, shows that rocks of the orogen core are overthrust in its southwestern sector. Examples include units of the Kular–Nera slate belt, which are overthrust by as much as 150 km westwards, overlying the craton margin. This value approximates to the width of the gravity gradient. Northwards, the width of the gradient decreases to 50–70 km, which may give evidence of a smaller-scale overthrust there.

The rear zone of the orogen

The rear zone includes the Ilin–Tas fold-and-thrust belt of northeastern vergence and the Zyryanka reardeep basin (Figs 2 and 3). The Ilin–Tas fold-and-thrust belt is made up of thick (> 6 km) Kimmeridgian–Volgian black shale of the backarc basin of the Uyandina–Yasachnaya magmatic arc (Parfenov 1995). Components of the orogen core are thrust northeastward onto the rocks of the backarc basin, which, in turn, overthrust the Lower Cretaceous–Neogene coal-bearing deposits of the Zyryanka (rear) basin, whose thickness reaches 5.5 km at the front of the fold-and-thrust belt. The detachment of the Zyryanka basin occurs at the bottom of Neocomian units (Gaiduk & Prokopiev 1999). The western and southwestern vergence of folding, characteristic of the outer zone and southwestern flank of the core orogen, is replaced by a northeastern vergence in the northeast of the inner zone and in the rear zone.

Deformation history of the orogen

Folding in the orogen core and in the rear of the outer zone began to form at the end of the Late Jurassic, advancing southwest- and northeastward, and terminated in the formation of systems of frontal thrusts along the boundary with the Siberian platform in the Late Cretaceous and of rear thrusts of the Ilin–Tas fold-and-thrust belt in the Cenozoic (Parfenov et al.,1995; Gaiduk & Prokopiev 1999). In the first (early collisional) stage of deformation (Tithonian–Neocomian), thrusts began to form in the orogen core and in the rear of the West Verkhoyansk sector of the Verkhoyansk

fold-and-thrust belt, where dislocations are intruded by collisional granites, whose age is given
as 150–134 Ma by $^{40}Ar/^{39}Ar$ dating (Layer et al.
2001). The Priverkhoyansk foreland basin and
syn-sedimentary folding began to form at the front
of the orogenic belt outer zone during that time
(Parfenov et al., 1995). In the rear zone, sediments
accumulated, transported from the orogen core
into the Ilin–Tas backarc basin. The Ilin–Tas
backarc basin is supposed to have formed as a
'pull-apart basin' structure in the course of
development of large strike-slip faults along the
northeastern boundary of the orogen core
(Prokopiev 1998). Late Mesozoic grabens in the
orogen inner zone could have formed similarly. At
the beginning of the Neocomian, the Zyryanka
rear basin began to form. The stage terminated in
the formation of thrusts and strike-slip structures
in the South Verkhoyansk sector, which in the east
are overlain by Neocomian volcanites of the Uda
belt (Fig. 5).

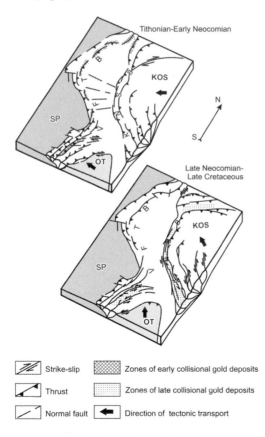

Fig. 5. Geodynamic conditions of collision-related gold
ore formation in the Verkhoyansk–Chersky orogen. SP,
Siberian platform; KOS, Kolyma–Omolon superterrane;
OT, Okhotsk terrane; VFTB, Verkhoyansk fold-and-
thrust belt; KNSB, Kular–Nera slate belt.

In the second (late collisional) stage (Barre-
mian–Late Cretaceous), fold-thrust structures and
frontal thrusts of the West Verkhoyansk sector of
the Verkhoyansk fold-and-thrust belt were formed.
A southwestern advance of folding at 132 to
98 Ma, according to $^{40}Ar/^{39}Ar$ dating (Layer et al.
2001), is marked by a rejuvenation in the same
direction as the trend of intruding granite plutons.
At the end of this stage, the thrusts in the orogen
core and in the rear of the outer zone are
transformed into sinistral strike-slip faults, which
is probably linked to the change in the direction
of the movement of the Kolyma–Omolon superterrane. The thrusts and strike-slip faults along the
southwestern margin of the superterrane are synchronous with them and are intruded by collisional granites ($^{40}Ar/^{39}Ar$ age of 127–120 Ma;
Parfenov 1995). In the rear zone, the formation of
the Zyryanka basin was by then in progress, and
the front of the thrusts of the Ilin–Tas fold-and-
thrust belt was shifted northeastwards. The formation of the Ilin–Tas fold-and-thrust belt could
occur at the expense of strike-slip displacements
along its axial part with the appearance of a
transpressional structure of a 'flower' or 'palm
tree' pattern, which is indicated by distribution of
faults with composite thrust – strike-slip kinematics (Gusev 1979).

Yana–Indigirka metallogenic belt

The 500–600 km wide Yana–Indigirka metallogenic belt extends in a northwest direction for
1200 km, encompassing central and western parts
of the Verkhoyansk–Chersky orogen. Its formation
was related to a collision between the Kolyma–
Omolon superterrane and the North Asia craton,
which occurred in the latest Late Jurassic to Early
Neocomian.

A metallogenic belt is a basic unit in the
zonation of a region. A belt includes all the
mineral deposits and occurrences that formed in a
particular geodynamic environment. Metallogenic
belts are subdivided into metallogenic zones,
including mineral deposits and occurrences of
common genesis. A zone is characterized by the
short time of its formation (within 10 million
years) and an irregular distribution of mineral
deposits. Metallogenic zones tend to group into
ore fields, which combine deposits of similar
composition and genesis. Recognition of metallogenic zones and ore districts makes it possible to
conduct the zonation of metallogenic belts and to
reveal the dynamics of their formation in time and
space (Parfenov et al. 1999).

Early collisional group of gold fields and deposits

The early collisional group of gold fields and deposits is localized in the the Allakh–Yun, Verkhoyansk, and Kular metallogenic zones within the western part of the Verkhoyansk fold-and-thrust belt and on the northwestern flank of the Kular–Nera slate belt (Fig. 1). The early group consists of gold–quartz deposits, the type of 'Au deposits in shear zonse and quartz veins' (after Nokleberg et al. 1996).

The Allakh–Yun metallogenic zone extends north–south for 300 km in the central part of the South Verkhoyansk sector of the Verkhoyansk fold-and-thrust belt (Fig. 1). It is confined to the Minor–Kiderikin zone of highly deformed rocks within the western South Verkhoyansk synclinorium. In the northern part of the South Verkhoyansk synclinorium, the Minor–Kiderikin zone consists of a system of en-echelon tight folds, while in the south there is a high-angle monocline separating compressive folds of Upper Carboniferous rocks on the west and simple open folds of Permian deposits on the east. The Minor–Kiderikin zone makes up an eastern termination of the Sette–Daban transpression shear zone (Fridovsky & Prokopiev 1997). The metamorphogenic gold–quartz veins, characteristic in the zone, are older than large granitic plutons of the South Verkhoyansk synclinorium, which are dated by ^{40}Ar/^{39}Ar at 120–123 Ma (Layer et al. 2001) and cut the ore bodies (Silichev & Belozertseva 1980).

The main ore bodies of the metallogenic zone are concordant and cross-cut veins occurring in the hinges and limbs of minor folds. The concordant veins thin out to form concordant stockwork-like bodies. There are also tabular ore bodies confined to tension fractures in sandstone beds. They are oriented sub-perpendicular to the contacts of the beds and parallel to the fold hinges. The concordant veins exhibit a zoned structure. The early thin-banded quartz of grey colour on the periphery of the veins grades towards the centre into the late milk-white massive quartz (Amuzinsky 1975; Silichev & Belozertseva 1980; Konstantinov et al. 1988). The early quartz occurs in the areas of bedding-plane slip and bedding-plane slaty cleavage. It replaced schistose rocks and has a thinly banded structure. The veins normally contain angular fragments of the host rocks measuring a few tens of centimetres across. Maximum thickness of the ore bodies is 3–4 m, and length ranges up to a few kilometres.

Principal minerals of the ore bodies are quartz (90–95%) and carbonate (5–8%). Pyrite, arsenopyrite, pyrrhotite, sphalerite, galena, and chalcopyrite constitute 1–2%. The early quartz–pyrite–arsenopyrite assemblage is followed by a quartz assemblage, then by productive quartz–gold–galena–sphalerite assemblage, and a late carbonate–quartz assemblage (Konstantinov et al. 1988). Gold from the productive assemblage is pure (830‰ fine) and coarse-grained (0.5–1 mm) (Seminsky et al. 1987; Konstantinov et al. 1988). Significant gold concentrations (up to several grams per tonne) also occur in pyrite and arsenopyrite of the early mineral assemblage. Maximum Au values (up to 3 kg t^{-1}) are found in galena and sphalerite (Kokin 1994).

The Allakh–Yun metallogenic zone exhibits a characteristic mineralogical, temperature, and geochemical zonality along strike (Kokin 1994). From south to north, the amount of arsenopyrite decreases and that of carbonates increases, decrepitation temperature of productive mineral assemblages goes down, and content of Ag, Sb, and Hg rises to high values. This zonality pattern is attributable to general upwarping of the structures in this direction and to ore formation at higher hypsometric levels. The zone includes the Yursky–Brindakit and the Bular–Onocholokh ore fields (Fig. 1).

The Yursky–Brindakit gold ore field can be traced for 36 km from the Yudoma river in the south to Brindakit creek in the north, within the western limb of the Minor Kiderekin zone of strongly deformed rocks, Figure 6. The boundaries of the district are defined by the outcrops of Upper Carboniferous and Lower Permian ore-bearing rocks of the Surkechan and Khalyya formations which are subdivided into five units, each up to a few hundred metres thick. Mineralization is confined to the base of the units and occurs in concordant veins. The lower parts of the units are made of sandstones, gritstones, conglomerate, and tuffaceous diamictites, which grade up section into siltstones. Total thickness of the ore-bearing deposits of the Surkechan and Khalyya formations ranges from 1200 m to 2200 m. The fold structure of the ore district is determined by open linear synclines and anticlines. The fold hinges are gently inclined (5–20°) towards the ore field. The largest fold structures up to 1 km wide are traced throughout the ore field. Vergence of the folds changes from western in the north and south of the ore district to eastern in the centre. The anticlines are asymmetric, with short western and longer eastern limbs. Fracture cleavage is developed, oriented parallel to axial surfaces of the concentric folds, and is well seen in aleuropelitic rocks and almost indistinguishable in psephitic and psammitic varieties. There are some localities with more compressive folding, including isoclinal and compressed folds which in plan form lens-

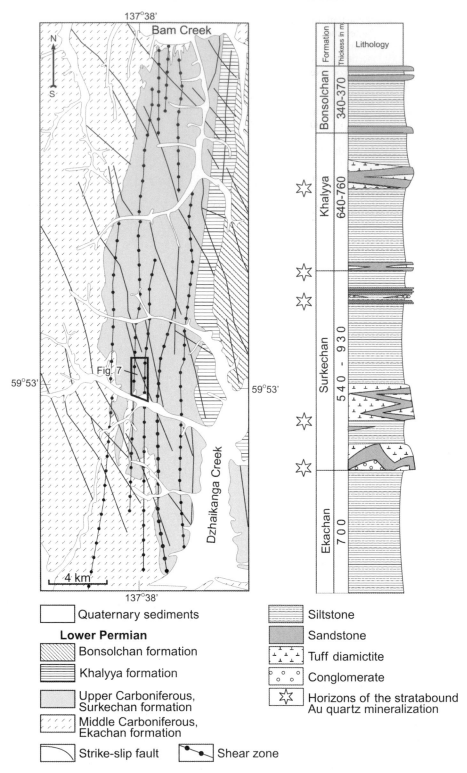

Fig. 6. Schematic geological structure of the Yursky–Brindakit gold ore field.

and-band patterns, non-persistent in width. Compressive folding is characteristic of shear zones which are the main ore-controlling structures of the Yursky–Brindakit ore field (Fridovsky 1998). Within the ore field, the Yurskoe, Nekur, and Duet deposits are known.

The Yurskoe gold–quartz deposit occurs within the limits of a low-angle monocline complicated by minor (with an amplitude of 100 m) folds and the ore bodies make up part of these folds (Fig. 7). Mineralization occurs in the third and fourth horizons of the Upper Carboniferous rocks. The ore bodies are massive in the centre and banded around the periphery. The thickness varies from 0.5 m to 5 m. The ore bodies occur at the base of sandstone units as thick as 1–20 m. Several north–south shear zones are recognized here. The rocks near the shear zones are draped into compressed and isoclinal folds filled with saddle reefs parallel to the cleavage and axial surfaces of the shear folds.

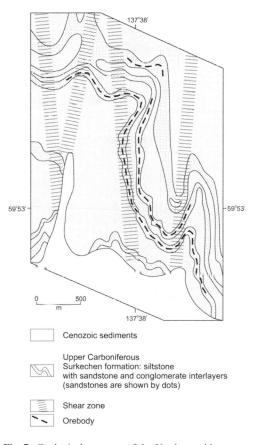

Fig. 7. Geological structure of the Yurskoe gold–quartz deposit, Yursky–Brindakit field.

The Verkhoyansk metallogenic zone extends as a narrow (up to 100 km) band for 1200 km along the western margin of the northern and central sectors of the Verkhoyansk fold-and-thrust belt (Ivensen *et al.* 1975). It is made largely of Carboniferous and Permian terrigenous rocks metamorphosed at greenschist facies. Metamorphism is thought to be related to thrust zones (Arkhipov *et al.* 1981), regional metamorphism (Kossovskaya & Shutov 1955) or to unexposed granitoid plutons (Yapaskurt & Andreev 1985). Early authors proposed the relation of gold mineralization with high-grade metamorphism of greenschist facies (Ivensen *et al.* 1975). Later on it was established that gold content is low in higher-grade rocks of the biotite sub-facies and that best gold values occur in the muscovite–chlorite subfacies (Andreev *et al.* 1990).

The zone consists of several ore fields corresponding to culminations of major anticlinoria (Amuzinsky 1975; Ivensen *et al.* 1975; Fridovsky 1998). From north to south these are the Dyandi–Okhonosoi, Meichan, Dzhardzhan, Balagannakh, Sereginsky, Aialyr, Sudyandalakh, Imtachan, Kygyltas, Sagandzha, Kitin, Barain, and Kharangak gold ore fields (Fig. 1). The main ore bodies of the zone are concordant veins complicated by cross veinlets clustering into stockworks in sandstone beds. Mineral composition of the ore bodies remains unchanged in all of the ore districts of the zone (Amuzinsky 1975). Quartz and carbonates predominate (98–99%). Ore minerals are pyrite, gold, pyrrhotite, galena, arsenopyrite, and chalcopyrite. Hydrothermal alterations include silicification, argillization, carbonatization, and sulphidization.

The Dyandi–Okhonosoi gold ore field is recognized in the north of the Verkhoyansk fold-and-thrust belt. It is made of undifferentiated Middle–Upper Carboniferous siltstones (with sandstone beds) and Lower Permian siltstones, sandstones and slates (Fig. 8). The ore field extends approximately north–south for 90 km and is 10–15 km wide. Gold–quartz mineralization occurs at three stratigraphic horizons. The ore bodies include concordant and cross veins, as well as stockworks in sandstone strata (Abel' & Slezko 1988). The Dyandi gold field and a suite of gold occurrences are known here. The structure of the Dyandi gold–quartz deposit area is determined by linear overturned folds and thrusts. Flow cleavage is clearly defined, following orientation of the thrusts. The deposit consists of stockworks, veins, and mineralized breccias controlled by approximately north–south high-angle faults. The stockworks are up to 900 m long and 100 m wide (avg. 20 m). Concordant and cross-cutting veins are present, ranging up to 80 m in length and 3 m in

Cenozoic sediments

Upper Carboniferous
Surkechen formation: siltstone
with sandstone and conglomerate interlayers
(sandstones are shown by dots)

Shear zone

Orebody

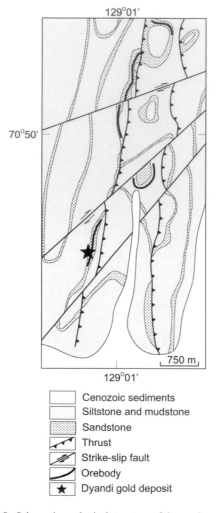

Cenozoic sediments
Siltstone and mudstone
Sandstone
Thrust
Strike-slip fault
Orebody
★ Dyandi gold deposit

Fig. 8. Schematic geological structure of the southern Dyandi–Okhonosoi gold–quartz ore field.

thickness. The veins and stockworks are accompanied by mineralized breccias. The highest Au values occur in the stockworks—up to 4.3 g t^{-1}. Ag content of the stockworks is up to 1 g t^{-1}. The gold is 700–900‰ fine and occurs as grains up to 2–3 mm in size.

The Kular metallogenic zone is located on the northwestern flank of the Kular–Nera slate belt. It extends northeastward for 150 km and is 30–40 km wide (Fig. 1). The zone is formed from deep-marine black slates of Permian–Triassic age which are intruded by granites dated at 103 Ma by ^{40}Ar/^{39}Ar (Layer *et al.* 2001). Early authors considered the zone to be a fault–fold uplift with simple box and slit-shaped folds (Ivensen *et al.*

1975; Gusev 1979). Detailed structural studies conducted at a later time revealed a complex fold-and-thrust structure of the zone with a wide distribution of refolded recumbent isoclines (Parfenov *et al.*, 1989; Fridovsky & Oxman 1997). The rocks are metamorphosed to greenschist facies (muscovite–chlorite and biotite subfacies).

Metamorphogenic gold–quartz veins form the Kyllakh, Burguat and Dzhuotuk ore fields recognized within the limits of antiforms composed of Permian rocks. The ore bodies consist of quartz, carbonates (ankerite, calcite), chlorite, muscovite and albite. The early pyrite–arsenopyrite assemblage was followed by the later productive Au pyrrhotite–chalcopyrite–sphalerite–galena one (Sustavov 1995). The gold is 750–850‰ fine.

The Kyllakh gold ore field is the most intensively studied (Ivensen *et al.* 1975; Fridovsky & Oxman, 1997). It is delineated in the axial part of the Ulakhan–Sis antiform. The structure of the district is defined by two systems of low-angle thrusts of opposite vergence which control gold mineralization. The ore bodies occur in concordant veins and as mineralized crush zones. The veins are several hundreds of metres long and up to 4 m thick. They contain 1% Pb, 1% Zn, and 0.01% Ag (Ivensen *et al.* 1975). The mineralized crush zones are confined to thrusts or their feathering fractures. Within the ore field, the Emelyanovskoye and Kyllakh deposits are known. The Emelyanovskoye gold–quartz deposit consists of concordant ore bodies. Stratabound saddle, lenticular, and sheet-like veins are typical. Closely spaced veins and veinlets form concordant stockworks. Most of the veins and veinlets follow the orientation of cleavage structures and some of them occupy S-shaped shear fractures. Both up and down dip, the veinlets pass into concordant veins or are truncated by decollement faults. The ore bodies are a few hundreds of metres long and up to 1.5 m thick. They consist mainly of quartz and carbonate, along with subordinate pyrite, galena, sphalerite, gold, pyrrhotite, arsenopyrite, fahlore and chalcopyrite. Gold grains are 3–4 mm in the longest dimension. The Kyllakh gold–quartz deposit consists of three mineralized zones: West, Central and East (Fig. 9). Morphology of the veins is defined by the degree of deformation and metamorphism of the host rocks (Fridovsky & Oxman 1997). The West zone occurs in an area of highly deformed rocks of greenschist facies (biotite) and is characterized by en-echelon quartz veins of various shape (saddle, beaded and lenticular), ranging in thickness from a few centimetres to 6 m. The Central zone occurs within the less deformed rocks of the stilpnomelane subfacies. Saddle veins here coexist with those developed along the thrust planes ranging up to 1.5 m

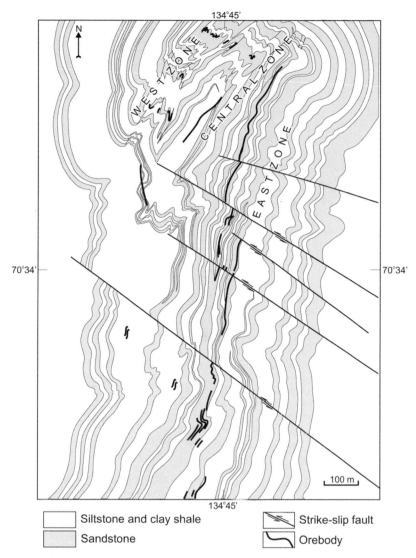

134°45'

70°34' — — 70°34'

134°45'

| | Siltstone and clay shale | | Strike-slip fault |
| | Sandstone | | Orebody |

Fig. 9. Schematic geological map of the Kyllakh gold–quartz deposit (modified from Fridovsky & Oxman 1997).

in thickness. The East zone is delineated in monoclinal rocks of the chlorite–muscovite subfacies and consists of tabular ore bodies about 1 m thick. Gold concentration does not exceed a few grams per tonne.

Late collisional gold fields and deposits

The late collisional group of gold fields and deposits is localized in the Adycha–Nera and South Verkhoyansk metallogenic zones within the eastern part of the Verkhoyansk fold-and-thrust belt and at the centre of the Kular–Nera slate belt (Fig. 1). The late group includes deposits of gold–

quartz type ('Au deposits in shear zones and quartz veins', after Nokleberg *et al.* 1996).

The Adycha–Nera metallogenic zone extends over the central and southwestern sectors of the Kular–Nera slate belt formed from Permian and Triassic deep-water black slates and the adjacent part of the Verkhoyansk fold-and-thrust belt made of Upper Triassic and, locally, Lower Jurassic shelf deposits. The zone extends northwesterly for 600 km and is 150 km wide (Fig. 1). It includes several hundred gold–quartz vein deposits and occurrences of different morphology. A long history of ore deposition is presumed here, beginning with accumulation of disseminated gold in the Upper Palaeozoic and Lower Mesozoic black slate units in distal parts of the Verkhoyansk passive continental margin and its subsequent mobilization during metamorphism and emplacement of granitoids in the course of Late Jurassic–Early Neocomian collision between the northeastern margin of the North Asia craton and the Kolyma–Omolon superterrane (Fridovsky 1998). The zone includes the Tuora–Tas, Badran, Yakutsky, Saninsky, Kurun–Agalyk, Sokh, Ol'chan, Talalakh, Bazovsky and Zhdaninsky ore fields (Fig. 1).

The Tuora–Tas gold ore field is about 40 km^2 in area, located in the central sector of the Kular–Nera slate belt. The district is composed of Lower Norian and Carnian siltstones and slates with rare sandstone beds. The rocks are metamorphosed at lowest greenschist facies. There are several andesite–basalt dykes up to a few hundred metres long, stiking NE. The dykes exhibit silicification, carbonatization, and chloritization. Small-sized massifs of biotite porphyry granite occur 7 km northwest of the ore district border and hornfels are found at a distance of 3.5 km. The main fold structures of the district trend northeasterly, parallel to the orientation of major north–south faults and cleavage. They seem to be superposed on the earlier isoclinal folds. The ore bodies normally have NE and, more rarely, east–west strike. There are concordant veins (Sokhatinoe deposit, Shyrokoe occurrence), mineralized crush zones, and stockworks (Venera and Sokh–Bar occurrences) (Fig. 10). The concordant veins occur at the contacts of beds with contrasting physico-mechanical properties. They often change into high-angle cross-cutting veins. The veins are up to 100 m long, ranging in thickness from a few tens of centimetres up to 2.5 m. The mineralized crush zones can be traced for a distance of 1.5 km and have a variable thickness. The stockworks are confined to areas of en-echelon mineralizing faults, ranging up to 100 m in thickness. The ore bodies are composed of quartz, carbonates, and ore minerals such as pyrite and arsenopyrite (early assemblage), along with subordinate sphalerite, chalcopyrite, and galena (late assemblage). The gold is free and is 792‰ fine (Rozhkov et al. 1971).

The Sokhatinoe gold–quartz deposit consists of concordant quartz veins up to 200 m long and 2 m thick accompanied by cross-cutting veinlets and veins up to 0.5 m thick. Mineralization is confined to a 1.5 km long mineralized crush zone in Carnian sandstones and siltstones (Fig. 10a). Gold content runs to a few tens of grams per tonne. The Venera gold–quartz occurrence is restricted to areas of en-echelon left-lateral strike-slip faults on

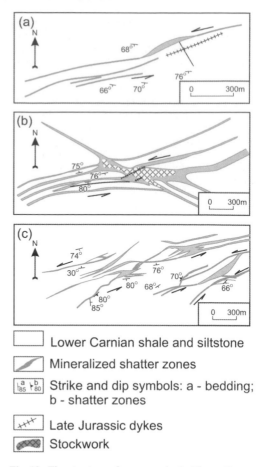

Lower Carnian shale and siltstone

Mineralized shatter zones

Strike and dip symbols: a - bedding; b - shatter zones

Late Jurassic dykes

Stockwork

Fig. 10. The structure of ore zones in the Tuora–Tas ore field. (**a**) Sokhatinoe deposit. Ore occurrences: (**b**) Venera, (**c**) Sokh–Bar.

the southern limb of an approximately east–west-striking fold. The ore body is a stockwork 100 m wide and 200 m long consisting of numerous quartz veinlets and rare master quartz veins up to 2 m thick (Fig. 10b). Gold content ranges from trace amounts to a few grams per tonne.

The Badran gold ore field is composed of Upper Triassic sandstones and siltstones deformed into wide low-angle anticlines and narrow slit-like synclines. Gold mineralization is controlled by northwesterly thrusts and strike-slip faults dipping to the NE, which are complicated by east–west- and NE-striking offset faults. The Badran gold–quartz deposit is confined to the plane of the Badran–Egelyakh strike-slip fault with a reverse fault component on which the amount of horizontal displacement is 800 m. The footwall of the fault is formed from Norian clastics, whereas the hanging wall is made of largely Carnian rocks

(Fig. 11). There is no apparent relationship of the deposit to magmatic rocks. The nearest granitoid rocks occur 30 km to the SE of the deposit. Quartz veins and veinlets tend to occur in mineralized crush zones within the Badran–Egelyakh strike-slip fault plane and are traced for a distance of 6 km to a depth of 800 m (Fridovsky 1999). The quartz veins in the fault zone are 200 m long and up to 4.2 m thick in swells. They are accompanied by thin quartz veinlets, most abundant in places where the veins pinch out. Along with the veins and veinlets, disseminated gold is found (mineralized boudins and tectonites). Maximum gold concentrations occur in the massive quartz veins. The ore bodies mainly consist of quartz, calcite, and dolomite. Ore minerals include pyrite, goethite, arsenopyrite, galena, sphalerite, tetrahedrite, along with minor ((1%) chalcopyrite, stibnite, bournonite, and free gold (Amuzinsky et al. 1989; Anisimova 1993). Gold is lumpy and inter-stitial. Fineness varies from 689‰ to 1000‰ (Anisimova 1993).

The South Verkhoyansk metallogenic zone occurs in the South Verkhoyansk sector of the Verkhoyansk fold-and-thrust belt. It is bounded, on the west, by the Minor–Kiderikin fault and, on the east, by the Yudoma fault. The zone is traced in a north–south direction for about 300 km. It is made of Upper Carboniferous to Middle Jurassic terrigenous rocks. Fold structures of the zone are characterized by wide gentle crests with smoothly undulating hinges. Northward, northeast striking strike-slip faults are important, with horizontal displacements of up to 10 km and vertical displacements ranging up to 1 km (Korostelev 1981). Magmatic rocks are represented by large polyphase plutons, as well as stocks, dykes, and

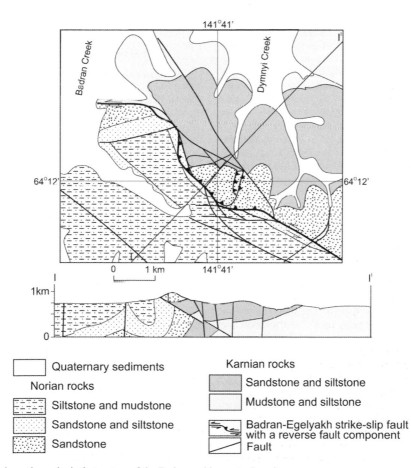

Quaternary sediments

Norian rocks

Siltstone and mudstone

Sandstone and siltstone

Sandstone

Karnian rocks

Sandstone and siltstone

Mudstone and siltstone

Badran-Egelyakh strike-slip fault with a reverse fault component

Fault

Fig. 11. Schematic geological structure of the Badran gold–quartz deposit.

subvolcanic bodies. There are gold–quartz veins and crush zones and gold–rare metal deposits within and above the apices of granitoid plutons. The zone includes the Nezhdaninskoe and Zaderzhinsky ore fields (Fig. 1). The Nezhdaninskoe gold ore field occurs in the northern part of the Allakh–Yun tectonic zone and is composed of Permian and Triassic shallow-marine sediments deformed into large linear folds of north–south trend. The major fold is the Dyby anticline about 10 km wide which extends throughout the metallogenic zone for 60 km. It has a gentle crest and rather steep (up to 40°) limbs, with well-defined cleavage of the axial plane. The folds are faulted by diagonal right-lateral northeasterly strike-slips with horizontal displacements of a few hundred metres. Two large (5–7 km^2) granite stocks are known. These are the Dyby stock in the north of the district dated by ^{40}Ar/^{39}Ar at 121 Ma and the Kurumsk stock in the south dated at 92–97 Ma by the same method (Layer et al. 2001). Dykes of the district range widely in composition and age. The oldest are NW-trending pre-granitoid dykes of lamprophyre and gabbro–diorite dated at 157–138 Ma (K–Ar) (Indolev 1979). Quartz diorite, granodiorite-porphyry, and plagiogranite-porphyry dykes have younger K–Ar ages (140–110 Ma). The youngest are quartz-porphyry and rhyolite dykes dated at 81–79 Ma which strike NE and approximately north–south for 15 km. The Nezhdaninskoe ore field includes the Nezhdaninka gold–silver–quartz deposit which is the largest in Yakutia. The Nezhdaninka gold–silver–quartz deposit occurs in the crest of the Dyby anticline and is confined to the intersection of four regional fault systems: approximately north–south, northeasterly, northwesterly and approximately east–west. The north–south system hosts the main mineralized crush zones of the deposit. The diagonal fault systems control the location of offset veins (Fig. 12). The deposit consists of mineralized crush zones, rather persistent along strike and down dip, and smaller veins. A total of 117 bodies with ore-bearing potential have been recognized at the deposit, of which only 12 have been thoroughly investigated. The richest is an ore zone in which two thirds of the ore reserves are contained. It is localized in a 15 km long and 1–40 km wide crush zone. The mineralized intervals extend continuously for 1 km along the strike of the zone to a depth of a few hundred metres. The host rocks within the ore field underwent the beresite facies metasomatism and the accompanying synberesite sulphidization. Alteration halos are about 50 m thick around individual zones, while in places where the zones are in close proximity to each other the thickness of the halos ranges up to a few hundred metres. Metasomatic sulphides

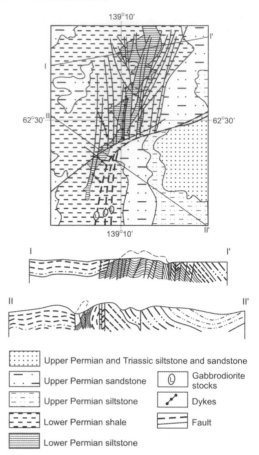

Fig. 12. Geology of the Nezhdaninskoe deposit. Modified after Shour (1985).

are highly auriferous: 30–500 g t^{-1} gold in arsenopyrite and 10–150 g t^{-1} gold in pyrite. Gold is 560–900‰ fine, with most values falling into the 780–820‰ category. Mineralization occurs mainly in veinlets and as disseminations. Quartz lenses are no more than 100 × 50 m in size, with the amount of sulphides up to 5%. The feathering quartz veins are traced for 300–400 m along strike and down dip. Their sulphide content does not exceed 3%. Along with quartz, significant carbonate (up to 5–7%) and arsenopyrite (2–3% and in places of intensely sulphidized rocks up to 15–20%) are found. Pyrite is less important. Late sulphides are represented by sphalerite and lesser galena. A range of sulphosalts are present including fahlore, freibergite, geocronite, jamesonite, pyrargyrite, miargyrite and stephanite, but bournonite and boulangerite are prevalent. Some of the ore bodies contain lens-like accumulations of stibnite (Gamyanin *et al.* 2001;Parfenov & Kuz'-min 2001).

Discussion

The structure of the Late Jurassic–Cretaceous Verkhoyansk–Cherskiy collisional orogen (orogenic belt) shows virtually all major elements of the idealized model of a collisional orogenic belt, which make it a typical example of this a class of tectonic structure (see Fig. 3). Details of the structure of the accretionary wedge complex remain debatable due to insufficient previous coverage. The absence of substantial ophiolite outcrops may be evidence that the oceanic crust in the Oimyakon basin was nearly completely absorbed in the subduction zone. The direction of movement of the Kolyma–Omolon superterrane in the period of its accretion and collision still remains unclear. Available data (Oxman & Prokopiev 1995) permit us to presume that the collision of the superterrane with the craton occurred in the first stage not frontally but rather at an acute angle during the process of oblique collision. Incidentally, the superterrane, shifting in the northwestern direction, was a rigid indentor analogous to the Indian Plate during its collision with the Asian continent (Tapponnier *et al.* 1982). This is favoured by the fact that deformations are observed along the periphery of the orogen core with composite sinistral strike-slip kinematics, and the age of the collisional granites of the Main Batholith Belt rejuvenates northwestward from 150 to 134 Ma (Layer *et al.* 2001).

We have considered only gold–quartz deposits of the Yana–Indigirka metallogenic belt within the Verkhoyansk fold-and-thrust belt, which are related to two stages of collision between the North Asia craton and Kolyma–Omolon superterrane resultant in the formation of the Verkhoyansk–Chersky orogenic belt. The early collisional Verkhoyansk metallogenic zone is within the West Verkhoyansk sector of the Verkhoyansk fold-and-thrust belt and deposits are confined to axial parts of large anticlinoria in whose cores the oldest Carboniferous rocks of the Verkhoyansk terrigenous complex are exposed. The anticlinoria are located in culminations of large duplex structures with their roof thrusts traced at the base of the Verkhoyansk terrigenous complex. Deposits of the Verkhoyansk zone appear to be related to shear zones representing branches of these faults. It is supposed that deposits of the Allakh–Yun metallogenic zone are located in fan-like thrust zones, while deposits of the Kular metallogenic zone are related to shear zones confined to detachment at the base of the Permian in the Kular–Nera slate belt. The late collisional Adycha–Nera and South

Verkhoyansk metallogenic zones are located in the hinterland of the Verkhoyansk fold-and-thrust belt and related to transpressional faults of combined thrust and strike-slip kinematics which originated during the second stage of the collision.

On the eastern margin of the North Asia craton (Verkhoyansk fold-and-thrust belt) there are other deposits assigned to the pre-collisional and post-collisional stages of the development of the territory. Formation of the pre-collisional deposits was related to the development of the passive margin of the Siberian continent. Typical are stratiform lead–zinc deposits of Vendian and Cambrian age (Sardana, Urui). Development of the Verkhoyansk passive continental margin was disturbed by Mid–Late Devonian rifting processes which produced copper mineralization here of the type of cupreous sandstones and schists and native copper in basalts (Kurpandzha), as well as copper–complex metal occurrences and apatite–pyrochlore mineralization in alkali-ultrabasic rocks and carbonatites (Gornoe Ozero). Some authors also include into the belt stratified silver–complex metal (Mangazeika) and gold–silver (Kysyl–Tas) deposits supposing that they formed in the Late Palaeozoic–Early Mesozoic in the course of sedimentation processes on the passive margin of the continent (Parfenov & Kuz'min 2001).

The post-collisional deposits include gold–antimony–mercury, gold–antimony, silver–complex metal, arsenic, antimony, and mercury deposits, among them Kyuchus, Sarylakh, Sentachan, and Prognoz. They were formed in the Late Cretaceous and, possibly, Early Cenozoic, i.e. in general synchronously with the formation of the Okhotsk–Chukotka volcanic–plutonic belt.

The authors thank L. M. Parfenov for critical comments, Warren Nokleberg for helpful review and Nigel Cook for comments and corrections which were taken into consideration in the final manuscript preparation. The work was completed within the framework of the following projects: 'Integratsia' (N 18.1), 'Universities of Russia-fundamental research' (N 1727), RFFI–Arktika (00-05-96212, 01-05-96231), RFFI (01-05-65485, 00-05-65105) and grant Minobrazovanie (E00-9.0-16), and the International project 'Mineral Resources, Metallogenesis and Tectonics of Northeast Asia'.

References

ABEL', V.E. & SLEZKO, V.A. 1988. [On stratiform gold mineralization in the Kharaulakh anticlinorium]. *In: Stratiform mineralization of Yakutia.* YaNTs SO AN SSSR, Yakutsk, 110–117 [in Russian].

AMUZINSKY, V.A. 1975. [Low-sulfide gold-quartz formation of the Verkhoyansk meganticlinorium]. *In: Gold ore formations and geochemistry of gold from the Verkhoyansk-Chukotka fold area.* Nauka, Moscow, 121–153 [in Russian].

AMUZINSKY, V.A., BORSHCHEVSKIY, YU.A., FEDCHUK, V.YA. & MEDVEDOVSKAYA, N.I. 1989. [Isotope-geochemical characteristics of endogenous carbonates at the Badran deposit]. In: Geology and mineral deposits of the central part of the Main metallogenic belt in the northeast USSR. YaF SO AN SSSR, Yakutsk, 91–103 [in Russian].

ANDREEV, V.A., NATAPOV, L.M. & YAPASKURT, O.V. 1990. [Lithologic-petrographic method for prediction of gold ore deposits in carbonaceous quartz-albite-muscovite-chlorite schists]. In: New methods for searching and prediction of gold ore and placer deposits. MG SSSR, Moscow, 18–32 [in Russian].

ANISIMOVA, G.S. 1993. [Mineralogical criteria for local prediction of gold mineralization on the basis of topomineralogical mapping of the Badran ore field]. In: Mineralogical-genetic aspects of magmatism and mineralization in Yakutia. YaNTs SO RAN, Yakutsk, 49–53 [in Russian].

ARKHIPOV, YU.V., VOLKODAV, I.G., KAMALETDINOV, V.A. & YAN-ZHIN-SHIN, V.A. 1981. [Thrusts in the western part of the Verkhoyansk-Chukotka fold area]. Geotektonika, 2, 81–98. [in Russian].

BULGAKOVA, M.D. 1997. [Paleogeography of Yakutia in the Early-Middle Paleozoic]. YaNTs SO RAN, Yakutsk 72. [in Russian]

FRIDOVSKY, V.YU. 1998. [Structures of the early collisional gold deposits of the Verkhoyansk fold-and-thrust belt]. Tikhookeanskaya Geologiya, 6, 26–36. [in Russian].

FRIDOVSKY, V.YU. 1999. [Strike-slip fault duplexes at the Badran deposit]. Izvestiya vuzov. Geologiya i razvedka, 1, 60–65. [in Russian].

FRIDOVSKY, V.YU. & OXMAN, V.S. 1997. [Gold mineralization of the Emis thrust (northeast Yakutia)]. In: Problems of geology and mining in Yakutia. YaGU, Yakutsk, 93–104 [in Russian].

FRIDOVSKY, V.YU. & PROKOPIEV, A.V. 1997. [Structural-geodynamic control of gold mineralization in the outer zone of the Verkhoyansk-Chersky orogen]. In: Geology and mineral resources of the Sakha Republic (Yakutia). IGN SO RAN, Yakutsk, 29–32 [in Russian].

FRIDOVSKY, V.YU. & PROKOPIEV, A.V. 1998. Structure and gold mineralization of the of Verkhoyansk fold-and-thrust belt (northeast Russia). In: International Conference on Arctic margins. BGR, Celle, Germany, 6.

FUJITA, K. & NEWBERRY, J.T. 1982. Tectonic evolution of northeastern Siberia and adjacent regions. Tectonophysics, 89, 337–357.

GAIDUK, V.V. & PROKOPIEV, A.V. 1999. [Methods of investigation of fold-and-thrust belts]. Nauka, Novosibirsk [in Russian], 160.

GAMYANIN, G.N., BORTNIKOV, N.S. & ALPATOV, V.V. 2001. [Nezhdaninskoe gold ore deposit - a unique deposit of the Northeastern Russia]. GEOS, Moscow [in Russian], 230.

GUSEV, G.S. 1979. [Fold structures and faults of the Verkhoyansk-Kolyma Mesozoide system]. Nauka, Moscow [in Russian], 207.

HATCHER, R.D. 1990. Structural geology. Merrill Publishing Company, Columbus, Ohio.

INDOLEV, L.N. 1979. [Dykes in ore districts of eastern Yakutia]. Nauka, Moscow [in Russian], 194.

IVENSEN, YU.P., AMUZINSKY, V.A. & NEVOISA, G.G. 1975. [Structure, formation history, magmatism, and metallogeny of the northern Verkhoyansk folded zone]. Nauka, Novosibirsk [in Russian].

KOKIN, A.V. 1994. [Mineral types of gold deposits in southeast Yakutia]. Otechestvennaya Geologiya, 8, 10–17. [in Russian].

KONSTANTINOV, M.M., KOSOVETS, T.N., ORLOVA, G.YU., SHCHITOVA, V.I., ZHIDKOV, S.N. & SLEZKO, V.A. 1988. [Control of localization of gold-quartz stratiform mineralization]. Geologiya rudnykh mestorozhdeniy, 5, 59–69. [in Russian].

KOROSTELEV, V.I. 1981. [Geology and tectonics of the South Verkhoyansk region]. Nauka, Novosibirsk [in Russian], 216.

KOSSOVSKAYA, A.G. & SHUTOV, V.D. 1955. [Epigenetic zones in the Mesozoic-Upper Paleozoic terrigenous complex of the West Verkhoyansk region]. Doklady AN SSSR, 103(6), 1085–1088. [in Russian].

LAYER, P.W., NEWBERRY, R., FUJITA, K., PARFENOV, L.M., TRUNILINA, V.A. & BAKHAREV, A.G. 2001. Tectonic setting of the plutonic belts of Yakutia, northeast Russia, based on $^{40}Ar/^{39}Ar$ geochronology and trace element geochemistry. Geology, 29, 167–170.

NOKLEBERG, W.J. & BUNDTZEN, T.K. 1996. Significant metalliferous lode deposits and placer districts for the Russian Far East, Alaska, and the Canadian Cordillera. US Geological Survey Open-File Reports, 513-A.

NOKLEBERG, W.J. & BUNDTZEN, T.K. 1997. Mineral deposit and metallogenic belt maps of the Russian Far East, Alaska, and the Canadian Cordillera. US Geological Survey Open-File Reports, 97-161.

NORTON, I., PARFENOV, L.M. & PROKOPIEV, A.V. 1994. Gravity modeling of crustal-scale cross section across the eastern margin of the North Asian Craton, northeast Siberia. In: International Conference on Arctic margins. SVNTs RAS, Magadan, 82–83.

OXMAN, V.S., PARFENOV, L.M., PROKOPIEV, A.V., TIMOFEEV, V.F., TRET'YAKOV, F.F., NEDOSEKIN, Y.D., LAYER, P.W. & FUJITA, K. 1995. The Chersky Range ophiolite belt, Northeast Russia. Journal of Geology, 103, 539–556.

OXMAN, V.S. & PROKOPIEV, A.V. 1995. Structural-kinematic evolution of the arcuate orogenic belt of the Chersky mountain system. In: Curved orogenic belts: their nature and significance. Buenos-Aires, 140–146.

PARFENOV, L.M. 1991. Tectonics of the Verkhoyansk-Kolyma Mesozoides in the context of plate-tectonics. Tectonophysics, 139, 319–342.

PARFENOV, L.M. 1995. [Terranes and formation history of Mesozoic orogenic belts of east Yakutia]. Tikhookeanskaya Geologiya, 14(6), 32–43. [in Russian].

PARFENOV, L.M., OXMAN, V.S., PROKOPIEV, A.V., ROZHIN, S.S., TIMOFEEV, V.F. & TRET'YAKOV, F.F. 1989. [Detailed structural studies in the Verkhoyansk region, their significance for large-scale geological mapping]. In: Tectonic investigations in connection with medium- and large-scale geological mapping. Nauka, Moscow, 109–127 [in Russian].

PARFENOV, L.M., PROKOPIEV, A.V. & GAIDUK, V.V. 1995. Cretaceous frontal thrusts of the Verkhoyansk fold belt, eastern Siberia. *Tectonics*, **4**, 342–358.

PARFENOV, L.M., PROKOPIEV, A.V. & TARABUKIN, V.P. 1998. [Paleontological evidence for large thrust motions in the South Verkhoyansk region]. *Doklady RAN*, **361A**(6), 809–813. [in Russian].

PARFENOV, L.M. & VETLUZHSKIKH, V.G. 1999. Main metallogenic units of the Sakha Republic (Yakutia), Russia. *International Geology Review*, **41**, 425–457.

PARFENOV, L.M. & KUZ'MIN, M.I. (EDS) 2001. *[Tectonics, geodynamics, and metallogeny of the Sakha Republic (Yakutia)]*. MAIK 'Nauka/Interperiodica', Moscow [in Russian], **571**.

PROKOPIEV, A.V. 1989. *[Kinematics of Mesozoic folding in the western part of the South Verkhoyansk region]*. YaNTs SO AN SSSR, Yakutsk [in Russian], **128**.

PROKOPIEV, A.V. 1998. [The Verkhoyansk-Chersky collisional orogen]. *Tikhookeanskaya Geologiya*, **17**(5), 3–10. [in Russian].

ROZHKOV, I.S., GRINBERG, G.A., GAMYANIN, G.N., KUKHTINSKIY, YU.G. & SOLOVYEV, V.I. 1971. *[Late Mesozoic magmatism and gold mineralization of the Upper Indigirka region]*. Nauka, Moscow [in Russian], **238**.

SEMINSKY, ZH.V., FILONYUK, V.A. & CHERNYKH, A.L. 1987. [*Structures of ore deposits in the Siberia.*] Nedra, Moscow. [in Russian]

SHOUR, V.I. 1985. *[Atlas of structures of the ore fields of Yakutia]*. Nedra, Moscow **154**. [in Russian]

SILICHEV, M.K. & BELOZERTSEVA, N.V. 1980. [Relationship between structural-lithological and barometric controls of gold mineralization]. *In: Problems of geology, mineralogy, and geochemistry of gold mineralization in Yakutia.* YaF SO AN SSSR, Yakutsk, 50–59 [in Russian].

SPEKTOR, V.B. & DUDKO, E.A. 1983. [Methodics for interpretation of geological-geophysical data in the Primorsk lowlands]. *In: Methods of geological investigation in prospecting for hard minerals in Yakutia.* Knizhnoe isdatel'stvo, Yakutsk, 51–57 [in Russian].

SUSTAVOV, O.A. 1995. [Deformation of vein quartz during the formation of gold mineralization in black slate sequences (Kular region, east Yakutia)]. *Geologiya i Geofizika*, **4**, 81–87. [in Russian].

TAPPONNIER, T., PELTZER, G., LE DAIN, A.Y., ARMIJO, R. & COBBOLD, P. 1982. Propagating extrusion tectonics in Asia: New insight from simple experiments with Plasticine. *Geology*, **10**, 611–616.

TWISS, R.J. & MOORES, E.M. 1992. *Structural geology.* W. H. Freeman and Co., New York.

YAPASKURT, O.V. & ANDREEV, V.A. 1985. [Zonal metamorphism and thermal domes of the North Verkhoyansk region]. *Doklady AN SSSR*, **280**(3), 714–717. [in Russian].

1.05–1.01 Ga Sveconorwegian metamorphism and deformation of the supracrustal sequence at Sæsvatn, South Norway: Re–Os dating of Cu–Mo mineral occurrences

H. J. STEIN[1,2] & B. BINGEN[2]

[1]*AIRIE Program, Department of Earth Resources, Colorado State University, Fort Collins, CO 80523-1482, USA (e-mail: hstein@cnr.colostate.edu)*

[2]*Geological Survey of Norway, N-7491 Trondheim, Norway*

Abstract: Re–Os dating of a suite of nine molybdenite samples from two small Cu–Mo mineral occurrences in the epidote-amphibolite facies Sæsvatn supracrustal sequence provides a temporal record of Sveconorwegian metamorphism and deformation. The sequence is situated in the western allochthonous lithotectonic domain of the Sveconorwegian orogen, the Rogaland–Hardangervidda terrane of south Norway. Onset of metamorphism occurred at about 1047 ± 2 Ma, as recorded in small gash veinlets, followed by a deformational peak at about 1032 ± 2 Ma, as recorded in mineralized breccia and mineralized metagabbro comprising a ductile shear zone constituting the Langvatn deposit. Deformation waned significantly by about 1017 ± 2 Ma, based on mineralization hosted in a brittle fault zone within stratigraphically higher metabasalts exposed at the Kobbernuten deposit. The origin of the mineralization at both deposits is most probably metamorphic with ore constituents provided by metasomatism of hosting basalt and gabbro. A metamorphic origin is supported by an array of Re–Os ages that can be related to structural features and the stratigraphic sequence, the absence of plutons related in time and space, the confinement of ore occurrences to mafic sequences in a bimodal supracrustal package that includes rhyolites and clastic units, and clear evidence for Cu mobility in mafic units. The 'main' Sveconorwegian orogenic event, probably a continent–continent collision involving imbrication, stacking, and burial of terranes took place at about 1.05 Ga and thereafter. Peak deformation in the Sæsvatn supracrustal sequence in the western part of the Sveconorwegian orogen (South Norway) may correlate with thermal metamorphism in the Idefjorden terrane in the eastern part of the orogen (SW Sweden), with a timing of about 1.03 Ga for both regions. The results of this study indicate that comparatively low grade domains in the orogen (greenschist to epidote-amphibolite facies), corresponding to upper crust, were deformed in a ductile fashion at about 1.03 Ga and were affected by brittle deformation as early as 1.025–1.015 Ga. In the high-grade domains (amphibolite to granulite facies), corresponding to middle and lower crust, ductile deformation is younger, beginning at about 1.025 Ga and persisting until 0.97 Ga.

One of the outstanding problems in understanding the Sveconorwegian (Grenvillian) province in SW Scandinavia is the relation between major terranes and lithotectonic domains bounded by continental scale orogen-parallel shear zones (Berthelsen 1980; Romer & Smeds 1996; Åhäll et al. 1998; Bingen et al. 2001). Geochronology is key to documenting and linking events that are unique to, or shared by, different terranes. It includes estimation of the timing and duration of metamorphism and its relation to a complex tectonic and magmatic history. The Sveconorwegian province contains regions affected by high-grade metamorphism (amphibolite to granulite facies) and regions affected by comparatively low-grade metamorphism (greenschist to epidote-amphibolite fa-

cies). The stratigraphic and tectonic relations between the low- and high-grade regions have been controversial for decades (Dons 1960). Although the timing of metamorphism has been estimated with a variety of techniques in the high-grade regions (Kullerud & Machado 1991; Kullerud & Dahlgren 1993; Bingen & van Breemen, 1998a; Cosca et al. 1998; Johansson et al. 2001), direct dating of metamorphism is virtually lacking in low-grade region. Attempts to date low-grade metamorphism using Rb–Sr and Sm–Nd have yielded spurious or unreliable results (Priem et al. 1973; Brewer & Menuge 1998). Since low-grade regions were not involved in a protracted middle- to lower-crustal metamorphic cycle, dating metamorphism in these regions provides im-

From: BLUNDELL, D.J., NEUBAUER, F. & VON QUADT, A. (eds) 2002. *The Timing and Location of Major Ore Deposits in an Evolving Orogen.* Geological Society, London, Special Publications, **204**, 319–335.
0305-8719/02/$15.00 © The Geological Society of London 2002.

portant information on the tectonic evolution of the orogenic belt.

We present a new approach to these problems, focusing on a low-grade region in the western allochthonous Rogaland–Hardangervidda terrane in southern Norway. Using carefully documented molybdenite samples from two well-studied mineral prospects that record the local development of ductile and brittle structures, we have tracked the evolution of metamorphism and deformation. The Re–Os dating of molybdenite provides a robust chronometer whose value reaches far beyond the dating and understanding of ore genesis (Stein *et al.* 1998, 2001). However, in order to obtain meaningful and accurate ages, the molybdenite mineral separates must be representative, and the data plotted appropriately (Stein *et al.* 2000, 2001). Molybdenite is unique among geochronometers in that new growth apparently does not seek existing grains. Indeed, unlike other commonly dated minerals using the U–Pb method (e.g. zircon, monazite), molybdenite overgrowths are exceedingly rare (Stein *et al.* 2001). And, although earlier-formed molybdenite may recrystallize during subsequent orogenic events, the chronometer remains intact, as chalcophile–siderophile Re and Os will not move from sulphide to silicate sites. The Re–Os chronometer in molybdenite has been shown to survive granulite facies metamorphism (Raith & Stein 2000; Bingen & Stein 2001), making it an ideal tool to address events in the Archaean and Proterozoic. Since large regions of the Earth's crust are locally endowed with molybdenite (e.g. southern Norway and Fennoscandia), the Re–Os dating of this mineral serves to accurately pinpoint and preserve the age of orogenic pulses related to its deposition. The high precision possible using ID-NTIMS for Re–Os dating of molybdenite allows confident discrimination of events that may be closely spaced in time.

The Sveconorwegian orogen

The Sveconorwegian orogen is situated at the southwestern margin of the Fennoscandian shield (Berthelsen 1980; Gorbatschev & Bogdanova 1993). It is made up of four main Palaeoproterozoic to Mesoproterozoic lithotectonic domains, bounded by orogen-parallel shear zones (Fig. 1). The orogen is partly covered by Caledonian nappes of Early Palaeozoic age and is cut obliquely by the Late Palaeozoic Oslo rift. The easternmost domain (Eastern Segment) is parautochthonous relative to the foreland and consists mainly of reworked granitoids of the 1.85–1.65 Ga Transcandinavian igneous belt (TIB) foreland (Christoffel *et al.* 1999; Söderlund *et al.*

1999). The other domains are allochthonous (Park *et al.* 1991; Romer & Smeds 1996; Stephens *et al.* 1996; Möller 1998; Andersson *et al.* 2002). From east to west, these are the Idefjorden terrane, the Telemark–Bamble terrane, and the Rogaland–Hardangervidda terrane (Fig. 1). These terranes are further divided into sectors. Major accretion of crust took place during the 1.7–1.5 Ga Gothian period in these terranes (Åhäll & Larson 2000). Stacking of terranes took place during the 'main' Sveconorwegian event between 1.05 and 0.97 Ga (Bingen & van Breemen 1998a; Johansson *et al.* 2001; Andersson *et al.* 2002; this work), probably as a result of a continent–continent collision between the Fennoscandian shield and an unknown continent (possibly Amazonia; Romer & Smeds 1996). Evidence for early Sveconorwegian metamorphism is recorded in the Bamble sector, where granulite-facies metamorphism took place at about 1.15–1.10 Ga (Kullerud & Machado 1991; Cosca *et al.* 1998). Late to post-orogenic granitoids are particularly abundant in the western part of the Svenconorwegian province and intruded between 1.00 and 0.92 Ga, with several distinct magmatic pulses in this window (Eliasson & Schöberg 1991; Kullerud & Machado 1991; Schärer *et al.* 1996).

Supracrustal rocks

Supracrustal sequences of rhyolite, basalt, and metasedimentary units that were affected by greenschist- to epidote-amphibolite-facies metamorphism are especially well exposed over a large area in the Telemark sector (Dons 1960). In this region, the Telemark supracrustals range in age from about 1.5 to 1.1 Ga (Fig. 1) and have been divided into four groups. Depositional ages and relationships between the groups are not fully established and are not transferable to other terranes. Several recent studies bracket the age of local supracrustal packages either by dating volcanic rocks, interleaving sills and dykes, or detrital zircons contained in metasedimentary units (Dahlgren & Heaman 1991; Menuge & Brewer 1996; de Haas *et al.* 1999). In the Rogaland–Hardangervidda terrane, several isolated supracrustal sequences have been described (Sigmond 1975, 1978, 1998; Ragnhildstveit *et al.* 1998). One of these, the Sæsvatn–Valldal supracrustal sequence, is interpreted as a continental rift sequence beginning at about 1.27 Ga (Prestvik & Vokes 1982; Bingen *et al.* 2002). The well-exposed Sæsvatn segment of this sequence contains the two Cu–Mo mineral occurrences described in this paper.

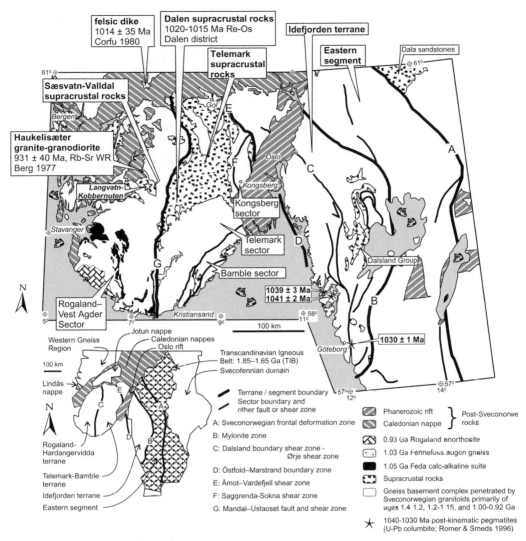

Fig. 1. The Sveconorwegian orogen in southern Norway and southwest Sweden. Modified from Bingen *et al.* (2002). Ages cited are U–Pb on zircon or monazite, unless otherwise indicated.

The Sæsvatn Synclinorium of supracrustal rocks

The supracrustal rocks at Sæsvatn (Fig. 2) have been described as a small outlier belonging to the Precambrian Telemark Suite (Prestvik & Vokes 1982). The basement complex discordantly underlying the Sæsvatn–Valldal supracrustals is composed mainly of augen gneiss with a magmatic intrusion age of 1.48 ± 0.06 Ga, as estimated by a Rb–Sr whole-rock errorchron (recalculated from 22 points on augen gneiss and amphibolite; MSWD = 40; initial $^{87}Sr/^{86}Sr = 0.703$; data from Berg 1977). This is similar to a U–Pb zircon age of 1468 ± 12 Ma for migmatitic gneissic base-

ment associated with supracrustal quartzite rafts in the Hardangervidda area, although in this area the age was interpreted as reflecting a migmatization event (Sigmond *et al.* 2000). The Sæsvatn outlier is about 15 by 10 km (150 km^2) and the stratigraphic sequence is clearly dominated in the lower part by rhyolite (Breive group, *c.* 2000 m thickness) with a basal sheet of sub-volcanic quartz porphyry, and in the upper part by basalt and clastic sedimentary rocks (Skyvatn group, *c.* 3000 m thickness). The change from felsic to mafic magmatism is abrupt and there is local discordance reported at the boundary, which has been debated (Bingen *et al.* 2002). Chemically, the basalts are transitional between tholeiitic and

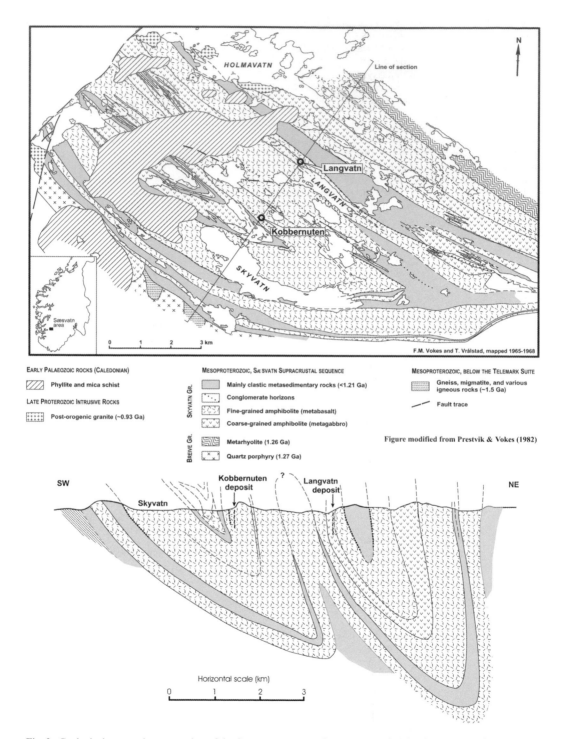

Fig. 2. Geological map and cross-section of the Sæsvatn supracrustal sequence, underlying basement gneisses, and overthrust Caledonian nappes. Locations of the Langvatn and Kobbernuten Cu–Mo deposits are shown. Modified from Prestvik & Vokes (1982). Stratigraphy and U–Pb ages cited for the Sæsvatn supracrustal rocks from Bingen *et al.* (2002).

alkaline (Prestvik & Vokes 1982). Within the massive basalt sequences, intrusions of gabbro are present as concordant lenses but locally displaying cross-cutting relations to the hosting basalts. All units have been folded into an asymmetric, partly overturned synclinorium in association with uppermost greenschist facies (epidote-amphibolite facies) metamorphism.

The age of the Sæsvatn supracrustal sequence is younger than 1275 ± 8 Ma, based on U–Pb dating of the basal quartz porphyry; the immediately overlying rhyolite sequence is 1264 ± 4 Ma (Bingen et al. 2002). The age of the metabasalts and sill-like metagabbros (Skyvatn group), host rocks for Langvatn–Kobbernuten mineralization, is not directly constrained, but is younger than 1.26 Ga. From the uppermost part of the Sæsvatn sequence, above the majority of the basaltic magmatism, U–Pb dating of detrital zircon from the Svartepodd formation (Skyvatn group) in the core of the synclinorium yields three main populations at 1.79, 1.56, and 1.22 Ga, with two Archaean grains detected. The youngest analysed zircon yielded an age of 1211 ± 18 Ma, which indicates that sedimentation took place after this time. This broad age distribution suggests that the Sæsvatn basin received detritus from felsic rocks of a variety of ages in a continental to pericontinental environment (Bingen et al. 2002).

The metagabbros and metabasalts are the relevant units for work reported here and, together with a few metasedimentary clastic and conglomeratic units, they constitute the outcrop area surrounding the Langvatn–Kobbernuten mineral occurrences. These units are strongly folded along NW–SE trending axes (Fig. 2). The metabasalts are dark green massive to schistose amphibolite with little preservation of original textures. One particularly striking variety of metabasalt, occurring locally, displays a mottled texture produced by disseminated 2–4 mm dark green ovoids of amphibole aggregate. All metabasalts are characterized by the presence of widespread lenses and pods of epidote + quartz + calcite, and at the thin section scale these secondary minerals are found in abundance (Prestvik & Vokes 1982). The sill-like metagabbros vary from seemingly unaltered coarse-grained massive varieties to foliated coarse- to medium-grained units. Like the metabasalts, widespread epidotization and uppermost greenschist facies mineral assemblages are characteristic. Regionally, the metagabbro unit containing the Langvatn deposit (Fig. 2) displays increasing deformation and epidotization as the ductile shear zone hosting the mineralization is approached. Some original textures and sedimentary structures are still preserved in the clastic units in the SW part of the Sæsvatn area (Prestvik & Vokes 1982).

The last magmatic event in the Sæsvatn area is exposed in the NW along the shores of Holmavatn where several post-orogenic (non-foliated) granite intrusions have pierced the supracrustal sequence, clearly cross-cutting all units. Prestvik & Vokes (1982) suggest that these granites form the southeastern margin of a large granite batholith that extends to the NW for about 60 km. It has also been proposed that a similar granite intrusion may be present beneath the mineralization at Langvatn, thereby providing a magmatic origin for the deposit (Prestvik & Vokes 1982). However, there is no field evidence for the existence of a buried granite intrusion. The Holmavatn granites have not been dated but they resemble other granites widespread in southern Norway with U–Pb ages of about 0.93 Ga. The Holmavatn granites are most probably of similar age and origin to the nearby Haukelisæter granite that intrudes the Sæsvatn–Valldal supracrustal rocks between the Valldal and Sæsvatn segments, less than 10 km north of Langvatn (Fig. 1). The Haukelisæter granite provides a Rb–Sr age of 931 ± 40 Ma (Fig. 3; recalculated from 21 data points on the granodiorite and medium-grained facies; MSWD = 64; initial $^{87}Sr/^{86}Sr$ = 0.7045; data from Berg 1977). Applying the maximum 2-sigma uncertainty in the older direction (931 + 40 = 971 Ma) yields an age that does not overlap the significantly older Re–Os ages for mineralization reported in this study.

The Langvatn and Kobbernuten Cu–Mo deposits

The Langvatn–Kobbernuten deposits are small, non-economic mineral occurrences of copper and molybdenum. The Kobbernuten prospect has not been described in the literature, but is noted as a small Cu prospect (Prestvik & Vokes 1982, plate 1). The Langvatn deposit, mined from 1911 to 1913, was evaluated for its mineral potential in the years 1965–1968, but geochemical and geophysical investigations followed by subsequent drilling failed to discover additional mineralization, along strike or at depth (Vokes et al. 1975).

The Langvatn Cu–Mo deposit lies in a steeply dipping (70–80° SW) to vertical ductile shear zone within a concordant metagabbro unit 200–300 m wide (Fig. 2). The zone of mineralization is characterized by extremely coarse-grained masses of gangue minerals hosting locally rich concentrations and disseminations of Cu–Mo mineralization. Massive calcite and epidote, with generally subordinate quartz, dominate the gangue mineralogy, and prominent ore minerals are chalcopyrite and molybdenite. Vokes et al. (1975) describe

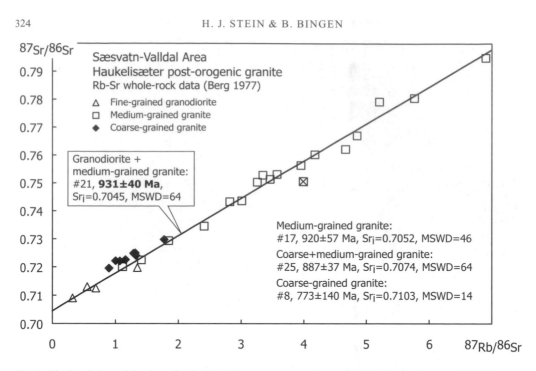

Fig. 3. Rb–Sr whole-rock isochron for the Haukelisæter post-orogenic granite intruding the Sæsvatn–Valldal supracrustal sequence. Rb–Sr data from Berg (1977). One data point (square with x) has been excluded. Even using the less robust Rb–Sr method, age calculations based on regressions of the isotopic data provide no indication that post-orogenic granites in this region are older than about 930 Ma.

minor bornite, plus traces of pyrite and pyrrhotite at Langvatn. Considerable magnetite is present. At one of the Langvatn adit entrances (UTM$_{VGS84}$: 32V 0398921 6614172) an earlier mineralization can be seen in calcite + quartz breccia clasts caught and deformed in the ductile shear zone (Fig. 4a). In addition, mineralization is present in the intensely deformed metagabbro within the shear zone and is parallel to the foliation, suggesting that it is at least as old as the ductile deformation in the shear zone (Fig. 4b, e). Mineralization occurring in coarse-grained and undeformed gangue includes occurrences in openspace vugs that must postdate this deformation.

The adit for the Kobbernuten Cu–Mo prospect (UTM$_{VGS84}$: 32V 0397682 6612390) is located at the base of a high angle planar fault zone, with two sub-parallel structures about 5 m apart, that reaches a height of about 10 m (Fig. 4c). Mineralization appears to have been localized between the two sub-parallel faults as noted by the bright blue Cu staining remaining. Epidote is present in the fault zone. In contrast to the metagabbro host rock at Langvatn (coarse-grained amphibolite), the Kobbernuten prospect is situated in massive fine-grained metabasalt (fine-grained amphibolite). Whereas mineralization was abundant at Langvatn, mineralization at Kobbernuten is scarce and

generally confined to remaining mine dump material. Kobbernuten mineralization is dominated by bornite with much subordinate chalcopyrite and occurs in clots and as irregular and discontinuous stringers and veinlets with calcite, epidote, amphibole, and quartz gangue. Relative to Langvatn, molybdenite mineralization at Kobbernuten is very scarce. It may be directly associated with the bornite clots, but is more common a few cm removed as mm slivers and tiny bladed rosettes penetrating the metabasalt (Fig. 4d). Open-space textures associated with massive coarse-grained material as observed at Langvatn are absent in the ore samples observed on the Kobbernuten mine dump. Bornite-bearing stringers and veinlets of calcite with delicate undeformed acicular needles of amphibole growing inward from the vein wall, document post-deformational growth of gangue and ore minerals at Kobbernuten.

The country rock and gangue mineral assemblages dominated by calcite + epidote + magnetite + actinolite + quartz suggest that mineralization conditions were associated with moderately high oxygen fugacity. Furthermore, ore assemblages with limited pyrite and pyrrhotite, but dominated by bornite, chalcopyrite, magnetite, and molybdenite support moderately high oxygen fugacity, and relatively low sulphur activity.

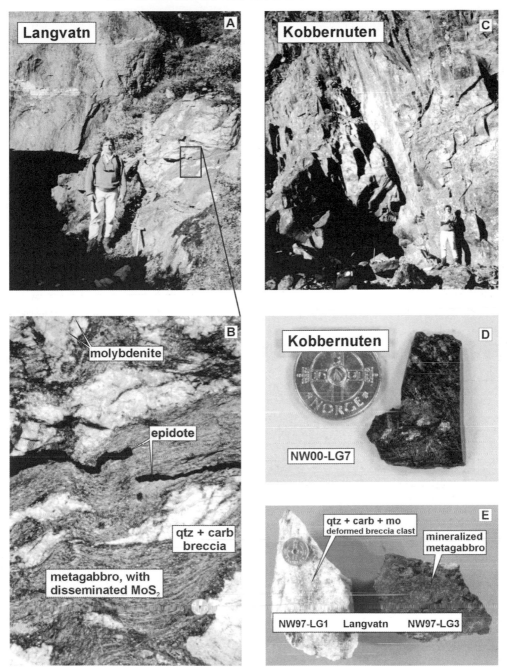

Fig. 4. The adits and ores at Langvatn–Kobbernuten. (**a**) Entrance to one of the adits at Langvatn located in metagabbro exhibiting ductile deformation. Cu staining is abundent above and to the right of the hammer. (**b**) Earlier mineralized calcite + quartz breccia clasts appear as boudinaged material in a Cu–Mo mineralized ductile shear zone at Langvatn. (**c**) Entrance to adit at Kobbernuten located in a brittle fault zone cutting a sequence of massive, fine-grained metabasalt. Cu–Mo mineralization was localized within the fault zone, and is also present as small clots and veinlets in the metabasalt. (**d**) Randomly oriented tiny blades and rosettes of late molybdenite cut the strong foliation in fine-grained metabasalt at Kobbernuten. (**e**) Analysed Langvatn molybdenite samples depict an earlier mineralization event in a deformed breccia clast (NW97-LG1), and a subsequent mineralization episode associated with ductile deformation in the metagabbro (NW97-LG3). Coin diameter is 2.0 cm.

Samples dated in this study

A total of nine different molybdenite samples were collected for this study, five from Langvatn and four from Kobbernuten. All samples were taken from the mine dumps, and reflect different types of ore–gangue–host rock relationships. The samples are also representative of mineralized outcrop at both mine adits.

Brief descriptions of the hand samples used for Re–Os dating are provided here. These are augmented by thin section descriptions and the photomicrographs in Figure 5. Samples NW97-LG1, NW97-LG2, and NW97-LG3 all represent gabbro-hosted ore samples from within the zone of ductile deformation at Langvatn. In particular, NW97-LG1 contains mineralized breccia clasts with rolled pyrite exhibiting well-developed pressure shadows, presumably developed as the breccia fragments were caught in the ductile deformation. Molybdenite and chalcopyrite are abundant. The NW97-LG1 molybdenite analysed in this study represents an occurrence confined to a carbonate

+ quartz breccia fragment (Fig. 4e). Massive and discontinuous fans and bladed patches of molybdenite, intergrown with chalcopyrite, are characteristic (Fig. 5a). Samples NW97-LG2 and NW97-LG3 are representative of Cu + Mo mineralization developed within ductile-deformed metagabbro (Fig. 4e). In all three samples, molybdenite is commonly kinked and recrystallized within single blades and laths, and gangue minerals quartz + calcite have marked undulatory extinction in thin section. Sample NW97-LG2 clearly shows that mineralization occurred prior to and in association with brecciation (Fig. 5b). Langvatn sample NW00-LG1 is representative of disseminated molybdenite associated with late, massive, coarse-grained, pink calcite. Langvatn sample NW00-LG2 is a segregation of epidote + molybdenite in the metagabbro.

The samples from Kobbernuten include NW00-LG4, NW00-LG5, and NW00-LG7, which consist of massive fine-grained metabasalt with small clots to irregular veinlets of calcite + quartz + epidote + amphibole ± pyrite ± chalcopyrite, up

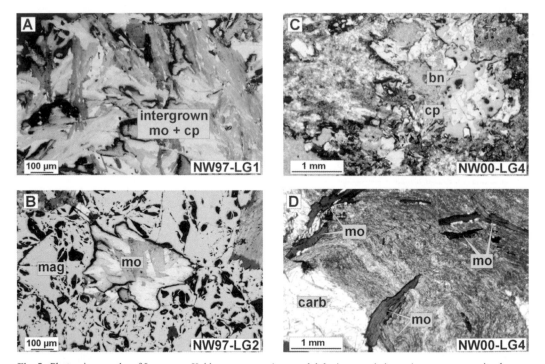

Fig. 5. Photomicrographs of Langvatn–Kobbernuten ores (mo, molybdenite; cp, chalcopyrite; mag, magnetite; bn, bornite). (**a**) Massive mo + cp intergrowths depicting early mineralization hosted in a carbonate + quartz breccia clast at Langvatn (plane polarized light). (**b**) Sulphide breccia clasts of mo ± cp hosted by massive mag ± cp mineralization at Langvatn (ppl). In this view, the cp is barely visible in the upper middle margin of a single clast. (**c**) Undeformed bn + cp blebs with poorly developed myrmekitic texture (barely visible) in metabasalt containing fine-grained chlorite and epidote patches. Cp is generally minor and bn blebs are more characteristic of the mineralization at Kobbernuten (ppl). (**d**) Blades of mo are locally parallel but mostly cross-cut a strong and complex foliation in metabasalt at Kobbernuten (crossed polars).

to 3 cm blebs of massive bornite ± chalcopyrite (Fig. 5c), and scattered 2–5 mm molybdenite blades. The molybdenite blebs may occur within the Cu-bearing veins, but they are more common as monomineralic occurrences in the metabasalt, but in close proximity to the veins. The molybdenite in NW00-LG4 and NW00-LG5 can be seen both closely parallel to and cross-cutting the pronounced foliation, with the cross-cutting occurrence dominating (Fig. 5d). The molybdenite in NW00-LG7 clearly cross-cuts foliation and post-dates deformation, as minute delicate blades and small rosettes (Fig. 4d). Thin section examination shows that a fourth Kobbernuten sample NW00-LG6 is in fact a metasedimentary clastic rock, containing minor disseminated chalcopyrite, and molybdenite as 1–2 mm platelets along the margin of a quartz + carbonate veinlet.

Analytical methods

The Re–Os chronometer in molybdenite provides robust single mineral ages, provided the mineral separates are representative and homogeneous (Stein *et al.* 2001). Dating is based on the beta decay of ^{187}Re (about 62% of total Re) to ^{187}Os. Since there is no measurable initial (common) Os in molybdenite (and common ^{187}Os is only about 1.5% of total Os in bulk Earth), essentially all Os is radiogenic ^{187}Os, whose abundance is dependent on Re concentration in the molybdenite and time for in-growth. Molybdenites typically have Re concentrations in the ppm range.

Re and ^{187}Os concentrations were determined at the AIRIE molybdenite laboratory at Colorado State University (CSU). For this study, a Carius-tube digestion was used, whereby molybdenite is dissolved and equilibrated with ^{185}Re and ^{190}Os spikes in HNO_3–HCl (inverse *aqua regia*) by sealing in a thick-walled glass ampoule and heating for 12 hours at 230 °C. The Os is recovered by distilling directly from the Carius tube into HBr, and is subsequently purified by micro-distillation. The Re is recovered by anion exchange. The Re and Os are loaded onto Pt filaments and isotopic compositions are determined using NTIMS on NBS 12-inch radius, 68° and 90° sector mass spectrometers at CSU. Two in-house molybdenite standards, calibrated at AIRIE, are run routinely as an internal check (Markey *et al.* 1998; Stein *et al.* 1997). The age is calculated by applying the equation $^{187}Os = {}^{187}Re$ $(e^{\lambda t} - 1)$, where λ is the decay constant for ^{187}Re and t is the calculated age. The ^{187}Re decay constant used is 1.666×10^{-11} a^{-1} with an uncertainty of 0.31% (Smoliar *et al.* 1996). Uncertainties in age calculations include error associated with (i) ^{185}Re and ^{190}Os spike calibrations, 0.05% and 0.15%, respectively,

(ii) magnification with spiking, (iii) mass spectrometric measurement of isotopic ratios, and (iv) the ^{187}Re decay constant (0.31%). Molybdenites rarely require a blank correction. In the AIRIE molybdenite laboratory, blanks are <20 pg for Re and <3 pg for ^{187}Os.

Analytical results

The Re–Os data and associated ages, derived from a Carius tube sample dissolution and ID-NTIMS, are presented in Table 1. Common Os and blank corrections have been applied, but are insignificant. Replicate ages are shown for two samples (two different mineral separates each from NW00-LG6 and NW00-LG7), and their agreement is excellent. These data are provided as a demonstration that the analytical results are reproducible at the geological (hand-sample) scale and can be used to argue that the different ages for different samples have geological meaning. It is typical to find slightly different Re concentrations with replicate runs of different mineral separates from the same samples, but with the proper preparation of mineral separates (Stein *et al.* 2001), the Re and ^{187}Os concentrations will be coupled such that ages are reproducible.

It is significant that the Re concentrations are highly distinct for the two deposits. At Langvatn, Re concentrations fall in a narrow range of 90–130 ppm. This suggests a fluid whose Re composition was stable and homogeneous throughout deposition of Langvatn molybdenite. In contrast, the Re concentrations for Kobbernuten molybdenite are markedly higher, ranging from 700 to 1000 ppm. We suggest that this concentration difference can be attributed to the volume-dilution effect (Stein *et al.* 2001), whereby a markedly smaller amount of molybdenite must take in all available Re present in the mineralizing fluid at Kobbernuten. We suggest that the ore fluids at both Langvatn and Kobbernuten were otherwise similar in composition and origin.

Presentation of the Re–Os results for this study is limited to a data compilation (Table 1). There are no data sets that are appropriate for construction of an ^{187}Re–^{187}Os molybdenite isochron, the most robust treatment possible for time-related samples (Stein *et al.* 1998, 2001). Although several molybdenite samples represent the timing of brecciation and ductile deformation at Langvatn, their Re concentrations have insufficient variation for an isochron plot (Table 1; NW97-LG1, NW97-LG2, NW97-LG3). If a single isochron for all the Langvatn–Kobbernuten Re–Os data is attempted, it is immediately apparent that this is the wrong approach, as a resulting age of 1032 ± 14 Ma based on 9 points is accompanied

Table 1. *Re–Os data for the Langvatn–Kobbernuten Cu–Mo deposits, Sæsvatn supracrustal sequence, Rogaland–Hardangervidda terrane, southern Norway*

AIRIE Run	Sample No.	Total Re, ppm	^{187}Os, ppb	Age, Ma
Langvatn				
CT-200	NW97-LG1	110.02 (5)	1200 (2)	1032 ± 4
CT-227	NW97-LG2	91.29 (4)	993 (2)	1030 ± 4
CT-233	NW97-LG3	133.34 (6)	1455 (2)	1033 ± 4
			Weighted mean = 1032 ± 2	
CT-457	NW00-LG1	106.18 (8)	1149.0 (8)	1025 ± 3
CT-458	NW00-LG2	99.14 (7)	1087.3 (7)	1038 ± 3
Kobbernuten				
CT-285	NW00-LG4	980.7 (6)	10618 (15)	1025 ± 4
CT-286	NW00-LG5	697.8 (5)	7564 (11)	1026 ± 4
			Weighted mean = 1025 ± 2	
CT-287	NW00-LG6	957.5 (9)	10595 (15)	1047 ± 4
CT-293 rep	NW00-LG6	790.7 (6)	8736 (6)	1046 ± 4
			Weighted mean = 1047 ± 2	
CT-304	NW00-LG7	685.8 (7)	7358 (8)	1016 ± 3
CT-459 rep	NW00-LG7	831.7 (9)	8937 (11)	1017 ± 3
			Weighted mean = 1017 ± 2	

Replicate runs (rep) on same sample represent two different mineral separates. All sample dissolutions by Carius tube, and isotopic ratio measurements by ID-NTIMS. Common ^{187}Os and blank corrections for molybdenites are insignificant. Errors (in parentheses) for concentration data are at 2-sigma for last digit(s) indicated. Ages calculated by ^{187}Os $= ^{187}$Re $(e^{\lambda t} - 1)$ and age uncertainty is absolute at 2-sigma. Weighted mean age uncertainties are less than the ^{187}Re decay constant uncertainty (± 0.31%).

by a MSWD of 2.5 × 10^6, a clear statement of excess geological scatter. Therefore, the correct approach for the Langvatn–Kobbernuten data set is to calculate weighted mean ages for related samples and replicate runs (Table 1). These weighted mean ages are used in the discussions presented in this paper.

Sæsvatn Cu–Mo mineralization and metamorphic evolution of supracrustal rocks

We suggest that Sæsvatn supracrustal sequences experienced metamorphism and associated deformation over a period of at least 40 million years from about 1.05 to 1.01 Ga, and that Cu–Mo mineralization occurred episodically for 30 million years during this period. A model for the mineralization is presented in Figure 6. The earliest indication of metamorphism (*c.* 1047 ± 2 Ma) is recorded at Kobbernuten where millimetre to centimetre scale gash veinlets of calcite ± quartz in a metasedimentary rock locally contain molybdenite and copper mineralization. These veinlets are not continuous over large distances, but may reflect local domains of extension, perhaps better developed in clastic units where fluids may migrate more easily, in a terrane that is embarking

on a metamorphic and folding cycle. These small gash veinlets might also be concentrated along fold axes where incipient extension in a compressional regime is possible, or they may be localized in association with pressure-solution and the development of spaced cleavage domains in clastic rocks (e.g. Wen *et al.* 1999). In either case, we suggest that they are an early feature of regional metasomatism. A sample containing an aggregate of molybdenite and epidote in metagabbro from Langvatn may also reflect early metasomatic segregation with its age of 1038 ± 3 Ma. With increasing metamorphism, culminating at uppermost greenschist facies (epidote-amphibolite facies), local shear zones developed at deeper levels within the supracrustal sequence at Sæsvatn, and ore deposition associated with Langvatn indicates that mineralization pre-dated, occurred during, and post-dated the development of a ductile shear zone. Dating of differently hosted ores at Langvatn, suggests that brecciation associated with mineralization and the continuation of mineralization during ductile shear zone development in the gabbro occurred during a remarkably brief geological period (1032 ± 2 Ma). This was followed by deposition of open-space mineralization associated with masses of extremely coarse crystalline undeformed calcite (1025 ± 3 Ma). Simultaneously with the open-space mineralization at Langvatn,

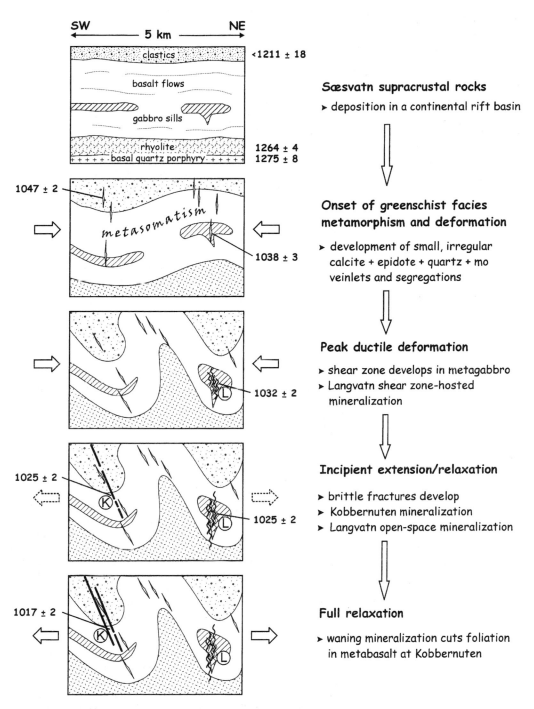

Fig. 6. Metamorphic–metasomatic model for Cu–Mo mineralization at Langvatn (L) and Kobbernuten (K) based on Re–Os dating and structural–paragenetic occurrence of the molybdenites dated. Stratigraphy is highly simplified; for example, metasedimentary clastic rocks are intercalated with basalt units.

the Kobbernuten deposit began to develop in apparently overlying basalt sequences. At Kobbernuten, mineralization associated with post-peak deformation (Fig. 5d) is recorded in two different molybdenite samples dated at 1025 ± 4 and 1026 ± 4 Ma. By about 1017 ± 2 Ma, Kobbernuten mineralization bears no relationship to the strong metamorphic fabric in the basalts (Fig. 4d), and single blades show no petrographic evidence for deformation. The general ore hosting structure at the Kobbernuten deposit reflects a brittle break that cuts a strong foliation in the metabasalt (Fig. 4c). Therefore, outcrop-scale features at the prospect site, coupled with petrographic observations, indicate that main Kobbernuten ore deposition occurred post-peak deformation. This is strongly supported by the younger Re–Os age data for Kobbernuten, using the 1032 ± 2 Ma Langvatn age as a time marker for peak deformation. Evidence of ductile shear development is clearly absent at Kobbernuten, and we suggest that its apparently higher stratigraphic position, coupled with a difference in host rock rheology (metagabbro at Langvatn versus metabasalt at Kobbernuten) may account for the distinct structural setting of these two related deposits.

We suggest that the origin of mineralization at both Langvatn and Kobbernuten is related to the host rocks themselves on a regional scale that includes perhaps a large part of the Sæsvatn supracrustal mafic sequence (Skyvatn group). In rhyolitic melts, molybdenum is sequestered with the crystallization of ferromagnesium or titanium-bearing phases, such as magnetite and ilmentite (Tacker & Candela 1987; Candela & Bouton 1990). Relative to a basaltic silicate melt, molybdenum is preferentially partitioned into Fe–S liquid, and this partitioning becomes very large for systems with Fe/S ratios that are above unity (Lodders & Palme 1991). A typical copper concentration in normal greenstones is in the order of one hundred ppm or more (Condie *et al.* 1977). Yet chemical analyses of the metabasalts and metagabbros in the Sæsvatn supracrustal sequence generally indicate only tens (most <50) of ppm Cu (Fig. 7), although chalcopyrite + pyrite ± pyrrhotite are petrographically described as opaque accessory minerals in the Sæsvatn mafic supracrustal rocks (Prestvik & Vokes 1982). Post-magmatic mobility of copper in the sequence is corroborated by the lack of correlation in binary diagrams featuring Cu and the incompatible trace element Zr, compared with a clear correlation between the two incompatible trace elements Y and Zr (Fig. 7), a feature typical for a magmatic differentiation process. We suggest that the Cu (and Mo) and the gangue epidote + calcite + quartz were derived from the mafic sequences that

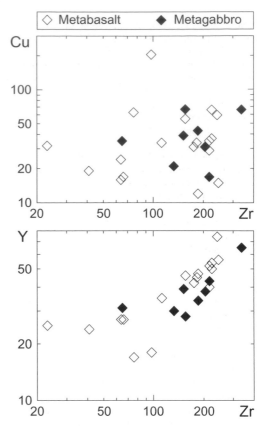

Fig. 7. Binary trace element diagrams in ppm for metabasalt and metagabbro whole rock samples of the Sæsvatn supracrustal sequence. Data from Prestvik & Vokes (1982) and Brewer & Atkin (1987).

host the deposits. Mobilization of Cu during alteration of mafic volcanic rocks is a well-documented phenomenon (e.g. Humphris & Thompson 1978; Condie *et al.* 1977), and metamorphic processes have been attributed to the formation of native copper deposits in metabasalts (e.g. Jolly 1974; Lincoln 1981), and to copper vein deposits in the Old Red Sandstone in Ireland, but within a very different stratigraphic setting (Wen *et al.* 1999). Epidosite, a rock type consisting of epidote + quartz, may arise during the metasomatic alteration of basalt, and large copper depletions may be found in epidosite and in epidotized greenstones (Kish & Stein 1989). In the Sæsvatn metabasalts, Prestvik & Vokes (1982) describe lens-shaped bodies of coarse-grained epidote + quartz ± calcite, and widespread epidotization of metabasalts. Epidote is also described in association with the metagabbro. Collectively, these observations suggest that alteration is regional and not a local effect in the vicinity of the mineralized shear-fault zones. The low and scattered Cu values associated with Sæsvatn metabasalt and metagabbro (Prestvik & Vokes 1982; Fig. 7), the stratigraphic confinement of mineralization to mafic units, the absence of spatially associated plutons of the ages obtained for mineralization, and Re–Os ages that correspond to structural, stratigraphic, and paragenetic positions for molybdenite support mineralization in the context of a metamorphic event lasting 30 million years.

There are relatively few examples of Cu–Mo deposits or occurrences whose origin is specifically attributed to greenschist- to epidote-amphibolite-facies metamorphism. In western Norway, calcite + quartz veins containing chalcopyrite ± pyrite have been recognized in the Solund basin, containing 6–7 km of Devonian sedimentary molasse floored by a late Caledonian detachment zone (Svensen *et al.* 2001). Unlike two other nearby basins (Hornelen and Kvamshesten) examined in their study, where the sedimentation is representative of continentally derived alluvial and fluvial deposits (sandstones, siltstones, and conglomerates), the Solund basin is dominated by coarse conglomerates and locally abundant metavolcanic material. The mineralization is concentrated in the central and basal parts of the Solund basin as veins associated with a shear zone and fracture networks. Based on fluid inclusion studies, vein formation is attributed to deep burial metamorphism at temperatures of 305–330 °C, at a pressure of 3.4 ± 0.2 kbar corresponding to a depth of 13.4 ± 0.6 km (Svensen *et al.* 2001). Based on the mineralogy cited for Solund, conditions in the lower part of the basin approached lower greenschist facies. Similar to Langvatn, Cu was apparently mobilized by metamorphic fluids interacting with metavolcanic material (detritus).

In the Carolina Slate Belt (southeastern USA), the Virgilina copper deposits, consisting of metre-wide quartz + calcite vein systems with bornite + chalcocite + hematite ± chalcopyrite are attributed to metasomatism associated with the waning stages of Acadian (Devonian) metamorphism (Kish & Stein 1989). Analogous to Langvatn-Kobbernuten, the Virgilina copper deposits are also situated in the core of a synclinorium. The Virgilina synclinorium contains Neoproterozoic felsic to mafic volcanic and epiclastic rocks, intruded by granitoid and gabbroic plutons, all metamorphosed to greenschist facies. The Virgilina copper veins are restricted to one of the stratigraphic units consisting of metabasalts. Regionally, the metabasalts have been epidotized, and are characteristically marked by lenses, pods, veins, and masses of epidosite. Although plutonic rocks were present, they hold no spatial association with the copper veins, and those dated by U–

Pb were found to be Neoproterozoic. Still, some workers attributed the deposits to magmatic fluids from unknown plutons (Linden 1981). Kish & Stein (1989) determined that the age of the Virgilina copper deposits was at least 235 million years younger than the hosting crustal sequence of mafic and felsic volcanic and associated clastic rocks, and that the Virgilina veins formed at about 340 Ma in association with post-Acadian fracturing that channelled metasomatizing fluids that scavenged Cu from greenstones.

Copper removal during epidotization of greenstone is sufficient to generate small vein-type deposits. This removal may have both vertical (structurally controlled) and lateral (stratigraphically controlled) components, depending on the local geologic conditions. We suggest that this type of metamorphic copper deposit may be more common than realized, but its economic value will be limited by the local stratigraphy, the extraction ability of metamorphic–metasomatic fluids, the ability to focus fluids in ductile shear zones and fractures, and the likely vertical geometry of the ore occurrences. The metamorphic model for Cu is not unlike the widely applied orogenic Au model (Cameron 1988; Groves *et al.* 1998), but also has affinities with models where pre-existing accessory magmatic sulphide may have supplied metal through liquid-state chemical transfer (Larocque & Hodgson 1995*a*, *b*; Marshall *et al.* 2000). At Langvatn-Kobbernuten, the present day exposures coupled with the Re–Os age results suggest that copper mineralization is not only late metamorphic (recorded at Kobbernuten and similar to Virgilina), but also occurred during peak deformation in association with possibly deeper level ductile deformation (Langvatn). In addition, gash veinlets of quartz ± calcite, with or without mineralization, record prograde metamorphism at Kobbernuten, whereas prograde metamorphism is reflected in early epidote-rich metasomatic segregations at Langvatn.

Regional implications

The geological evolution in the Sæsvatn region includes deposition of a supracrustal sequence from 1.27 Ga to at least 1.21 Ga, in a pull-apart or rift basin. The basin possibly formed at some distance inboard from a subduction trench in a continental setting possibly on the Fennoscandia plate (Bingen *et al.* 2002). There is little else in the geological record during this interval and, with the exception of a few gabbro–tonalite complexes in the Telemark–Bamble terrane (Pedersen & Konnerup-Madsen 2000), extensive magmatic activity during this period has not yet been recognized in southern Norway. Regional metamorphism in the Sæsvatn region is recorded by Re–Os dating of molybdenite from the Langvatn–Kobbernuten Cu–Mo deposits (Table 1). It took place from about 1047 ± 2 to 1017 ± 2 Ma, a 30 million years interval of time. Based on the sample suite in this study, earliest metasomatism is recorded in small gash veinlets and segregations (1047 ± 2 Ma at Kobbernuten and 1038 ± 3 Ma at Langvatn) at the onset of metamorphism and folding. Ductile deformation at epidote-amphibolite-facies conditions peaked at 1032 ± 2 Ma. The development of open-space mineralization and the onset of brittle faulting began at 1025 ± 2 Ma, with full relaxation by 1017 ± 3 Ma.

Several robust age estimates fall within the 1.05–1.01 Ga metamorphism and deformation window established for the Sæsvatn supracrustal sequence. In the southern part of the Rogaland–Hardangervidda terrane, significant 1050^{+2}_{-8} Ma high-K calc-alkaline syn-kinematic granite magmatism (Feda suite) is believed to represent the youngest magmatism with a subduction-related signature (Bingen & van Breemen 1998a, b). These granite plutons have a generally narrow and highly elongate outcrop pattern that in some places hugs terrane boundaries and parallels shear zones, suggesting that their intrusion may have been localized along margins of Sveconorwegian terranes that were converging and/or undergoing transpression (Fig. 1). In the Rogaland–Hardangervidda terrane, metamorphic grade increases southwestwards from epidote-amphibolite facies in the Sæsvatn region to upper amphibolite and granulite facies in the southwestern coastal region. In this region, monazite ages in the 1024–970 Ma range, associated with the Feda suite and other charnockitic gneisses (Bingen & van Breemen 1998a), record amphibolite to granulite-facies regional metamorphism.

About 125 km south of the Sæsvatn area, the Fennefoss feldspar porphyritic augen gneiss just inside the Telemark sector (Fig. 1) is well dated by U–Pb at 1035^{+2}_{-3} to 1031 ± 2 Ma, and is noted to have a geochemical signature transitional between high-K calc-alkaline and A-type (Bingen & van Breemen 1998b; Pedersen & Konnerup-Madsen 2000). About 170 km north of the Sæsvatn area, U–Pb ages for exposures of Proterozoic rocks underlying the Caledonian Jotun nappe include a 1014 ± 35 Ma felsic dyke that cuts a hosting high-grade gneiss complex of 1518 ± 17 Ma (Corfu 1980), the same age as the basement underlying the Sæsvatn–Valldal supracrustal rocks (1.48 ± 0.06 Ga; Berg 1977). This 1.01 Ga intrusive event is distinct from two nearby granites dated at 932 ± 10 Ma (Corfu 1980).

Perhaps the most intriguing and supporting data is found in recently dated molybdenite from quartz + molybdenite + bornite + chalcopyrite + galena + pyrite + hematite + tourmaline + calcite + barite + epidote veins hosted in metabasalt in the Dalen district in the Telemark sector (Hasan 1971). Like the Sæsvatn sequence, supracrustal rocks in the Dalen district (quartz-schist, quartzite, alkali metabasalt, metarhyolite of the Bandak group) are deformed into steeply plunging folds, which were metamorphosed to upper greenschist (epidote-amphibolite) facies. Quartz + calcite + epidote veins and massive segregations occur in abundance in the area, and intrusions are absent (Hasan 1971). The mineralized veins in the Dalen district occupy fractures perpendicular to regional folds. Although the veins are slightly deformed, their formation clearly post-dates the main phase of regional deformation in this region. Re–Os ages of about 1015 Ma for the Askom deposit (Stein & Bingen, unpublished data) indicate that supracrustal rocks in the Telemark sector (Bandak group) were affected, at least in part, by the same metamorphic event(s) recorded in the Sæsvatn area in the Rogaland–Hardangervidda sector, suggesting that they began to share a common metamorphic history no later than about 1.015 Ga. This study shows that small ore occurrences in molybdenum-endowed regions of the Earth's crust are capable of unleashing important age information that bears on the metamorphic history and tectonic assembly of major orogens.

Available data in the Rogaland–Hardangervidda terrane suggest that this terrane was caught in the 'main' Sveconorwegian collision event at or shortly after 1.05 Ga, as indicated by intrusion of calc-alkaline syn-kinematic granites (Feda suite; Bingen & van Breemen 1998a, b) and the oldest metamorphic age on molybdenite in the Sæsvatn supracrustal sequence. In low-grade regions, peak ductile deformation and metamorphism took place at about 1.03 Ga (Sæsvatn) and was followed by brittle deformation at about 1.025–1.015 Ga (Sæsvatn and Dalen). In the high-grade regions (southwestern Rogaland–Hardangervidda), ductile deformation and high-grade metamorphism is significantly younger at about 1.025–0.97 Ga as recorded in monazite geochronology. The succession of dates in the Rogaland–Hardangervidda terrane supports a model where early deformation (1.05–1.03 Ga) linked to 'main' Sveconorwegian convergence was distributed over the entire crust in this terrane, whereas later deformation (1.025–0.97 Ga) was confined to the middle and lower crust, with the upper crust behaving in a brittle fashion.

We suggest that 1.05 to 1.045 Ga metamorphism may record the early stages of terrane imbrication and stacking associated with the 'main' Sveconorwegian collision across a wide area of the Sveconorwegian province. In the eastern part of the province, the comparatively low-grade Idefjorden terrane (greenschist to amphibolite facies) is situated in the hanging wall of the Mylonite zone and is separated from the parauthochthonous high-grade Eastern Segment by this continental-scale shear zone (Park *et al.* 1991; Stephens *et al.* 1996; Romer & Smeds 1996; Möller 1998; Andersson *et al.* 2002). In the Idefjorden terrane, intrusion of 1041 ± 2 to 1030 ± 1 Ma rare-mineral post-kinematic pegmatites are associated with a magmatic and metamorphic event in the underlying rocks and related to stacking and imbrication of terranes in this part of the orogen (Romer & Smeds 1996). In the Idefjorden terrane, the tectonic regime probably became extensional well before 1.00 Ga, whereas in the underlying (footwall) Eastern Segment, high-pressure granulites and eclogite-facies rocks record compressional tectonics as young as 0.96 Ga (Johansson *et al.* 2001). The latest Sveconorwegian history is characterized by prolific magmatic activity that produced a range of undeformed granitoid suites, and an anorthosite suite (with associated thermal metamorphism) over a wide region of the Sveconorwegian orogen from about 1.00 to 0.92 Ga (Eliasson & Schöberg 1991; Kullerud & Machado 1991; Schärer *et al.* 1996). These suites are related to relaxation of the orogen.

This work was supported by a Fulbright research fellowship awarded to HS for studies based at the Geological Survey of Norway in Trondheim (NGU), and the support of director Arne Bjørlykke is gratefully acknowledged. The analytical work was made possible by combining funds from NGU and the U.S. National Science Foundation (NSF, EAR-0087483). The work was also carried out in the framework of the EU-funded GEODE programme. Frank Vokes provided an initial set of samples for this study and we appreciate our many discussions with him, his review comments, and his contribution to the geology at Langvatn–Kobbernuten. We thank Richard Markey, AIRIE-Colorado State University, for the Re–Os analytical work. An early draft of this manuscript benefited from the thoughtful comments of Judith Hannah and Anders Scherstén. Helpful comments from journal reviewers Johann Raith and David Alderton are also appreciated.

References

ÅHÄLL, K.-I., CORNELL, D.H. & ARMSTRONG, R. 1998. Ion probe zircon dating of metasedimentary units across the Skagerrak: new constraints for early Mesoproterozoic growth of the Baltic Shield. *Precambrian Research*, **87**, 117–134.

ÅHÄLL, K.-I. & LARSON, Å. 2000. Growth-related 1.85-1.55 Ga magmatism in the Baltic shield; a review addressing the tectonic characteristics of Svecofen-

nian, TIB 1-related, and Gothian events. *GFF*, **122**, 193–206.

ANDERSSON, J., MÖLLER, C. & JOHANSSON, L. 2002. Zircon chronology of migmatite gneisses along the Mylonite Zone (S Sweden): a major Sveconorwegian terrane boundary in the Baltic Shield. *Precambrian Research*, **114**, 121–147.

BERG, Ø. 1977. *En geokronologisk analyse av prekambrisk basement i distriktet Røldal-Haukeliscæter-Valldalen ved Rb-Sr whole-rock metoden, dets plass i den sørvestnorske Prekambriske Provinsen, og en vurdering av denne provinsen i lys av det generelle geokronologiske mønster i det Nord-Atlantiske Prekambriske området.*. Master's thesis, University of Oslo, Norway.

BERTHELSEN, A. 1980. Towards a palinspastic tectonic analysis of the Baltic shield. *International Geological Congress. Colloquim*, **C6**, 5–21.

BINGEN, B. & VAN BREEMEN, O. 1998a. U-Pb monazite ages in amphibolite- to granulite-facies orthogneisses reflect hydrous mineral breakdown reactions: Sveconorwegian province of SW Norway. *Contributions to Mineralogy and Petrology*, **132**, 336–353.

BINGEN, B. & VAN BREEMEN, O. 1998b. Tectonic regimes and terrane boundaries in the high-grade Sveconorwegian belt of SW Norway, inferred from U-Pb zircon geochronology and geochemical signature of augen gneiss suites. *Journal of the Geological Society, London*, **155**, 143–154.

BINGEN, B. & STEIN, H. 2001. Re–Os dating of the Ørsdalen W-Mo district in Rogaland, S Norway, and its relationship to Sveconorwegian high-grade metamorphism. *In: Ilmenite Deposits in the Rogaland Anorthosite Province, South Norway.* NGU Reports, **2001.042**, 15–18.

BINGEN, B., BIRKELAND, A., NORDGULEN, Ø. & SIGMOND, E.M. 2001. Correlation of supracrustal sequences and origin of terranes in the Sveconorwegian orogen of SW Scandinavia: SIMS data on zircon in clastic metasediments. *Precambrian Research*, **108**, 293–318.

BINGEN, B., SIGMOND, E.M.O., MANSFELD, J. & STEIN, H. 2002. Pre- to early-Grenvillian development of supracrustal basins in the Sveconorwegian orogen: geochronology of the Sæsvatn-Valldal sequences, S Norway. *Canadian Journal of Earth Sciences*, **39**, in press.

BREWER, T.S. & ATKIN, B.P. 1987. Geochemical and tectonic evolution of the Proterozoic Telemark supracrustals, southern Norway. *In:* PHARAOH, T.C., BECKINSALE, R.D. & RICKARD, D. (eds) *Geochemistry and Mineralization of Proterozoic Volcanic Suites.* Geological Society, London, Special Publications, **33**, 471–487.

BREWER, T.S. & MENUGE, J.F. 1998. Metamorphic overprinting of Sm-Nd isotopic systems in volcanic rocks: the Telemark Supergroup, southern Norway. *Chemical Geology*, **145**, 1–16.

CAMERON, E.M. 1988. Archean gold: relation to granulite facies formation and redox zoning in the crust. *Geology*, **16**, 109–112.

CANDELA, P.A. & BOUTON, S.L. 1990. The influence of oxygen fugacity on tungsten and molybdenum partitioning between silicate melts and ilmenite. *Economic Geology*, **85**, 633–640.

CHRISTOFFEL, C.A., CONNELLY, J.N. & ÅHÄLL, K.-I. 1999. Timing and characterization of recurrent pre-Sveconorwegian metamorphism and deformation in the Varberg-Halmstad region of SW Sweden. *Precambrian Research*, **98**, 173–195.

CONDIE, K.C., VILJOEN, M.J. & KABLE, E.D. 1977. Effects of alteration on element distributions in Archean tholeiites from the Barberton greenstone belt, South Africa. *Contributions to Mineralogy and Petrology*, **64**, 75–89.

CORFU, F. 1980. U-Pb and Rb-Sr systematics in a polyorogenic segment of the Precambrian shield, central southern Norway. *Lithos*, **13**, 305–323.

COSCA, M.A., MEZGER, K. & ESSENE, E.J. 1998. The Baltica-Laurentia connection: Sveconorwegian (Grenvillian) metamorphism, cooling, and unroofing in the Bamble sector, Norway. *Journal of Geology*, **106**, 539–552.

DAHLGREN, S. & HEAMAN, L. 1991. U-Pb constraints for the timing of Middle Proterozoic magmatism in the Telemark region, southern Norway. *Geological Association of Canada - Mineralogical Association of Canada Program with Abstracts*, **16**, A28.

DE HAAS, G.J.L.M., ANDERSEN, T. & VESTIN, J. 1999. Detrital zircon geochronology: new evidence for an old model for accretion of the southwest Baltic shield. *Journal of Geology*, **107**, 569–586.

DONS, J.A. 1960. Telemark supracrustals and associated rocks. *In:* HOLTEDAHL, O./eds;*Geology of Norway.* Norges Geologiske Undersøkelse Bulletin, **208**, 49–58.

ELIASSON, T. & SCHÖBERG, H. 1991. U-Pb dating of the post-kinematic Sveconorwegian (Grenvillian) Bohus granite, SW Sweden: evidence of restitic zircon. *Precambrian Research*, **51**, 337–350.

GORBATSCHEV, R. & BOGDANOVA, S. 1993. Frontiers in the Baltic shield. *Precambrian Research*, **64**, 3–21.

GROVES, D.I., GOLDFARB, R.J., GEBRE-MARTIN, M., HAGEMANN, S. & ROBERT, F. 1998. Orogenic gold deposits: a proposed classification in the context of their crustal distribution and relationship to other gold deposit types. *Ore Geology Reviews*, **13**, 7–27.

HASAN, Z.-U. 1971. Supracrustal rocks and Mo-Cu bearing veins in Dalen. *Norsk Geologisk Tidsskrift*, **51**, 287–309.

HUMPHRIS, S.E. & THOMPSON, G. 1978. Trace element mobility during hydrothermal alteration of oceanic basalts. *Geochimica et Cosmochimica Acta*, **42**, 127–136.

JOHANSSON, L., MÖLLER, C. & SÖDERLUND, U. 2001. Geochronology of eclogite facies metamorphism in the Sveconorwegian province of SW Sweden. *Precambrian Research*, **106**, 261–275.

JOLLY, W.T. 1974. Behavior of Cu, Zn, and Ni during prehnite-pumpellyite metamorphism of Keweenawan basalts, northern Michigan. *Economic Geology*, **69**, 1118–1125.

KISH, S.A. & STEIN, H.J. 1989. Post-Acadian metasomatic origin for copper-bearing vein deposits of the Virgilina district, North Carolina and Virginia. *Economic Geology*, **84**, 1903–1920.

KULLERUD, L. & DAHLGREN, S.H. 1993. Sm-Nd geo-

chronology of Sveconorwegian granulite facies mineral assemblages in the Bamble shear belt, south Norway. *Precambrian Research*, **64**, 389–402.

KULLERUD, L. & MACHADO, N. 1991. End of a controversy: U-Pb geochronological evidence for significant Grenvillian activity in the Bamble area, Norway. *Terra Abstracts, supplement to Terra Nova*, **3**, 504.

LAROCQUE, A.C.L. & HODGSON, C.J. 1995*a*. Effects of greenschist-facies metamorphism and related deformation on the Mobrun massive sulfide deposit, Quebec, Canada. *Mineralium Deposita*, **30**, 439–448.

LAROCQUE, A.C.L. & HODGSON, C.J. 1995*b*. Ion-microprobe analysis of pyrite, chalcopyrite and pyrrhotite from the Mobrun VMS deposit in northwestern Quebec: evidence for metamorphic remobilization of gold. *The Canadian Mineralogist*, **33**, 373–388.

LINCOLN, T.N. 1981. The redistribution of copper during low-grade metamorphism of the Karmutsen volcanics, Vancouver island, British Columbia. *Economic Geology*, **76**, 2147–2161.

LINDEN, M.A. 1981. *Mineralogical and chemical characteristics of gold occurrences in the Virgilina district, Halifax County, Virginia*. Masters thesis, Virginia Polytechnic Institute State University, Blacksburg, Virginia.

LODDERS, K. & PALME, H. 1991. On the chalcophile character of molybdenum: determination of sulfide/silicate partition coefficients of Mo and W. *Earth and Planetary Science Letters*, **103**, 311–324.

MARKEY, R.J., STEIN, H.J. & MORGAN, J.W. 1998. Highly precise Re–Os dating of molybdenite using alkaline fusion and NTIMS. *Talanta*, **45**, 935–946.

MARSHALL, B., VOKES, F.M. & LAROCQUE, A.C.L. 2000. Regional metamorphic remobilization: upgrading and formation of ore deposits. *In:* SPRY, P.G., MARSHALL, B. & VOKES, F.M. (eds) *Metamorphosed and Metamorphogenic Ore Deposits.* SEG Reviews in Economic Geology, **11**, 19–38.

MENUGE, J.F. & BREWER, T.S. 1996. Mesoproterozoic anorogenic magmatism in southern Norway. *In:* BREWER, T.S. (ed.) *Precambrian Crustal Evolution in the North Atlantic Region.* Geological Society, London, Special Publications, **112**, 275–295.

MÖLLER, C. 1998. Decompressed eclogites in the Sveconorwegian (-Grenvillian) orogen of SW Sweden: petrology and tectonic implications. *Journal of Metamorphic Geology*, **16**, 641–656.

PARK, R.G., AHÄLL, K.-I. & BOLAND, M.P. 1991. The Sveconorwegian shear-zone network of SW Sweden in relation to mid-Proterozoic plate movements. *Precambrian Research*, **49**, 245–260.

PEDERSEN, S. & KONNERUP-MADSEN, J. 2000. Geology of the Setesdalen area, south Norway: implications for the Sveconorwegian evolution of south Norway. *Bulletin of the Geological Society of Denmark*, **46**, 181–201.

PRESTVIK, T. & VOKES, F.M. 1982. Amphibolites and metagabbros from the Proterozoic Telemark suite of Setesdalsheiene, south-central Norway. *Norges Geologiske Undersøkelse Bulletin*, **378**, 49–63.

PRIEM, H.N., BOELRIJK, N.A., HEBEDA, E.H., VERDURMEN, E.A. & VERSCHURE, R.H. 1973. Rb-Sr investigations on Precambrian granites, granitic gneisses and acid metavolcanics in Central Telemark: metamorphic resetting of Rb-Sr whole-rock systems. *Norges Geologiske Undersøkelse Bulletin*, **289**, 37–53.

RAGNHILDSTVEIT, J., NATERSTAD, J., JORDE, K. & EGELAND, B. 1998. *Geologisk kart over Norge, berggrunnskart Haugesund, 1:250000*. Norges Geologiske Undersøkelse.

RAITH, J.G. & STEIN, H.J. 2000. Re–Os dating and sulfur isotope composition of molybdenite from tungsten deposits in western Namaqualand, South Africa: implications for ore genesis and the timing of metamorphism. *Mineralium Deposita*, **35**, 741–753.

ROMER, R.L. & SMEDS, S.-A. 1996. U-Pb columbite ages of pegmatites from Sveconorwegian terranes in southwestern Sweden. *Precambrian Research*, **76**, 15–30.

SCHÄRER, U., WILMART, E. & DUCHESNE, J.-C. 1996. The short duration and anorogenic character of anorthosite magmatism: U-Pb dating of the Rogaland complex, Norway. *Earth and Planetary Science Letters*, **139**, 335–350.

SIGMOND, E.M.O. 1975. *Geologisk kart over Norge, berggrunnskart Sauda, 1:250000.*. Norges Geologiske Undersøkelse.

SIGMOND, E.M.O. 1978. Beskrivelse til det berggrunnsgeologiske kartbladet Sauda 1:250000. *Norges Geologiske Undersøkelse Bulletin*, **341**, 1–94.

SIGMOND, E.M.O. 1998. *Geologisk kart over Norges, berggrunnskart Odda, 1:250000.*. Norges Geologiske Undersøkelse.

SIGMOND, E.M.O., BIRKELAND, A. & BINGEN, B. 2000. A possible basement to the Mesoproterozoic quartzites on Hardangervidda, south-central Norway: zircon U-Pb geochronology of a migmatitic gneiss. *Norges Geologiske Undersøkelse Bulletin*, **437**, 25–32.

SMOLIAR, M.I., WALKER, R.J. & MORGAN, J.W. 1996. Re–Os ages of group IIA, IIIA, IVA and IVB iron meteorites. *Science*, **271**, 1099–1102.

SÖDERLUND, U., JARL, L.-G., PERSSON, P.-O., STEPHENS, M.B. & WALGREN, C.-H. 1999. Protolith ages and timing of deformation in the eastern, marginal part of the Sveconorwegian orogen, southwestern Sweden. *Precambrian Research*, **94**, 29–48.

STEIN, H.J., MARKEY, R.J., MORGAN, J.W., DU, A, &

SUN, Y. 1997. Highly precise and accurate Re–Os ages for molybdenite from the East Qinling molybdenum belt, Shaanxi Province, China. *Economic Geology*, **92**, 827–835.

STEIN, H.J., SUNDBLAD, K., MARKEY, R.J., MORGAN, J.W. & MOTUZA, G. 1998. Re–Os ages for Archaean molybdenite and pyrite, Kuittila-Kivisuo, Finland and Proterozoic molybdenite, Kabeliai, Lithuania: testing the chronometer in a metamorphic and metasomatic setting. *Mineralium Deposita*, **33**, 329–345.

STEIN, H.J., MORGAN, J.W. & SCHERSTÉN, A. 2000. Re–Os dating of low-level highly-radiogenic (LLHR) sulfides: the Harnäs gold deposit, southwest Sweden records continental scale tectonic events. *Economic Geology*, **95**, 1657–1671.

STEIN, H.J., MARKEY, R.J., MORGAN, J.W., HANNAH, J.L. & SCHERSTÉN, A. 2001. The remarkable Re–Os chronometer in molybdenite: how and why it works. *Terra Nova*, **13**, 479–486.

STEPHENS, M.B., WAHLGREN, C.-H., WEIJERMARS, R. & CRUDEN, A.R. 1996. Left lateral transpressive deformation and its tectonic implications, Sveconorwegian orogen, Baltic shield, southwestern Sweden. *Precambrian Research*, **79**, 261–279.

SVENSEN, H., JAMTVEIT, B., BANKS, D.A. & KARLSEN, D. 2001. Fluids and halogens at the diagenetic-metamorphic boundary: evidence from veins in continental basins, western Norway. *Geofluids*, **1**, 53–70.

TACKER, R.C. & CANDELA, P.A. 1987. Partitioning of molybdenum between magnetite and melt: a preliminary experimental study of partitioning of ore metals between silicic magmas and crystalline phases. *Economic Geology*, **82**, 1827–1838.

VOKES, F.M., SINDRE, A., EIDSVIG, P. & BØLVIKEN, B. 1975. Geological, geophysical and geochemical investigations of the molybdenum-copper deposit at Langvatn, Setesdalsheiene, south-central Norway. *In:* JONES, M.I./eds;*Prospecting in Areas of Glaciated Terrain*. Institute of Mining and Metallurgy, London, 32–40.

WEN, N., IXER, R.A., KINNAIRD, J.A., ASHWORTH, J.R. & NEX, P.A. 1999. Ore petrology and metamorphic mobilization of copper ores in red beds, southwest County Cork, Ireland. *Transactions Institute of Mining and Metallurgy (Section B, Applied Earth Sciences)*, **108**, B53–B63.

Fluorine in orthoamphibole dominated Zn–Cu–Pb deposits: examples from Finland and Australia

LEENA RAJAVUORI[1] & LEO M. KRIEGSMAN[1,2]

[1]Department of Geology, University of Turku, 20014 Turku, Finland
[2]National Museum of Natural History, PO Box 9517, 2300 RA Leiden, the Netherlands
(e-mail: kriegsman@naturalis.nnm.nl)

Abstract: Volcanogenic massive sulphide (VMS) deposits commonly occur within much larger fluorine (F)-bearing hydrothermal systems, where cordierite–orthoamphibole rocks are characteristic pathfinder assemblages. Here we report whole-rock and mineral F contents for orthoamphibole bearing rocks and associated rock types from Zn–Cu–Pb deposits in Finland (Orijärvi, Iilijärvi, Pyhäsalmi, Mullikkoräme) and central Australia (Oonagalabi). Textural and mineralogical data suggest that F influx predates peak metamorphism in these deposits.

The Mullikkoräme whole-rock data show positive correlations between F and X_{Mg} = molar $Mg/(Mg + Fe)$ and between F and elements of relatively low mobility (Ti, Al, Mg) and negative correlations between Fe and these elements. This suggests that iron was leached from silicate rocks by F-bearing fluids and was transferred to the Fe-rich sulphide ore. When normalized to immobile elements, F correlates positively with total metal content (Cu + Pb + Zn + Fe), consistent with the commonly observed increase of F content towards ore bodies.

Combining all microprobe data, hydrous minerals show the following order of decreasing F/(F + OH) ratio when coexisting: apatite > chondrite > biotite > gedrite > (hornblende, muscovite, anthophyllite) > chlorite. The low- to medium-grade Finnish samples (Mullikkoräme: 500–560 °C at 1–3 kbar; Orijärvi: 550–650 °C at 3–5 kbar; Pyhäsalmi: 600–700 °C at 5–7 kbar) contain mainly F-poor anthophyllite, whereas F-rich gedrite is dominant in the higher grade Oonagalabi deposit (750 800 °C, 8–9 kbar). Temperature seems to have a significant, but X_{Mg} a negligible influence on F partitioning between biotite and orthoamphiboles.

Fluid–rock interaction has become one of the main issues addressed by metamorphic geologists over the past few years. In order to evaluate the potential of crustal fluids in transporting and concentrating elements in different types of ore deposits, it is critical to constrain their composition and source. Mineral chemical equilibria (e.g., papers in Humphris *et al.* 1995; Wood & Samson 1998), fluid inclusion and stable isotope studies (e.g, Heaton & Sheppard 1977; Barnes 1979) have been widely applied for this purpose. An alternative approach uses the capability of high-grade minerals to act as a sink for fluids in crustal rocks and hence to act as a monitor of fluid–rock interaction.

Magmatic volatiles, such as H_2O, F, Cl and Br, play a crucial role in magmatic–hydrothermal ore formation, and F (as well as Cl) can transport metals as aqueous metal halide complexes. Many volcanogenic massive sulphide (VMS) deposits occur in F-bearing hydrothermal systems, which are usually more voluminous than these ore deposits and therefore more easily recognizable (Lavery 1985). Hence, a geochemical exploration strategy is to use F as a pathfinder element to trace hydrothermal systems and then to locate possible ore deposits within the system. The average F content in volcanic rocks is usually below 500 ppm; in F-enriched rocks it can easily rise to several thousand ppm. Anomalous F values often define a distinctive zone stratigraphically above and below the sulphide ores, and F contents are usually highest near ore bodies (Lavery 1985).

Available studies on the behaviour and distribution of F in natural rocks and fluids suggest that (i) F is strongly partitioned into the solid phases during fluid–rock interaction (Anfilogov *et al.* 1977; Willner *et al.* 1990; Zhu & Sverjensky 1991; Willner 1993); (ii) F stabilizes relict biotite and amphiboles during high-grade metamorphism and partial melting (Foley 1991; Peterson *et al.* 1991; Hensen & Osanai 1994; Tareen *et al.* 1995; Dooley & Patiño-Douce 1996); (iii) high F contents are common in A-type granites (Collins *et al.* 1982; Kovalenko & Kovalenko 1984; Rogers & Satterfield 1994; Hogan & Gilbert 1997); and

(iv) high F contents are common in rocks associated with massive sulphide deposits (Lavery 1985; Jiang *et al.* 1994) and Au deposits (Matthäi *et al.* 1995). Hence, it seems that pre-metamorphic F may persist in high-grade metamorphic rocks. If metamorphism is unrelated to F influx, this could allow one to 'see through' the various metamorphic events and investigate the role of F in the mineralizing fluids during earlier hydrothermal processes.

Orthoamphibole–cordierite rocks are rare, but characteristic rock types that have been studied intensely ever since the first detailed petrographical descriptions of the Orijärvi outcrops in Finland (Eskola 1914, 1915). Many of these rocks are associated with sulphide ore deposits (Eskola 1914; Froese 1969; Zaleski & Peterson 1995; Treloar *et al.* 1981; Wolter & Seifert 1984; Moore & Waters 1990; Dobbe 1994; Pan & Fleet 1995), which partly explains the interest in these possible pathfinder assemblages. Alteration suites around ore deposits often exhibit vertical and lateral zonation, which is expressed as sericification, silicification, chloritization and a cordierite–orthoamphibole association (Vallance 1967; Sánchez-España *et al.* 2000). The stabilizing effect of F in orthoamphiboles is, however, not yet well documented. In addition, orthoamphibole–cordierite rocks have a high potential for the study of phase relations and the derivation of the medium- to high-grade segments of P–T paths (Robinson *et al.* 1981; Schumacher & Robinson 1987; Visser & Senior 1990; Schneiderman & Tracy 1991). Orthoamphibole–cordierite-bearing rocks show a different mineralogy at higher P–T conditions. In transitional granulites, orthoamphibole may coexist with orthopyroxene and cordierite (West Uusimaa, Finland: Schreurs & Westra 1985); with orthopyroxene and garnet (altered felsic rocks in Oonagalabi, central Australia: Raith & Kriegsman 1998) or with spinel, corundum, chondrodite and sapphirine (same area, altered mafic to ultramafic rocks). The high-pressure, high-temperature Oonagalabi rocks occur in an orthoamphibole-dominated unit, which hosts a small, presently sub-economic, Cu–Zn–Pb deposit (Stewart & Warren 1977), underscoring the similarity with low-pressure, medium-temperature orthoamphibole–cordierite rocks.

The present study uses whole-rock and mineral chemistry to assess the role of F in (i) hydrothermal activity leading to strongly altered felsic and mafic sequences and associated sulphide ore deposits and (ii) medium- to high-temperature metamorphism producing orthoamphibole bearing assemblages. The emphasis is on samples from Finland (Orijärvi, Iilijärvi, Pyhäsalmi, Mullikkoräme), but samples from Oonagalabi are used to provide additional constraints on F partitioning data between hydrous phases.

Methods employed

Whole-rock analyses were made on 20 selected samples from Orijärvi, which were crushed and powdered at the Department of Geology, University of Turku. Powders were analysed at Genalysis Laboratory Services Ltd., Perth, Australia. ICP-AES techniques were used for Na, Ca, Mn, P, Zn, Cr, Ni, Cu and V (acid digest) and for K, Fe, Mg, Al, Ti, S and B (peroxide fusion). ICP-MS techniques were used for REE, Ba, Rb, Sr, Th, Y, U, Pb, As, Te, Be and W (acid digest) and for Nb, Ta, Zr and Hf (peroxide fusion). F and LOI (loss on ignition) were also measured except for sample LMR 49. For F extraction, the sample was fused with a flux of sodium carbonate, potassium carbonate, zinc oxide and silicic acid. The melt was then leached with water to dissolve the fluoride. International standards were fused with each set of samples and a calibration curve plotted each time the measuring electrode was used.

Whole-rock F contents varied from 150 to 3550 ppm (detection limit: 50 ppm; average: 979 ppm). SiO_2 was not analysed for Orijärvi samples, so their SiO_2 content was calculated by subtracting other element concentrations from 100% (Table 1). Existing geochemical databases (>1000 whole-rock analyses, 52 of which include F) on the Iilijärvi, Pyhäsalmi and Mullikkoräme deposits in central Finland, kindly provided by Outokumpu Mining Ltd, were used as additional sources of information and were critical for improving the statistical relevance of observed correlations. Some representative analyses are also given in Table 1. The total weight of some Mullikkoräme samples exceeds 100% because a considerable amount of Fe is in sulphide rather than oxide form (see geochemistry section). The total weight has also been corrected for oxygen in baryte and CO_2 (Table 1).

After petrological analysis (recognition of minerals in thin sections, assessment of their growth relationships and relative timing; qualitative evaluation of P–T conditions and possible P–T paths), minerals from polished thin sections of selected samples were analysed for major elements and F, using the JEOL JXA 8600 electron microprobe at the Department of Earth Sciences, University of Bristol. The detection limit for F is routinely put at 1000 ppm (= 0.1 wt%) but, with special care, analyses down to 50–100 ppm give reproducible results. Representative results are given in Tables 2 and 3.

Table 1. Representative whole-rock analyses, including fluorine, from samples of the Orijärvi and Mullikkoräme deposits

Area	ORI	ORI	ORI	ORI	ORI	ORI	MUL	MUL	MUL	MUL	MUL	MUL	MUL	MUL	MUL	MUL
Sample	LMR 7	LMR 34c	LMR 42	LMR 45	LMR 40a	LMR 54	MUR-273	MUR-144	MUR-273	MUR-144	MUR-293	MUR-314	MUR-293	MUR-144	MUR-283	MUR-326
Rock type	Grt–Bt gneiss	Crd–Ath rock	Crd–Bt gneiss	Mafic volcanite	Granodiorite	Felsic vulcanite	Mafic vulcanite	Felsic vulcanite	Skarn	Tlc skarn	Tlc–Chl skarn	Phl–Chl gneiss	Plg–Uralite porphyrite	Sphalerite–galena ore	Pyrite ore	Pyrite ore
SiO_2	56.54	64.90	60.61	52.69	72.58	73.35	37.99	61.50	32.60	30.00	43.10	53.50	40.80	17.80	7.95	18.30
TiO_2	1.5	0.33	0.72	0.7	0.3	0.2	0.42	0.33	0.20	0.06	0.06	0.28	0.02	0.02	0.02	0.01
Al_2O_3	14.74	12.09	16.25	15.87	13.6	12.09	8.73	11.60	9.14	2.60	2.86	11.00	1.24	1.15	0.91	0.86
FeO	11.58	7.72	6.43	9.78	2.44	3.15	21.90	6.41	16.56	21.60	12.96	13.86	15.57	14.31	15.48	29.69
MnO	0.1	0.14	0.05	0.17	0.05	0.04	0.65	0.07	0.14	0.14	0.204	0.10	0.061	0.252	0.209	0.171
MgO	5.31	11.61	5.47	8.95	0.9	0.93	17.20	4.35	11.60	22.70	23.5	7.46	22.6	21.4	11	13.4
CaO	2.17	0.17	2.1	6.58	2.8	1.75	1.55	3.24	7.62	1.28	4.65	1.54	1.39	4.54	18.9	4.23
Na_2O	2.7	0.16	2.97	3.57	4.58	3.98	0.10	0.73	0.85	–	–	1.41	–	–	–	0.24
K_2O	3.49	1.04	3.98	0.51	1.87	2.59	2.65	0.89	1.35	–	0.017	1.72	0.104	0.007	0.018	0.004
P_2O_5	0.46	0.05	0.12	0.09	0.07	0.03	0.09	0.10	0.18	0.107	0.123	0.09	0.079	0.044	0.34	0.165
LOI	0.88	1.46	0.85	0.89	0.58	1	–	–	–	–	–	–	–	–	–	–
F	1000	2450	3550	300	650	150	n.a.	n.a.	n.a.	n.a.	n.a.	n.a.	n.a.	n.a.	n.a.	n.a.
Cl	–	–	–	–	–	–	4200	1000	2800	3400	4700	1300	4400	3500	900	1400
C	–	–	–	–	–	–	180	–	90	200	–	100	–	160	–	–
S	2700	100	0	500	500	7200	100	200	200	200	700	100	100	1700	55100	3900
Zn	125	265	62	125	27	68	58600	45300	124000	177000	85600	81900	117000	167000	136000	227000
Pb	12	42	8	32	8	14	22919	26304	23465	71354	49307	12053	75158	223330	98474	96970
Cu	15	19	2	50	4	37	6986	11268	4631	1292	3300	533	9872	22677	3733	7525
Ni	0	2	4	64	3	3	–	12289	2390	1348	2132	641	1322	215	5743	34186
Cr	70	130	80	280	160	108	–	–	–	–	–	–	–	–	–	–
V	155	24	114	220	14	0	101	–	–	–	–	21	–	23	–	49
Ba	920	64	330	94	490	900	5158	13414	5804	–	210	4492	120	–	6956	21
Rb	64	21.5	92	9.6	42	36	56	31	42	20	–	27	16	177	–	–
Sr	106	7.4	58	235	140	74	22	369	231	–	17	78	14	25	117	–
Y	13.5	8.6	9	12.5	27	21.5	15	29	19	–	–	20	–	59	–	–
Zr	72	112	94	42	145	165	100	203	106	31	29	147	12	14	15	–
Nb	7.5	11.5	7.5	3.5	12	13	–	19	10	–	–	–	–	–	–	–
Ce	38	40	54	17	64	56	76	200	78	–	–	92	37	–	–	13
Th	4	6	6.6	2.2	8.2	9	–	–	–	–	–	–	–	81	20	–
U	2.1	2.2	2.1	1.0	3.1	4.1	–	49	11	14	14	–	–	13	–	13
Total*	100	100	100	100	100	100	101.09	100.28	96.63	103.94	102.07	101.10	102.66	101.41	85.53	104.18

*Orijärvi samples: SiO_2 calculated as 100 − total of other elements

wt% Fe in sulphides (calculated)							4.02	2.40	9.62	12.35	5.30	6.52	6.84	4.70	7.45	15.53
Real FeO (wt%)							17.43	3.32	4.19	5.71	6.14	5.47	6.76	8.27	5.90	9.72
Oxygen in baryte and CO_2							0.27	0.68	0.32	0.05	0.20	0.24	0.03	0.46	15.02	1.04
Total corrected for Fe in sulphides and oxygen in baryte and CO_2							100.91	100.27	94.20	100.45	100.75	99.47	100.74	100.53	98.42	100.77

Table 2. *Representative microprobe analyses and recalculated mineral formulae of amphiboles*

Area	ORI	IIL	IIL	IIL	OONA	ORI	IIL	ORI	OONA	OONA	OONA	OONA	OONA	OONA	OONA	OONA	IIL
Sample	62b	214	266	219	AR227b	62b	214	62b	AR126a	AR126a	AR126b	AR126c	AR126d	AR129	AR131b	AR245a	266
Mineral	Ath	Ath	Ath	Ath	Ath	Oam	Oam	Ged	Ged	Ged	Ged	Ged	Ged	Ged	Ged	Ged	Cum
ID	65	178	435	527	767	51	183	55	716	216	38	94	444	18	178	198	439
SiO_2	51.65	50.71	51.22	53.36	55.03	50.09	50.70	42.90	45.15	45.15	43.25	43.62	45.00	45.01	45.25	49.08	51.46
Al_2O_3	2.23	3.00	2.48	1.76	1.53	4.04	4.18	14.01	16.92	16.92	18.64	19.07	17.25	13.45	14.82	9.61	1.48
TiO_2	0.02	0.05	0.09	0.02	0.03	0.03	0.14	0.18	0.12	0.12	0.32	0.16	0.15	0.29	0.08	0.02	0.00
FeO	26.97	24.63	23.69	21.61	18.72	26.25	25.29	26.38	7.69	7.69	8.51	10.87	7.66	17.80	13.79	14.81	23.47
MnO	0.87	1.18	1.01	0.87	0.14	0.92	1.15	0.97	0.26	0.26	0.71	0.42	0.30	0.16	0.23	0.25	0.98
MgO	14.88	15.47	15.74	18.37	21.55	14.43	14.70	10.67	24.74	24.74	23.00	21.68	24.85	17.77	21.41	22.05	15.39
CaO	0.15	0.16	0.23	0.23	0.25	0.19	0.25	0.44	0.45	0.45	0.73	0.29	0.43	0.06	0.11	0.08	1.07
Na_2O	0.15	0.17	0.23	0.00	0.04	0.43	0.40	1.33	2.18	2.18	2.46	2.40	2.31	1.59	1.94	1.56	0.06
K_2O	0.01	0.03	0.01	0.00	0.02	0.00	0.00	0.00	0.00	0.00	0.02	0.02	0.01	0.00	0.01	0.01	0.02
ZnO	–	–	0.27	0.61	–	–	–	–	–	–	–	–	–	–	–	–	0.13
Cl	0.001	0.000	0.000	0.002	0.014	0.006	0.000	0.000	0.014	0.010	0.011	0.017	0.000	0.003	0.020	0.017	0.005
F	0.077	0.036	0.025	0.088	0.000	0.049	0.044	0.107	0.625	0.620	0.218	0.272	0.765	0.217	0.579	0.653	0.034
Total wt.	97.00	95.43	95.01	96.91	97.33	96.45	96.85	96.99	98.15	98.14	97.88	98.81	98.73	96.35	98.23	98.14	94.10
$-O\equiv F,Cl$	-0.03	-0.02	-0.01	-0.04	0.00	-0.02	-0.02	-0.05	-0.26	-0.26	-0.09	-0.11	-0.32	-0.09	-0.24	-0.28	-0.01
H_2O+ (calc.)	1.96	1.96	1.97	2.00	2.10	1.96	1.98	1.92	1.87	1.87	2.05	2.03	1.82	1.95	1.84	1.80	1.94
$-FeO\equiv Fe_2O_3$	0.04	0.04	0.00	0.00	0.00	0.02	0.00	0.03	0.24	0.24	0.23	0.11	0.25	0.01	0.24	0.06	0.00
Corr. wt	98.97	97.42	96.96	98.88	99.42	98.41	98.81	98.90	100.00	100.00	100.06	100.84	100.47	98.22	100.06	99.73	96.03
Cations / 23 oxygens (total cations = 15+Na+K)																	
Si	7.753	7.675	7.829	7.829	7.859	7.564	7.587	6.513	6.203	6.202	5.993	6.043	6.159	6.577	6.372	6.947	7.878
Al^{IV}	0.247	0.325	0.171	0.171	0.141	0.436	0.413	1.487	1.797	1.798	2.007	1.957	1.841	1.423	1.628	1.053	0.122
Al^{VI}	0.147	0.210	0.205	0.134	0.117	0.284	0.325	1.020	0.942	0.942	1.037	1.156	0.941	0.893	0.831	0.550	0.145
Fe^{3+}	0.050	0.048	0.000	0.000	0.002	0.019	0.000	0.036	0.249	0.250	0.237	0.118	0.255	0.015	0.249	0.066	0.000
Ti	0.003	0.005	0.010	0.002	0.004	0.004	0.016	0.020	0.012	0.012	0.034	0.017	0.015	0.031	0.008	0.002	0.000
Mg	3.330	3.490	3.554	4.019	4.589	3.248	3.278	2.415	5.065	5.066	4.751	4.476	5.070	3.871	4.494	4.653	3.511
Fe^{2+}	3.336	3.069	3.003	2.652	2.234	3.297	3.165	3.313	0.634	0.633	0.749	1.141	0.621	2.160	1.375	1.687	3.004
Mn	0.110	0.151	0.130	0.108	0.016	0.117	0.145	0.125	0.030	0.030	0.084	0.049	0.034	0.019	0.027	0.030	0.127
Ca	0.024	0.026	0.038	0.036	0.039	0.031	0.040	0.072	0.067	0.066	0.109	0.043	0.063	0.010	0.016	0.012	0.176
Na	0.042	0.051	0.067	0.000	0.012	0.125	0.117	0.391	0.580	0.581	0.662	0.645	0.613	0.451	0.530	0.430	0.018
K	0.002	0.005	0.002	0.000	0.003	0.001	0.000	0.000	0.000	0.000	0.004	0.004	0.002	0.001	0.002	0.003	0.004
Zn	0.000	0.000	0.031	0.066	0.003	0.001	0.000	0.000	0.003	0.000	0.000	0.000	0.000	0.001	0.005	0.004	0.015
Cl	0.000	0.000	0.000	0.000	0.000	0.001	0.000	0.000	0.003	0.002	0.002	0.004	0.000	0.001	0.005	0.004	0.001
F	0.036	0.017	0.012	0.041	0.000	0.024	0.021	0.051	0.271	0.269	0.095	0.119	0.331	0.100	0.258	0.292	0.016
X_F	0.018	0.009	0.006	0.020	0.000	0.012	0.010	0.026	0.136	0.135	0.048	0.060	0.167	0.050	0.130	0.146	0.008
X_{Mg}	0.496	0.528	0.542	0.602	0.672	0.495	0.509	0.419	0.852	0.851	0.828	0.780	0.853	0.640	0.735	0.726	0.539

Table 3. *Representative microprobe analyses and recalculated mineral formulae of sheet silicates*

Place	IIL	IIL	IIL	ORI	OONA	OONA	OONA	OONA	OONA	OONA	OONA	OONA	ORI	MUR	ORI	ORI
Sample	214	219	266	62b	AR126a	AR126b	AR126c	AR126d	AR129	AR131b	AR245a	AR227b	55	142	62b	62b
Mineral	Bt	Bt	Bt	Bt	Bt	Bt	Bt	Bt	Bt	Bt	Bt	Bt	Musc	Musc	Chl incl.	Chl
ID	179	509	445	73	240	46	101	371	22	170	204	287	7	334	87	59
SiO_2	36.79	38.94	35.88	37.14	42.50	40.42	39.67	42.22	38.78	40.73	39.97	37.64	45.32	44.81	24.89	24.45
Al_2O_3	17.02	16.12	17.53	18.17	13.67	15.86	16.84	13.82	15.40	14.39	14.49	15.97	28.20	35.58	21.93	21.66
TiO_2	1.51	0.98	1.61	1.20	0.52	1.13	1.15	0.41	2.13	0.72	0.64	3.45	0.68	0.08	0.07	0.08
FeO	17.73	12.34	15.58	17.30	2.32	3.77	5.33	2.29	7.97	5.53	5.10	10.45	4.66	0.00	22.47	23.04
MnO	0.10	0.07	0.16	0.02	0.02	0.04	0.04	0.02	0.02	0.02	0.04	0.01	0.15	0.00	0.19	0.24
MgO	12.58	16.71	12.58	13.07	26.19	24.49	22.65	26.53	20.13	23.80	24.02	17.42	4.67	0.37	15.97	15.95
CaO	0.00	0.00	0.10	0.01	0.02	0.00	0.00	0.01	0.01	0.00	0.00	0.00	0.07	0.03	0.02	0.01
Na_2O	0.35	0.33	0.29	0.25	1.04	0.71	0.73	1.12	0.75	0.53	1.11	0.23	0.11	0.74	0.03	0.00
K_2O	8.79	9.23	9.51	8.49	8.63	8.88	9.19	8.61	8.49	9.33	8.44	8.84	10.66	11.10	0.00	0.02
ZnO	—	0.16	0.18	0.00	—	—	—	—	—	—	—	—	—	—	0.00	0.00
Cl	0.00	0.01	0.00	0.01	0.02	0.02	0.04	0.03	0.03	0.12	0.12	0.02	0.004	0.005	0.000	0.000
F	0.24	0.63	0.13	0.20	1.76	0.87	0.75	1.50	0.90	1.66	1.63	0.11	0.053	0.049	0.034	0.006
Total wt.	95.11	95.54	93.55	95.87	96.68	96.20	96.39	97.36	94.61	96.81	95.55	94.13	94.58	93.90	85.61	85.46
−O=F,Cl	−0.10	−0.27	−0.06	−0.08	−0.74	−0.37	−0.31	−0.80	−0.38	−0.70	−0.69	−0.04	−0.02	−0.02	−0.01	−0.00
H$_2$O+ (calc.)	3.86	3.77	3.85	3.93	3.45	3.85	3.88	3.40	3.68	3.40	3.37	4.01	4.32	4.39	11.25	11.18
−FeO=Fe$_2$O$_3$	0.00	0.00	0.00	0.00	0.00	0.00	0.00	0.00	0.00	0.00	0.00	0.00	0.52	0.13	0.02	0.02
Corr. wt.	98.86	99.04	97.35	99.72	99.39	99.69	99.96	99.96	97.91	99.51	98.24	98.10	99.40	98.40	96.84	96.65
Nr. of oxygens	22	22	22	22	22	22	22	22	22	22	22	22	22	22	28	28
Si	5.561	5.732	5.494	5.526	5.938	5.678	5.606	5.873	5.657	5.798	5.745	5.547	6.176	6.069	5.302	5.241
Al IV	2.439	2.268	2.506	2.474	2.062	2.322	2.394	2.127	2.343	2.202	2.255	2.453	1.824	1.931	2.698	2.759
Al VI	0.592	0.529	0.658	0.713	0.188	0.302	0.412	0.139	0.305	0.212	0.199	0.320	2.706	3.748	2.807	2.712
Fe^{3+}	—	—	—	—	—	—	—	—	—	—	—	—	0.531	0.129	—	0.024
Ti	0.172	0.108	0.185	0.135	0.054	0.120	0.122	0.043	0.234	0.077	0.069	0.383	0.069	0.008	0.011	0.024
Mg	2.834	3.666	2.871	2.900	5.454	5.128	4.773	5.583	4.376	5.049	5.146	3.827	0.949	0.075	5.070	5.094
Fe^{2+}	2.242	1.520	1.995	2.153	0.272	0.443	0.630	0.267	0.972	0.658	0.613	1.288	0.000	0.000	4.002	4.106
Mn	0.013	0.009	0.021	0.002	0.003	0.005	0.005	0.003	0.002	0.002	0.004	0.001	0.017	0.000	0.035	0.044
Ca	0.000	0.000	0.017	0.002	0.002	0.000	0.000	0.001	0.002	0.000	0.000	0.000	0.010	0.004	0.004	0.003
Na	0.103	0.093	0.087	0.072	0.281	0.194	0.199	0.302	0.211	0.145	0.310	0.066	0.030	0.195	0.013	0.000
K	1.696	1.733	1.857	1.611	1.538	1.591	1.656	1.529	1.580	1.694	1.548	1.663	1.854	1.917	0.000	0.004
Zn	0.000	0.018	0.020	0.000	0.000	0.000	0.000	0.000	0.000	0.000	0.000	0.000	0.000	0.000	0.000	0.000
Cl	0.000	0.004	0.001	0.002	0.005	0.005	0.010	0.007	0.006	0.028	0.028	0.004	0.001	0.001	0.000	0.000
F	0.113	0.295	0.064	0.094	0.777	0.389	0.334	0.835	0.415	0.747	0.741	0.049	0.023	0.021	0.023	0.004
Total cations	15.651	15.675	15.711	15.587	15.792	15.783	15.797	15.866	15.681	15.838	15.888	15.548	14.166	14.075	19.941	20.000
X_F	0.028	0.074	0.016	0.023	0.194	0.097	0.083	0.209	0.104	0.187	0.185	0.012	0.006	0.005	0.001	0.001
X_{Mg}	0.558	0.707	0.590	0.574	0.953	0.921	0.883	0.954	0.818	0.885	0.894	0.748	0.641	0.369	0.559	0.552

Orijärvi area, southern Finland

Regional background

The Orijärvi area is located within the Svecofennian orogenic belt of SW Finland (Figs 1 and 2). The bedrock of the area is composed of supracrustal and infracrustal rocks that are of Palaeoproterozoic age (Mäkelä 1989). During the Svecofennian orogeny, the supracrustal rocks were deformed during a main foliation-forming event (D_1) and a subsequent refolding event (D_2), regionally metamorphosed and intruded by granodiorites (Schneiderman & Tracy 1991). Metamorphism in the area started during the early stages of the Svecofennian orogeny (1900–1870 Ma: Neuvonen et al. 1981) and reached peak temperatures (550–650 °C at 3–5 kbar: Schreurs & Westra 1985) at about 1830 Ma (Ploegsma & Westra 1990). After cooling, some shear zones were reactivated during the last deformation event at about 1530 Ma (Ploegsma & Westra 1990).

The Orijärvi area belongs to a metallogenic zone that extends from the southwestern part of

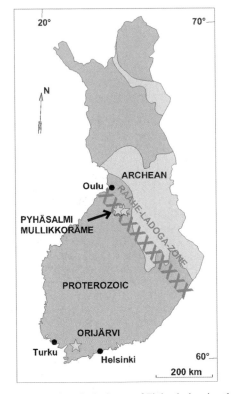

Fig. 1. General geological map of Finland, showing the locations of the Orijärvi, Pyhäsalmi and Mullikkoräme Zn–Cu–Pb deposits

Finland to central Sweden. There are several principal ore deposits in the Finnish section of that zone: the Aijala Cu–Zn deposit (Mäkelä 1983), the Metsämonttu and Attu Zn–Pb deposits, the Orijärvi Cu–Zn deposit (Latvalahti 1979) and the Iilijärvi Zn–Pb–Ag(–Au) deposit located 1.3 km northwest from the Orijärvi mine. The Orijärvi and Iilijärvi deposits lie within the same zone (Fig. 2) of altered rocks. The zone is composed of several elongated lenses of cordierite–anthophyllite rocks; these lenses are tens to hundreds of metres in length and less than 300 m in width. The ores and their host rocks are intensively deformed and metamorphosed (Colley & Westra 1987). The Orijärvi deposit is located within a zone that is composed of andalusite–cordierite–muscovite gneisses, cordierite–muscovite–biotite gneisses with garnet porphyroblasts and cordierite–anthophyllite gneisses. Mineral deposits are associated with strongly hydrothermally altered rocks (Eskola 1914; Latvalahti 1979) that occur in two separate zones (Mäkelä 1989). The Zn–Pb–Cu ore in the main ore zone occurs in narrow, elongated bodies. It is hosted in a diopside–tremolite skarn. Common ore minerals are pyrite, chalcopyrite, sphalerite and galena (Mäkelä 1989; Colley & Westra 1987; Papunen 1986). The Cu–Zn ore lies 50 m south of the main zone and is hosted in cordierite–anthophyllite rocks. Common ore minerals in this ore deposit are sphalerite, pyrrhotite, chalcopyrite and galena (Mäkelä 1989; Colley & Westra 1987; Papunen 1986).

Petrography

Petrographic studies indicate that quartz, plagioclase, zircon, titanite, fluorite, apatite and opaque minerals usually predate D_1 deformation whereas phlogopite, gedrite, anthophyllite, cordierite and cummingtonite crystallized pre- to syn-kinematically with respect to a first deformation phase (D_1). Peak temperatures (550–650 °C) in the Orijärvi area were reached syn- to post-kinematically with respect to a second deformation event (D_2) and garnets, often associated with cummingtonite, have probably grown at that stage (Bleeker & Westra 1987). In many places, amphiboles have recrystallized at this stage and have coarsened and randomly overgrown earlier, well-aligned amphiboles.

Fluorite is a common accessory mineral in most rock types in the Orijärvi area with the exception of the amphibolites and granodiorites. As fluorite and F-rich apatite commonly form inclusions in porphyroblastic minerals (cordierite, anthophyllite, garnet), they must have crystallized before peak metamorphism. This suggests that F-bearing fluids entered relatively early into the system, consistent

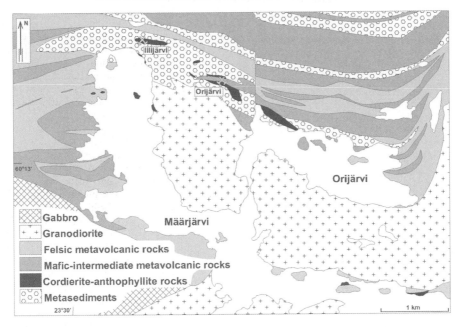

Fig. 2. Geological sketch map of the Orijärvi area (after data kindly provided by M. Väisänen)

with conclusions of earlier workers (Eskola 1914; Latvalahti 1979; Mäkelä 1983).

Geochemistry and mineralogy

The range in the whole-rock F contents for all Orijärvi samples is 150–3550 ppm (Fig. 3, Table 1). Cordierite–anthophyllite rocks contain 950–2450 ppm F and have slightly enriched rare earth elements (REE) with negative Eu anomalies. The F content in minerals from sample LMR 34c was

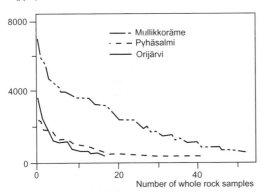

Fig. 3. F contents in rock samples (not as geographical trend) from the Orijärvi, Pyhäsalmi and Mullikkoräme Zn–Cu–Pb deposits

2.71–2.74 wt% in apatite, 0.57–0.83 wt% in phlogopite, 0.08–0.20 wt% in anthophyllite and <0.13 wt% in chlorite.

Cordierite–biotite and garnet–biotite gneisses contain 400–3550 ppm F. Cordierite–biotite gneisses show more enriched and fractioned light REE (LREE) than garnet–biotite gneisses. The F content in samples from a concordant amphibolite unit (previously described as subvolcanic mafic sill: Mäkelä 1983) was 200–450 ppm. These samples have low Cu, Zn and Pb contents, with the exception of one sample from the Orijärvi mine area. In the amphibolite samples ($n = 3$), which are also characterized by low metal contents, F correlates positively with FeO ($r^2 = 1.00$) and K_2O ($r^2 = 0.95$), but negatively with Na_2O ($r^2 = 0.98$) and SiO_2 ($r^2 = 0.91$). The amphibolite data also show a positive correlation between F and total metal content (Cu + Zn + Pb; $r^2 = 0.96$). Hence, hydrothermal activity of an F-rich fluid may have caused the variation in major elements as well as the local enrichment in Cu, Zn and Pb. These preliminary results are augmented by the extensive geochemical databases of the Outokumpu Mining Ltd (see next section).

Mineral assemblages in probed samples from Orijärvi are: anthophyllite + chlorite + Zn-rich spinel; anthophyllite + biotite + cordierite + quartz ± Zn-rich spinel; biotite + hornblende + muscovite + quartz + plagioclase; and biotite + anthophyllite quartz + plagioclase ± chlorite.

Sample ORI62b contains the last assemblage with additional gedrite and minor supersolvus orthoamphibole intermediate between anthophyllite and gedrite (Table 2). The order of decreasing X_F (normalized to biotite) in biotite-bearing samples is (see also Tables 2 and 3): biotite (X_F range: 0.011–0.106, $n = 49$) > gedrite (0.006–0.026, $n = 11$) > intermediate supersolvus orthoamphibole (0.009–0.020, $n = 12$) > anthophyllite (0.006–0.026, $n = 26$) > hornblende (0.003–0.010, $n = 6$) ~ muscovite (0.006, $n = 1$) > chlorite (0.001–0.012, $n = 11$). Surprisingly, the highest recorded F content is for anthophyllite (X_F range: 0.047–0.132) in sample LMR35e, which lacks biotite.

Mineral assemblages in probed samples from Iilijärvi are (see also Tables 2 and 3): anthophyllite + biotite ± intermediate supersolvus orthoamphibole ± cummingtonite (± quartz ± plagioclase). The order of decreasing X_F (normalized to biotite) is: biotite (X_F range: 0.006–0.074, $n = 31$) > intermediate supersolvus orthoamphibole (0.002–0.027, $n = 8$) ~ anthophyllite (0.002–0.029, $n = 16$).

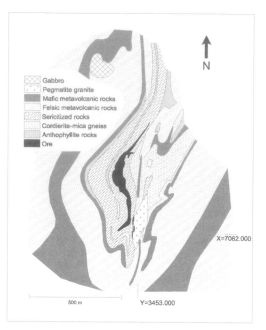

Fig. 4. Geological sketch map of the Pyhäsalmi area (after Ekberg & Penttilä 1986)

Pyhäsalmi & Mullikkoräme deposits, central Finland

Regional background

The Pyhäsalmi and Mullikkoräme mines are situated in central Finland, c.160 km SSE of Oulu (Figs 1 and 4), within the Raahe–Ladoga Zone. The Raahe–Ladoga Zone, previously called the Main Sulphide Ore Belt, has been described as a collisional boundary zone between Archaean and Proterozoic crustal domains (Lahtinen 1994). It is the main metallogenic zone in Finland where five principal phases of mineralization have been recorded (Ekdahl 1993; Weihed & Mäki 1997). The Pyhäsalmi and Mullikkoräme mines occur within the Pyhäsalmi volcanic complex, a local unit of the larger Pyhäsalmi Island Arc, which forms a 10–30 km wide and c. 350 km long zone including over 60 significant mineral deposits. According to Ekdahl (1993), ore deposits in the Pyhäsalmi Island Arc are characterized by (i) medium- to high-grade metavolcanic zones and complexes, (ii) submarine, tholeiitic and calc-alkaline island arc volcanism, (iii) vertical or lateral hydrothermal alteration characterized by sericitization, silicification, chloritization, the presence of garnet–cordierite–anthophyllite rocks and (iv) the Mg, Fe, H_2O and S enrichment and Si, Ca, Na and K depletion towards the ore bodies.

The Pyhäsalmi type ore deposits are thought to have formed during the early orogenic stage of the Svecofennian orogen at c.1.92 Ga (Ekdahl 1996). Volcanism is bimodal in style: most of the felsic

rocks are calc-alkaline, low-K rhyolites, whereas mafic volcanic rocks are sub-alkalic, resembling present day low-K tholeiitic basalts to basaltic andesites (Kousa & Lundqvist 2000). The lower stratigraphic levels are dominated by felsic mass flows or tuffaceous and pyroclastic rocks with mafic intercalations. Mafic lavas and pyroclastic rocks, pillow lavas and breccias become more abundant at higher stratigraphic levels (Kousa & Lundqvist 2000). Lahtinen (1994) suggested that the felsic rocks and associated VMS deposits represent an aborted rifting stage in an island arc environment. According to Weihed & Mäki (1997), volcanism started in an extensional continental margin with felsic volcanism and continued in a rifted marine environment with mafic volcanism. According to these authors, large-scale hydrothermal alteration and ore formation occurred near the centres of mafic volcanism. The metavolcanic rocks have later been intruded by syn-tectonic intrusions (Kousa et al. 1994; Lahtinen & Huhma 1997; Kousa & Lundqvist 2000).

In the Pyhäsalmi mine area, the volcanogenic host rocks of the ore bodies have undergone amphibolite facies metamorphism and pervasive deformation resulting in sericite schists and quartzites, cordierite and sillimanite gneisses and cordierite-, anthophyllite- and garnet-bearing rocks (Huhtala 1979). Peak metamorphic conditions reached 600–700 °C and 5–7 kbar (A.-P. Tapio, pers. comm. 2000). The Pyhäsalmi ore deposit is a typical VMS deposit surrounded by a hydrothermal alteration halo. All rock types are most intensely altered near the ore bodies, but the width of the alteration zone differs from 300 m at the surface to a few metres at deeper levels (Weihed & Mäki 1997). The ore contains *c.* 70% sulphides, mainly pyrite, chalcopyrite, sphalerite and pyrrhotite; common accessory minerals are galena, magnetite and arsenopyrite (Helovuori 1979; Weihed & Mäki 1997). Production from the start-up date (1962) to the present day has been about 33.4 Mt with levels of 2.48% Zn, 0.8% Cu and 35.2% S. A new ore body at greater depth was discovered at the end of 1996 with reserves indicating 17 Mt with levels of 2.1% Zn, 1.1% Cu and 34.2% S (Weihed & Mäki 1997; Mäki & Puustjärvi 2002).

The Mullikkoräme mine is located in one of three satellite deposits, 7 km east of the main Pyhäsalmi mine. Host rocks in mineralized zones are mainly chloritized and sericitized cordierite–amphibole gneisses and other altered felsic to mafic volcanites. Locally, felsic volcanic rocks with talc–chlorite schist intercalations are abundant. The metamorphic grade reaches greenschist facies, which is considerably lower than in Pyhäsalmi (Weihed & Mäki 1997). Based on the reported mineral assemblage, andalusite + biotite + chlorite, our temperature estimate is 500–560 °C at a pressure of 1–3 kbar, using published KFMASH grids (e.g. Spear & Cheney 1989). The mine was closed in 2000 after a total production of 1.2 Mt in its ten-year history, at an average grade of 0.23% Cu, 5.99% Zn, 0.62% Pb, 1.21 $g\,t^{-1}$ Au, and 33 $g\,t^{-1}$ Ag (Mäki & Puustjärvi 2002). Mineralized horizons at Mullikkoräme consist of massive sulphides, with pyrite, sphalerite, galena and chalcopyrite as common ore minerals; gangue minerals are quartz, talc, carbonate, Fe–Mg silicates and barite (Weihed & Mäki 1997).

Petrography

The most common silicate minerals in Mullikkoräme samples are biotite, chlorite and muscovite, garnet, andalusite and plagioclase. Cordierite and hornblende are present in some sections. Observed mineral assemblages are biotite + chlorite + garnet + plagioclase + quartz ± anthophyllite; biotite + chlorite + hornblende + plagioclase + quartz; biotite + andalusite + cordierite + quartz + tourmaline; and biotite + chlorite + andalusite + garnet + muscovite + staurolite + plagioclase + quartz. 'Skarn' samples (possibly metamarls or altered ultramafics) contain talc, muscovite, spinel, chlorite, tremolite and carbonate minerals. Beside sulphide minerals, our Pyhäsalmi samples contain the assemblages tourmaline + muscovite ± biotite; biotite + chlorite + hornblende + talc; biotite + hornblende ± apatite. A.-P. Tapio (pers. comm. 2000) also reported anthophyllite + Zn-rich staurolite ± sillimanite + quartz ± biotite ± cordierite.

Geochemistry and mineralogy

A significant number ($n = 52$) of F analyses exist of whole-rock samples from the Mullikkoräme mine. The F contents in those rocks are 300–7100 ppm (Table 1, Fig. 3), significantly higher than in Orijärvi. This difference is also reflected in the mineralogy: the average F content of biotites from Mullikkoräme is 0.22 wt%, whereas it is only 0.12 wt% in biotites from Orijärvi. Samples from the nearby Pyhäsalmi mine have lower whole-rock F contents, namely 55–2300 ppm (Fig. 3), which is a similar range as in Orijärvi. This may be due to mineralogical constraints (see below), as the Pyhäsalmi samples have higher ore contents.

The hydrous minerals in the Pyhäsalmi samples are biotite, chlorite, muscovite, hornblende, talc, tourmaline and apatite. The highest recorded F content is in apatite from Pyhäsalmi sample 208, ($X_F = 0.328$–0.340, $n = 2$), much higher than coexisting biotite ($X_F = 0.030 \pm 0.004$, $n = 4$). Tourmaline ($X_F = 0.078 \pm 0.041$, $n = 4$) has slightly higher F content than coexisting biotite ($X_F = 0.074 \pm 0.006$, $n = 10$) in sample 168. Talc ($X_F = 0.015 \pm 0.004$, $n = 10$) has slightly lower F content than coexisting biotite ($X_F = 0.018 \pm 0.010$, $n = 3$) in sample 181. X_F of Hbl is 10–30% lower than that of coexisting biotite. F contents are still lower in muscovite and negligible in chlorite.

The hydrous minerals at Mullikkoräme are biotite, chlorite, muscovite, hornblende, talc, anthophyllite, tourmaline and staurolite. F contents are low in chlorite ($X_F = 0.002 \pm 0.001$, $n = 7$), but significantly higher in biotite ($X_F = 0.015 \pm 0.006$, $n = 25$), muscovite ($X_F = 0.008 \pm 0.003$, $n = 11$) and hornblende ($X_F = 0.007 \pm 0.004$, $n = 5$). Hence, the last two seem to be the main F carrier minerals in view of the low fluorite abundance. Tourmaline ($X_F = 0.028$–0.033, $n = 2$) and staurolite ($X_F = 0.009$, $n = 1$) have fairly high fluorine contents, but they are relatively rare. No talc and anthophyllite analyses exist for this area. The highest F contents occur in talc-rich rocks with $X_{Mg} > 0.8$, which probably represent highly altered mafic rocks, so talc may also be a significant F carrier mineral (see Pyhäsalmi data). F contents in anthophyllite are expected to be as low as elsewhere.

The main S-bearing minerals in the Mullikkor-äme mine are pyrite, sphalerite, galena, chalco-pyrite and barite. Using these mineralogical constraints, the amount of Fe residing in sulphides can be estimated and the total weight corrected for the lower amount of iron oxide (Table 1). The results show that on average $60 \pm 16\%$ of the Fe resides in sulphides, giving a strong positive correlation between Fe and S contents of the samples (Fig. 5). Geochemical data from a 30 m long drill core through a metal rich horizon in the Mullikkoräme Zn–Cu–Pb deposit show a positive correlation between F and X_{Mg} [$= MgO/(MgO + FeO)$], without correcting for ferric iron (Fig. 6). This suggests that leaching of corrosive, F-bearing fluids has reduced the iron content of this rock suite. The data also show a moderately strong, negative correlation between F and total metal content (Pb + Zn + Cu + Fe; Fig. 6). Hence, this metal rich zone (metal sink) seems to show a trend that is opposite to the trend of metal poor zones (metal sources ?) in Orijärvi (see also below).

Considering all samples from Mullikkoräme for which F was measured ($n = 52$), the positive correlation ($r = 0.85$) between F and X_{Mg} (Fig. 7a) is confirmed. A plot of F against metal content (Fig. 7b) shows a similar, negative correlation as the drill core data (Fig. 6), but with considerable scatter. F also shows a negative correlation with S content (Fig. 7c). At first sight the data would therefore suggest that the higher the F flux through the system, the lower the content of sulphide ore, at odds with the commonly observed increase of F content towards ore bodies (e.g.,

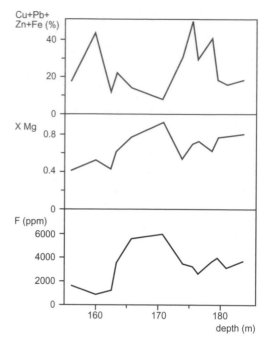

Fig. 6. Positive correlation between F (ppm) and X_{Mg} [$= MgO/(MgO + FeO)$] and moderately strong, negative correlation between F (ppm) and total metal content (wt%) in a drill core through the Mullikkoräme Zn–Cu–Pb deposit

Lavery 1985). However, this is only *apparent* because of the interplay of two factors: (i) the closure problem in geochemical data (e.g. Rollinson 1993) and (ii) the mineralogical con-straint that F is dominantly incorporated into hydrous silicates. The more sulphide a sample contains, the less silicates are present and the less F can be accommodated in the rock, even when F concentrations in F carrier minerals may be high. This may also explain why the ore-rich Pyhäsalmi samples have lower F contents (Fig. 3).

To circumvent the closure problem, it is useful to normalize element concentrations to elements of relatively low mobility such as Ti, Al (and Mg, to a lesser extent). Ti and Al show a strong positive correlation (Fig. 8a), suggesting similar geochemical behaviour. Combining TiO_2, Al_2O_3 and MgO as the total mass of relatively immobile elements not or very little affected by whole-rock alteration, some clear correlations become appar-ent. F correlates positively with the sum of these relatively immobile elements (Fig. 8b), whereas Fe shows a negative correlation (Fig. 8c). Hence, leaching by F-rich fluids probably led to lower Fe concentrations and an increase in the concentra-tion of immobile elements Ti, Al and Mg. This

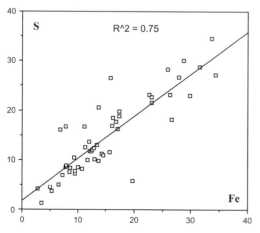

Fig. 5. Positive correlation between Fe and S (element wt%) in the Mullikkoräme samples, confirming that Fe occurs dominantly in the form of sulphides (mainly pyrite and chalcopyrite)

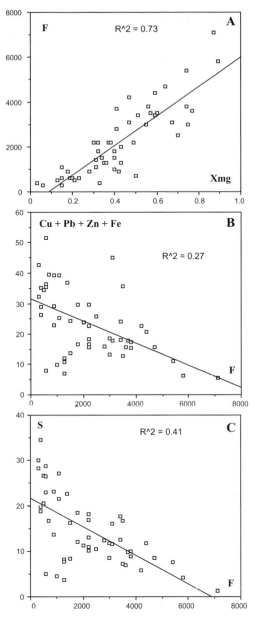

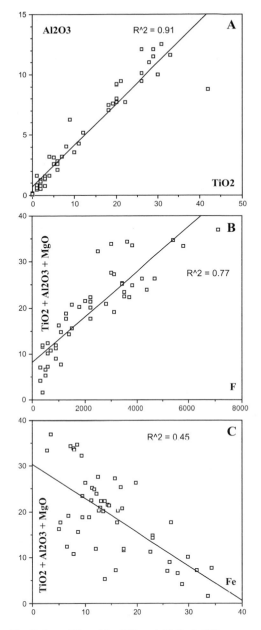

Fig. 7. Mullikkoräme samples show a strong positive correlation between F (ppm) and X_{Mg} [= MgO/(MgO + FeO)], where all iron is taken as FeO (**a**), and a negative correlation between F (ppm) and the total metal content (wt%), with considerable scatter (**b**). F (ppm) and S (wt%) show a negative correlation (**c**)

Fig. 8. Immobile oxides TiO_2 and Al_2O_3 (wt%) correlate positively (**a**), suggesting similar geochemical behaviour. F (ppm) shows a positive (**b**) and Fe (wt%) a negative (**c**) correlation with the sum of immobile elements (wt%)

strengthens the interpretation of the F versus X_{Mg} diagrams (Figs 6 and 7a). Normalizing to the immobile elements, F correlates positively with the total metal content (Fig. 9). This correlation is quite strong when normalized to TiO_2 (Fig. 9a)

and moderate when normalized to Al_2O_3 (Fig. 9b). Viewed in this way, the data seem to suggest that a higher F flux through the system corresponds to a higher content of sulphide ore, consistent with the increase of F content towards ore bodies (e.g., Lavery 1985).

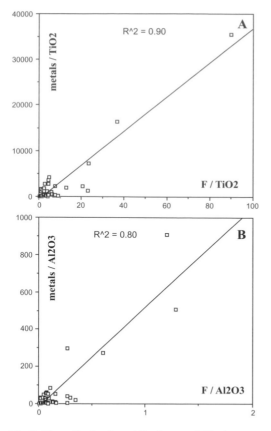

Fig. 9. Normalized to immobile elements (TiO$_2$ in **a**; Al$_2$O$_3$ in **b**), total metal content and F correlate positively (all data in wt%)

Oonagalabi domain, Arunta block, central Australia

Regional background

The Oonagalabi domain is located in the western Harts Range, about 50 km east of the Strangways Complex in the eastern Arunta Block, Central Australia. It comprises a succession of orthogneisses and subordinate supracrustal units (Sivell 1986) that were metamorphosed under transitional amphibolite to granulite facies conditions (Oliver *et al.* 1988; Raith & Kriegsman 1998). The Oonagalabi domain displays a fold nappe geometry (Ding & James 1985) and has been interpreted as a tectonic window to granulites underlying the amphibolite facies rocks of the Eastern Harts Range. All units record the same deformation stages, being top-to-the-SW thrusting and subsequent refolding into NE-plunging folds with NE-trending, subvertical axial planes. The Oonagalabi domain hosts a small, sub-economic Pb–Zn–Cu

deposit where most of our samples have been collected. Peak metamorphism, which also affected these orthoamphibole dominated units, culminated at *c.* 750–800 °C, 8–9 kbar (Raith & Kriegsman 1998).

Petrography

Aluminous lenses in a concordant orthoamphibole dominated unit show an early assemblage gedrite + spinel ± corundum, overgrown by sapphirine and garnet. The earliest recorded mineral assemblage in K-poor felsic rocks is gedrite + sillimanite + quartz, overprinted by garnet and cordierite. Locally, gedrite has broken down to orthopyroxene + garnet + quartz and the orthopyroxene is overgrown at the edges by small (retrograde) anthophyllite grains. Rocks closest to the Pb–Zn–Cu deposit are dominated by retrograde anthophyllite with orthopyroxene relics. Details are given in Raith & Kriegsman (1998) and a paper in preparation.

Geochemistry and mineralogy

The high F content of gedrite (0.3–0.8 wt%), phlogopite (0.9–2.1 wt%) and anthophyllite (up to 0.3 wt%), and the presence of fluorite in marble layers and chondrodite (with 3.3–3.8 wt% F) in some orthoamphibole dominated lenses, suggest a high F content of the hydrous fluids that have affected the deposit. Zn content is highest (up to 12.5 wt% ZnO) in spinel from chondrodite bearing lenses, strengthening the link between Zn mineralization and F. A reference sample several kilometres away from the deposit, with the assemblage orthopyroxene + garnet + quartz + biotite + plagioclase + rutile (+ retrograde anthophyllite), contains biotite with only *c.* 0.14 wt% F, much lower than near the ore deposit.

There are few whole-rock chemical analyses from this area, but available microprobe data and modal abundances can be used to estimate the F content of some samples. Sample AR125 contains *c.* 60 vol% gedrite (at 0.74 wt% F) and 10 vol% chondrodite (at 3.6 wt% F), yielding an estimated 8000 ppm F in the rock. Sample AR126c contains *c.* 40 vol% phlogopite (at 2 wt% F) and consequently has a similar F content. Sample AR126d contains about 80 vol% anthophyllite (at 0.15 wt% F), giving an estimated 1200 ppm F. Sample AR129 contains *c.* 2 vol% gedrite (at 0.7 wt% F) and 0.5 vol% biotite (at 1 wt% F), equating with an estimated 1450 ppm F. Sample AR245 contains *c.* 15 vol% gedrite (at 0.64 wt% F), corresponding to an estimated 960 ppm F in the rock; this is consistent with a whole-rock analysis of 950 ppm. The range of whole-rock F contents in Oonagalabi is therefore *c.* 1000–8000 ppm, quite similar to the range of values at Mullikkoräme mine.

F partitioning data and F carrier minerals

Figure 10 shows $X_F = [F/(F + OH)]$ data for coexisting biotite and orthoamphiboles. The data for anthophyllite in samples from Finland (Orijärvi and Iilijärvi) and Oonagalabi fall on different lines, with the slope of the Oonagalabi line closer to $K_D = 1$ (Fig. 10a). Within both areas, X_{Mg} increases with overall F content, without affecting the K_D of F partitioning. In view of the well-documented influence of X_{Mg} on the F content of hydrous minerals (e.g., Rosenberg & Foit 1977), the constant K_D for F partitioning implies that anthophyllite shows a similar Fe^{2+}/Fe avoidance as biotite. The data suggest that peak temperature (500–650 °C in Finnish samples: Schreurs & Westra 1985; Schneiderman & Tracy 1991; 750–800 °C in Oonagalabi: Raith & Kriegsman 1998) is the dominant factor controlling F partitioning. X_F data for biotite coexisting with gedrite (Orijär-

vi and Oonagalabi) show a similar pattern (Fig. 10b). In the Oonagalabi samples, X_{Mg} increases with overall F content, without affecting the K_D of F partitioning. The Oonagalabi line has a higher slope than the Finnish line, consistent with higher peak temperatures in the former.

The slope for biotite–gedrite is significantly higher than for biotite–anthophyllite in Oonagalabi; Fig. 10a, b). Hence, combining these data, the order of decreasing X_F for coexisting biotite and orthoamphiboles is: biotite > gedrite > anthophyllite. Intermediate, supersolvus orthoamphiboles (Finnish samples) plot close to the single Finnish gedrite point. As for other minerals, chlorites in Finnish samples have F contents close to 0 even in samples where biotite has significant F contents, showing that it is not an important F carrier mineral. An interesting observation, however, is that chlorite inclusions in garnet in Orijärvi sample 62b show higher F contents (0.034 wt%) than texturally late, retrograde, chlorite (0.006 wt%; Table 3), which suggests that the prevailing fluid was more F-rich prior to peak metamorphism than on the retrograde path. Hornblende and muscovite in Mullikoräme and Pyhäsalmi samples have much lower X_F than coexisting biotite, but higher F contents than chlorite. Talc analysed so far had F contents comparable to hornblende ($X_F = 0.015$ and 0.016, respectively, in Pyhäsalmi) but a sample from Mullikoräme with very high bulk F content (7100 ppm) is dominated by talc and we suspect it may be the main F carrier mineral there.

Apatite (Pyhäsalmi sample 208) and chondrodite (Oonagalabi sample AR125) have much higher F contents than coexisting phlogopite, with X_F up to 0.3. The relative F order of preference between apatite and chondrodite can be estimated indirectly. In sample AR125, X_F in chondrodite (0.322 ± 0.013) is twice as high as in gedrite (0.160 ± 0.012). This is the only Oonagalabi sample without biotite, but using the average K_D for F exchange between biotite and gedrite (from Fig. 10b), the X_F value for coexisting biotite is estimated at 0.24 ± 0.04. In Pyhäsalmi sample 208, X_F in apatite is 0.334 ± 0.009, whereas the biotite X_F is only 0.030 ± 0.004. This suggests that, when normalized to biotite X_F, apatite has a much higher preference for F than chondrodite.

Combining all data, the relative F partitioning order of hydrous minerals is apatite > chondrodite > biotite > gedrite > (hornblende, muscovite, anthophyllite) > chlorite. The F partitioning between hornblende, muscovite and anthophyllite is still unclear. Our data on talc, staurolite and tourmaline are also inconclusive because of the low number of analyses. The F partitioning data can be used to quickly identify the F carrier

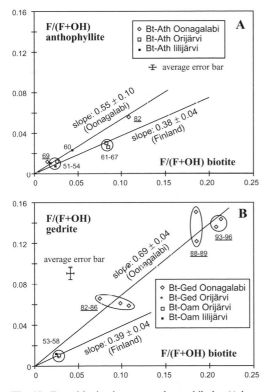

Fig. 10. F partitioning between orthoamphiboles (Ath, anthophyllite; Ged, gedrite; Oam, intermediate, supersolvus orthoamphibole) and biotite (Bt). Numbers near (groups of) average analyses indicate Mg numbers = 100 × X_{Mg} (underlined: Oonagalabi data)

mineral in any rock type, by estimating the modal abundance of each mineral in thin section or hand specimen. The Orijärvi samples contain principally anthophyllite, whereas gedrite is very common in Oonagalabi. This probably reflects a combination of different bulk compositions and different peak metamorphic temperatures. As our data show that anthophyllite has a much lower F content (up to 0.3 wt%) than gedrite (up to 0.8 wt%), much less F can be accommodated in orthoamphiboles at Orijärvi than at Oonagalabi.

For some samples or suites of samples, both mineral and whole-rock chemical analyses were obtained, allowing a more precise determination of the main F carrier minerals. For example, sample LMR 34c (cordierite–anthophyllite rock) from Orijärvi contains 2450 ppm (=0.245 wt%) F based on whole-rock analysis. Sample LMR 34b (cordierite–anthophyllite gneiss), which was sampled 10 m away in the same rock unit, contains biotite (0.77 wt% F: Table 2), anthophyllite (0.16 wt%: Table 2) and chlorite (0.13 wt%). The approximate F content for this sample (at 30% biotite, 30% anthophyllite, 30% cordierite, 1% chlorite and 9% anhydrous minerals) is 0.28 wt%, similar to the analytical value of the adjacent sample. Biotite contributes 0.23 wt% and is thus the main F carrier mineral in cordierite–anthophyllite rocks and gneisses. Other rock types usually have much less biotite, but plenty of fluorite, which suggests that fluorite is the most important F-bearing mineral in those medium-grade rock types. In contrast, orthoamphiboles are the dominant F-bearing minerals in the transitional granulite facies rocks at Oonagalabi, where fluorite is absent (except in marbles). Anthophyllite is the dominant mineral in the deposit and thus the main F carrier on the regional scale. However, chondrodite, gedrite and phlogopite are the main F carriers in other samples.

Discussion

F content and metamorphic grade

A comparison of whole-rock samples from Orijärvi, Pyhäsalmi and Mullikkoräme shows that the Mullikkoräme samples have exceptionally high F contents (Fig. 3). As Mullikkoräme is the only low-grade metamorphic area of the three, one could conclude that significant amounts of F have left the system during peak metamorphism in Orijärvi and Pyhäsalmi. However, the Oonagalabi samples (see above) show similar F contents as in Mullikkoräme, even though the peak metamorphic temperatures were c. 750–800 °C. As mineral–fluid partitioning coefficients for F indicate that F will remain in solids at least until advanced partial

melting (Anfilogov et al. 1977; Zhu & Sverjensky 1991), the whole-rock data probably reflect F contents prior to the peak metamorphic overprint.

Timing of F influx

High F contents of magmatic rocks are a feature typical of so-called anorogenic granitoids (e.g. Rogers & Satterfield 1994; Goodenough et al. 2000). In the Svecofennian orogen of Finland, this may be either early orogenic (e.g., Orijärvi back-arc basin: Väisänen 2001), late-orogenic (post-collisional: Eklund et al. 1998), or post-orogenic (e.g., rapakivi granites: papers in Haapala & Rämö 1999). As a result, ore deposits related to high F flux are probably either early orogenic or late- to post-orogenic. These relative timings are easy to distinguish, because early orogenic intrusions predate peak metamorphism, whereas late- to post-orogenic intrusions post-date peak metamorphism. Arguments for early F influx in the deposits studied here, are:

1 fluorite, apatite and oriented F-rich biotite form inclusions in peak metamorphic porphyroblasts (garnet, cordierite);
2 chlorite inclusions in garnet show higher F contents than texturally late, retrograde chlorite;
3 volcanics and related granitoids that are possible F sources, predate peak metamorphism (Orijärvi: Väisänen 2001; Oonagalabi: Sivell 1986); at Orijärvi, extrusion of rhyolite and coeval intrusion predate peak metamorphism by some 50–60 Ma.

Correlation of F with ore minerals

Our data superficially suggest that a higher F flux through a system results in lower contents of sulphide ore, which is inconsistent with the commonly observed increase of F content towards ore bodies (e.g. Lavery 1985). However, we have shown that this is due to a mineralogical constraint, F being dominantly incorporated into hydrous silicates, in combination with the closure problem in geochemical data. When normalized to relatively immobile elements (Ti, Al), F correlates positively with the total metal content, as expected.

A similar feature can be seen in the Crandon Cu–Zn deposit (Lavery 1985, fig. 4): the F content increases towards the deposit, but drops sharply at the point where Cu and Zn sulphides start to dominate the mineralogy; small Cu and Zn peaks within the deposit also correlate with low values of F. We believe that this is due to the same geochemical and mineralogical complications observed in our data and that it could be solved in a similar manner. Hence, geographical F trends may become clearer when normalized to immobile elements, which may improve geochemical exploration for Pb–Zn–Cu deposits.

We acknowledge the support of the European Community Access to Research Infrastructure action of the Improving Human Potential Programme, contract HPRI-CT-1999-00008 awarded to B.J. Wood (EU Geochemical Facility, University of Bristol), and are greatly indebted to its professional staff. L.R. acknowledges financial support from the K.H. Renlund Foundation in Finland. L.M.K. thanks J.G. Raith, Mining University of Leoben (Austria), for making the Oonagalabi field trip in 1997 possible. Thanks are also due to T. Mäki and H. Puustjärvi for allowing access to the geochemical databases of the Outokumpu Mining Company Ltd. and drill core material. Critical reviews by A. Mogessie and P. Tropper have substantially improved our thinking.

References

ANFILOGOV, V.N., BUSHLYAKOV, I.N., VILISOV, V.A. & BRAGINA, G.I. 1977. Distribution of F between coexisting biotite and amphibole and granitic melt at 780 °C and 1000 atm pressure. *Geochemistry International*, **14**, 95–98.

BARNES, H.L. (ED.) 1979. *Geochemistry of hydrothermal ore deposits*. Wiley Interscience, New York.

BLEEKER, W. & WESTRA, L. 1987. The evolution of the Mustio gneiss dome, Svecofennides of SW Finland. *Precambrian Research*, **36**, 227–240.

COLLEY, H. & WESTRA, L. 1987. The volcano-tectonic setting and mineralization of the early Proterozoic Kemiö-Orijärvi-Lohja belt, SW Finland. *In:* PHARAOH, T.C., BECKINSALE, R.D. & RICKARD, D. (eds) *Geochemistry and Mineralization of Proterozoic Volcanic Suites*. Geological Society, London, Special Publications, **33**, 95–107.

COLLINS, W.J., BEAMS, S.D., WHITE, A.J.R. & CHAPPELL, B.W. 1982. Nature and origin of A-type granites with particular reference to southeastern Australia. *Contributions to Mineralogy and Petrology*, **80**, 189–200.

DING, P. & JAMES, P.R. 1985. Structural evolution of the Harts Range area and its implication for the development of the Arunta Block, central Australia. *Precambrian Research*, **27**, 251–276.

DOBBE, R.T.M. 1994. Geochemistry of cordierite-anthophyllite rocks, Tunaberg, Bergslagen, Sweden. *Economic Geology*, **89**, 919–930.

DOOLEY, D.F. & PATIÑO-DOUCE, E. 1996. Fluid-absent melting of F-rich phlogopite + rutile + quartz. *American Mineralogist*, **81**, 202–212.

EKBERG, M. & PENTTILÄ, V.-J. 1986. The Pyhäsalmi Cu-Zn-pyrite deposit. *In:* GAÁL, G. (ed.) *Proterozoic mineral deposits in central Finland. 7th IAGOD Symposium and Nordkalott Project Meeting (Luleå, 1986): excursion guide no 5*. Sveriges Geologiska Undersökning, Series Ca, **63**, 20–25.

EKDAHL, E. 1993. *Early Proterozoic Karelian and Svecofennian formations and the evolution of the Raahe–Ladoga Ore Zone, based on the Pielavesi area, central Finland*. Geological Survey of Finland, Bulletin, **373**.

EKDAHL, E. 1996. The evolution and metallogenesis of the Raahe-Ladoga Zone. *In:* EKDAHL, E. & AUTIO, S. (eds) *Global Geoscience Transect/SVEKA: proceedings of the Kuopio seminar, Finland 25-26.11.1993*. Geological Survey of Finland, Research Reports, **136**, 13–17.

EKLUND, O., KONOPELKO, D., RUTANEN, H., FRÖJDÖ, S. & SHEBANOV, A.D. 1998. 1.8 Ga Svecofennian post-collisional shoshonitic magmatism in the Fennoscandian shield. *In:* LIÉGEOIS, J.-P. (eds) *Post-collisional magmatism*. Lithos, **45**, 87–108.

ESKOLA, P. 1914. On the petrology of the Orijärvi region in southwestern Finland. *Bulletin de la Commission Géologique de Finlande*, 40.

ESKOLA, P. 1915. Om sambandet mellan kemisk och mineralogisk sammansättning hos Orijärvitraktens metamorfa bergarter. *Bulletin de la Commission Géologique de Finlande*, **44**, 109–145.

FOLEY, S. 1991. High-pressure stability of the fluor- and hydroxy-endmembers of pargasite and K-richterite. *Geochimica et Cosmochimica Acta*, **55**, 2689–2694.

FROESE, E. 1969. Metamorphic rocks from the Coronation Mine and surrounding area. *In:* BYERS, A.R. (eds) *Symposium on the Geology of Coronation Mine, Saskatchewan*. Geological Survey of Canada Papers, **68-5**, 55–78.

GOODENOUGH, K.M., UPTON, B.G.J. & ELLAM, R.M. 2000. Geochemical evolution of the Ivigtut Granite, South Greenland; a F-rich 'A-type' intrusion. *Lithos*, **51**, 205–221.

HAAPALA, I. & RÄMÖ, O.T. (EDS) 1999. Rapakivi granites and related rocks. *Precambrian Research*, **95**, 1–167.

HEATON, T.H.E. & SHEPPARD, S.M.F. 1977. *In: Volcanic processes in ore genesis*. Institute of Mining and Metallurgy, London, 42–57.

HELOVUORI, O. 1979. Geology of the Pyhäsalmi ore deposit, Finland. *Economic Geology*, **74**, 1084–1101.

HENSEN, B.J. & OSANAI, Y. 1994. Experimental study of dehydration melting of F-bearing biotite in model pelitic compositions. *Mineralogical Magazine*, **58A**, 410–411.

HOGAN, J.P. & GILBERT, M.C. 1997. Intrusive style of A-type sheet granites in a rift environment; the Southern Oklahoma Aulacogen. *Geological Society of America, Special Papers*, **312**, 299–311.

HUHTALA, T. 1979. The geology and zinc-copper deposits of the Pyhäsalmi-Pielavesi district, Finland. *Economic Geology*, **74**, 1069–1083.

HUMPHRIS, S.E., ZIERENBERG, R.A., MULLINEAUX, L. & THOMSON, R.E. (EDS) 1995. *Seafloor hydrothermal systems; physical, chemical, biological, and geological interactions*. American Geophysical Union, Geophysical Monographs, **91**.

JIANG, S.Y., PALMER, M.R., XUE, C. & LI, Y. 1994. Halogen-rich scapolite-biotite rocks from the Tongmugou Pb-Zn deposit, Qinling, northwestern China; implications for the ore-forming processes. *Mineralogical Magazine*, **58**, 543–552.

KOUSA, J. & LUNDQVIST, T. 2000. Meso- and Neoproterozoic cover rocks of the Svecofennian Domain. *In:* LUNDQVIST, T. & AUTIO, S. (eds) *Description to the bedrock map of central Fennoscandia (Mid-Norden).* Geological Survey of Finland, Special Papers, **28**, 75–76.

KOUSA, J., MARTTILA, E. & VAASJOKI, M. 1994. Petrology, geochemistry and dating of Paleoproterozoic metavolcanic rocks in the Pyhäjärvi area, central Finland. *In:* NIRONEN, M. & KÄHKÖNEN, Y. (eds) *Geochemistry of Proterozoic supracrustal rocks in Finland.* Geological Survey of Finland, Special Papers, **19**, 7–27.

KOVALENKO, V.I. & KOVALENKO, N.I. 1984. Problems of the origin, ore-bearing and evolution of rare-metal granitoids. *In:* ALLEGRE, C.J. & DIDIER, J. (eds) *Granitoids.* Physics of the Earth and Planetary Interiors, **35**, 51–62.

LAHTINEN, R. 1994. *Crustal evolution of the Svecofennian and Karelian domains during 2.1-1.79 Ga, with special emphasis on the geochemistry and origin of 1.93-1.91 Ga gneissic tonalites and associated supracrustal rocks in the Rautalampi area, central Finland.* Geological Survey of Finland, Bulletin, **378**.

LAHTINEN, R. & HUHMA, H. 1997. Isotopic and geochemical constraints on the evolution of the 1.93-1.79 Ga Svecofennian crust and mantle in Finland. *Precambrian Research,* **82**, 13–34.

LATVALAHTI, U. 1979. Cu-Zn-Pb ores in the Aijala-Orijärvi area, southwest Finland. *Economic Geology,* **74**, 1035–1059.

LAVERY, N.G. 1985. The use of F as a pathfinder for volcanic-hosted massive sulfide ore deposits. *Journal of Geochemical Exploration,* **23**, 35–60.

MÄKELÄ, U. 1983. On the geology of the Aijala-Orijärvi area, Southwest Finland. *In:* LAAJOKI, K. & PAAKKOLA, J. (eds) *Exogenic processes and related metallogeny in the Svecokarelian geosynclinal complex.* Geological Survey of Finland, Guides, **11**, 140–159.

MÄKELÄ, U. 1989. *Geological and geochemical environments of Precambrian sulphide deposits in southwestern Finland.* Annales Academia Scientiarum Fennicae, Series A3, Geologica-Geographica, **151**.

MÄKI, T. & PUUSTJÄRVI, H. 2002. The Pyhäsalmi Massive Zn-Cu-pyrite Deposit, Middle Finland - a Paleoproterozoic VMS-class 'giant'. *In:* EARTH, C. (eds) *The Geology and Genesis of Europe's Major Base Metal Deposits.* Irish Association for Economic Geology, in press.

MATTHÄI, S.K., HENLEY, R.W., BACIGALUPO, R.S., BINNS, R.A., ANDREW, A.S., CARR, G.R., FRENCH, D.H., McANDREW, J. & KANANAGH, M.E. 1995. Intrusion-related, high-temperature gold quartz veining in the Cosmopolitan Howley metasedimentary rock-hosted gold deposit, Northern Territory, Australia. *Economic Geology,* **90**, 1012–1045.

MOORE, J.M. & WATERS, D.J. 1990. Geochemistry and origin of cordierite-orthoamphibole/orthopyroxene-phlogopite rocks from Namaqualand, South Africa. *Chemical Geology,* **85**, 77–100.

NEUVONEN, K.J., KORSMAN, K., KOUVO, O. & PAAVOLA, J. 1981. Paleomagnetism and age relations of the rocks in the Main Sulphide Ore Belt in central Finland. *Geological Society of Finland, Bulletin,* **53**, 109–133.

OLIVER, R.L., LAWRENCE, R., GOSCOMBE, B.D., DING, P., SIVELL, W.J. & BOWYER, D.G. 1988. Metamorphism and crustal considerations in the Harts Range and neighbouring regions, Arunta Inlier, central Australia. *Precambrian Research,* **40/41**, 277–295.

PAN, Y. & FLEET, M.E. 1995. Geochemistry and origin of cordierite-orthoamphibole gneiss and associated rocks at an Archaean volcanogenic massive sulphide camp: Manitouwadge, Ontario, Canada. *Precambrian Research,* **74**, 73–89.

PAPUNEN, H. 1986. Geology and ore deposits of southwestern Finland. *In:* GAÁL, G. (eds) *Metallogeny and ore deposits in South Finland.* Geological Survey of Finland, Guides, **16**, 8–18.

PETERSON, J.W., CHACKO, T. & KUEHNER, S.M. 1991. The effects of vapor-absent melting of phlogopite + quartz: Implications for deep-crustal processes. *American Mineralogist,* **76**, 470–476.

PLOEGSMA, M. & WESTRA, L. 1990. The early Proterozoic Orijärvi Triangle (Southwest Finland): a key area on the tectonic evolution of the Svecofennides. *Precambrian Research,* **47**, 51–69.

RAITH, J.G. & KRIEGSMAN, L.M. 1998. Granulite-amphibolite facies transitions in high-grade terranes and their relation to ore-genetic processes. *The New Petroleum World,* **11**, 19–23.

ROBINSON, P., SPEAR, F.S., SCHUMACHER, J.C., LAIRD, J., KLEIN, C., EVANS, B.W. & DOOLAN, B.L. 1981. Phase relations of metamorphic amphiboles: natural occurrence and theory. *In:* VEBLER, D.R. & RIBBE, P.H. (eds) *Amphiboles: Petrology and Experimental Phase Relations.* Mineralogical Society of America, Reviews in Mineralogy, **9B**, 1–228.

ROGERS, J.J.W. & SATTERFIELD, M.E. 1994. Fluids of anorogenic granites; a preliminary assessment. *Mineralogy and Petrology,* **50**, 157–171.

ROLLINSON, H.R. 1993. *Using geochemical data; evaluation, presentation, interpretation.* Longman Scientific & Technical, Harlow, UK, **352**.

ROSENBERG, P.E. & FOIT, F.F. JR 1977. Fe^{2+}-F avoidance in silicates. *Geochimica et Cosmochimica Acta,* **41**, 345–346.

SÁNCHEZ-ESPAÑA, J., VELASCO, F. & YUSTA, I. 2000. Hydrothermal alteration of felsic volcanic rocks associated with massive sulphide deposition in the northern Iberian Pyrite Belt (SW Spain). *Applied Geochemistry,* **15**, 1265–1290.

SCHNEIDERMAN, J.S. & TRACY, R.J. 1991. Petrology of orthoamphibole-cordierite gneisses from the Orijärvi area, southwest Finland. *American Mineralogist,* **76**, 942–955.

SCHREURS, J. & WESTRA, L. 1985. Cordierite-orthopyroxene rocks: the granulite facies equivalents of the Orijärvi cordierite-anthophyllite rocks in west Uusimaa, Southwest Finland. *Lithos,* **18**, 215–228.

SCHUMACHER, J.C. & ROBINSON, P. 1987. Mineral chemistry and metasomatic growth of aluminous enclaves in gedrite-cordierite-gneiss from southwestern New Hampshire, USA. *Journal of Petrology,* **28**, 1033–1073.

SIVELL, W.J. 1986. A basaltic-ferrobasaltic granulite

association, Oonagalabi gneiss complex, Central Australia: magmatic variation in an Early Proterozoic rift. *Contributions to Mineralogy and Petrology*, **93**, 381–394.

SPEAR, F.S. & CHENEY, J.T. 1989. A petrogenetic grid for pelitic schists in the system SiO_2-Al_2O_3-FeO-MgO-K_2O-H_2O. *Contributions to Mineralogy and Petrology*, **101**, 149–164.

STEWART, A.J. & WARREN, R.G. 1977. The mineral potential of the Arunta Block, central Australia. *BMR Journal of Australian Geology & Geophysics*, **2**, 21–34.

TAREEN, J.A.K., KESHAVA, P.A.V., BASAVALINGU, B. & GANESHA, A.V. 1995. The effect of F and titanium on the vapour-absent melting of phlogopite and quartz. *Mineralogical Magazine*, **59**, 566–570.

TRELOAR, P.J., KOISTINEN, T.J. & BOWES, D.R. 1981. Metamorphic development of cordierite-amphibole rocks and mica schists in the vicinity of the Outokumpu ore deposit, Finland. *Transactions of the Royal Society of Edinburgh, Earth Sciences*, **72**, 201–215.

VÄISÄNEN, M. 2001. *Tectonic evolution of the Svecofennian Orogen in SW Finland: structural, petrological and U-Pb zircon dating (SIMS) constraints*. Licentiate thesis, University of Turku, Finland, **87**.

VALLANCE, T.G. 1967. Mafic rock alteration and isochemical development of some cordierite-anthophyllite rocks. *Journal of Petrology*, **8**, 84–96.

VISSER, D. & SENIOR, A. 1990. Aluminous reaction textures in orthoamphibole-bearing rocks: the pressure-temperature evolution of the high-grade Proterozoic of the Bamble sector, south Norway. *Journal of Metamorphic Geology*, **8**, 231–246.

WEIHED, P. & MÄKI, T. (EDS) 1997. *Research and exploration - where do they meet? 4th Biennial SGA Meeting, 1997, Turku, Finland. Excursion guidebook A2*. Geological Survey of Finland, Guides, **41**.

WILLNER, A.P. 1993. *Grundlagen zur Dynamik der mittleren Erdkruste: P-T Pfad, Fluidentwicklung und Fluorverteilung in einer Modellregion im Namaqua Mobile Belt (Südafrika)*. Enke Verlag, Stuttgart, **191**.

WILLNER, A.P., SCHREYER, W. & MOORE, J.M. 1990. Peraluminous metamorphic rocks from the Namaqualand metamorphic complex (South Africa); geochemical evidence for an exhalation-related, sedimentary origin in a mid-Proterozoic rift system. *Chemical Geology*, **81**, 221–240.

WOLTER, H.U. & SEIFERT, F. 1984. Mineralogy and genesis of cordierite-anthophyllite rocks from the sulfide deposit of Falun, Sweden. *Lithos*, **17**, 147–152.

WOOD, S.A. & SAMSON, I.M. 1998. Solubility of ore minerals and complexation of ore minerals in hydrothermal solutions. *Reviews in Economic Geology*, **10**, 33–80.

ZALESKI, E. & PETERSON, V.L. 1995. Depositional setting and deformation of massive sulfide deposits, iron-formation, and associated alteration in the Manitouwadge greenstone belt, Superior Province, Ontario. *Economic Geology*, **90**, 2244–2261.

ZHU, C. & SVERJENSKY, D.A. 1991. Partitioning of F Cl–OH between minerals and hydrothermal fluids. *Geochimica et Cosmochimica Acta*, **55**, 1837–1858.

Index

Note: Page numbers in **bold** type refer to tables, those in *italic* type refer to illustrations.